The Seismic Wavefield

The Seismic Wavefield provides a guide to the understanding of seismograms in terms of physical propagation processes within the Earth. The focus is on the observation of earthquakes and man-made sources, from the near source region out to thousands of kilometres from the source, for both body waves and surface waves. The link between theory and observations is made at a number of levels, starting from a broad illustrative survey and then progressing to greater detail. Most existing advanced textbooks on seismology deal mainly with the theory of seismic wave propagation, and only a few deal with the observations: *The Seismic Wavefield* is a comprehensive text that links theory with observation.

Volume I of *The Seismic Wavefield* begins with a survey of the structure of the Earth and the nature of seismic wave propagation, using examples of observed seismograms. Many concepts are introduced here and then elaborated on later in a full development of the theoretical background for seismic waves. This material will be drawn on in Volume II of *The Seismic Wavefield* (to be published later), which will cover local and regional seismic events, global wave propagation, and the three-dimensional Earth.

The emphasis throughout is on waves in seismological applications and the selection of methods and techniques is designed to provide physical insight rather than concentrate on numerical efficiency. This book draws on numerous graduate courses given by the author, and the combination of observation and theoretical development with a strong visual focus will greatly appeal to graduate students in seismology. The book will also be valuable to researchers and professionals in academia and the petroleum industry.

Brian Kennett is a Professor in the Seismology Group at the Research School of Earth Sciences, The Australian National University and Deputy Director of the Australian National Seismic Imaging Resource Research Facility. Professor Kennett's research interests involve the development of interpretational techniques for seismic records, with the object of extracting detailed information about the nature of the seismic velocity distribution within the Earth, and the character of seismic sources. He is currently President of the *International Association of Seismology and Physics of the Earth's Interior (IASPEI)*, and was Vice-President from 1991–1999. He is a Fellow of the American Geophysical Union and the Australian Academy of Sciences, and an Associate of the Royal Astronomical Society. Professor Kennett was an Editor of *Geophysical Journal International* from 1979–1999, and since 1985 has been an Associate Editor of *Physics of the Earth and Planetary Interiors.* He is the author of *Seismic Waves in Stratified Media* (1983; Cambridge University Press), compiler of the *IASPEI 1991 Seismological Tables*, and over 160 research papers.

The Seismic Wavefield

Volume I: Introduction and Theoretical Development

B.L.N. KENNETT
Research School of Earth Sciences, The Australian National University

CAMBRIDGE
UNIVERSITY PRESS

PUBLISHED BY THE PRESS SYNDICATE OF THE UNIVERSITY OF CAMBRIDGE
The Pitt Building, Trumpington Street, Cambridge, United Kingdom

CAMBRIDGE UNIVERSITY PRESS
The Edinburgh Building, Cambridge CB2 2RU, UK
40 West 20th Street, New York, NY 10011–4211, USA
10 Stamford Road, Oakleigh, VIC 3166, Australia
Ruiz de Alarcón 13, 28014 Madrid, Spain
Dock House, The Waterfront, Cape Town 8001, South Africa

http://www.cambridge.org

First published 2001

Printed in the United States of America

Typefaces Lucida Bright and Lucida NewMath (Y&Y) *System* LaTeX2ε [author]

A catalogue record for this book is available from the British Library

Library of Congress Cataloging-in-Publication Data is available.

ISBN 0 521 80945 2 hardback
ISBN 0 521 00663 5 paperback

Contents - Volume 1

Preface

The inspiration for this book came from the remarkable series of articles written by Beno Gutenberg in the 1930's for the multi-volume *Handbuch der Geophysik* on the theory and observations of waves from earthquakes that captured most of what was known at the time.

With the subsequent growth of studies in Seismology, it would be rash for any individual to attempt to cover the full field. Nevertheless, this book has arisen from the need to provide a broad survey of the nature of seismic wave observations and the relation of the seismic wavefield to the structure of the Earth. In the last decade the volume of high quality digital observations has grown enormously with the development of high quality global networks as well as the widespread use of broad-band recording using portable instrumentation. Yet, at the same time, the switch from analogue to digital presentation means that more effort is required to obtain an overview of the seismic wavefield.

The focus of this book is on observations of earthquakes and man-made sources, from the near-source region out to thousands of kilometres from the source, for both body waves and surface waves. The emphasis is on frequencies above 10 mHz so that the development can be regarded as complementary to the free oscillation orientation of *Theoretical Global Seismology* by Dahlen and Tromp (1998).

The aim is to relate observations to the relevant physical processes. The link to theory is made at a number of levels, starting with a broad survey and then progressing towards more detail. The first part provides a survey of the structure of the Earth and the nature of seismic wave propagation with illustrative examples of observations. Many topics are introduced here and then elaborated in the second part which provides a full development of the theoretical background. This material is then drawn upon in Volume II, which is concerned with seismic observations at different distance ranges from the source and the way in which the seismic wavefield across the globe evolves with distance.

The emphasis throughout is on waves in seismological applications, and the selection of methods and techniques is designed to provide physical insight rather than concentrate on numerical efficiency. The book incorporates material developed for graduate level courses at the Australian National University and short courses given at Kyoto University and the University of Tokyo. I am grateful for the feedback that I have received from many people and particularly my research students, K. Yoshizawa and K. Marson-Pidgeon, who have helped to remove some of the blemishes from the work.

Once again, I would like to thank my wife, Heather, for many forms of support without which this volume would never have been finished.

B.L.N. Kennett

1

Introduction

The most visible effects of earthquakes are their impacts on natural and man-made structures at the Earth's surface. The transfer of energy from the source that causes the destruction takes place through the agency of seismic waves. These same waves travel out through the Earth to be recorded at distant locations using sensitive seismographs. In their passage through the Earth the waves acquire characteristics related to the properties of the regions through which they have travelled. Our knowledge of the interior structure of the Earth derives in large part from the unravelling of the different influences on the character of the seismic wavefield.

For the large scale structure of the Earth, the major source of information comes from the records of ground motion produced by the waves from earthquakes at seismic stations across the globe. At smaller scales man-made sources become important. Most of the knowledge of the seismic wavespeeds in the crust and the uppermost mantle is derived from observations of reflected and refracted waves on networks of seismic instruments specifically deployed for structural studies. In exploration work, particularly for petroleum, all the information comes from complex observational procedures in which attention is focussed on those parts of the seismic wavefield whose propagation paths are close to vertical.

The interpretation of seismic records requires an understanding of the generation and propagation of the seismic waves and the influence of the recording process, since each imposes its imprint on the seismogram. Improvements in the quality of seismic instrumentation mean that it is now possible to obtain a faithful rendition of the particle motion at the seismic sensor in digital form over a broad range of frequencies. The use of broad-band data requires careful attention to the nature of the disorganised component of seismic motion, commonly known as *noise*. It is against the background of seismic noise that the arrivals of interest have to be sought.

1.1 Seismic signals

The surface of the Earth is in constant slight movement which can be detected with a sensitive seismometer. The ground motion arises from both local effects, such as

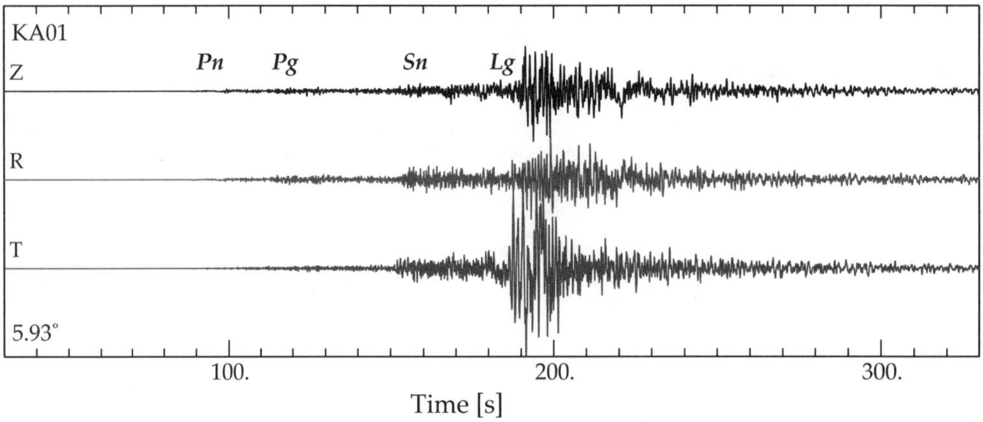

Figure 1.1. Unfiltered three-component velocity records from a portable broad-band recording system for a large local event (Mb = 6.3) in NW Australia at an epicentral distance $\Delta = 5.93°$. The horizontal component traces have been rotated to align along (R) and transverse (T) to the propagation path.

man-made disturbances or the rocking of trees in the wind, and vibrations induced by distant processes such as the microseisms generated by distant storms in the ocean.

From time to time the irregular pattern of ground motion is interrupted by an organised pattern of energy which rises above the background (figures 1.1, 1.2). Such well-defined wavetrains are induced by the excitation of seismic waves by a natural or artificial source and their subsequent propagation through the earth to the recording site. The two seismograms in figure 1.1 and figure 1.2 have been recorded at nearby sites in northwestern Australia in a deployment of portable broad-band seismic instrumentation. They show the major differences in the appearance of the seismic wavetrain with distance from the source. Figure 1.1 shows the three-orthogonal components of ground velocity at a distance of about 660 km from a relatively large local event (Mb 6.3); whereas figure 1.2 shows comparable records for a shallow event near the Fiji Islands at a distance of 5685 km from the source.

The characteristic form of the the wavetrain in each case includes a number of distinct arrivals associated with particular propagation paths, which have a more distinct pulse-like appearance at the larger distance. The distinct arrivals are accompanied by a lower level 'coda' arising from the superposition of many different processes and the influence of scattering. Following the initial group of body-wave phases such as *P*, *S*, the amplitude of the records increases as a sequence of waves arrives which have been guided by the presence of the Earth's surface (*Lg* in figure 1.1, *Love*, *Rayleigh* in figure 1.2). These surface waves have a more limited penetration into the Earth than the body waves and have somewhat lower frequencies. The difference in frequency is small for the shorter distance of propagation in figure 1.1 but is much more apparent in figure 1.2 where the long-period undulation associated with the

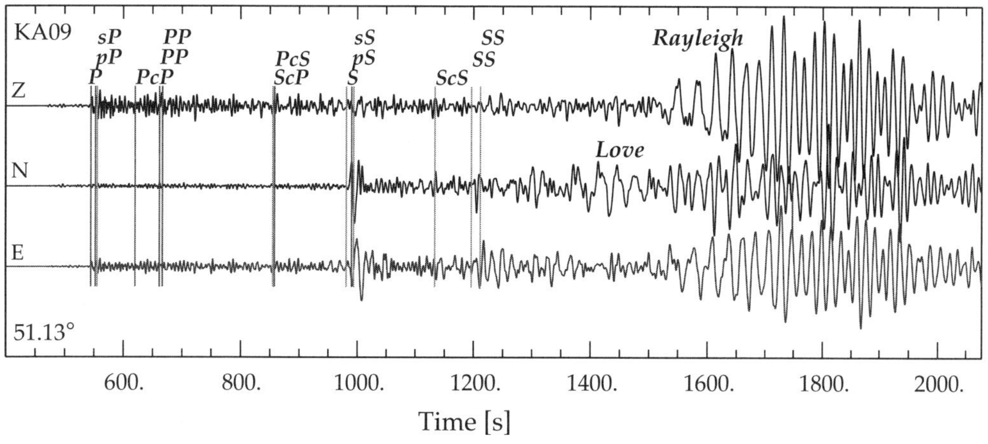

Figure 1.2. Unfiltered three-component velocity records from a portable broad-band recording system in NW Australia for a shallow event near the Fiji islands at an epicentral distance Δ = 51.13°. The event is almost due east of the receiver so the horizontal components are naturally polarised with transverse motion to the path on the N/S component.

surface waves is very distinct. In later chapters we will follow the development of the seismic wavefield and show how seismograms such as those displayed in figures 1.1 and 1.2 attain their form.

The most common natural generators of seismic waves are earthquakes associated with tectonic processes which can impart substantial energy in the form of seismic waves. In the period range from 0.001 Hz to 4 Hz the seismic waves from earthquakes can be detected at considerable ranges from the source (e.g. with a surface displacement of around 10^{-8} m at 9000 km for a surface wave magnitude of 4). For the largest earthquakes, wavetrains can be observed which have circled the globe a number of times.

The largest earthquakes have a tectonic character and involve substantial displacement on a fault surface. These relatively rare events are accompanied by very large numbers of smaller events. Nearly all large events are followed by a set of smaller earthquakes in the same region. These aftershocks appear to be related to the fault plane that slipped during the large event, but can also occur on nearby fault systems which have been activated by the redistribution of stresses following the main event. The aftershocks normally have significantly lower magnitude than the main event and decay fairly rapidly with time. The largest aftershocks, which occur soon after the event, can have a significant effect because of their interactions with structures which have been damaged by the large event. The distribution of aftershocks is often used to infer both the fault area and to provide information on the distribution of the slip in the rupture associated with the main event. Aftershocks often seem to concentrate around regions with lower slip which are termed asperities.

The most common natural generators of seismic waves are earthquakes associated

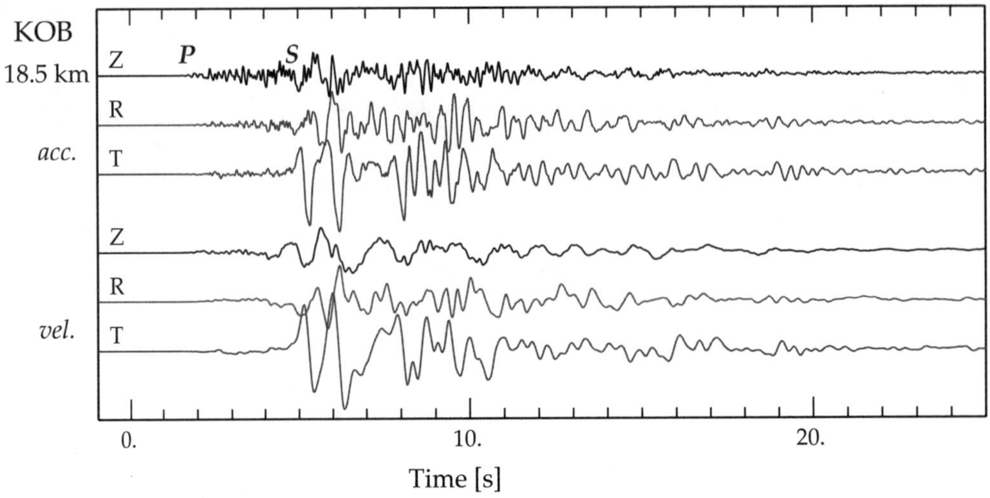

Figure 1.3. Three-component records for the KOB accelerometer in the city of Kobe, Japan for the 1995 Hyogo-ken Nanbu (Kobe) earthquake (Mw 6.9) which devastated the city. The station was about 19 km from the epicentre of the earthquake and the traces are rotated so that the horizontal components aligned along (R) and perpendicular (T) to the path from the epicentre. The ground velocity records obtained by integrating the accelerograms are also shown. The records are dominated by *S* waves with a long duration arising in part from the complex rupture pattern of the event. [Data courtesy of the Japan Meteorological Agency].

with subduction processes. However, sizeable events have occurred in continental regions well away from any tectonic boundary, as for example the sequence of three Mb 6 events within 12 hours in 1988 near Tennant Creek in the Northern Territory of Australia (see, e.g., Bowman et al, 1990). Many volcanoes also show associated earthquake activity which may, for example, arise from rapid magma transport or motions on faults above a magma chamber.

In the immediate neighbourhood of a large earthquake the Earth's surface suffers very large ground motion which is sufficient to overload many seismic instruments. In regions with a high incidence of earthquakes, such as California and Japan, specially emplaced accelerometer systems are installed to record these very large motions. The recordings are normally triggered by the arrival of the compressional *P* waves and modern systems have an extended data buffer, so that at close ranges the onset can be captured as well as the *S* waves and surface waves. In figure 1.3 we illustrate such strong ground motion with the record from the accelerometer operated by the Japan Meteorological Agency in the city of Kobe, Japan for the major earthquake in 1995 (Mj 7.2) which devastated the city. The station was about 19 km from the epicentre of the earthquake. The faulting pattern was complex with multiple subevents and rupture on both sides of the focus leading to prominent surface faulting in Awaji Island to the south and major destruction in the city of Kobe itself (Yoshida, 1996). Complex 3-dimensional structure beneath the city tended to reinforce the ground motion near

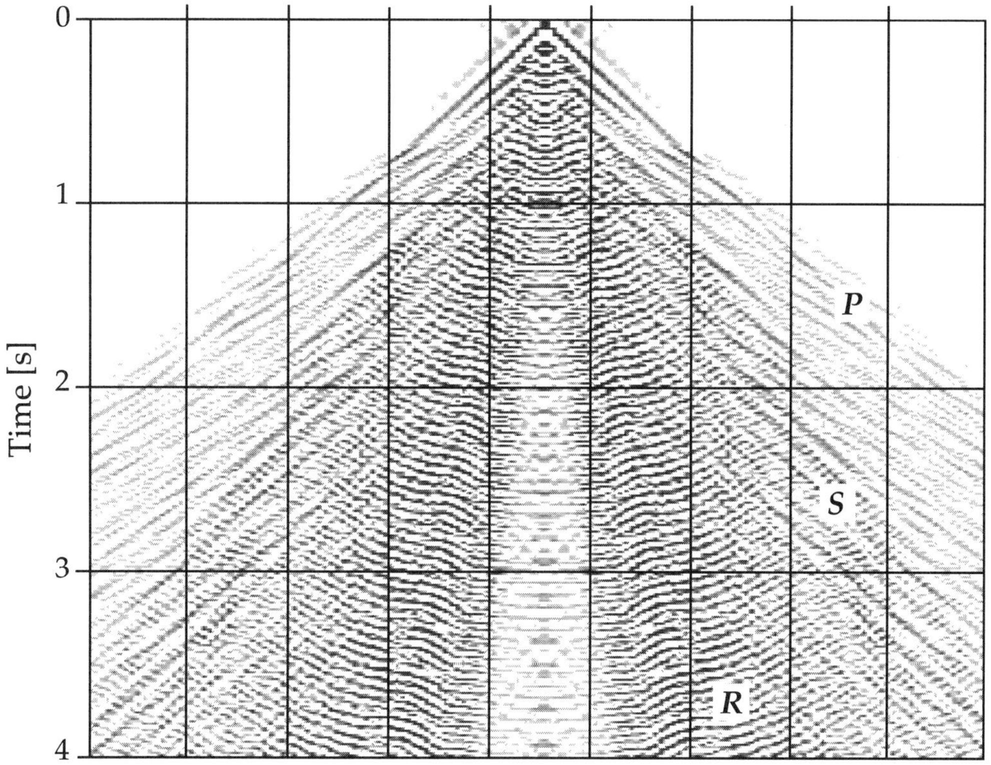

Figure 1.4. Land reflection spread on land from a vibrator source on a complex layered structure, showing prominent *P* reflections and refractions as well as *S* wave and Rayleigh wave (*R*) arrivals.

the fault line, leading to major damage (Koketsu & Furumura, 1998). Figure 1.3 shows both the original accelerometer records and the ground velocity derived by integration. The initial high frequency *P* waves are followed closely by large amplitude *S* waves which are particularly prominent on the tangential component to the path to the epicentre (T) on which horizontally polarised *SH* waves are expected. The complexity of the *S* wave records reflects the rupture process in the earthquake as well as the local structure.

Most man-made sources of seismic energy such as chemical explosions, surface vibrators or weight-dropping devices have a limited range over which they give detectable arrivals. This distance is about 2 km for a single surface vibrator and may be as large as 1000 km for a charge of several tons of TNT. Only large nuclear explosions rival earthquakes in generating seismic waves which are observable over a considerable portion of the Earth's surface.

Three examples of man-made seismic signals are shown in figures 1.4–1.6, from exploration seismology, large-scale refraction seismology and recordings of an underground nuclear test.

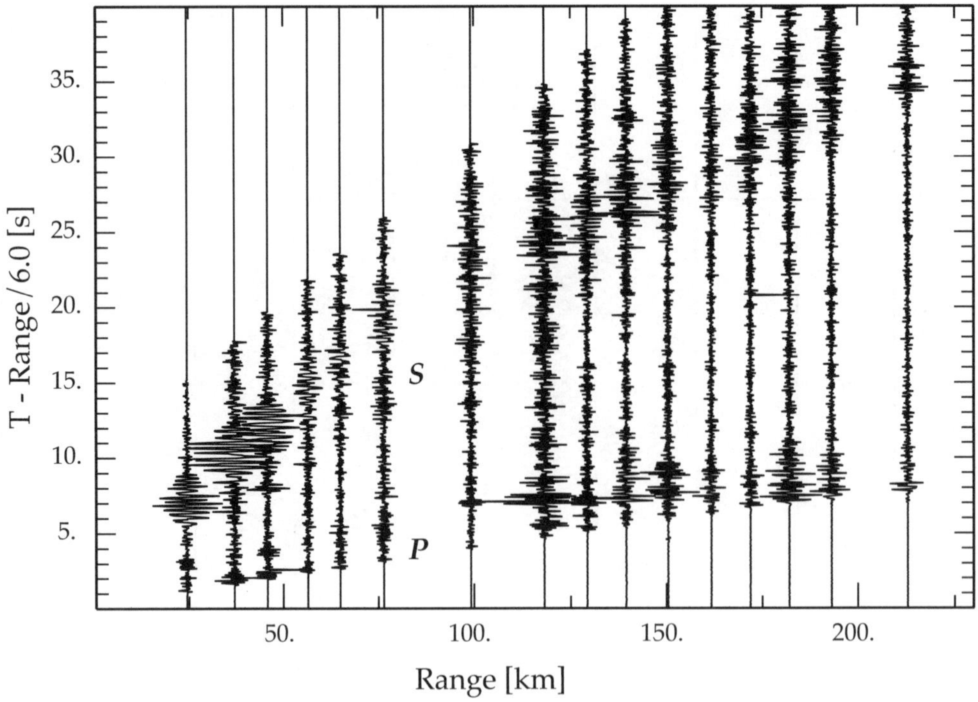

Figure 1.5. Radial component records for short distances from shotpoint B of the Fennolora seismic refraction experiment [courtesy of University of Karlsruhe].

The exploration example (figure 1.4) shows a 4 km spread from vibrator sources over complex layered horizons. The vibrators produce significant radiation in the form of vertically directed compressional (*P*) waves, but also shear waves radiated at about 45 degrees to the vertical. The source lies at the surface and so a large fraction of the radiated energy is carried as fundamental mode Rayleigh waves. These effects can be clearly seen in figure 1.4. The onset of *P* waves with prominent refractions is clear, and the associated reflections can be tracked back towards short offsets, but at larger times the *P* arrivals tend to be masked by other arrivals such as *S* waves and Rayleigh waves (*R*). The object of seismic processing in exploration is to extract the reflection signal which contains the major part of the geological information. The other arrivals whose properties are controlled by the shallow part of the layering are less used; although *P* information is sometimes used to build corrections for variations in the wavespeed in the near-surface layers.

The second example is taken from investigations of the structure of the Earth's crust and uppermost mantle where attention is concentrated on the body wave portion of the wavefield. Figure 1.5 shows radial component records from a small portion of the FENNOLORA project, a major long-range refraction profile through Sweden and northern Finland. The source for these records was an explosion in shallow water

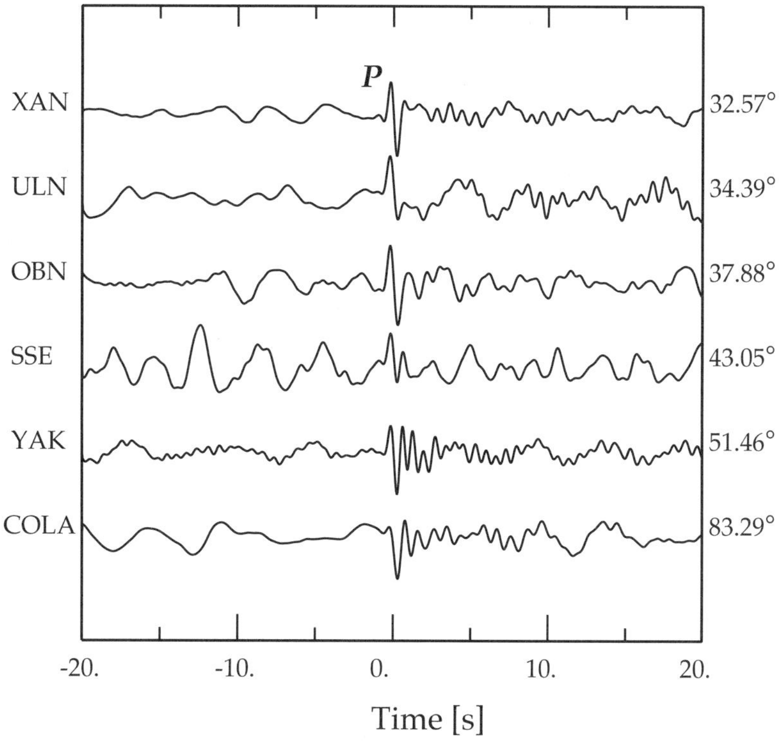

Figure 1.6. Vertical component *P* wave records from the Indian nuclear test of 1998 May 11 recorded at teleseismic broadband stations. The records are aligned on the onset of *P*.

and the seismometer component represents horizontal motion along the line of the profile. The group of records has been shifted in time with distance so that we can follow the earliest *P* waves, and reduce the total time span to be displayed. Very clear *P* and *S* wave signals are seen, with quite complex character associated with crustal propagation. There is a prominent *P* reflection from the crust-mantle boundary near a reduced time of 8 s for ranges beyond 100 km. The explosive source generates only *P* waves so all the *S* wave energy has been produced by wavetype conversion.

The final example (figure 1.6) shows the records of *P* waves produced by underground nuclear test in northwestern India on 1998 May 11 at teleseismic distances at a number of broadband stations around the globe (the epicentral distances in degrees are indicated at the right). There is a sharp simple *P* pulse which is very consistent between stations. No significant *S* waves can be found in the records; this is consistent with the simple model of an expanding pressure pulse source from the nuclear explosion.

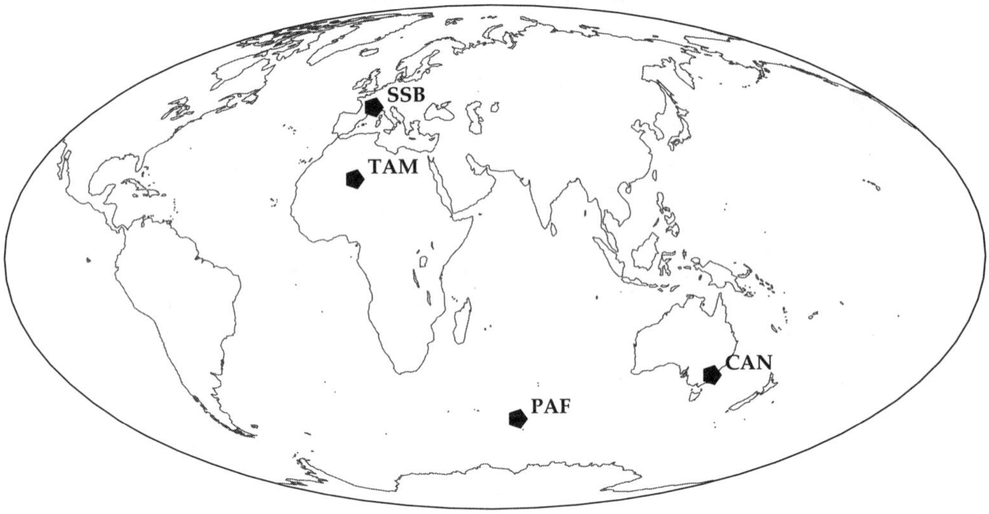

Figure 1.7. GEOSCOPE stations selected for display of the seismic noise spectrum.

1.2 Seismic noise

The nature of the seismic noise spectrum has had a profound influence on the nature of the instruments which have been developed to record earthquake signals and this in turn has affected the way in which seismic wave theory has developed.

There are a wide range of contributions to the background noise including the presence of tides, atmospheric pressure, diurnal effects mostly associated with temperature variation, and human-induced activity. Although efforts are made to place permanent seismic instruments in quiet locations away from sources of potential noise, long-established sites frequently suffer from increased noise due to encroachment of human activity as cities expand.

The noise pattern is dominated by a strong noise peak for frequencies from 0.09–0.18 Hz which arises from the influence of microseisms which mostly arise from standing waves induced by storms in the deep ocean. Wave surf can also be significant in the frequency range 0.05–0.10 Hz. The level of microseismic noise therefore depends on the season and is of most significance for stations close to the ocean. Mid-continental stations generally have conditions with somewhat lower noise levels.

We illustrate the influence of station location with noise spectra for the year 1995 taken from a number of stations in the GEOSCOPE network (figure 1.7). We display spectra taken at the mid-continental site TAM in Algeria, SSB in central France, CAN about 160 km from the coast of southeastern Australia and the island station PAF in the Kerguelen islands (figure 1.8). Each station uses Streckeisen STS-1 broad-band sensors with a comparable recording system. The power spectral density for acceleration is displayed for each of the three components at the stations, by combining results for different recording bands. The units are decibels, i.e. $10\log_{10}$(signal power), so that a variation of 20 dB corresponds to a factor of 10 variation in ground acceleration. The dashed lines shown in each of the frames of figure 1.8 represent the high- and low-noise models developed by Peterson (1993). In favourable circumstances the noise levels on the vertical component instruments can approach the low-noise floor.

The variation of the noise levels near the microseismic peak exceeds 36 dB, with nearly a factor of 100 in acceleration. The levels at this peak and in the high frequency range are normally comparable for each of the three-components. But, as can be seen from figure 1.8, the noise levels at long periods for horizontal component instruments are somewhat higher than for the vertical component. A major influence on the long period noise level is atmospheric pressure. Slight variations of pressure induce ground tilting which is only recorded on the horizontal components.

The energy release from earthquakes spans an enormous range. The waves generated by the smallest events lie well below the detection threshold of even the most sensitive seismometers. The largest events produce very strong ground motion in their immediate vicinity. A seismic recording system thus requires a large dynamic range to cope with the range of conditions that may be experienced.

The strong variation in noise level with frequency and station location means that the detection capability for seismic signals of interest differs dramatically depending on the nature of the disturbance. Long period surface waves from an event may well rise above the noise even though the shorter period body waves are submerged in the ambient ground motion. At an island station such as PAF, it will not be possible to detect seismic signals from events which would be readily recorded on a mid-continental station such as TAM at a comparable distance from the source.

The seismic noise spectrum is not static but varies on both a daily and seasonal basis as illustrated in figure 1.9 for station SSB in France. The daily variations lead to enhanced high frequency noise in daylight hours, much of which is due to increased human activity during the day, but also to the higher temperatures and larger atmospheric variations during the day. The seasonal variations correlate with the presence of stronger storm systems in the North Atlantic in the winter which raise the level of the microseismic peak noticeably.

Such seasonal variations can be particularly severe for stations with climates with a very strong contrast between winter and summer. Stations in the northern land masses can be very quiet in winter when the whole area is frozen and there is a snow blanket, but the spring thaw can bring much higher noise particularly when stations lie near a body of water.

Details of the method used for noise estimation and a discussion of noise conditions at the full range of GEOSCOPE stations for the year 1995 can be found in Stutzman, Roult & Astiz (2000).

The major noise peaks associated with the microseisms at 0.09–0.18 Hz caused

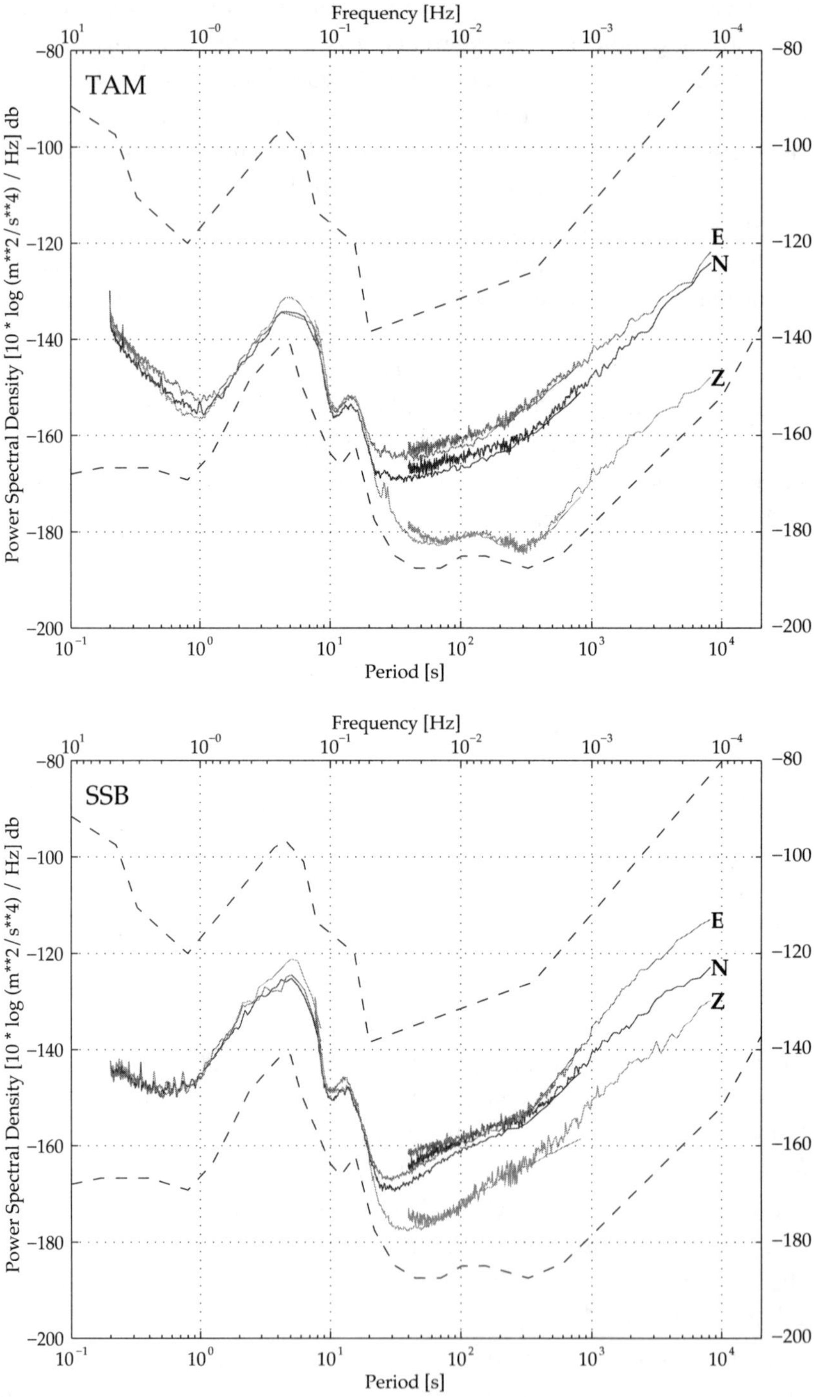

Figure 1.8. Power spectral density for acceleration at the Earth's surface. The upper and lower dashed curves represent the high- and low-noise model of Petersen (1981). (a) TAM: Tammanraset, Algeria, (b) SSB: St Saveur, France [courtesy of GEOSCOPE].

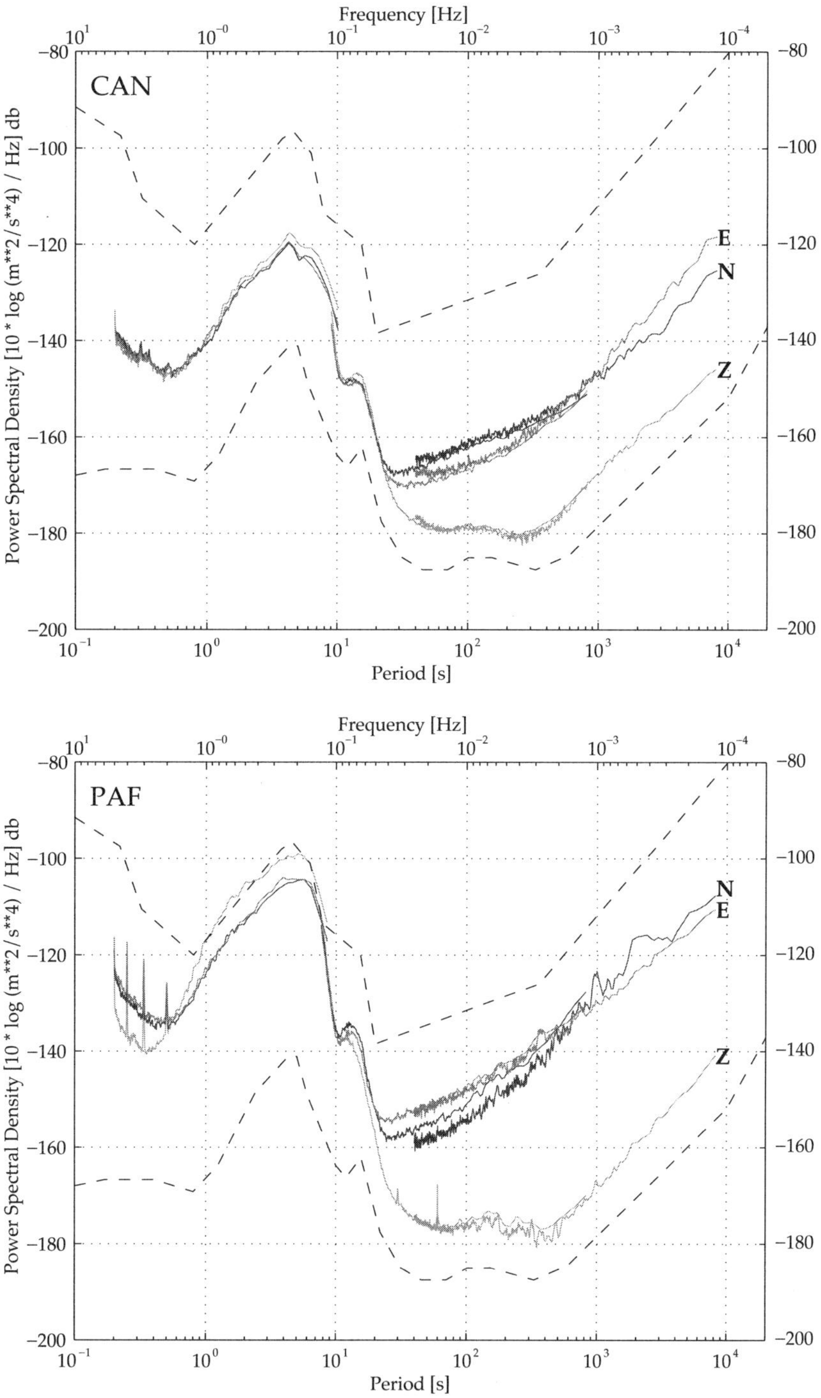

Figure 1.8. Power spectral density for acceleration at the Earth's surface. The upper and lower dashed curves represent the high- and low-noise model of Petersen (1981). (c) CAN: Canberra, Australia, (d) PAF: Port aux Français, Kerguelen Islands [courtesy of GEOSCOPE].

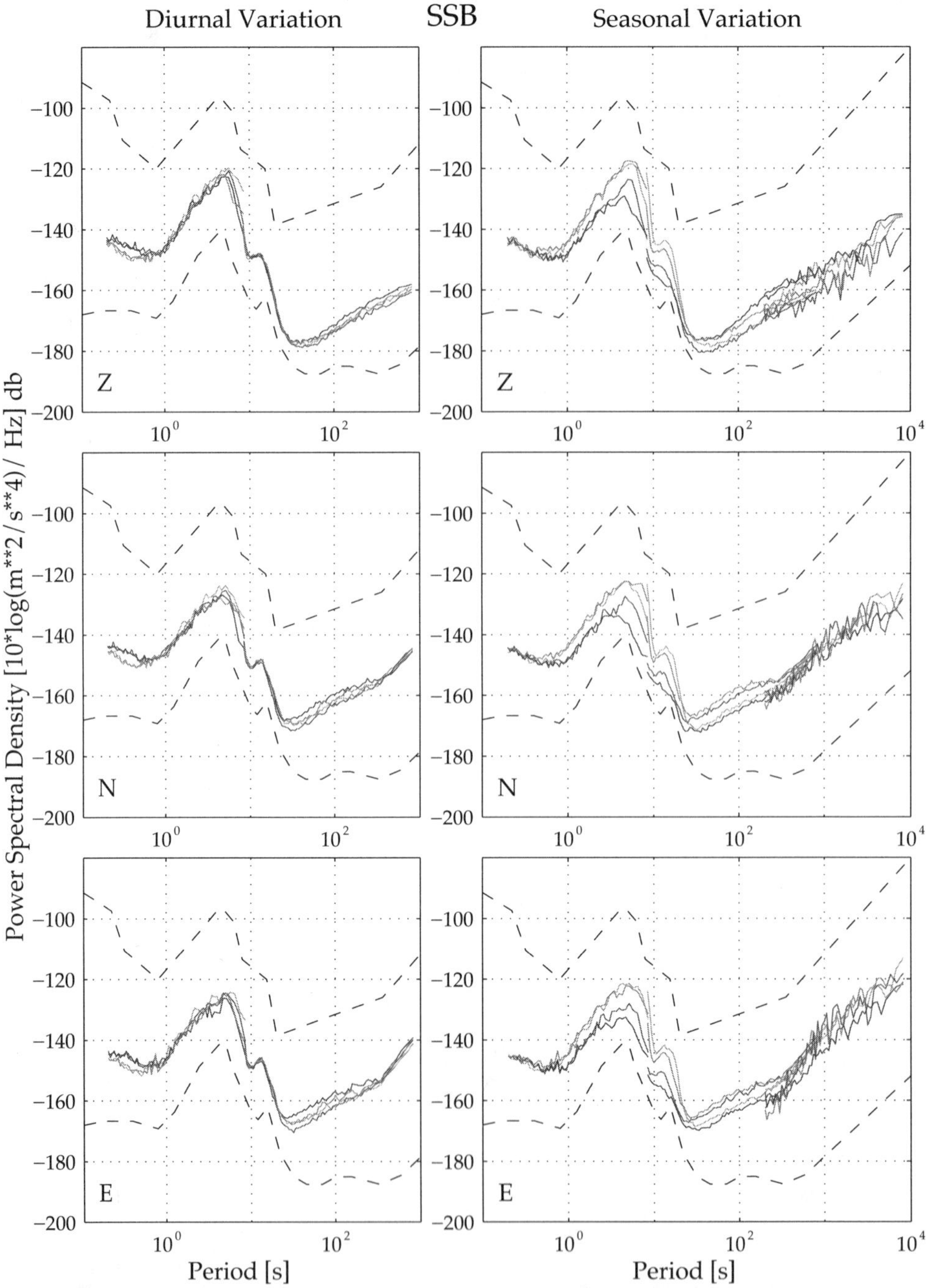

Figure 1.9. Diurnal and seasonal variations in the power spectral density for acceleration at the station SSB in France [courtesy of GEOSCOPE].

considerable difficulties for the analogue recording systems which were the primary means of recording seismic data up to the 1980's, since it is difficult to achieve adequate dynamic range to cope with the whole range of signals. For a photographic recording system there is an unavoidable limit to the smallest signal which can be discerned due to the width of the light beam and large signals simply disappear off scale. To avoid swamping the records of medium size events with microseismic noise, the commonest procedure was to operate two separate instruments with characteristics designed to exploit the relatively low noise conditions on the two sides of the noise peak. This was the approach followed in the World Wide Seismograph Network (WWSSN) which installed separate long-period and short-period instruments at over 100 sites around the world in the 1960's. The records from the long-period seismometers were then dominated by the fundamental modes of surface waves although some body waves were present. The short-period records were most useful for body waves. A similar arrangement was made for the digital recording channels of the SRO network which was designed to enhance the WWSSN system, and the instrumental responses for the short-period and long-period recording bands had a very rapid fall off in the neighbourhood of the noise peak so that even with digital techniques it was hard to recover information in this region.

At mid-continental stations well away from the oceans the microseismic noise levels are much reduced. It is then possible to follow the approach pioneered by Galitzin in the early 1900's and use a single instrument over the frequency band of interest. This style of broad-band recording was exploited in the Kirnos instrument used in the former Soviet Union. The sensor for the SRO system was also a broad-band seismometer, but the recording channels conformed to the scheme originally devised for analogue purposes.

The current systems of broad-band sensors are based on feed-back electronics designed to maintain a small test mass close to a fixed position. The output of the seismometer is derived from the feedback circuitry, and provides a stable response over a very broad range of frequencies (e.g., 360 s to 0.1 s period for the Streckeisen STS-1 seismometer based on a design by Wielandt). Further, because the movement of the mass is limited, the linearity of the seismometer is very high. Broad-band seismometer packages have been built for both vault and borehole installation. The advantage of a borehole is that the influence of wind noise on the seismometer can be significantly reduced compared with a near surface site. As a result boreholes have been used for newly established stations particularly those in the IRIS networks.

The advent of digital recording using 24-bit analogue to digital conversion means that sufficient dynamic range (140 db) is available in the recording system to cope with the expected variations in seismic signals and noise and in consequence high fidelity rendition of ground motion can be made. A variety of different recording modes can then be employed to capture different aspects of the seismic wavefield.

Figure 1.10 displays the frequency bands employed in 2000 for recording at the

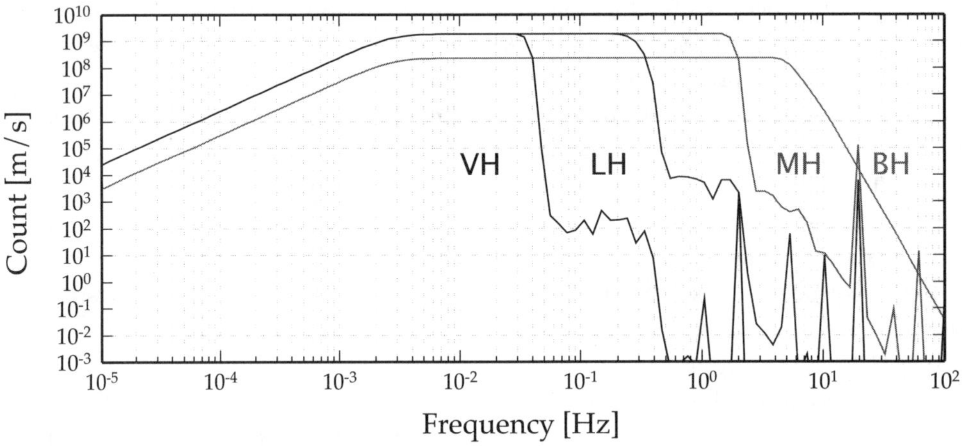

Figure 1.10. The set of recording bands employed during 2000 for the station CAN, Canberra, Australia [courtesy of GEOSCOPE].

station CAN in southeast Australia, which is part of the GEOSCOPE network. The different bands are recorded with different sampling rates and used for different styles of seismic analysis. The very-long period VH band is used for free-oscillation work and studies of global surface waves with a sampling rate of 0.1 samples/s and continuous recording. The long-period LH band reproduces much of the character of an analogue long-period instrument and is particularly useful for surface wave studies and uses a sampling rate of 1 sample/s, once again continuous recording is used. The full band-width of the seismometer and recording system are exploited in the BH band with the highest sampling rate of 20 samples/s used with an event detection scheme, so that 1 hour segments are preserved following a suitable trigger.

With digital recording such broad-band instruments can be used in a wide range of environments and digital filtering can be employed if it is necessary to suppress the microseisms. As a result of a number of initiatives, the network of digital broad-band stations provides reasonable coverage of the main land masses, and a design goal for the international community has been to achieve a station spacing of approximately 2000 km. Many stations have been deployed on oceanic islands which, as we have seen, have somewhat higher noise levels. However, large regions of the oceans have very limited coverage and work continues on the development of sea-floor observatories to enhance the sampling of the globe (Stephen et al, 1999).

The noise spectra shown in figures 1.8 and 1.9, determine the minimum size of events at each site for which useful records can be obtained. The expected spectra for a moderate earthquake (magnitude 5.0) and a great earthquake (magnitude 9.5) at 30° epicentral distance are compared with a model for the lower limit of the Earth Noise spectrum (Peterson, 1993) and the instrument pass-band for the IRIS Global Seismic Network in figure 1.11. Except for effects associated with the radiation pattern

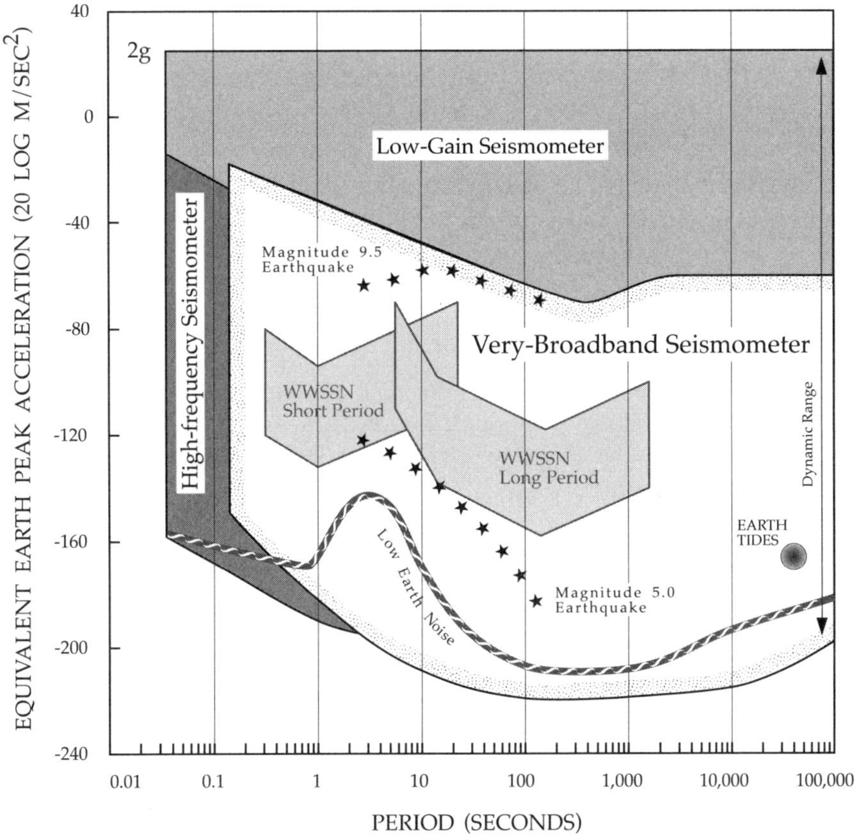

Figure 1.11. The instrumentation bands employed in the IRIS GSN system compared with the earlier WWSSN network [courtesy of IRIS].

from the source, it should therefore be possible to maintain teleseismic recordings for significant earthquakes within the dynamic range of very broadband seismometers with 24-bit recording systems, and in addition recover Earth tides. The limited dynamic range associated with the older analogue systems is very clearly seen for the WWSSN short-period and long-period systems in figure 1.11.

The distribution of permanent broad-band seismic stations around the world is now quite extensive and mutual cooperation between different agencies through the Federation of Digital Seismographic Networks (FDSN) has done much to improve the standards of instrumentation and data exchange. The rapid availability of high quality data from an extensive global network is having a significant impact on the way in which seismology is carried out.

Broad-band seismometers are now available in compact packages that are capable of field deployment and in conjunction with 24-bit recording systems can be used in temporary deployments to supplement the permanent station network. Solar power provides operational capability in remote areas. Major deployments of portable instruments have been carried out in many parts of the world, e.g., Tibet (Owens et al., 1993), Australia (van der Hilst et al., 1994). The data quality that be achieved with such field deployments is high and can match many observatory environments (cf. figures 1.1, 1.2). The advent of high-quality recording from portable instruments allows a change in character of seismological investigations, since it is now possible to aim for a selective increase in data density in a particular region to focus attention on a specific aspect of the seismic wavefield.

1.3 Understanding the seismic wavefield

The aim of this book is to provide a link between the features of observed seismograms and the processes which generate them, so that it is possible to develop an improved understanding of the structure of the Earth and the nature of the seismic source.

For most of the first century of seismic recording, seismology has been an intensely visual subject with attention focussed on the details of seismograms. As digital recording and analysis has supplanted analogue records and manual interpretation, there have been major gains in the fidelity of the representation of ground motion. However, the expertise built up through systematic reading and interpretation of seismograms needs to be preserved and transferred into the digital era.

This work therefore presents a wide variety of examples of different aspects of the seismic wavefield through displays of individual and multiple records, accompanied by a description of the propagation processes which dictate the nature of the seismograms. There is a strong link between the seismogram displays and the theoretical development so that the character of the wavefield in different distance ranges from the source can be clearly represented.

Part I provides a general discussion of the structure of the Earth and the way in which a seismogram is built up by the sequence of processes between wave generation at the source and arrival at the receiver. The treatment is largely descriptive with cross-referencing to the more comprehensive and mathematical analysis in Part II. The aim is to provide a clear development in terms of physical processes which can then be applied explicitly to different aspects of the seismic wavefield in Volume II (Parts III–V).

Part II presents a treatment of the main elements of the theory of seismic wave propagation, employing a range of different representations of the wavefield and the generation of waves by the seismic source for both body waves and surface waves. The material is linked both to the general introduction in Part I and the more detailed treatment of the nature of the seismic wavefield as revealed in observed seismograms in the second volume.

Part III is concerned with the development of the seismic wavefield out to about 3000 km from the source, where the nature of the seismogram is dominated by propagation in the Earth's crust and upper mantle. Close to the source, ground motion can be significant and is strongly affected by local conditions. Further away from the source, the character of the records is controlled by the nature of crustal structure with prominent guided waves such as *Lg*, which are sensitive to structural complexity along the propagation path. Waves travelling in the uppermost mantle become relatively more important as the range from the source increases. The structure of the transition zone in the upper mantle with discontinuities in seismic wavespeeds imposes a complex structure on the arrivals of *P* and *S* waves out to 3000 km. These complexities are repeated in the surface multiples at larger distances.

In contrast, Part IV is concerned with more distant events for which propagation processes can involve much of the Earth. Analysis of teleseismic disturbances requires a clear understanding of the influence of both mantle and core structure. Propagation of *P* and *S* waves through the lower mantle is dominated by simple refraction until waves penetrate into the complex D" region just above the core-mantle boundary. Even at relatively short ranges the presence of the Earth's core is significant through reflected phases such as *PcP*, *ScP*, and *ScS*. Beyond 9000 km from the source the influence of the core becomes most apparent, leading to a variety of ways in which seismic energy can travel from source to receiver. The distinctive styles of propagation of *PKP* and *SKS* waves provide very different sampling of the core and can be used in diverse ways to examine core structure.

Part V is concerned with mapping three-dimensional variations in Earth structure building on the observations and techniques introduced in the earlier sections. Seismic tomography is covered on a variety of scales from crustal to global using body waves, surface waves and the free oscillation of the Earth with a discussion of the nature of the various methods together with their associated assumptions and limitations. Frequently 3-D structure is introduced as a perturbation to 1-D reference structures but in many regions the levels of variation are large enough that direct methods for modelling 3-D structures are needed and these are drawn from the developing area of numerical seismology.

Part I

Seismic Waves and the Structure of the Earth

2

Earthquakes and Earth Structure

2.1 The distribution of seismic sources

Interest in earthquakes as a scientific study was stimulated by the very destructive Lisbon earthquake of 1755, which produced effects at considerable distances from the source with oscillations of water level in lakes (*seiches*) observed in Loch Lomond in Scotland and Lake Geneva in Switzerland.

The general pattern of occurrence of earthquakes had been deduced from combining felt reports from around the globe in the middle of the 19th century. However, the recognition that the earthquake activity is concentrated in relatively narrow belts required the development of a broad spread of seismic instrumentation around the globe, and good international collaboration to combine information from different observatories to determine the location of the source. The difficulty faced in the early instrumental studies was that the determination of source location requires knowledge of the structure of the Earth, which in turn depends on the quality of source information.

From the 1920s as the number of seismic stations and the quality of time keeping improved, the times of arrival of seismic phases were analysed to produce sets of *travel time tables* relating the time of passage of seismic phases to the distance from the source to the receiver. Such tables could then be used to improve event location. By the early 1930s Jeffreys and Bullen had produced a good table for P waves and in 1940 published a comprehensive set of tables (Jeffreys & Bullen, 1940) which are still in use as the reference for event location by major agencies such as the International Seismological Centre and the National Earthquake Information Centre of the U.S. Geological Survey.

In the early days of instrumental seismology the common assumption was that all earthquakes were shallow, but the improvements in travel time curves soon forced the realisation that events must occur at depth. In the island arcs around the Pacific, Benioff and Wadati identified the presence of inclined zones in which the earthquake depth increased systematically and these zones are now recognised to indicate the presence of subduction zones.

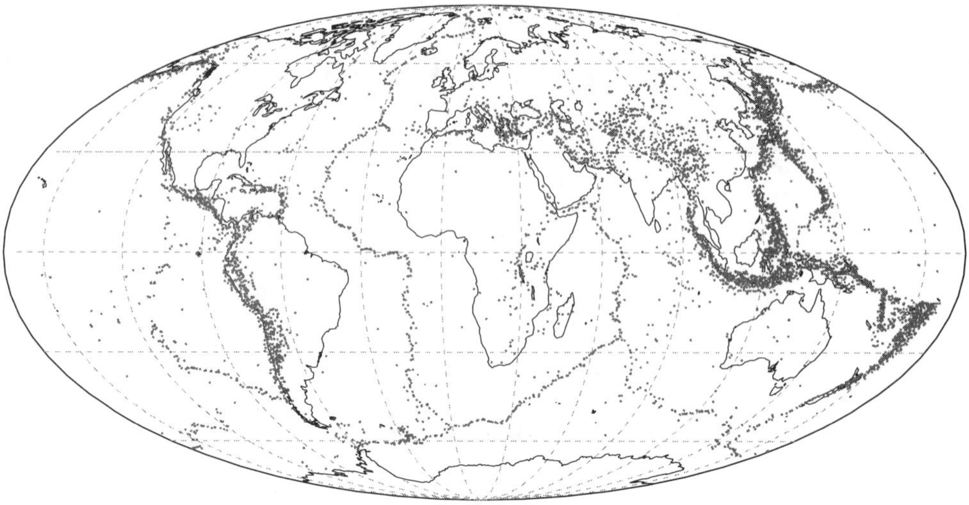

Figure 2.1. The distribution of seismic sources across the globe which gives a clear indication of the mid-ocean ridge system.

Gutenburg and Richter (1949) collated all the available information to produce an excellent summary of the main distribution of seismic events. Their work was only supplanted once much denser global networks of seismographs were established in the 1960s. In particular the presence of relatively narrow zones of seismicity traversing the oceans was recognised and correlated with improved bathymetry information to outline the mid-ocean ridge system (see figure 2.1).

The present distribution of seismic stations is understandably dense in those regions which suffer earthquakes but provides quite a good coverage of the whole globe. Even moderate size events, such as a Mb 5.4 event in China, can have several hundred phase arrival times reported. The result is that the pattern of seismicity across the Earth is now well determined with more than a 30-year span of high quality observations. The epicentre associated with a event is the projection of the actual location on to the Earth's surface and a display of epicentres is a convenient means of presenting event locations in map view.

A selection of 6000 of the best determined earthquake epicentres are shown in figure 2.1 and give a very clear indication of the differences between the earthquake distribution in oceanic and continental regions. The distribution of events within the continents is generally diffuse, as in the complex patterns in the Alpine-Himalayan mountain system. The rift valley system of East Africa is well outlined by its seismicity, but many continental areas have very few events.

A very distinctive feature of oceanic seismicity is the narrow zone of earthquakes marking the mid-oceanic ridge system. Many of these events lie on the transform faults

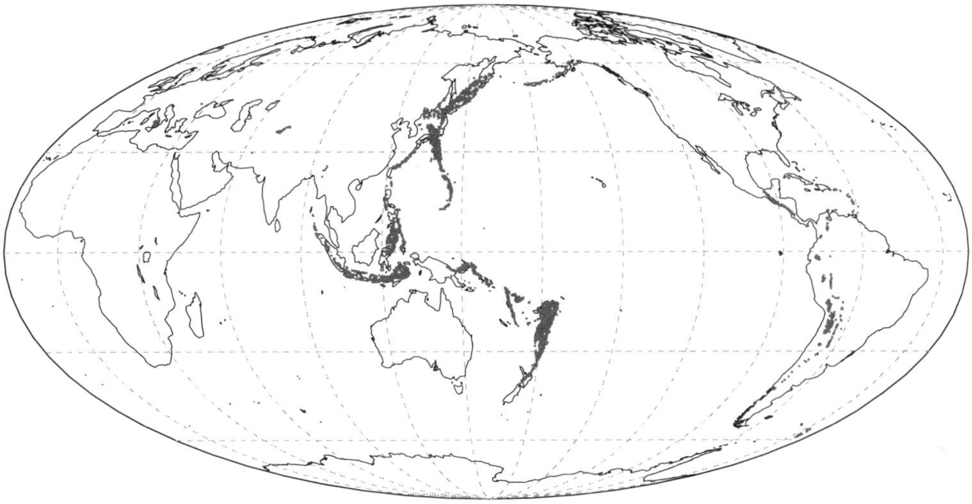

Figure 2.2. The distribution of seismic sources with depths greater than 150 km across the globe which indicates the presence of subduction zones.

that offset the ridge system. A good example can be seen in the equatorial Atlantic in figure 2.1.

The densest assemblage of earthquakes occurs beneath and around the island arcs of the Pacific, where the events in the subduction zones extend to depths of the order of 600 km. The subduction zones are clearly outlined when we confine attention to those seismic events whose depth lies greater than 150 km (figure 2.2). There are minor clusters of events associated with the Calabrian arc in Italy and the Aegean arc in Greece, but the majority of the activity occurs around the Pacific rim. A nearly continuous zone of earthquakes extends from the Aleutians in the north through Kamchatka and the Kuriles to central Japan, where the pattern bifurcates. One branch extends south through the Izu-Bonin and Mariana islands, whilst the other veers to the south-west through the Ryukyus and the Philippines to connect with the east-west system through the Indonesia Arc, from which events continue to the west of the Andaman islands and up into Burma. From New Guinea a further skein of earthquakes extends though the Solomon Islands and Vanuatu to Fiji and Tonga, and then south through the Kermadec Islands into New Zealand. Other significant subduction occurs in the Caribbean, through central America, along the length of the South American Cordillera and also in the Scotia Arc between South America and Antarctica. There are also some isolated pockets of deep events in the Vrancea region of Roumania and beneath the Hindu-Kush.

The distribution of earthquakes and the nature of seismic source mechanisms played a major role in the development of the plate tectonic concept for the dynamics of the outer part of the Earth. This model is based on the idea of the tessallation of the surface

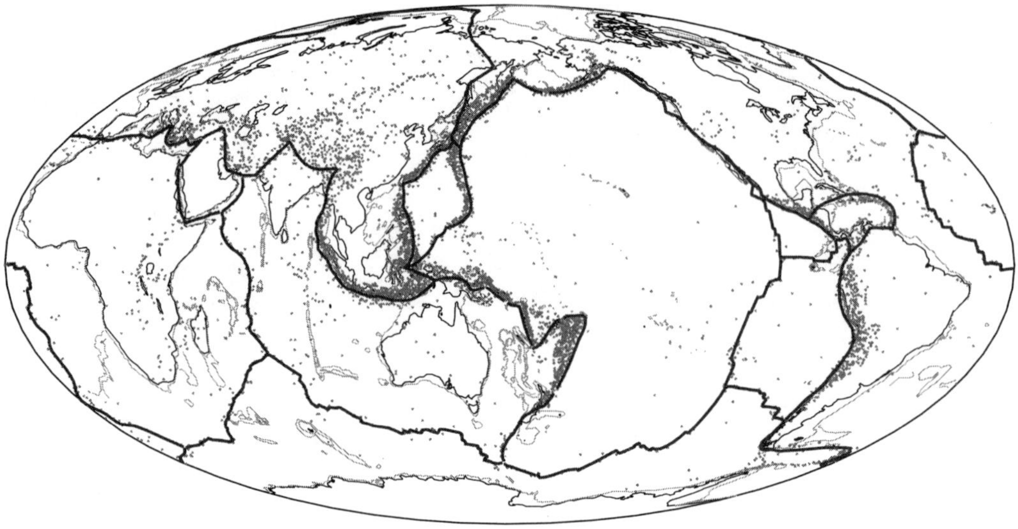

Figure 2.3. The major plate-boundaries superimposed on the distribution of seismic sources across the globe.

of the earth by a set of rigid plates, with accretion of new material at the mid-ocean ridge system and destruction of former ocean floor in the subduction zones. Relative motion between two of the set of rigid plates is described by rotation about a pole. By balancing the available constraints on all the sets of plate pairs it is possible to deduce a consistent set of plate motions (e.g., DeMets et al, 1990, 1994). The portion of the outer shell of the Earth that translates with plate motion is referred to as the *lithosphere* and is somewhat thicker beneath continental regions than beneath the oceans (see Section 2.2.1).

The pattern of the major plate boundaries is shown in figure 2.3 superimposed on the global earthquake distribution. In addition to the coastline, the main continental shelves and main marine features are indicated in light grey. The plate boundaries vary in character and their nature ties closely to the earthquake patterns. Subduction zones represent the main style of convergent boundary where lithosphere is extracted from the surface by sinking into the mantle.

Divergent boundaries such as mid-ocean ridges and continental rifts represent zones where two plates are moving apart. New lithosphere is produced at the mid-ocean ridges and we have already noted the narrow zone of seismicity outlining the ocean ridges. As the lithosphere moves away from the ridge it gets progressively thicker through cooling, and the ocean depth increases. Both increase approximately as the square root of the age of the lithospheric material. In the continental rift zones, such as the East African rift, the lithosphere is being thinned and eventually a new ocean ridge may be produced, as has happened relatively recently in the Red Sea.

The segments of the mid-ocean ridge where active spreading are occurring are linked by transform faults, which accommodate the relative motion between the two plates. These shear faults follow small circles about the pole of rotation between the two plates, and the morphology of transform faults is one of the major constraints on the pattern of relative plate motions.

Other transcurrent fault systems connect different combinations of plate boundaries. The most famous of these strike-slip systems is the San Andreas Fault in California, on which a significant fraction of the relative motion between the Pacific and North American plates is taken up. However the boundary between these plates is quite complex, with motion being accommodated over a significant zone and on a number of faults. The San Andreas system links the sea-floor spreading in the Gulf of California to the subduction occurring beneath the Cascade ranges of Oregon and Washington. A similar pattern of behaviour is seen in the Alpine fault complex in the South Island of New Zealand, where a dominantly strike slip fault grades into a extensive fault complex linking to the subduction zone to the north. Another major transcurrent system is the Anatolian fault in northern Turkey which accommodates the deformations associated with the collision of Africa and Eurasia.

Intraplate earthquakes occur away from the main plate boundaries. Such events may be associated with rifting processes or with ongoing continental collision such as the continuing motion of India to the north indenting the Asian plate and producing the broad sweep of seismicity in southern Asia. Isolated events in other continents can be due to be the build up of stresses as a result of plate boundary interactions; for example the events in central Australia have thrusting from the south, which is consistent with the northward movement of Australia away from Antarctica being impeded by the subduction zones to the north in Indonesia and New Guinea.

The character of subduction zones has been a subject of extensive study in recent years and much effort has been made to use the variations in the arrival times of seismic phases to build up three-dimensional images of the structure in these zones. The process is aided by the presence of intermediate depth (70–300 km) and deep earthquakes (> 300 km) within the zone. Figure 2.4 represents a very clear example of a cross-section through a subduction zone beneath the northern part of Honshu in Japan. In the lower panel the subducted material is evident through the high seismic wavespeeds associated with the cold descending lithosphere. The cartoon in the upper panel shows the relation of the earthquakes and the arc volcanoes to the subduction zone.

There is considerable variety in the character of subduction zones, both in the dip of the zones and in the distribution of seismicity with depth. The variations arise from the character of the material being subducted, for example, old cold lithosphere is much more dense than younger and more buoyant material. Also the pattern of plate motions is not constant in time but changes in response to the evolution of the plates

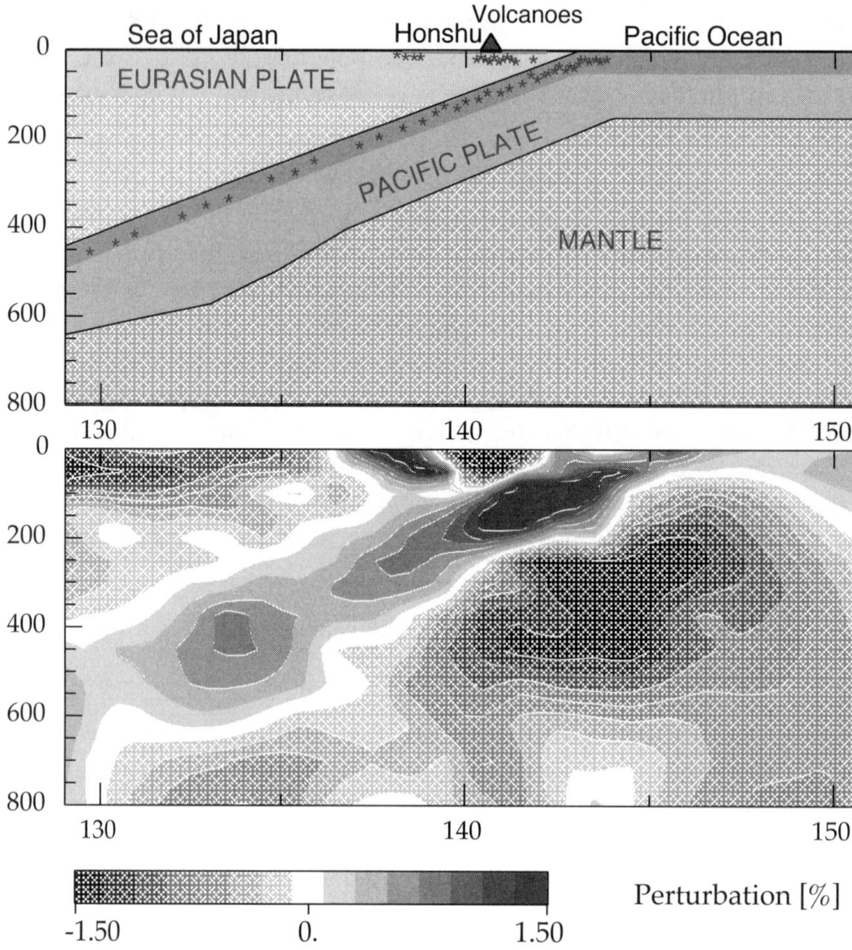

Figure 2.4. Detail of a *P* wavespeed tomographic model for the northwest Pacific: slice across Japan at 40°N with a clear representation of the subduction zone associated with the Pacific plate. The scale represents deviations from the AK135 reference model.

themselves. Entire plates have been eliminated by progressive subduction, e.g. the Farallon plate has disappeared beneath North America leaving traces in the mantle.

We can illustrate the nature of current subduction with east-west cross sections through parts of the subduction zones in the western Pacific (figure 2.5). The upper panel of figure 2.5 displays sections in the Northwest Pacific; all three of these sections show the subduction of the Pacific plate, but in the southernmost section we also see the subduction of the Philippine plate beneath the Asian plate in the Ryukyu arc. In the Kuriles to the north (50°N), we can follow the plate readily to about 600 km depth, and we also see shallow seismicity associated with the plate boundary to the west of

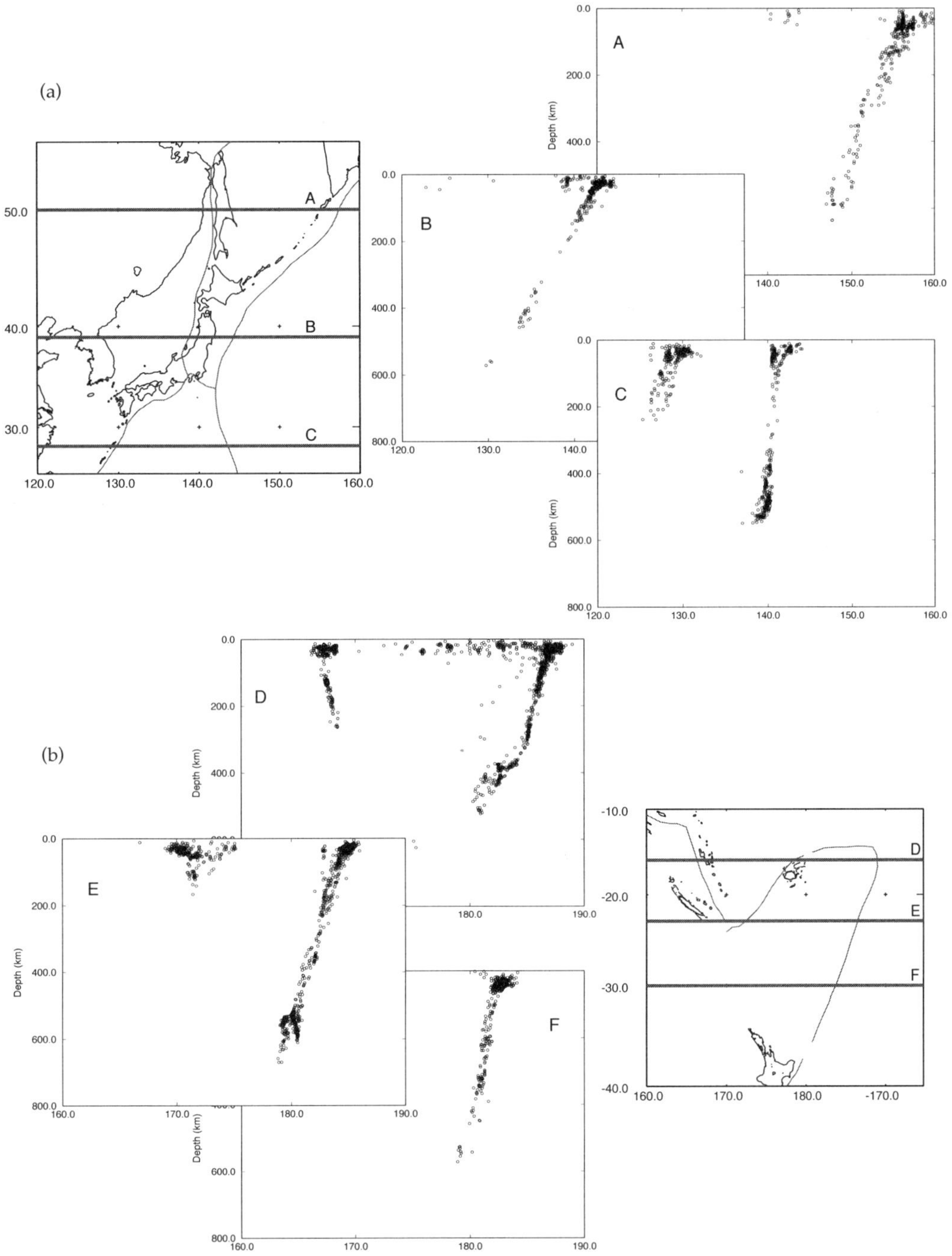

Figure 2.5. Cross sections across the subduction zones in the western Pacific. In each case the section includes all events lying within 1° in latitude of the line of section displayed in the map view. (a) Northwestern Pacific: [A] Kurile arc 50°N, [B] Northern Honshu 39°N, [C] Ryukyu and Izu-Bonin subduction zones 28°N. (b) Southwest Pacific: [D] Vanuatu and Tonga subduction zones 16°S, [E] Tonga subduction zone 23°S, [F] Tonga-Kermadec subduction zones 30°S.

Sakhalin. Beneath northern Honshu (39°N), there is a very clear dipping structure but little activity at depths between 200 and 300 km, the continuation of the subduction zone is clear in the zone from 300 to 450 km and links to the isolated deep events near 580 km depth (note that this section lies close to the line of the tomographic image in figure 2.4). The two subduction zones in the southern section (28°N) have somewhat different character: the Izu-Bonin arc in the east has a steep dip with significant activity at depths below 400 km, whereas the Ryukyu zone, which is cut more obliquely, has only activity to 200 km depth.

The lower panel of figure 2.5 illustrates sections across the subduction zones of the SW Pacific. The northern section (16°S) cuts across the Vanuatu arc as well as the northern end of the Tonga-Kermadec zone. The westward dipping structure in Vanuatu is clear to below 200 km. Shallow activity occurs near Fiji and in the Lau Basin but the main concentration of events is in the subduction zone dipping to the west; there is a distinct change in dip near 400 km depth. At 23°S the section cuts across the southern end of the Vanuatu arc and the most prominent feature is the Tonga subduction zone where the deep seismicity appears to bifurcate near 600 km depth, which may be associated with buckling of the slab as it impinges on the 660 km discontinuity in the mantle. Further south (30°S), the subduction zone appears to have a nearly constant dip.

2.2 The major elements of Earth structure

The Earth is an oblate spheroid with a flattening of 1/298 so that the polar radius is 6357 km and the equatorial radius is 6378 km. For most purposes we are able to work with a spherical model of the Earth with a mean radius of 6371 km, and use perturbation methods to account for the ellipticity of figure, e.g., via corrections to the travel times of seismic phases depending on source and receiver location.

There is now clear evidence of pervasive three-dimensional structure in the solid portions of the Earth, but the dominant variation of seismic properties is with radius. It is is therefore possible to work with a reference model for seismic structure which varies only with radius and to describe three-dimensional variations as deviations from the reference model.

The major radial subdivisions within the Earth (figure 2.6), have been inferred from the patterns of seismic wave propagation. The base of the crust was seen in the analysis of the Kupatal earthquake of 1909 by Mohorovičić. Since that time seismic methods have been used extensively to determine the properties of the crust using both natural and man-made sources. Beneath the crust lies the silicate *mantle* which extends to a depth of 2890 km, and this is separated from the metallic *core* by a major change of seismic properties that has a profound effect on global seismic wave propagation. The outer core behaves as a fluid at seismic frequencies and does not allow the passage of shear waves, while the inner core appears to be solid.

The need for a core was recognised at the end of the last century by Oldham from

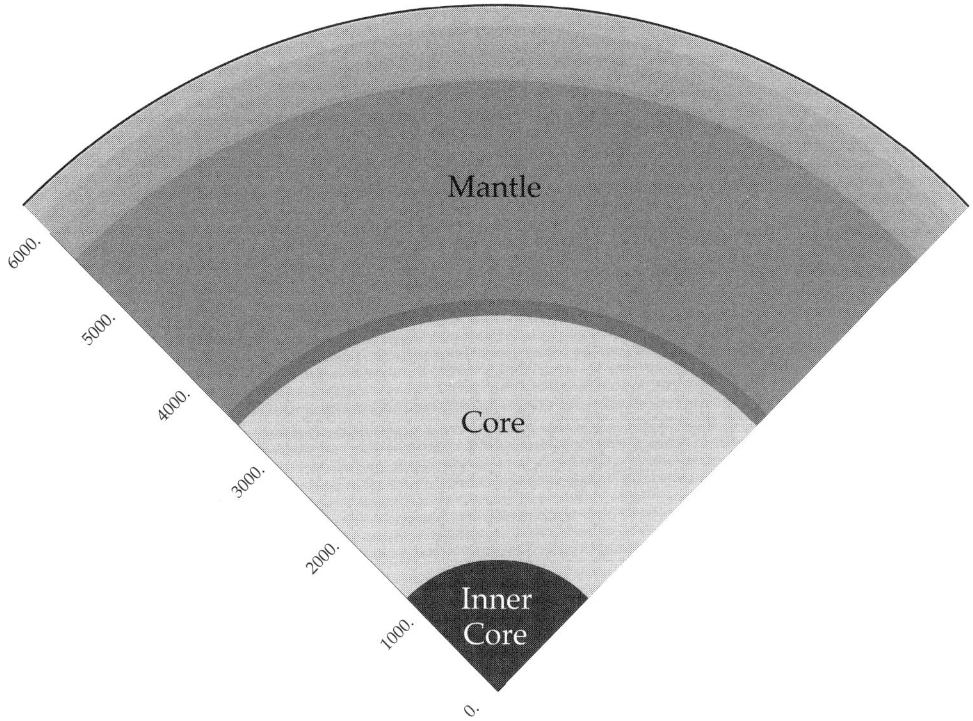

Figure 2.6. The major divisions of the radial structure of the Earth.

the great Assam earthquake of 1890 by the presence of a shadow zone in seismic observations. By 1914 Gutenburg had obtained an estimate of core radius which is quite close to the current value. The presence of the inner core was inferred by Inge Lehmann in 1932 from careful analysis of arrivals within the shadow zone, which had to be reflected from some substructure within the core.

The mantle also shows considerable variation in properties with depth, with strong gradients in seismic wavespeed in the top 800 km. The presence of structure in the upper mantle was recognised from the change in the slope of the travel time variation with distance near 20° (Jeffreys, 1939) but evidence for significant discontinuities in wavespeed first came from observations at seismic arrays (e.g., Johnson, 1967). Subsequent studies have demonstrated the global presence of discontinuities near 410 and 660 km depth, but also significant variations in seismic structure within the upper mantle (for a review see Nolet, Grand & Kennett, 1994).

The times of arrival of seismic phases constrain the variations in P and S wavespeed, and can be used to produce models of the variation with radius. The increase in the volume and quality of arrival time data from around the world collected by the International Seismological Centre over the last 35 years has enabled the development

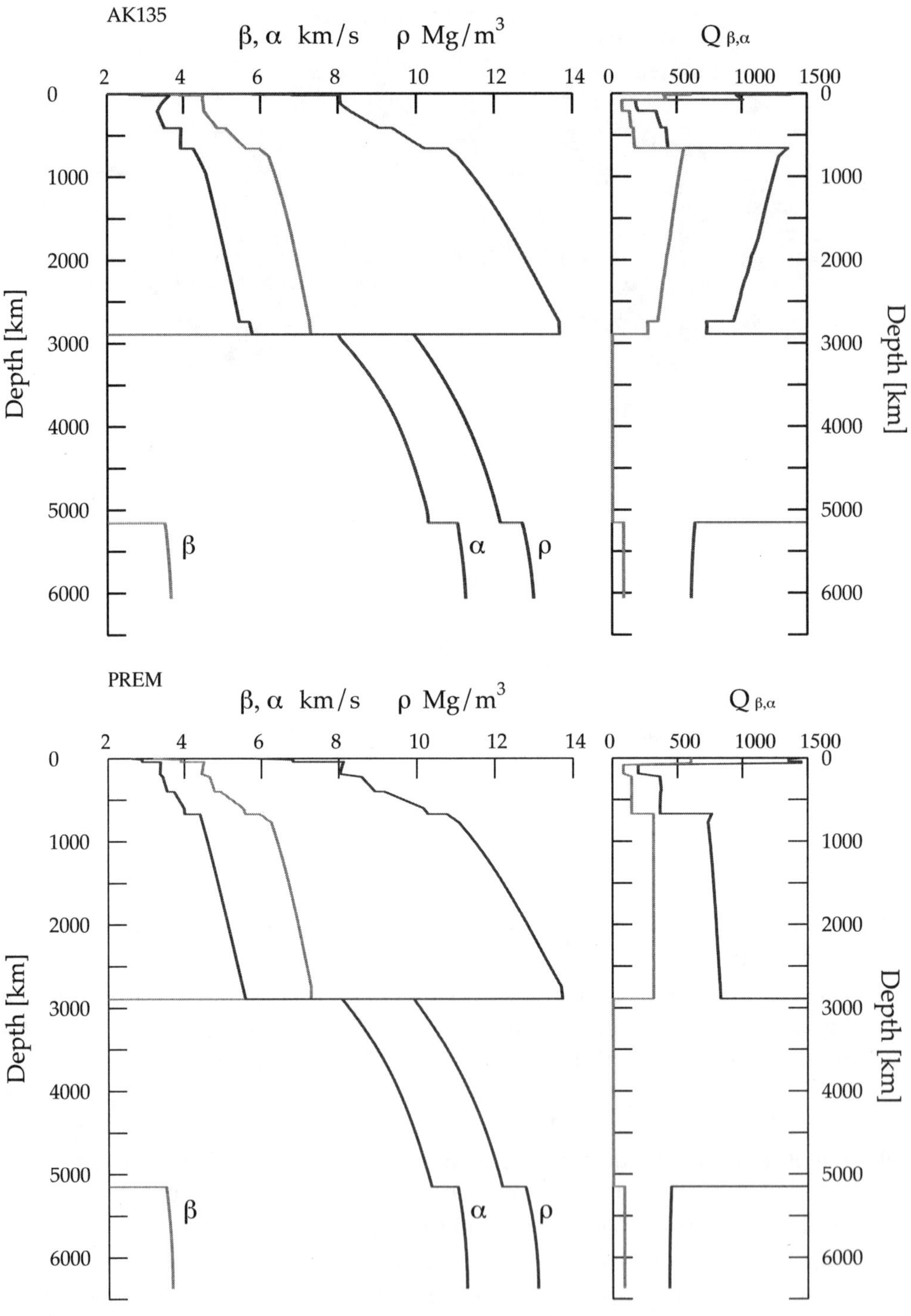

Figure 2.7. Radial reference earth models: (a) AK135, seismic wavespeeds: Kennett et al (1995), attenuation parameters, density: Montagner & Kennett (1996) (b) PREM, Dziewonski & Anderson (1981).

of high quality travel time tables, which can in turn be used to improve the locations of events. Kennett & Engdahl (1991) developed the IASP91 model based on the major phases *P* and *S*, with associated software for digital implementation. In a parallel effort, Morelli & Dziewonski (1993) produced the SP6 model with improved constraints on core structure. The IASP91 model has been adopted as the global reference model for the International Data Centre in Vienna established under the 1996 Comprehensive Nuclear-Test-Ban Treaty (CTBT).

Subsequently Engdahl and co-workers have reworked the arrival time information for all available events of high quality to improve the identification of time picks for later phases and produce new empirical relations between travel time and epicentral distance for a wide range of phases. This work enabled Kennett, Engdahl & Buland (1995) to produce a new reference model for both *P* and *S* wavespeeds, which gives a good fit to mantle and core phases. This model AK135 has since been used for a further reprocessing of the arrival time information (Engdahl, van der Hilst & Buland, 1998). The reprocessed data-set and the AK135 reference model have formed the basis of much recent work on high-resolution travel-time tomography to determine three-dimensional variations in seismic wavespeed.

The AK135 reference model is shown in figure 2.7(a) and this model will be used in many of the illustrations of the characteristics of seismic wave propagation throughout the book.

The use of the times of arrival of seismic phases constrains the *P* and *S* wavespeed, but more information is needed to provide a full model for Earth structure. The major constraints on the density distribution come from the mass and moment of inertia of the Earth. The mean density of the Earth can be reconciled with the moment of inertia if there is a concentration of mass towards the centre of the Earth; which can be associated with a major density jump going from the mantle into the outer core and a smaller density contrast at the boundary between the inner and outer cores (Bullen, 1975). With successful observations of the free-oscillations of the Earth following the great Chilean earthquake of 1960, additional information could be extracted from the frequencies of oscillation on both the seismic wavespeeds and the density. A major effort to produce a spherically symmetric reference model culminated in the PREM model of Dziewonski & Anderson (1981) shown in figure 2.7(b). The PREM model forms the basis for much current global seismology using quantitative exploitation of seismic waveforms at longer periods (see, e.g., Dahlen & Tromp, 1998).

In order to reconcile the both free oscillation and travel time observations, it is necessary to take account of the influence of anelastic attenuation within the Earth. A consequence of attenuation is a small variation in the seismic wavespeeds with frequency, so that waves with frequencies of 0.01 Hz (at the upper limit of free-oscillation observations) travel slightly slower than the 1 Hz waves typical of the short period observations used in travel time studies.

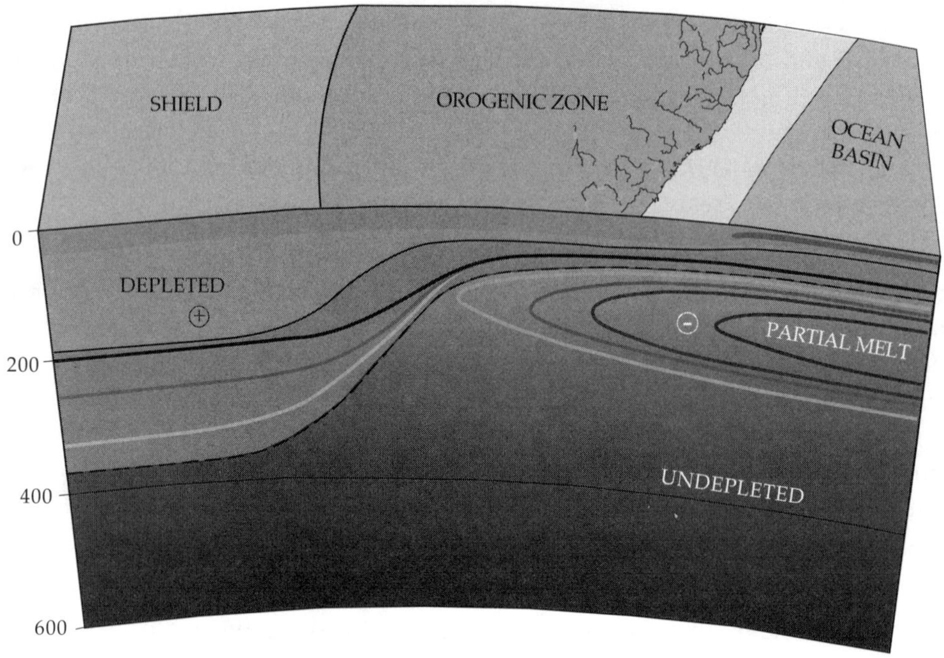

Figure 2.8. Schematic representation of lithospheric structure and associated velocity variations based on the "tectosphere" hypothesis of Jordan (1975, 1978).

2.2.1 Lithosphere and uppermost mantle

The crust and the upper part of the mantle which are entrained with plate motion form the *lithosphere* of the Earth, and there are substantial differences in the character of the lithosphere beneath continental and oceanic regions.

Oceanic lithosphere is continually being created at the mid-ocean ridges from the mantle by melt processes and is subsequently recycled back into the mantle in subduction zones. The result is that there is no very old oceanic lithosphere, the oldest current material in the northwest Pacific has an age of about 140 Ma. Oceanic lithosphere has a relatively dense basaltic crust about 7 km thick, underlain by ultramafic cumulates with lherzolitic material beneath the Moho underlain by harzburgite at depth. In the oldest, and coldest, oceanic lithosphere the thickness may reach 70 km.

In contrast, continental lithosphere has somewhat lower average density and once again has been produced by chemical segregation from the mantle involving melt extraction. The density contrast with the depleted mantle from which the lithosphere has been extracted has been sufficient to segregate the material. As a result the continental lithosphere is rather long-lived. The continental crust has an average thickness of about 35 km, but is thickened beneath mountain belts reaching nearly

70 km thick beneath the Tibetan plateau. There is considerable variation in crustal structure but some general correlations with crustal age and tectonic setting. Two recent studies have developed models for crustal variation. Mooney, Laske & Masters (1998) have produced the model CRUST 5.1 in which the structure is represented in $5 \times 5°$ blocks; structure from seismic refraction experiments is used where available, otherwise structures are inferred from analogous geologic environments. The 3SMAC model of Nataf & Ricard (1995) is more ambitious, it not only provides a crustal model at a 2° scale but also an upper mantle model based on geologic reasoning and thermal modelling.

Based on information from heat flow, chemical evidence and the relative delay times of S waves in different environments, Jordan (1975,1978) proposed the "tectosphere" model in which the zone moving with plate motion beneath old continental cratons is expected to be be very thick, so that contrasts in seismic wavespeeds between continental and oceanic environments should persist to significant depth. Figure 2.8 shows the general aspects of this model with application to the eastern coast of Australia, where Phanerozoic orogenic belts abut onto the Precambrian craton to the west and the Tasman Sea to the east. The expected variations in seismic wavespeed are indicated schematically in figure 2.8, with fast seismic wavespeeds beneath the cratons $\oplus$ and comparatively low seismic wavespeeds on the oceanic side $\ominus$. The oceanic lithosphere is thin and beneath it is a well developed zone of lower seismic wavespeeds which is also strongly attenuative, and may well be associated with the presence of partial melt.

Clearly the upper part of the mantle is expected to have substantial deviations from radial stratification, and the presence of major differences is confirmed in a wide variety of studies using different classes of seismological probes, from variations in the passage times of body wave phases, to analysis of the waveforms of seismic surface waves. High seismic wavespeeds are found to considerable depth (up to about 250 km) beneath cratonic regions, and low seismic wavespeeds beneath the mid-ocean ridges and rift zones.

2.2.2 Mantle

The character of mantle structure shows distinctive properties in different depth ranges, and it is convenient to make a distinction into four main zones (see, e.g., Jackson & Ridgen, 1998)

Upper Mantle (depth $z < 350$ km), with a high degree of variability in seismic wavespeed, as noted above, (exceeding ± 4%) and relatively strong attenuation.

Transition Zone ($350 < z < 800$ km), including significant discontinuities in P and S wavespeeds and generally high velocity gradients with depth.

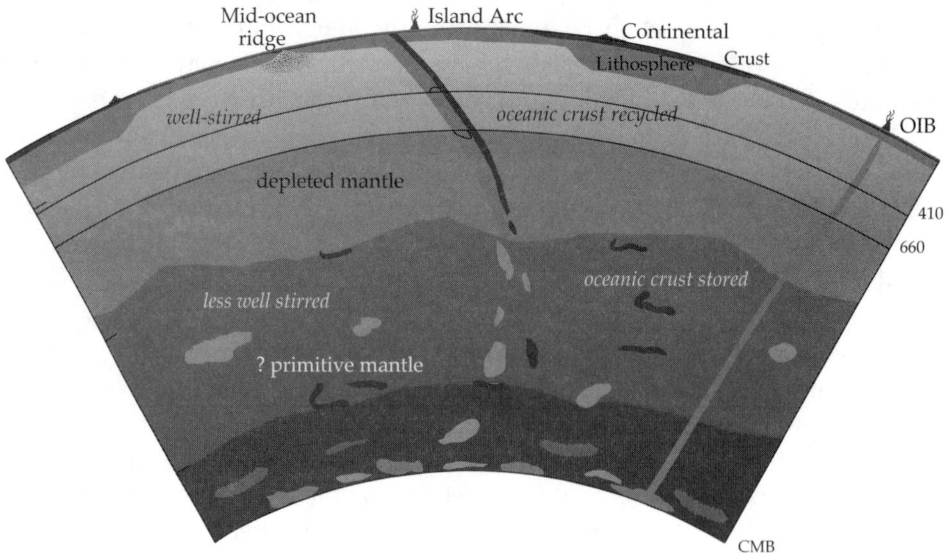

Figure 2.9. Schematic representation of processes and structures in the mantle, the features are drawn at approximately true scale.

Lower Mantle (800 < z < 2600 km) with a smooth variation of seismic wavespeeds with depth that is consistent with adiabatic compression of a chemically homogeneous material.

D" layer (2600 < z < 2900 km) with a significant change in velocity gradient and evidence for strong lateral variability and attenuation.

The presence of two major discontinuities in seismic wavespeeds near depths of 410 and 660 km are well established from a variety of classes of study, for example, array studies (e.g., Johnson, 1967), the interpretation of refracted seismic waves (e.g., Nolet, Grand & Kennett, 1994), and minor arrivals imaged in stacks of global seismic records (e.g., Shearer, 1991). These discontinuities are associated with phase transformations in silicate minerals induced by the effects of increasing pressure, and represent changes in the organisation of the oxygen coordination with the silicon atoms. The change in seismic wavespeed across these discontinuities occurs quite rapidly, and they are seen in both short-period and long-period observations. Other minor discontinuities have been proposed, but only the one near 520 km appears to have some global presence in long-period stacks but not short-period data; the transition may occur over an extended zone, e.g, 30–50 km, so that it still appears sharp for long-period waves with wavelengths of 100 km or more. Jackson & Ridgen (1998) provide a broad ranging review of the interpretation of seismological models for the transition zone and their reconciliation with information from mineral physics.

Frequently the lower mantle is taken to begin below the 660 km discontinuity, but strong gradients in seismic wavespeeds persist to depths of the order of 800 km and it seems appropriate to retain this region in the transition zone. However, the lower mantle between 800 km and 2600 km has, on average, relatively simple properties which would be consistent with the adiabatic compression of a mineral assemblage of constant chemical composition and phase. Near the base of the mantle, in the D" layer, the character of the seismic wavespeeds changes significantly, the velocity gradient with depth is reduced and there is extensive evidence for lateral heterogeneity (see, e.g., Young & Lay, 1990).

Superimposed on the main radial variation of seismic properties the mantle shows widespread three-dimensional variation which must reflect the range of processes which have shaped the evolution of the structure. Geochemical evidence requires the presence of a number of distinct geochemical reservoirs within the mantle. Figure 2.9 gives a schematic representation of these reservoirs (see, e.g., O'Neill & Palme, 1998), and the geodynamic processes which link them. Some of the features such as the thick continental lithosphere and the presence of subduction zones have a clear expression in seismic properties. Others, such as the plume features associated with e.g. ocean island basalts, have marginal observability with current seismological probes.

We can understand the main features of the seismic wavefield without needing to dwell on the detail of three-dimensional variation in the mantle, but the variations are important and will contribute to the background against which we try to recognise seismic arrivals.

2.2.3 Core

The core-mantle boundary at about 2890 km depth marks a substantial change in physical properties in the transition from the silicate mantle to the fluid metallic core (see figure 2.7). There is a significant jump in density, and a dramatic drop in P wavespeed from 13.7 to 8.0 km/s. The major change in wavespeed arises from the absence of shear strength in the fluid outer core, which means that no shear waves are transmitted through this region.

The process of core formation requires the segregation of heavy iron-rich components in the early stages of the accretion of the earth (see, e.g., O'Neill & Palme, 1998). The core is believed to be largely composed of an iron-nickel alloy, but its density requires the presence of some lighter elemental components. A wide variety of candidates have been proposed for the light components, but it is difficult to satisfy the geochemical constraints on the nature of the bulk composition of the Earth. The inner core appears to be solid and formed by crystallisation of material from the outer core, but may include some entrained fluid. The shear wavespeed inferred from free-oscillation studies is very low and the the ratio of P to S wavespeeds is comparable to a slurry-like material. The structure of the inner core is both anisotropic and shows three-dimensional variation (see, e.g., Creager, 1999).

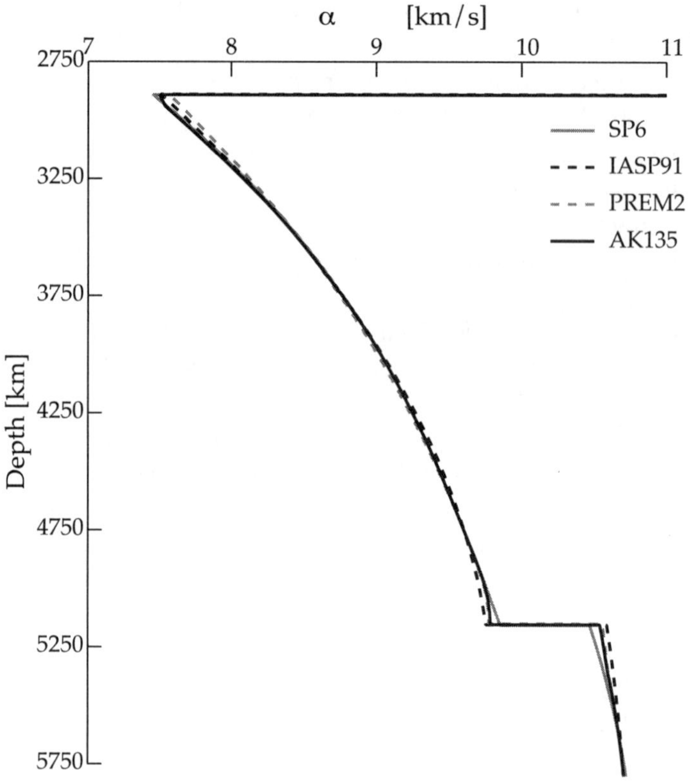

Figure 2.10. Comparison of *P* wavespeed models for the core: IASP91, Kennett & Engdahl, 1981; SP6, Morelli & Dziewonski, 1982; AK135, Kennett et al, 1995; PREM2, Song & Helmberger, 1993.

The fluid outer core is conducting and motions within the core create a self-sustaining dynamo which generates the main component of the magnetic field at the surface of the Earth. The dominant component of the geomagnetic field is dipolar but with significant secondary components. Careful analysis of the historic record of the variation of the magnetic field has lead to a picture of the evolution of the flow in the outer part of the core (see, e.g., Bloxham & Gubbins, 1989). The presence of the inner core may well be important for the action of the dynamo, and electromagnetic coupling between the inner and outer cores could give rise to differential rotation between the two parts of the core (Glatzmaier & Roberts, 1996). Efforts have been made to detect this differential rotation using the time history of different classes of seismic observations but the results are currently inconclusive.

The main features of the variation in seismic wavespeeds in the core are well established, but there are noticeable differences in the details of models near the core-mantle boundary and the boundary between the inner core and outer core (figure 2.10).

3

Seismic Waves

For most of the Earth's mantle a convenient approximation is that the polycrystalline aggregate of silicate minerals has properties close to isotropic. In the core we would expect the convection in the fluid outer core to maintain a nearly isotropic condition, but the segregation process as the inner core grows may give rise to some class of anisotropy. For convenience we will base our general discussion in Part I on isotropic media and will introduce concepts from anisotropy only where they are explicitly needed. In the more detailed treatment in Part II we will frequently employ a full anisotropic representation both for generality and for concise exposition.

3.1 Body waves

In a uniform isotropic medium two types of elastic waves are possible (figure 3.1):

***P* waves** which are compressional waves which have longitudinal particle motion aligned with the propagation direction. The deformation produced by the passage of these waves involves change in the volume of a material element and is directly analogous to the propagation of sound waves in a fluid.

***S* waves** which induce shearing deformation with no change in volume, the particle motion is transverse to the propagation path (as in electromagnetic waves). *S* waves do not propagate through perfect fluids because they cannot sustain shear deformation.

The properties of these waves can be developed in terms of the equations of motion for an elastic solid and the relation between stress and strain (see chapters 7 and 8). In terms of the bulk modulus κ, the shear modulus μ, and the density ρ, the speeds of propagation for the two wave types are:
for P waves

$$\alpha = \sqrt{\frac{\kappa + \frac{4}{3}\mu}{\rho}}, \qquad (3.1.1)$$

and for S waves

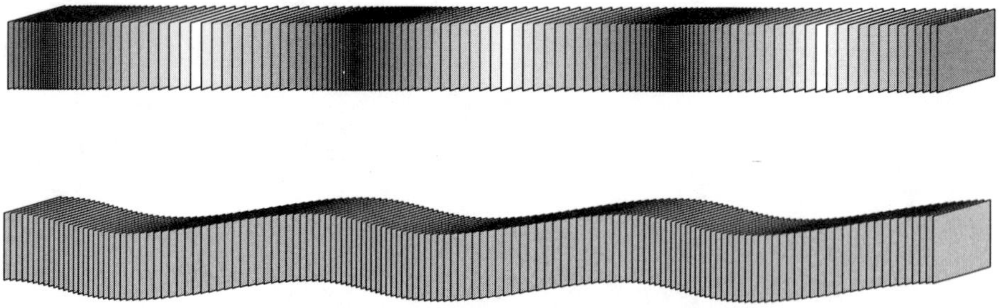

Figure 3.1. Elastic waves in a uniform medium: P waves with longitudinal motion and S waves with transverse (shear) motion.

$$\beta = \sqrt{\frac{\mu}{\rho}}. \tag{3.1.2}$$

The P waves travel with a faster propagation speed than the S waves. The first motion arriving from a source will therefore be associated with the P wave.

For a source consisting of a point force in a uniform medium, Stokes (1849) demonstrated that the resulting disturbance consisted of two dominant elements propagating at the P and S wave speeds linked by a contribution which decays more rapidly with distance away from the source. At large distances from the source the "far-field" contribution consists of two disturbances spreading spherically from the source which decay with distance from the source R as R^{-1}, the spherical wavefronts are modulated by a radiation pattern which depends on the orientation of the source (see Section 4.2). The "far-field" P wave contribution represents a wave spreading spherically about the source propagating with the wavespeed α with the radial dependence

$$\frac{1}{4\pi\rho\alpha^2 R}\mathcal{F}\left(t - \frac{R}{\alpha}\right). \tag{3.1.3}$$

The S waves take a similar form.

$$\frac{1}{4\pi\rho\beta^2 R}\mathcal{F}\left(t - \frac{R}{\beta}\right). \tag{3.1.4}$$

The decay as R^{-1} is required by the conservation of energy on the spherically expanding waves.

The spherical wavefronts of (3.1.3), (3.1.4) may alternatively be represented as a superposition of plane waves with differing frequencies and directions of propagation. The way in which plane wave elements can combine to describe the spatial characteristics of a wave front is displayed in figure 3.2(a), in which the shape of the wavefront is built up by the envelope of plane wave segments. In a full plane wave representation, including the amplitude characteristics, there would be

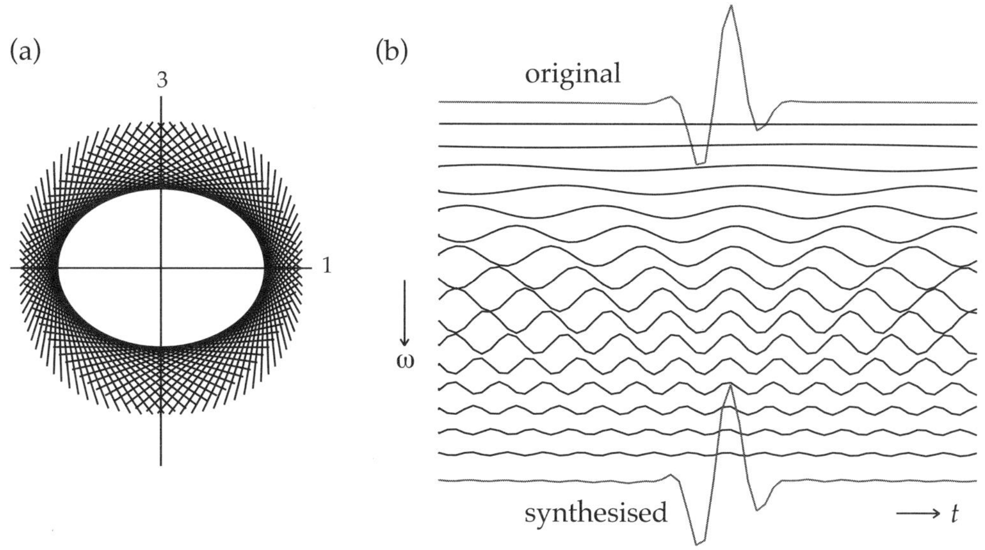

Figure 3.2. (a) Representation of a wavefront by the envelope of plane wave segments. (b) Synthesis of a seismic waveform by superposition of frequency components. The upper trace represents the original pulse and the lower the sum of 16 sinusoidal traces with steadily increasing angular frequency (ω).

constructive interference in the vicinity of the zone where the linear segments coincide, and destructive interference where they are separated.

When we make a plane wave decomposition of the wavefield, the mathematical form is a Fourier representation in terms of frequency and wavenumber components. We can illustrate the process by the synthesis of a waveform from a set of discrete frequency components [figure 3.2(b)]. The various sinusoidal terms interfere both constructively and destructively to reconstruct the original waveform.

The formal mathematical representation involves a continuous distribution over frequency

$$f(t) = \frac{1}{\sqrt{2\pi}} \int_{-\infty}^{\infty} \mathrm{d}\omega\, \bar{f}(\omega) \exp(-\mathrm{i}\omega t), \qquad (3.1.5)$$

where the transform $\bar{f}(\omega)$ is a complex variable containing the amplitude and phase shift information for the frequency ω which is complemented by the complex exponential $\exp(-\mathrm{i}\omega t)$ that contains the main phase information.

There is a significant advantage in examining the properties of seismic waves in the frequency domain because it provides a means of including the attenuation of the waves. The rocks through which seismic waves propagate are not perfectly elastic but the effects of loss are normally not large in one propagation cycle. As a result the effect of the wide range of different processes which convert elastic to thermal energy can be represented via introducing an imaginary part to the elastic moduli at each frequency. This means that the elastic wavespeeds are also complex: the real part corresponds to the propagating term whilst the imaginary part introduces a slight decay to take account of the energy transfer processes. With this complex representation the equations for the seismic waves at each frequency mirror those for an elastic medium. Details of the treatment of attenuation can be found in section 8.3.

A plane wave is a convenient mathematical abstraction describing a disturbance with fixed frequency ω and a wavenumber vector $\mathbf{k}$,

$$\mathbf{u} = \mathbf{v}(\mathbf{k}, \omega) \exp(\mathrm{i}\mathbf{k} \cdot \mathbf{x} - \mathrm{i}\omega t), \tag{3.1.6}$$

where the complex wave vector $\mathbf{v}(\mathbf{k}, \omega)$ includes amplitude, phase shift and polarisation information.

It is common to separate out the dependence on the horizontal coordinates x_1, x_2 and work in terms of a horizontal slowness vector

$$\mathbf{p} = [p_1, p_2, 0] = \frac{1}{\omega}[k_1, k_2, 0] \tag{3.1.7}$$

The horizontal slowness components represent the inverse of the apparent propagation speed along the corresponding coordinate axis. As we shall see later, Snell's law for horizontal boundaries corresponds to the preservation of the horizontal slowness $\mathbf{p}$. The extraction of the horizontal terms leaves the dependence on the vertical direction x_3, and this is convenient for layered media where we will need to deal with both upgoing and downgoing waves. We introduce a vertical slowness $q = k_3/\omega$ and can then rewrite the expression for the plane wave in the form

$$\mathbf{u} = \mathbf{v}(\mathbf{k}, \omega) \exp(\mathrm{i}\omega[\mathbf{p} \cdot \mathbf{x}_\perp + q x_3 - t]), \tag{3.1.8}$$

where the horizontal position vector $\mathbf{x}_\perp = [x_1, x_2, 0]$.

The complex wave vector $\mathbf{v}$ will have different forms depending on the polarisation of the wave-type. For simplicity we will consider waves travelling in the x_1 direction in a coordinate system with the x_3 axis directed vertically downwards as in the common seismological system [x_1 - North, x_2 - East, x_3 - Down].

For a P wave travelling at an angle i to the vertical the horizontal slowness

$$p = p_1 = \frac{\sin i}{\alpha}, \tag{3.1.9}$$

and the corresponding displacement takes the form

$$\mathbf{u}^P = \begin{pmatrix} \sin i \\ 0 \\ \cos i \end{pmatrix} A \exp(\mathrm{i}\omega[p x_1 + q_\alpha x_3 - t]), \tag{3.1.10}$$

where the vertical slowness q_α is given by

$$q_\alpha = \left(\frac{1}{\alpha^2} - p^2\right)^{1/2} = \frac{\cos i}{\alpha}, \tag{3.1.11}$$

and A is a complex quantity including the amplitude and phase behaviour.

When the horizontal slowness p exceeds the P wave slowness α^{-1}, we have an *evanescent* wave in which there is energy transport in the x_1 direction but not in x_3. The vertical slowness satisfies

$$q_\alpha^2 = \frac{1}{\alpha^2} - p^2, \tag{3.1.12}$$

and the square root to extract q_α imposes a branch cut in the complex p-plane. We choose the convention for the radical so that $\operatorname{Im} \omega q_\alpha \geq 0$. For a perfectly elastic medium the evanescent wave then has the form

$$\exp\left(\mathrm{i}\omega[px_1 - t] + \mathrm{i}\omega q_\alpha x_3\right) = \exp\left(\mathrm{i}\omega[px_1 - t] - \omega|q_\alpha|x_3\right), \tag{3.1.13}$$

with an exponential decay with depth.

Such evanescent waves need to be included to provide a full 'plane-wave' description of the wavefield, and can play an important role in many situations particularly for surface waves.

For S waves in an isotropic medium the polarisation is perpendicular to the propagation direction and we are free to choose two orthogonal directions to describe the wavefield. The conventional choice is to take one axis horizontal, designated *SH*, and the other in the vertical plane, designated *SV*. For an S wave travelling at an angle j to the vertical, the horizontal slowness

$$p = p_1 = \frac{\sin j}{\beta}, \tag{3.1.14}$$

with associated vertical slowness q_β,

$$q_\beta = \left(\frac{1}{\beta^2} - p^2\right)^{1/2} = \frac{\cos j}{\beta}. \tag{3.1.15}$$

For *SH* waves travelling in the x_1 direction, the displacement is entirely in the x_2 direction

$$\mathbf{u}^H = \begin{pmatrix} 0 \\ 1 \\ 0 \end{pmatrix} C \exp(\mathrm{i}\omega[px_1 + q_\beta x_3 - t]), \tag{3.1.16}$$

with a complex amplitude C.

The displacement for *SV* waves lies in the vertical $\{13\}$ plane

$$\mathbf{u}^V = \begin{pmatrix} \cos j \\ 0 \\ \sin j \end{pmatrix} B \exp(\mathrm{i}\omega[px_1 + q_\beta x_3 - t]), \tag{3.1.17}$$

with complex amplitude B.

The S waves also display evanescent behaviour when the horizontal slowness p exceeds the S waveslowness β^{-1}; we again impose the condition $\operatorname{Im} \omega q_\beta \geq 0$ on the choice of radical.

For a common horizontal slowness p the inclination of the P and S waves to the vertical are related by a form of Snell's law

$$p = \frac{\sin i}{\alpha} = \frac{\sin j}{\beta}. \tag{3.1.18}$$

In a region with uniform properties we can build up the full seismic wavefield by superposition of different plane wave components, and this approach is also suitable for anisotropic media (Section 8.2.2). Plane waves therefore provide a convenient simple representation to examine propagation processes.

In an isotropic medium with smoothly varying seismic wavespeeds there are still just two types of seismic waves but the behaviour loses the simplicity of the plane waves. However, for high frequency disturbances, we can examine the evolution of seismic wavefronts and introduce seismic rays for which the local slowness vector lies normal to the wavefront (see Chapter 9). The evolution of the seismic rays can be characterised via the progressive use of local plane wave representations along the ray path.

Plane wave and ray solutions provide a very different perspective on the seismic wavefield, but both are useful in linking together different aspects of the propagation process. We will use plane wave representations (or their equivalents) when we discuss stratified media. Ray representations will be employed when attention is focussed on the expected arrival times of seismic phases, and for more complex propagation scenarios.

3.1.1 Interaction of seismic waves with the Earth's surface

The surface of the Earth represents a strong contrast between the properties of the solid material and the low density air above, and commonly the presence of the atmosphere can be neglected and instead be replaced by a vacuum. The conditions at the boundary depend on the components of the stress tensor (see Chapter 7). With a upper vacuum, the surface becomes a boundary at which the surface traction components τ_{13}, τ_{23}, and τ_{33} must vanish.

Thus when a wave disturbance impinges on this free surface, a new set of reflected waves must be created so that the interaction with the incident field annihilates the surface tractions. The processes can be illustrated conveniently by looking at the behaviour of plane wave solutions which we can characterise by their horizontal slowness p or alternatively through the inclination of the P, S waves to the vertical, cf. (3.1.18).

The simplest case is the incidence of an *SH* wave at the surface [figure 3.3(a)] since this class of waves has just τ_{23} as the only non-zero traction. The surface conditions can be satisfied by adding in a wave with equal and opposite τ_{23} which can be achieved by having a reflected wave of the same amplitude travelling at the same angle j to the vertical but propagating away from the boundary, as discussed in detail in Section

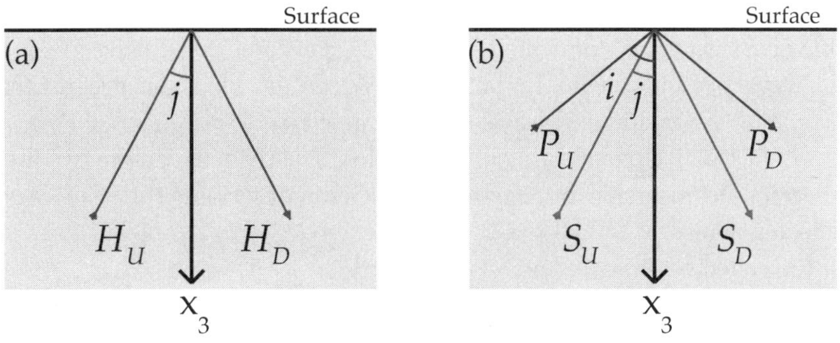

Figure 3.3. (a) *SH* wave reflection at the free-surface, (b) *P-SV* wave reflection at the free-surface.

13.1. Total reflection occurs with the incident and reflected waves having the same amplitude, which can be described by a unit free-surface reflection coefficient for *SH* waves which is independent of slowness.

$$\mathrm{R}_F^{HH}(p) = 1. \tag{3.1.19}$$

Although the tractions are equal and opposite, the displacements of the incident and reflection waves are additive, and so the surface displacement is twice that in the incident wave. We can describe this behaviour by introducing a surface amplification factor for the transverse displacement in the horizontal plane,

$$\mathrm{W}_F^{TH}(p) = 2, \tag{3.1.20}$$

which is again independent of slowness.

The situation becomes more complex when we consider either *P* or *SV* waves for which the particle motion is confined to the {13} plane, with displacement on both the 1- and 3-components [figure 3.3(b)]. Both *P* and *SV* waves have associated τ_{13} and τ_{33} tractions but unlike the *SH* case these tractions cannot be eliminated by superposing just a reflected wave of the same type. The *P* and *SV* waves are linked together by the boundary condition and so conversion between the wave types must occur at the boundary. Both incident *P* and *SV* waves will produce reflected *P* and *SV* waves.

In order that the plane wave solutions corresponding to the *P* and *SV* waves can be combined at the surface we require both waves to have the same dependence on the x_1 coordinate through a common complex exponential term

$$\exp(i\omega[px_1 - t]), \tag{3.1.21}$$

so that the waves have a common slowness p. The inclination of the waves to the vertical is given by

$$p = \frac{\sin i}{\alpha_0} = \frac{\sin j}{\beta_0}, \tag{3.1.22}$$

where α_0 and β_0 are the *P* and *S* wavespeeds at the surface. The reflected waves in the same wavetype have the same angles of incidence and reflection, and the inclination of the converted waves is controlled by (3.1.22). Because the P wavespeed is larger than the S wavespeed the inclination of the P wave to the vertical is greater than for S.

If we consider an incident P wave this will give rise to a reflected P wave and a reflected SV wave with a partitioning of the incident energy between the two wavetypes. We can therefore introduce two reflection coefficients describing the relative character of the reflected waves:

R_F^{PP} for the reflected P wave, and

R_F^{SP} for the reflected S wave.

In a similar way for an incident SV wave we produce reflected SV and P waves with corresponding reflection coefficients R_F^{SS} and R_F^{PS}. The detail of the mathematical derivation is presented in Section 13.1. With a normalisation of the wave amplitudes with respect to energy transport in the x_3 direction, the reflection coefficients show strong symmetries:

$$\mathrm{R}_F^{PP}(p) = \mathrm{R}_F^{SS}(p), \qquad \mathrm{R}_F^{SP}(p) = \mathrm{R}_F^{PS}(p). \tag{3.1.23}$$

The surface tractions associated with the various waves depend on the slowness and so the free-surface reflection coefficients R_F^{SS}, R_F^{PS} have a significant slowness variation as illustrated in figure 3.4. We work with slowness rather than angle of incidence because this enables us to use a common reference for both P and S waves.

At vertical incidence ($p = 0$) R_F^{SS} is unity and diminishes as p increases, falling to a minimum just before the P slowness α_0^{-1} (0.167 s/km) where the inclination $i = 90°$ for P and $j = \sin^{-1}(\beta_0/\alpha_0) = 34.52°$ for S. Beyond this critical slowness P waves become evanescent, the vertical slowness q_α is imaginary and the amplitude of the P wave diminishes exponentially away from the surface; in such an evanescent wave there is no energy propagation in the x_3 direction but energy can be transported in the x_1 direction. Once the P waves are evanescent, the R_F^{SS} reflection coefficient has a unit amplitude until the S waves themselves become evanescent ($p > 0.294$ s/km). For $p > \alpha_0^{-1}$ the R_F^{SS} coefficient is complex so that the reflected plane S waves acquire a phase shift relative to the incident wave.

The reflection coefficient R_F^{PS}, involving conversion of wavetypes, starts from zero at vertical incidence and the efficiency of conversion increases with increasing slowness until just before the P waves become evanescent. At slowness α_0^{-1} there is a null which is followed by a regime in which propagating S waves and evanescent P waves are coupled at the surface. For this slowness range, the R_F^{PS} coefficient does not have a direct physical interpretation and can exceed unity. Note that the S waves are totally reflected so $|\mathrm{R}_F^{SS}| = 1$.

The interaction of the incident and reflected waves (including conversion) has the effect of eliminating the traction at the surface, and also by the interaction of the displacements in the different wave components of altering the displacement field from that in the incident wave. As in the SH case we can introduce the idea of surface amplification factors for the incident waves, but now have to take account

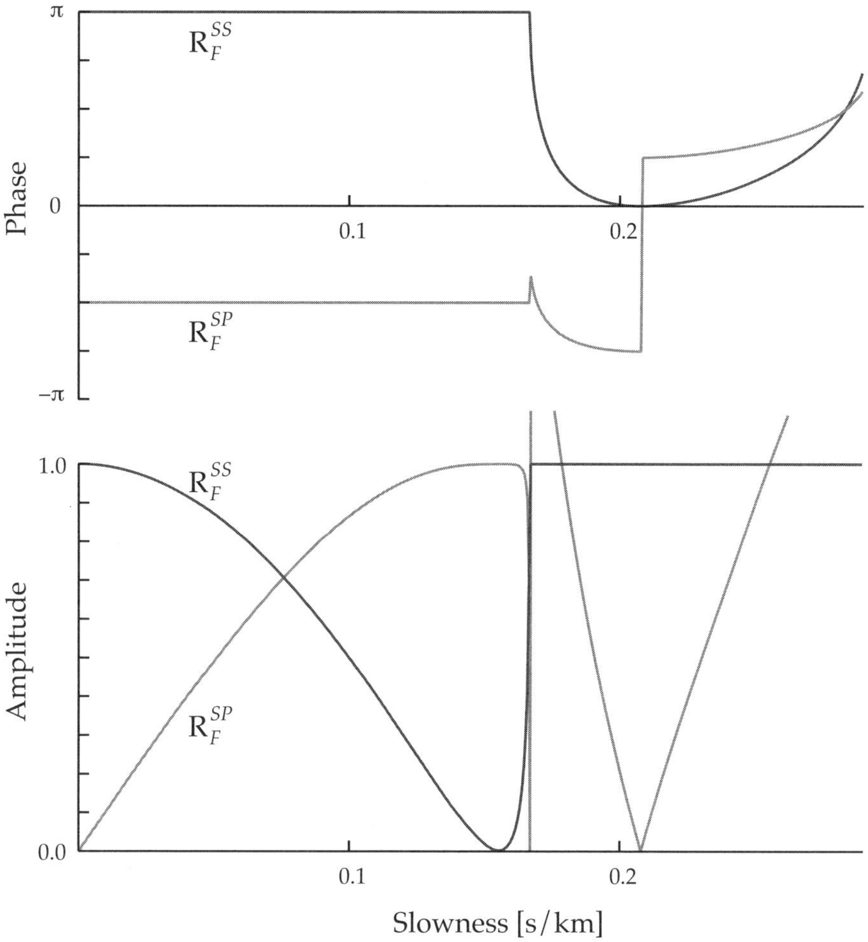

Figure 3.4. The variation of the free-surface reflection coefficients as a function of slowness: $\alpha_0 =$ 6.0 km/s, $\beta_0 = 3.4$ km/s. Critical slowness for *P*, 0.167 s/km.

of the distribution of the displacement between the 1- and 3-components. We need to introduce four coefficients W_F^{ZP}, W_F^{ZS}, W_F^{RP}, W_F^{RS} to describe the behaviour by representing the contributions of the *P* and *S* waves to the vertical and horizontal components of ground motion. In order to have a portable notation for waves travelling in different directions in the horizontal plane we have written *Z* for the 3-component and *R* for the 1-component.

The amplification of the displacements relative to that in the incident waves can be represented in terms of just two coefficients C_1 for W_F^{ZP}, W_F^{RS} and C_2 for W_F^{RP}, W_F^{ZS}. The variation of these amplification factors with slowness is illustrated in figure 3.5. The amplification factors are close to 2 for the slowness range in which *P* waves are

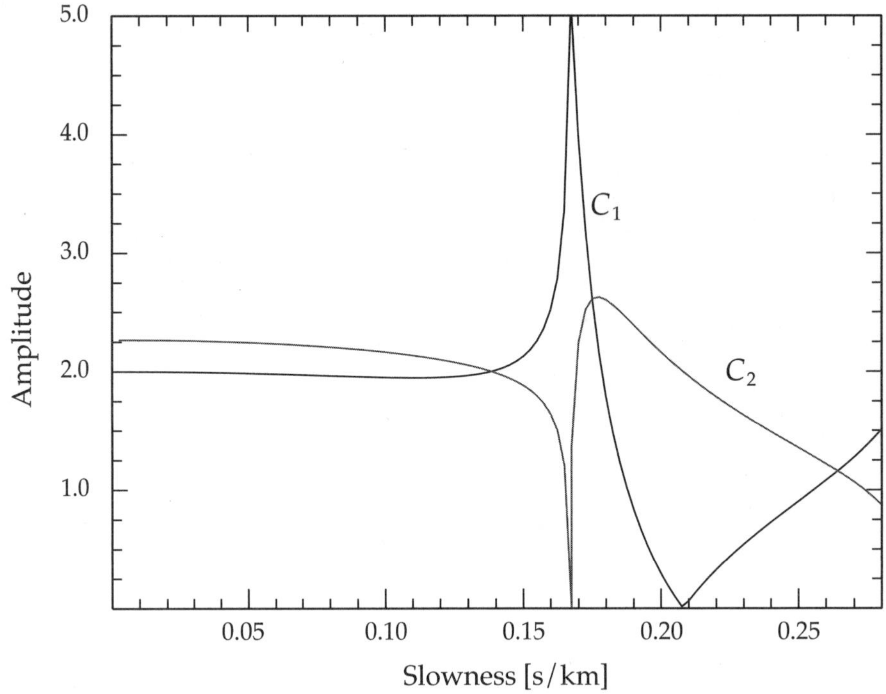

Figure 3.5. The variation of the surface amplification factors C_1, C_2 at the free-surface as a function of slowness: $\alpha_0 = 6.0$ km/s, $\beta_0 = 3.4$ km/s.

propagating. However C_1 is singular at $p = \alpha_0^{-1}$ and C_2 goes to zero. For larger slowness the amplification factors are complex and include phase shifts (only the amplitudes are shown in figure 3.5).

The plane wave representation can describe the angular dependence of the reflection process but does not give direct insight into the way the wavefield develops. A useful alternative viewpoint is to look at evolution of wavefront patterns with time, as waves interact with the free surface. In figure 3.6 we display sets of wavefronts emerging from sources of both *P* and *S* waves at the same depth beneath the free surface. For the *P* source the pattern is fairly simple: initially circular wavefronts for *P* undergo mirror reflection at the surface whereas the reflected *S* wavefront is hyperbolic and links to the reflection point for *P* on the surface. In contrast, for an *S* source, the wavefront patterns are more complex because propagating *S* waves can link to evanescent *P*. The main pattern of incident and reflected waves for *S* retains the free surface as a mirror, but now a *P* wavefront can break away from the *S* reflection point and outstrip the *S* waves. This wavefront is associated with a refracted *P* wave propagating along the free-surface with a higher wavespeed than for *S*. The surface disturbance creates a conical wavefront which links across to the reflected *S* front, and represents *S* waves travelling with an apparent horizontal velocity α_0, i.e. slowness $p = \alpha_0^{-1}$.

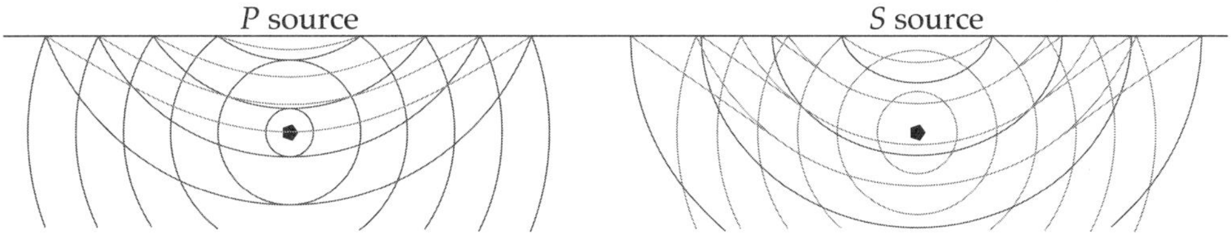

Figure 3.6. Wavefronts from P and S wave sources impinging on the free surface, time interval 2 s for P and 3 s for S: $\alpha_0 = 6.0$ km/s, $\beta_0 = 3.4$ km/s.

We can introduce a pattern of seismic rays via the normals to the wavefronts at each time instant, and we will often find this a convenient way to visualise seismic wave propagation. In a high frequency approximation we can attach amplitudes to the rays and the reflection when a ray impinges on the surface can be taken to be described by the plane wave coefficient for that angle of incidence.

3.1.2 *The influence of an internal boundary*

The example of the free-surface introduces the concept of reflection including the possibility of conversion of wavetypes, but at an internal boundary between two solid media we have also transmitted waves and still the possibility of conversion.

We require the upper and lower sides of the interface to remain in contact as a wave impinges on the boundary and hence that the displacement and the traction on the horizontal plane is continuous. For a plane wave representation we require the phase variation of all waves to have the same dependence on x_1 and hence a common slowness. This simple requirement imposes Snell's law on the inclination of the waves to the vertical, so that designating angles in the upper medium by subscript 1 and the lower medium with subscript 2,

$$p = \frac{\sin i_1}{\alpha_1} = \frac{\sin i_2}{\alpha_2} = \frac{\sin j_1}{\beta_1} = \frac{\sin j_2}{\beta_2}. \tag{3.1.24}$$

The configuration of the reflected and transmitted waves are shown in figure 3.7 for both incident P and SV waves for the material contrast we will use in all our illustrations. The simpler configuration of incident SH waves has the same geometry as the incident SV case but without the P conversions. Figure 3.7 also displays the notation for the reflection and transmission coefficients for the different incident wave types. We need to keep track of both the incident and generated waves as well as the direction of incidence onto the interface (in this case downwards as indicated by the subscript D).

The relatively complicated boundary condition means that the derivation of the reflection and transmission coefficients is not straightforward, but is much simpler for the SH case than for the coupled P-SV wave system. The detail is presented in Section 13.2, but here we will concentrate on the results. The consequence of Snell's

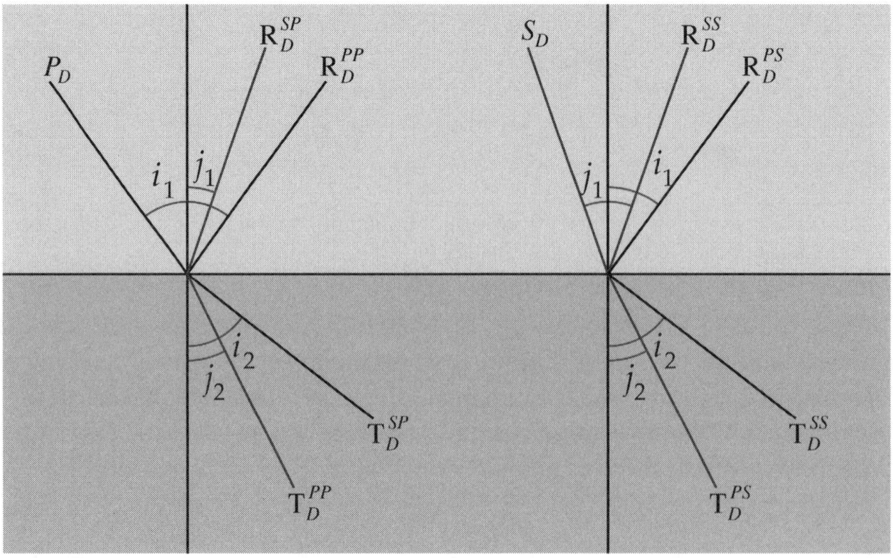

Figure 3.7. The configuration of the reflected and transmitted waves at an internal interface for $p = 0.1$ s/km: $\alpha_1 = 6.0$ km/s, $\beta_1 = 3.4$ km/s, $\rho_1 = 2.7$ Mg/m^3, $\alpha_2 = 8.0$ km/s, $\beta_2 = 4.6$ km/s, $\rho_2 = 3.3$ Mg/m^3.

law is that the inclination to the vertical increases as waves pass from a medium (or wavetype) with lower wavespeed to a medium (or wavetype) with higher wavespeed. Thus for P waves incident on a jump to higher wavespeeds from above, there will be a critical angle of incidence at which the refracted angle i_2 in the second medium becomes 90° with propagation along the boundary. For angles of incidence larger than the critical angle

$$i_{c\alpha} = \sin^{-1}\left(\frac{\alpha_1}{\alpha_2}\right), \tag{3.1.25}$$

corresponding to $p > \alpha_2^{-1}$, the P wave in the second medium is evanescent. As the incidence angle approaches critical the width of a transmitted beam in the second medium becomes steadily narrower (figure 3.8). The ratio of beam width is governed by $\cos i_2 / \cos i_1$ and hence shrinks to zero as evanescence is approached.

There is a comparable critical angle for incident S waves

$$i_{c\beta} = \sin^{-1}\left(\frac{\beta_1}{\beta_2}\right), \tag{3.1.26}$$

corresponding to the slowness $p = \beta_2^{-1}$ at which the S waves become evanescent in the lower medium. When there is a very large contrast in properties between the two media, we can have the situation of P waves incident from the upper medium giving rise to evanescent S waves in the lower medium. The slowness is the same but the critical incidence angle will involve the wavespeeds α_1 and β_2.

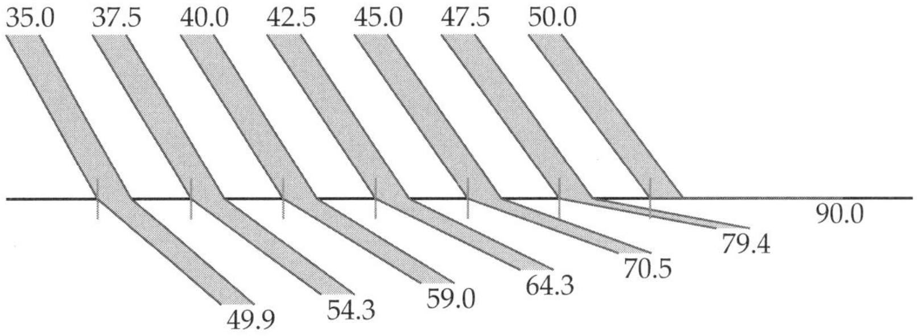

Figure 3.8. The variation in the transmission of a beam of *P* waves with increasing angle of incidence. The incident and transmitted angles are indicated in each case.

As an illustration of the interface coefficients we have been discussing, we display in figures 3.9 and 3.10 the amplitude and phase behaviour of the reflection and transmission coefficients for a plane wave incident from above on the interface illustrated in figure 3.7.

The behaviour of the reflection coefficients is governed by the relative sizes of the wave slownesses for *P* and *S* in the media on the two sides of the interface. When $p < \alpha_2^{-1}$ (here 0.125 s/km) the reflection coefficients R_D^{PP}, R_D^{SS} and R_D^{HH} are real. At vertical incidence the amplitudes of the S wave coefficients $|\mathrm{R}_D^{SS}|$, $|\mathrm{R}_D^{HH}|$ are equal and there is no conversion from P to S waves. The behaviour of R_D^{SS} and R_D^{HH} as slowness increases is very different: the *SH* wave coefficient is fairly simple, but the *SV* wave coefficient is profoundly influenced by the *P* wave behaviour.

At $p = \alpha_2^{-1}$, *P* waves are travelling horizontally in the lower medium and we have reached the critical slowness for *P* waves. For $\alpha_2^{-1} < p < \alpha_1^{-1}$, *P* waves are reflected at the interface and give rise to only evanescent waves in the lower half space. Once $p > \alpha_2^{-1}$ all the reflection coefficients for the *P-SV* system become complex. The phase of R_D^{PP} and R_D^{PS} change fairly rapidly with slowness, but R_D^{SS} has slower change.

For $p > \alpha_1^{-1}$ (here 0.167 s/km) *P* waves become evanescent in the upper medium, but we can still extend the mathematical representations of the reflection coefficients for these evanescent incident waves. The amplitude of R_D^{PS} drops to zero at $p = \alpha_1^{-1}$ and then recovers before falling to zero again at $p = \beta_2^{-1}$. The amplitude of R_D^{SS} has an inflection at $p = \alpha_1^{-1}$ and the character of the phase variation changes at this slowness.

The critical slowness for *S* waves is β_2^{-1} (0.217 s/km) and for *p* greater than this value both *SV* and *SH* waves are totally reflected. The phase for the SV wave coefficient for this interface varies more rapidly with slowness than that for *SH* waves.

The properties in transmission are controlled by the same set of critical slowness values. For $p < \alpha_2^{-1}$ transmission is efficient for both incident *P* and *S* with only a small component of the energy reflected. The T_D^{PP} coefficient dips as the reflection of *P* waves become more effective close to the critical slowness. For $p > \alpha_2^{-1}$ the transmission terms T_D^{PP}, T_D^{PS} link to evanescent waves and are just the mathematical continuation of

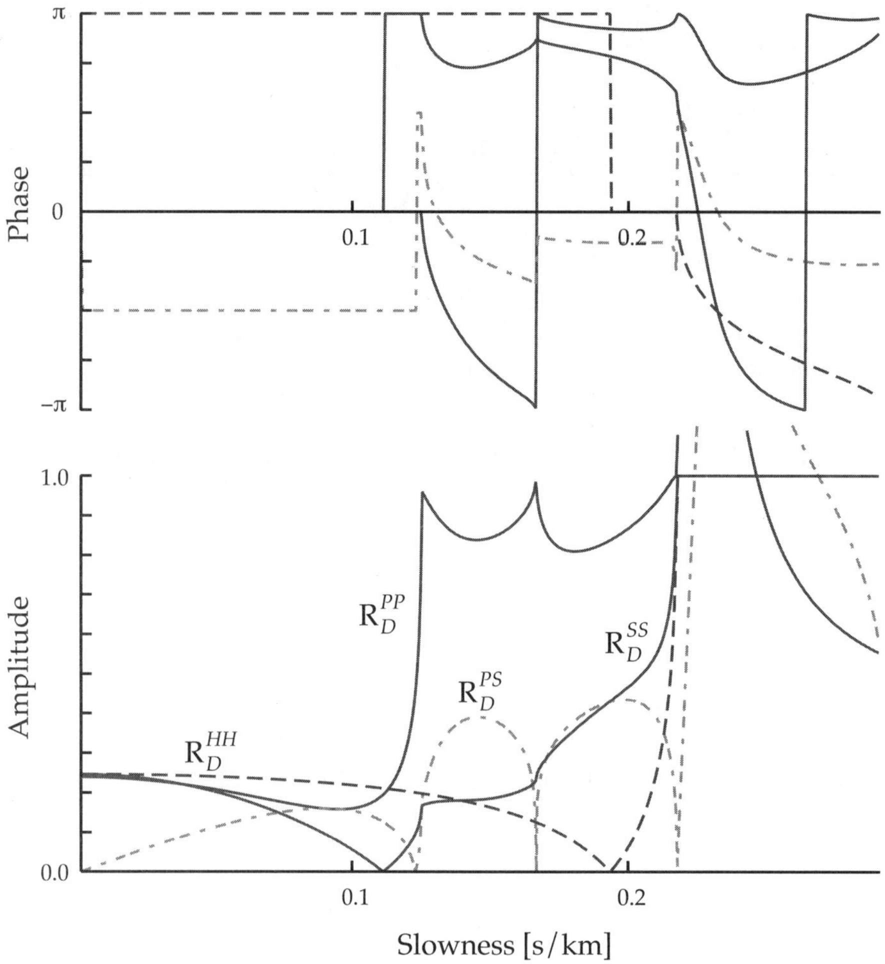

Figure 3.9. The variation of the reflection coefficients at the interface as a function of slowness.

the functional relationships that have physical meaning for smaller slownesses. The T_D^{HH}, T_D^{SS}, T_D^{SP} coefficients retain a physical interpretation until $p > \beta_2^{-1}$. The behaviour is simpler for the *SH* wave system without conversion than for incident *SV*.

When the reflection and transmission coefficients are real, an incident plane wave pulse with slowness p is merely scaled in amplitude on reflection or transmission. Once the coefficients become complex, the shape of the pulse in the generated wave is modified (see, e.g., Hudson, 1962). The real part of the coefficient gives a scaled version of the original pulse and the imaginary part introduces a scaling of the Hilbert transform of the pulse, which for an impulse has precursory effects. For post-critical reflection $p > \alpha_2^{-1}$ there will therefore be a distortion of the reflected and transmitted pulses (see also Section 13.2).

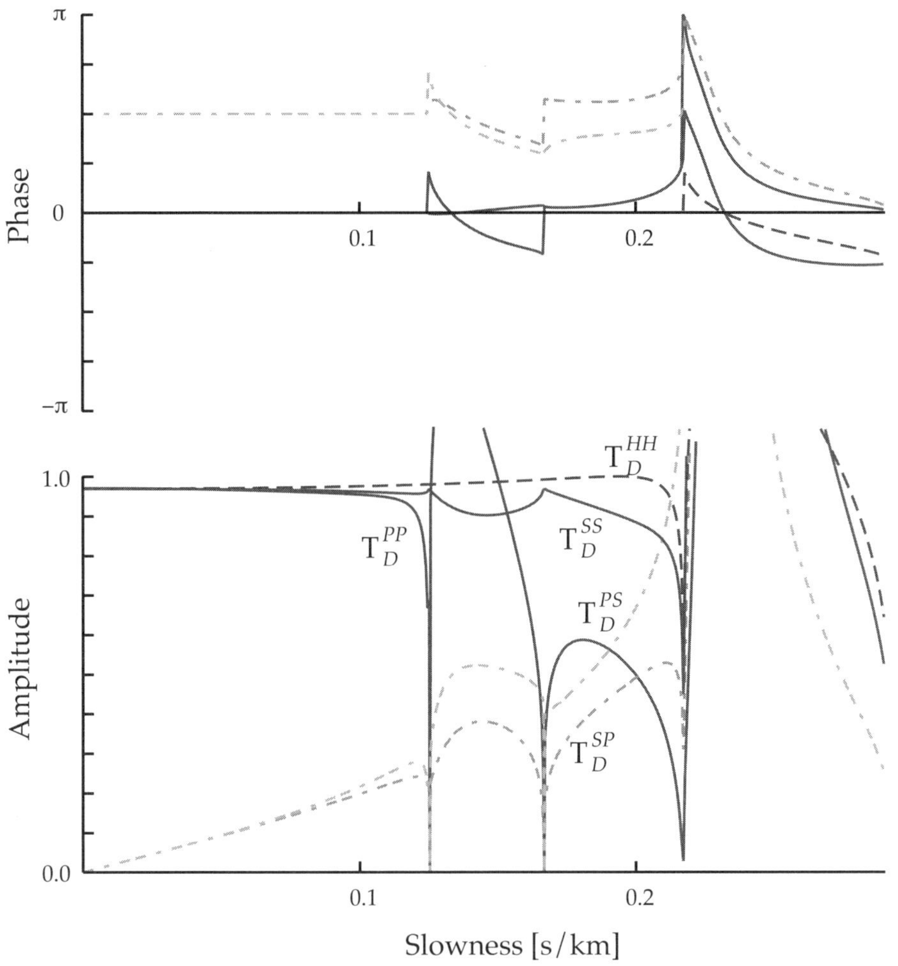

Figure 3.10. The variation of the transmission coefficients at the interface as a function of slowness.

As in the case of reflection at the free surface we gain a useful alternative view of the propagation processes at the interface by examining the evolution of wavefronts with time. In figure 3.11 we display the wavefront pattern from a *P* wave source in the upper medium. The circular wavefronts from the *P* source are reflected in the interface, as in a mirror, and so retain a simple configuration. The disturbance created by the incident wave on the interface links also into the reflected *S* wave and the transmitted *P* and *S* waves. It is the combination of all these waves which is needed to satisfy the boundary conditions of continuity of displacement and traction at the interface. Because the *P* wavespeed in medium 2 is faster than that in medium 1 where the source is situated, as time progresses the *P* wavefront in the lower medium separates from the reflection point. In order to link this disturbance with wavespeed α_2 along the interface with the

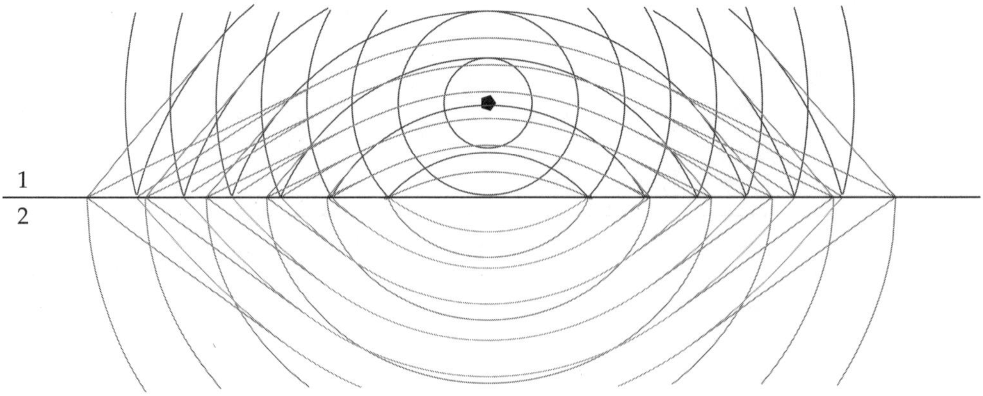

Figure 3.11. Wavefronts from a P wave source in the upper medium.

other components of the wavefield, a set of conical wavefronts are formed which link to the points on the reflected and transmitted wavefronts where the apparent horizontal velocity is also α_2. These 'head' waves heal the discontinuity in the wavefronts that would otherwise exist between the transmitted P wave and the wavefronts tied to the P reflection point. Thus linked to the reflected P wave we have a P head wave with slowness α_2^{-1} returned into medium 1, together with an S head wave with the same slowness, and thus a smaller inclination to the vertical, linked to the reflected S wavefront. There is a third conical S wavefront tied to the transmitted S wave in the lower medium. In the plane wave viewpoint these head waves link propagating waves to evanescent P with slownesses $p > \alpha_2^{-1}$.

The predictions of this simple analysis in terms of wavefronts are confirmed by direct numerical calculations. Figure 3.12 shows snapshots of the seismic wavefield for both *P-SV* waves and *SH* waves calculated using the pseudospectral method (Furumura & Takenaka, 1996) for an earthquake source at a depth of 5 km in a model with a crust overlying a stratified mantle with the inclusion of the free surface. Head waves *Pn*, *Sn* generated by the faster propagation in the mantle are clearly seen for both wave types. We also see the influence of multiple reflections from the free surface in generating a complex wavefield in the crust, with many reflections for P waves and a clear group of two distinct S wavefronts associated with *Lg* wavegroup.

When a P wave source is placed in the lower medium the wavefront pattern is substantially simplified because there are no conical wave contributions. In figure 3.13 we see the approach of the P wave to the interface, followed by mirror reflection for P in the lower medium. The incident P wave generates wavefronts of reflected S in medium 2 and transmitted P, S waves in medium 1. These elliptical wavefronts are linked to the current reflection point for P, where the incident and reflected wavefronts coincide.

However, if we consider waves from an S wave source in the lower medium the

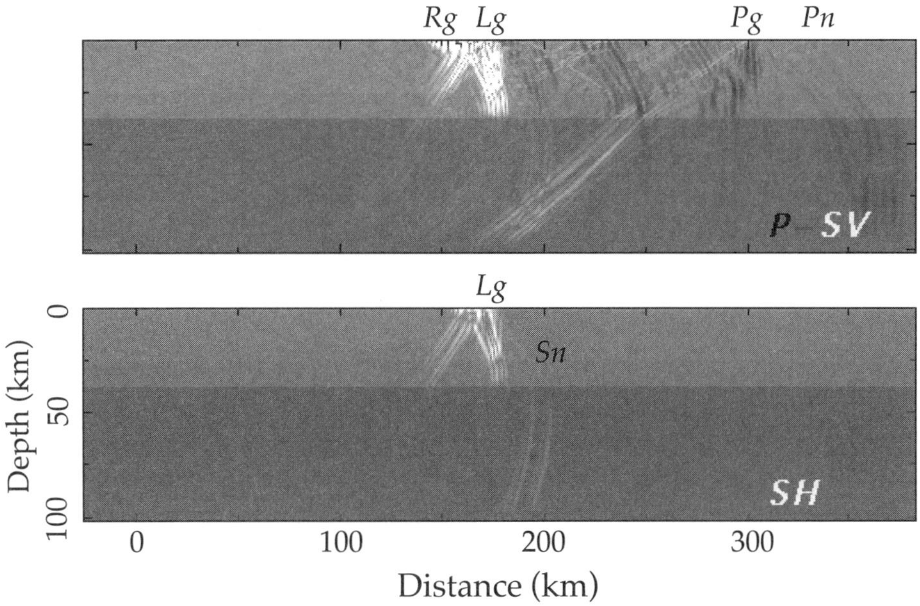

Figure 3.12. Snapshots of the seismic wavefield 50 s after initiation, for both *P-SV* waves and *SH* waves illustrating the development of head waves *Pn*, *Sn* and multiply reflected waves, e.g., for *P* in the upper panel, and *Lg* in both panels. *P* waves are indicated in dark grey and *SH* waves in white.

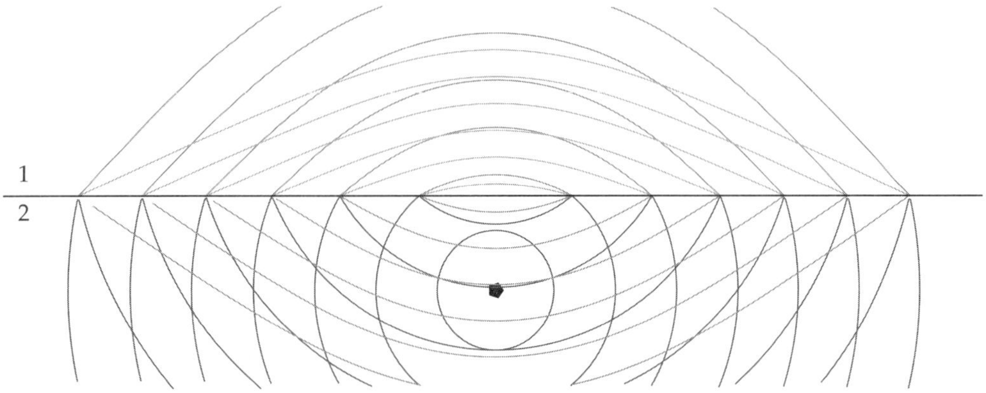

Figure 3.13. Wavefronts from a *P* wave source in the lower medium.

situation would be more complicated. The reflection and transmission of *S* will remain simple in character, with mirror reflection and an elliptical wavefront for transmitted *S*. But, the wavespeeds for *P* on both sides of the interface exceed the wavespeed for *S* in medium 2 and so there will be conical waves linked to the hyperbolic wavefronts for both reflected and transmitted *P* waves.

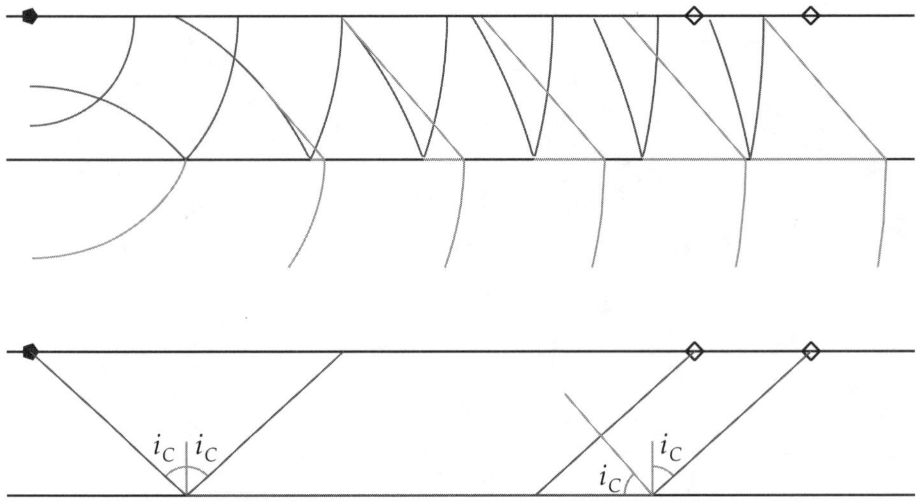

Figure 3.14. Wavefront and ray pictures for P waves illustrating the generation of a head wave at the interface between two media, with higher seismic wavespeed below the interface.

3.1.3 Ray representations of seismic propagation

As previously indicated we can introduce seismic rays whose trajectories lie normal to the wavefronts at each instant. The relationship between the ray and wavefront representations can be conveniently illustrated by looking at the generation of a head wave, as shown for P waves in figure 3.14. In the upper panel we see that the fast wavefront in the lower medium drags a disturbance along the interface which generates a head wave with slowness α_2^{-1} in the upper medium, which propagates away from the interface ultimately reaching a receiver. The corresponding ray picture is shown in the lower panel. No head wave reaches the surface until the distance corresponding to the ray at the critical angle $i_{c\alpha}$ (3.1.25), For greater distances, in addition to a post-critical reflection from the interface there will be a head wave which appears to originate from the interface, since the ray will lie perpendicular to the conical wavefront.

We can readily extend the analysis of rays to a model with horizontal stratification for which the slowness p for a ray remains a constant (see section 10.1). The total travel time and distance for a particular ray path can then be constructed by combining the contributions from individual layers, invoking Snell's law for reflection or refraction at any interfaces.

For a single ray leg traversing a layer from z_a to z_b, the horizontal distance increment $\Delta X_{ab}(p)$ and time increment $\Delta T_{ab}(p)$ are given by

$$\begin{aligned} \Delta X_{ab}(p) &= \int_{z_a}^{z_b} \mathrm{d}z \, \frac{p}{[v^{-2}(z) - p^2]^{1/2}}, \\ \Delta T_{ab}(p) &= \int_{z_a}^{z_b} \mathrm{d}z \, \frac{v^{-2}(z)}{[v^{-2}(z) - p^2]^{1/2}}. \end{aligned} \tag{3.1.27}$$

Here $v(z)$ is the local elastic wave speed for the wave mode in which the ray is currently travelling. The total horizontal distance travelled and the corresponding travel time are then to be found by summing over the contributions from all the requisite ray legs

$$X^r(p) = \sum_i \Delta X_i, \quad T^r(p) = \sum_i \Delta T_i. \tag{3.1.28}$$

where the superscript indicates the particular seismic phase.

For such a stratified medium the horizontal slowness p has also the significance of the slope of the travel time-distance curve $p = \mathrm{d}T/\mathrm{d}X$, and so the intercept time $\tau(p)$ for the local tangent to the travel time curve is given by

$$\tau^r(p) = T^r(p) - pX^r(p) = \sum_i [\Delta T_i - p\Delta X_i]. \tag{3.1.29}$$

The individual layer contributions to $\tau^r(p)$ can be derived from (3.1.27) as

$$\Delta\tau_{ab}(p) = \int_{z_a}^{z_b} \mathrm{d}z\, [v^{-2}(z) - p^2]^{1/2}. \tag{3.1.30}$$

The derivative of $\tau^r(p)$ with respect to slowness p is simply related to the horizontal distance travelled,

$$\frac{\mathrm{d}\tau^r}{\mathrm{d}p} = -X^r(p). \tag{3.1.31}$$

For an individual seismic phase $\tau^r(p)$ will be a monotonically decreasing function of increasing slowness and is single valued in the absence of any velocity inversions.

An important special case is a velocity distribution which is locally a linear function of depth:

$$v(z) = v_a + b(z - z_a), \tag{3.1.32}$$

with a constant velocity gradient b with depth z. The increments in horizontal distance and time for a ray leg for this case are then

$$\Delta X_{ab}(p) = \frac{1}{pb}(\cos i_a - \cos i_b),$$
$$\Delta T_{ab}(p) = \frac{1}{b} \ln \left[\frac{\sin i_b(1 + \cos i_b)}{\sin i_a(1 + \cos i_a)} \right], \tag{3.1.33}$$

where i is the inclination of the ray to the vertical

$$\sin i(z) = pv(z).$$

In such a situation with a constant vertical gradient of velocity, the ray paths are arcs of circles and as the velocity increases the inclination to the vertical becomes larger and will be 90° when $v(z) = 1/p$. For such a turning ray the ray leg will not penetrate to the base of the layer but will return to the level z_a and the contributions from the layer will be

$$\Delta X_{ab}(p) = \frac{2}{pb} \cos i_a,$$

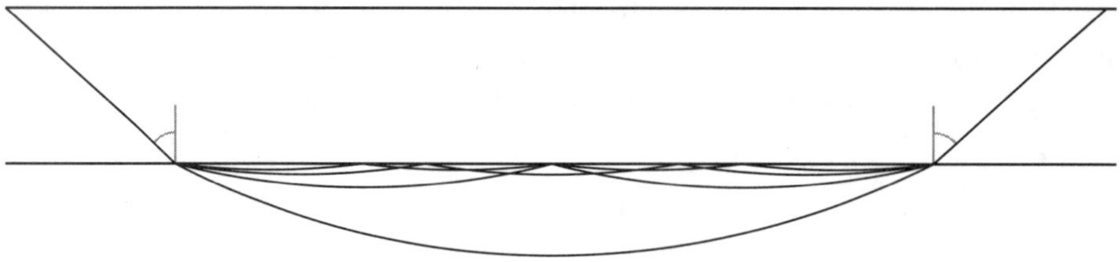

Figure 3.15. Illustration of the formation of an interference head wave.

$$\Delta T_{ab}(p) = \frac{2}{b} \ln\left[\frac{\sin i_b}{\sin i_a(1+\cos i_a)}\right]. \tag{3.1.34}$$

The refracted waves observed on seismic records will generally be composed of such turning rays. The behaviour is illustrated in figure 3.15. Consider an interface overlying a layer with a weak vertical gradient in velocity. A ray with slowness just less than $1/v_a$ will be turned back by the structure without penetrating far beneath the interface. The time-distance behaviour will be almost identical to a straight line path along the interface, as in the classical 'head wave' in a uniform medium illustrated above. With a gradient, more energy can be transported in the refracted arrival by the formation of an interference head wave. The turning ray is closely followed by rays which correspond to multiple reflections beneath the interface; there is little variation in time since the paths are very similar but the amplitude is significantly augmented.

As an illustration of the geometric properties of reflections and refractions in the time (T) – distance (X) and intercept time (τ) – slowness (p) domains we consider the upper part of a simple velocity model (figure 3.16). Each of the layers in figure 3.16 has a very small velocity gradient (0.005/s) so that refracted waves are modelled by turning rays as discussed above.

We consider just P wave propagation and look at the way in which the P waves interact with the first two interfaces. A schematic representation of the ray

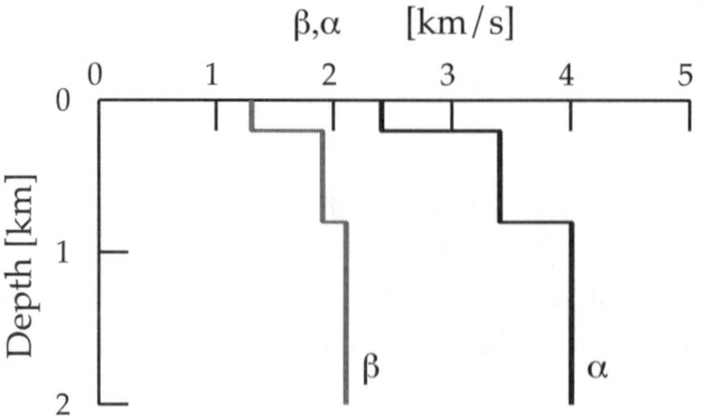

Figure 3.16. Upper part of simple velocity model used for ray calculations.

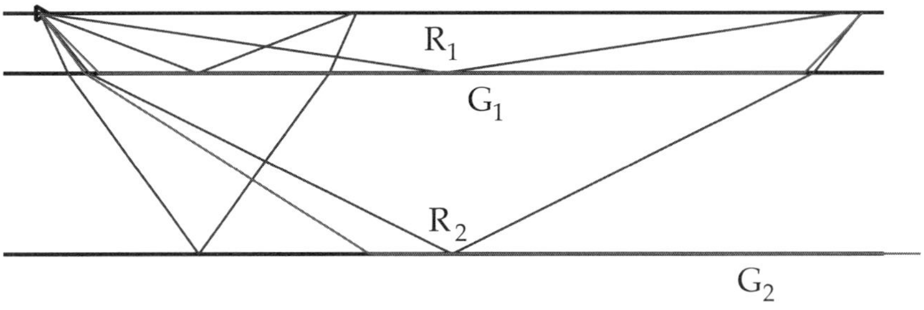

Figure 3.17. Ray configuration for reflected and refracted waves in the simple model.

configuration is shown in figure 3.17 and the trajectories for the rays in the T–X domain and τ–p domain are compared in figure 3.18.

The direct waves from the source D spread out across the surface and sweep across the surface at constant velocity and so map to a single point in the τ–p domain. The next arrival at short offsets is the reflection R_1 from the first interface with two way time at vertical incidence of $2h_0/\alpha_0$. The trajectory for this reflection in the time-offset domain will be a simple hyperbola,

$$\alpha_0^2 t^2 - x^2 = 4h_0^2, \tag{3.1.35}$$

whereas the corresponding trajectory in the tau–slowness domain is elliptical

$$\alpha_0^2(\tau^2 + 4h_0^2 p^2) = 4h_0^2. \tag{3.1.36}$$

At larger distances from the source the very wide angle reflections reach the first interface at near grazing incidence and arrive just behind the direct wave. For slowness less than 0.294 s/km, rays can be transmitted through the first interface and be reflected from deeper interfaces. However at this critical slowness the ray refracted into the second medium propagates parallel to the interface and so we have a refracted wave G_1 travelling with an apparent velocity equal to the P wave speed just below the interface (3.40 km/s). This refracted wave coalesces to a point in the slowness-time display. The critical point for the bifurcation of the travel time curve into refracted and post-critically reflected branches occurs at 460 m offset. From the behaviour of the reflection and transmission coefficients we have discussed above we expect the largest amplitude in reflection to occur for slownesses close to critical. The amplitude maximum is displaced just beyond the critical point as a result of the finite frequency of the waves.

Reflections from the second interface R_2 span the slowness range from 0.0 to 0.294 s/km from vertical to grazing incidence. The shape of the reflection trajectory is approximately hyperbolic in the time-distance domain. The very wide angle reflections asymptotically approach the time curve for the first refraction G_1 with increasing offset. The critical point for this reflection (corresponding to slowness 0.25 s/km) lies beyond the 2.5 km span of the display in figure 3.18. Ultimately the refraction

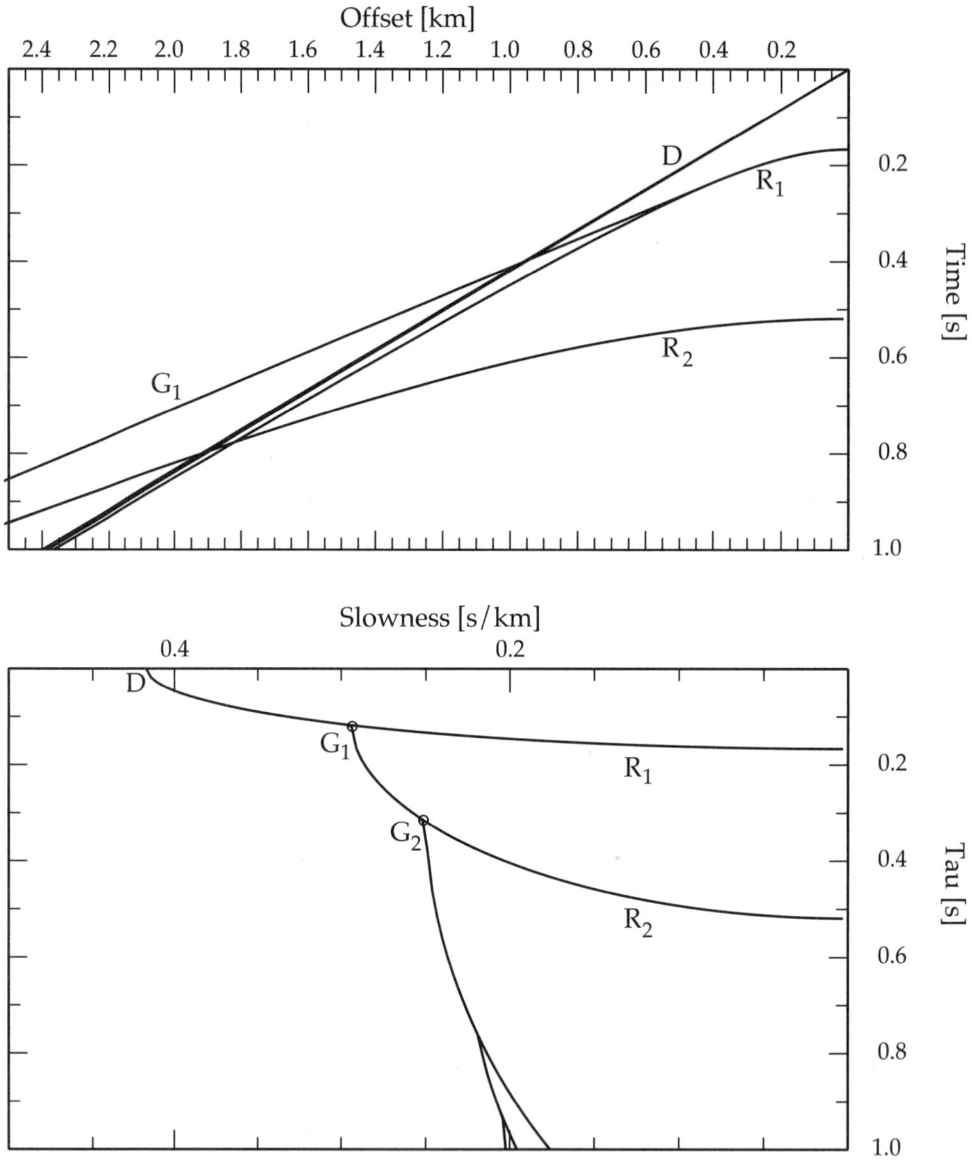

Figure 3.18. Ray calculations for the simple velocity model presented in (a) the time - distance domain and (b) the intercept time (τ) - slowness (p) domain.

G_2 will overtake the refraction from the first interface G_1. The reflection branch has an approximately ellipsoidal shape in the slowness-time domain and terminates at the critical slowness for the first interface 0.294 s/km - the intersection marks the mapping of the refraction G_1. In a similar way the third reflection joins the R_2 branch

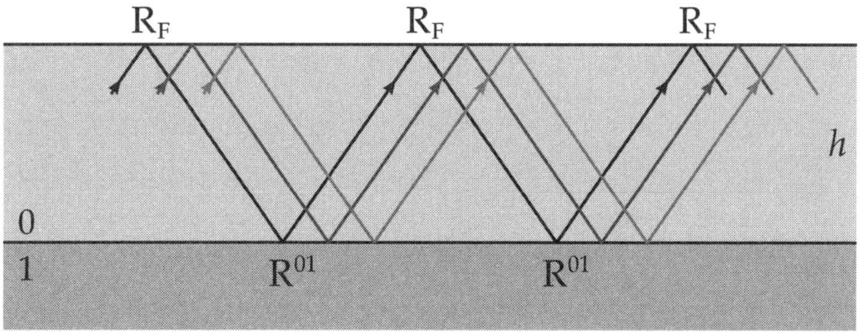

Figure 3.19. Constructive interference of *SH* waves to form Love wave modes.

at the refraction point for G_2 (0.217 s/km). The additional branches in the slowness time display come from even deeper horizons.

The contrasting behaviour of seismic phases in the time - distance and intercept time - slowness domains can often be used to unravel the nature of the processes giving rise to different features in the wavefield.

3.2 Guided and surface waves

Earlier in this chapter we have seen how seismic waves can be reflected by both the free surface and by internal interfaces, and further how a refracted wave can be turned back by a velocity gradient.

A dominant feature of the structure of the Earth is the generally strong increase in seismic velocity with depth in the top 1000 km with a major jump at the crust-mantle boundary. As a result we get the possibility of seismic energy from a source being reflected back from the structure in the crust and upper mantle to impinge on the free surface. The efficiency of reflection of *S* waves at the surface allows the generation of multiply reflected waves bouncing back and forth between the shallow structure and the free surface. The interference of many multiple reflections generates complex wavetrains in which the apparent slowness varies with frequency which may alternatively be represented in terms of the modes of the near-surface waveguide (see also Chapter 16). Figure 3.12 shows how a complex wavefield can be quite quickly established in the crust.

Such modes are solutions of the seismic wave equations which simultaneously satisfy the free-surface boundary condition of vanishing traction and also the requirement of decaying displacement at great depth. This gives rise to a constructive interference condition which can be illustrated for *SH* waves in a single layer over a half space (figure 3.19). For such *SH* waves the free surface reflection coefficient R_F^{HH} is unity and if we consider slownesses p between the wave slowness in the lower medium β_1^{-1} and the wave slowness in the upper medium β_0^{-1} we will have total reflection from the interface at depth h. For a given slowness p the reflected wave will acquire a phase shift $\chi^{01}(p)$ in the process of reflection. There will be a further phase shift of

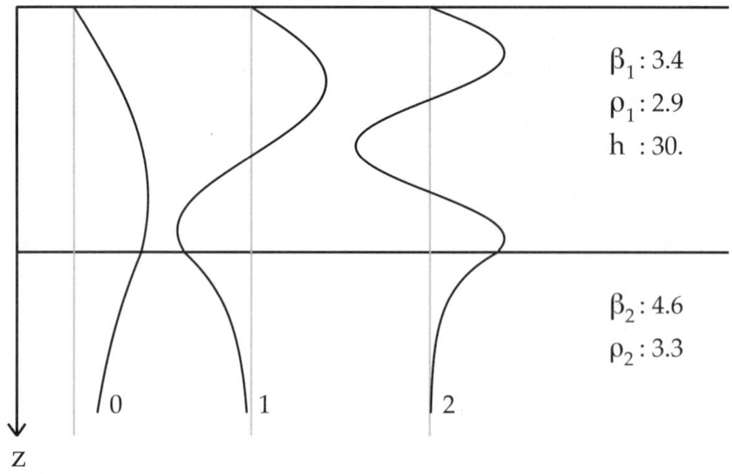

Figure 3.20. A simple model of a waveguide for Love waves, with the mode shape for the fundamental and first two higher modes.

$\exp[2i\omega q_{\beta 0}h]$ acquired in two-way passage through the layer. If the total effect of the reflection and propagation process is such that that the waves are in phase with the incident wave on the interface, we have a self-sustaining situation corresponding to the existence of a Love wave mode. There will therefore be a sequence of frequencies ω_n for which

$$\exp[i\chi^{01}(p)]\exp[2i\omega_n q_{\beta 0}(p)h] = 1 = \exp[2in\pi]. \tag{3.2.1}$$

On equating the phases on the two sides of the equation, we find

$$2i\omega_n q_{\beta 0}(p)h + i\chi^{01}(p) = 2in\pi. \tag{3.2.2}$$

The set of frequencies at the fixed slowness p are

$$\omega_n = \frac{n\pi - \frac{1}{2}\chi^{01}(p)}{h(\beta_0^{-2} - p^2)^{1/2}}, \tag{3.2.3}$$

with an equal frequency increment between successive ω_n.

Along each branch of the dispersion curves the frequency varies with slowness with a pattern controlled by the phase $\chi^{01}(p)$ of the reflection coefficient at the interface. The mode number n gives a direct indication of the number of zero-crossings in the depth dependence of the guided wave mode (figure 3.20).

The horizontal slowness p for each guided mode varies between β_1^{-1} at the low-frequency end and β_0^{-1} in the high frequency limit. Because of the frequency dispersion the apparent slowness for energy propagation g is no longer the slowness p.

We can provide a heuristic justification for the behaviour of energy transport when the slowness varies with frequency by looking at the properties of two sinusoidal

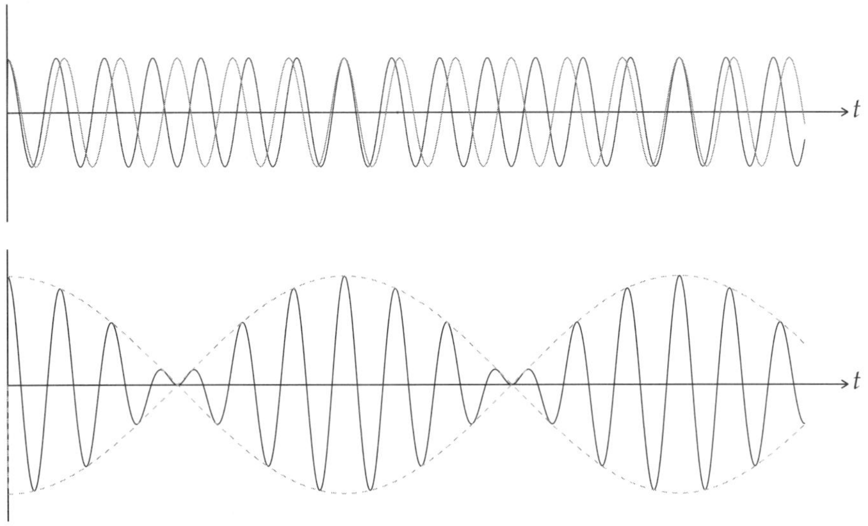

Figure 3.21. A linear combination of two sinusoidal waves with the same amplitude and nearly the same wavenumber. The envelope encloses 'packets' of waves that travel with the slowness (3.2.5).

disturbances with the same amplitude and nearly the same slowness (figure 3.21). The addition formula for cosines shows that

$$\cos(k_1x - \omega_1t) + \cos(k_2x - \omega_2t) = \\ \{2\cos[\tfrac{1}{2}(k_2 - k_1)x - \tfrac{1}{2}(\omega_2 - \omega_1)t]\}\cos[\tfrac{1}{2}(k_2 + k_1)x - \tfrac{1}{2}(\omega_2 + \omega_1)t]. \quad (3.2.4)$$

The factor in curly bracket is a slowly varying modulation, with small wavenumber $\frac{1}{2}(k_2 - k_1)$, for the oscillations with much larger wavenumber $\frac{1}{2}(k_2 + k_1)$ represented by the second cosine. The wave combination therefore takes the form of a series of wave 'packets' travelling with slowness

$$(k_2 - k_1)/(\omega_2 - \omega_1), \quad (3.2.5)$$

the packets are isolated from each either by the nodal points at which the amplitude is zero. We can achieve a more isolated 'packet' of waves by the superposition of a number of sinsusoids with very similar properties. The condition for maintaining constructive interference and hence the integrity of the wave 'packet' is that it travels with the group slowness (inverse of group velocity)

$$g(\omega) = \frac{\partial k}{\partial \omega} = \frac{\partial}{\partial \omega}(\omega p), \quad (3.2.6)$$

where we have expressed the wavenumber k in terms of the horizontal slowness p.

The wave packet associated with a particular mode will travel with the group slowness

$$g(\omega) = p(\omega) + \omega\frac{\partial}{\partial \omega}p(\omega). \quad (3.2.7)$$

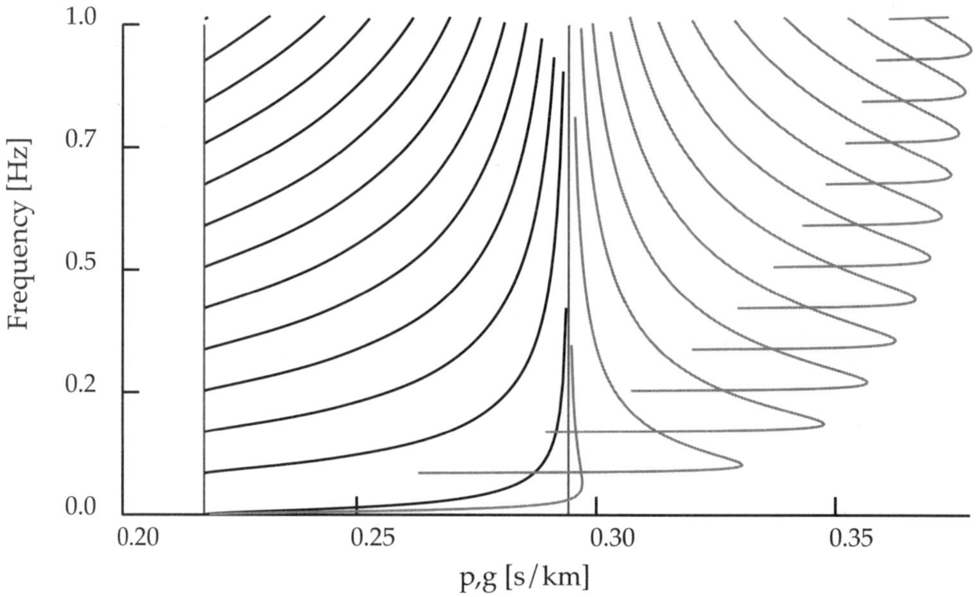

Figure 3.22. Dispersion curves for phase slowness p and group slowness g (grey tone) for Love waves in a layer over a halfspace using the model of figure 3.20.

The group slowness has some advantages over the more commonly used group velocity because arrivals on a seismogram are ordered by group slowness. A group slowness dispersion curve can therefore be mapped directly into the pattern of expected arrivals with frequency.

The behaviour of the group and phase slownesses as a function of frequency for the first 12 modes for Love waves in a single layer over a half space are illustrated in figure 3.22. At high frequencies both the group and phase slownesses asymptote to the surface slowness β_0^{-1}. However the group slowness is usually greater than the slowness in the upper layer and the guided wave energy will therefore travel slower than 3.4 km/s. The main contribution from each mode will come from the frequency at which the group slowness is a maximum and this is known as the Airy phase for the mode. For a shallow source the principal excitation is usually into the fundamental Love mode $n = 0$.

The dispersion relation (3.2.3) for Love waves can be extended to situations where there is further structure beneath the first interface for which the phase of the reflection coefficient χ will become frequency dependent; the resulting equation can be solved recursively for ω_n (cf. Kennett & Clarke, 1983).

The displacement in a Love wave in an isotropic medium is purely horizontal and transverse to the propagation path, whereas the analogous Rayleigh waves for the coupled *P-SV* wave system have elliptical polarisation in the vertical plane through the propagation path (figure 3.23).

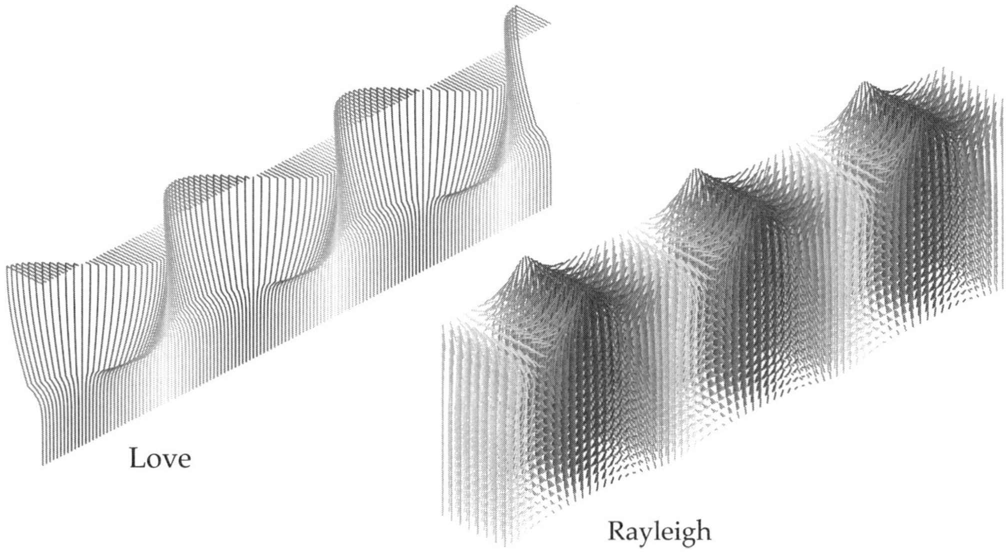

Figure 3.23. The particle motion for fundamental Love and Rayleigh wave modes.

We were able to derive a simple analytic expression for the dispersion of Love waves for a single layer over a half space because we were dealing with a single wavetype. However, when we consider *SV* waves, we have to take account of the linkage to *P* waves through the reflection at the free-surface and, to a lesser extent, in the structure beneath. Although the conditions are more complex we can find Rayleigh wave modes which satisfy both the surface condition and the requirements for decaying displacement at depth (see Chapter 16).

The higher modes of Rayleigh waves have a similar dispersion character to the Love wave modes and are restricted to slownesses $p < \beta_0^{-1}$. The fundamental mode has a somewhat different character because a solution exists in the slowness range $p > \beta_0^{-1}$ which couples evanescent *SV* and *P* waves through the free-surface condition. This form of wave is the generalisation to a stratified medium of the non-dispersive solution found by Lord Rayleigh (1885) for a uniform half space. The high frequency asymptote in slowness of the fundamental mode Rayleigh wave is the slowness p_{R0} corresponding to the uniform half space case since then only the properties just below the surface are sampled. For typical materials $p_{R0} \approx 1.1\beta^{-1}$.

In figure 3.12 the *Lg* wavegroup can be described in terms of higher mode surface waves and the following *Rg* arrival, in the *P-SV* calculation, with disturbance confined close to the free surface represents the fundamental mode Rayleigh wave.

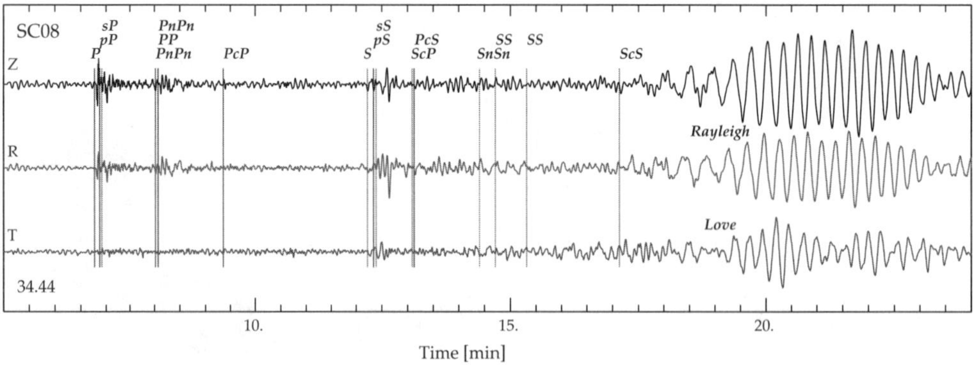

Figure 3.24. An example of a three-component seismogram from a shallow source at an epicentral distance $\Delta = 34°$ showing very prominent surface waves.

3.3 Seismic wavetrains

As the distance from the source increases, body waves spread out into the volume of the Earth but the surface wave field is guided along the surface with a much lower spreading rate. As a result the surface wave field tends to become more pronounced and play a major role in the character of the seismic wave train for shallow sources (figure 3.24).

Surface waves can also be quite important relatively close to the source since they carry much of the energy associated with strong ground motion from major earthquakes, and also commonly occur in exploration records on land. In reflection seismology, the portion of the wavefield of interest are the weak reflections from depth and the large amplitude surface waves generated by vibrator sources ('ground roll' - Telford et al., 1976) tend to obscure the desired information. The recording configuration commonly uses signals derived from adding the outputs of a spatial pattern of geophones designed to suppress the slowest surface waves.

When high frequency information is sought in reflection work, single geophones or very tight clusters may be employed to avoid problems with lateral variations in near-surface properties. In this case records show very clearly the onset of compressional P wave energy and the Rayleigh wave energy giving rise to the ground-roll (figure 3.25). The emergence of refracted P waves can be seen through the change in the slope of the first arrivals. The second refraction which just emerges at the furthest offset traces links back to a clear reflection with an intercept time of 0.12 s. S waves are not seen very clearly on vertical component geophones but the onset of the S arrivals can just be discerned on figure 3.25. At larger offsets S waves reflected by the near-surface layering separate from the ground-roll, and in figure 3.25 there is also some indication of higher mode surface waves. Normally the ground-roll and S waves would be suppressed by some form of velocity filtering in processing, but

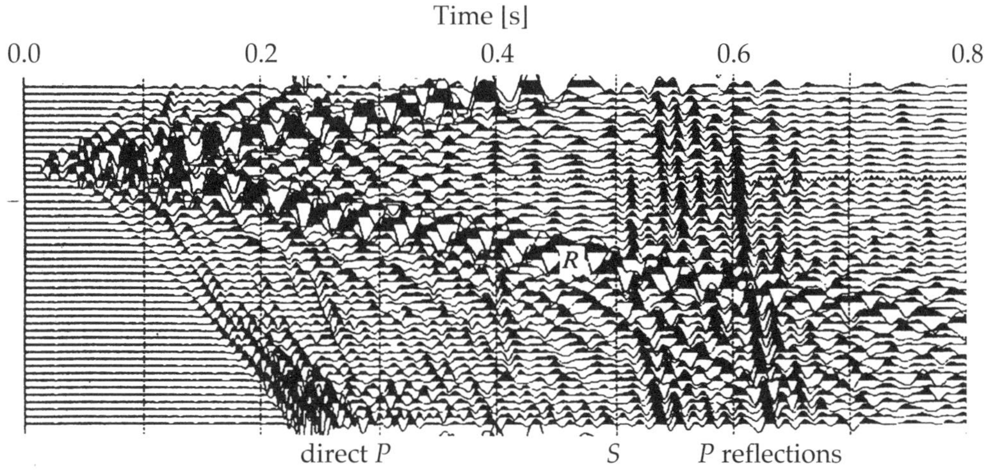

Figure 3.25. Single spread for shallow reflection work showing *P* refractions and reflections, and prominent Rayleigh waves (*R*).

they do contain useful information about shallow structure which can complement *P* wave information.

4

Seismic Sources

For a seismic wave to be recorded at a seismometer it has to be generated by the action of a source such as an earthquake or explosion. The seismic record itself bears the imprint of the propagation processes within the earth, but the patterns of variation in the behaviour of seismic phases are to a large extent imposed by the nature of the source. As a result it is possible to infer the character of a seismic source from relatively sparse observations.

4.1 Nature of seismic sources

All sources of seismic waves involve the sudden release of some form of potential energy within the Earth or at its surface. The radiation of seismic waves is a secondary phenomenon which removes only a fraction of the original energy from the vicinity of the source.

In shallow earthquakes the stored energy is normally associated with the strain built up in a region by continuing deformation, and is usually released on a pre-existing fault system. Some major events break new rock, e.g., by linking a set of faults and a substantial amount of energy is consumed in the process. For an existing fault substantial energy is required to overcome frictional resistance leading to melting of material on the fault surface (McKenzie & Brune, 1972). Earthquakes in the shallower part of subduction zones also correspond to energy release in slippage on fault systems, but the mode of failure in the deepest events is still not clear and may be initiated by the release of configurational energy in metastable upper mantle minerals via a phase transition.

In a explosion either chemical or nuclear energy is released in the detonation process, leading to the production of a cavity around the source point and compaction of the zone around the original charge. In a nuclear explosion much energy is dissipated through the melting of the host rock. Surface impact sources such as a weight drop or vibrator utilise stored mechanical energy and energy is lost through surface deformation and compaction.

In the immediate neighbourhood of a seismic source, a shock wave spreads out into the medium and the stress is non-linearly related to the considerable displacements. Eventually, at some distance from the point of initiation the displacements associated with the disturbance become small enough for the state to be described in terms of simple seismic waves. The size of the region for which non-linear effects are important will scale with the size of the source.

As a result the representations that we employ for seismic sources are simplified and are designed to match the radiation characteristics well away from the source; they do not attempt to describe the details of the source process. Thus, for many purposes, we are able to adopt the simple model of a point source for an earthquake or explosion with radiation characteristics derived from the nature of the energy release process. For larger earthquakes a better model is often a representation with several sub-events or a distribution of slip across a fault surface.

The representations which we employ for seismic sources make use of equivalent force systems designed to produce the same radiation as the actual source. The development of source descriptions and equivalent sources are discussed in detail in Chapter 11.

For a source within the Earth, such as an explosion or an earthquake, we require that there should be no net change in angular momentum and as a result the equivalent force system cannot exert a net moment. This condition can be satisfied by describing the source in terms of a set of weights for a system of point dipoles and couples. The weighting terms are referred to as the components of the "Moment tensor" M_{ij} for the source. The condition of no net moment imposes the symmetry condition $M_{ij} = M_{ji}$.

The radiation induced by the source can be conveniently described by considering the point moment-tensor source to lie in a uniform medium with the properties at the source point (P wavespeed α_S, S wavespeed β_S and density ρ_S). The far-field contributions to the displacement field at a point $\mathbf{x}$ depend on the distance R from the source point $\mathbf{h}$ and the rate of change of the moment-tensor components with time

$$4\pi\rho_S u_i(\mathbf{x}, t) = \gamma_i\gamma_j\gamma_k \frac{1}{\alpha_S^3 R}\partial_t M_{jk}(t - \frac{R}{\alpha_S}) \tag{4.1.1}$$

$$-(\gamma_i\gamma_j - \delta_{ij})\gamma_k \frac{1}{\beta_S^3 R}\partial_t M_{jk}(t - \frac{R}{\beta_S}). \tag{4.1.2}$$

where $R = |\mathbf{x}-\mathbf{h}|$ and the γ_i are the direction cosines of $\mathbf{x}$ relative to $\mathbf{h}$. The contribution (4.1.1) to the displacement corresponds to P wave radiation and (4.1.2) to S wave radiation.

The diagonal elements of the moment tensor M_{11}, M_{22}, M_{33} correspond to dipoles (see figure 4.1) and have a bilobed radiation pattern for both P and S. The off-diagonal elements $M_{ij}, i \neq j$, weight couples with a four-lobed radiation pattern for P and a bilobate pattern for S. When we take account of the symmetry property of the moment tensor and combine M_{ij} and M_{ji} to create a double couple without moment we produce a four-lobed radiation pattern for both P and S (see figure 4.3).

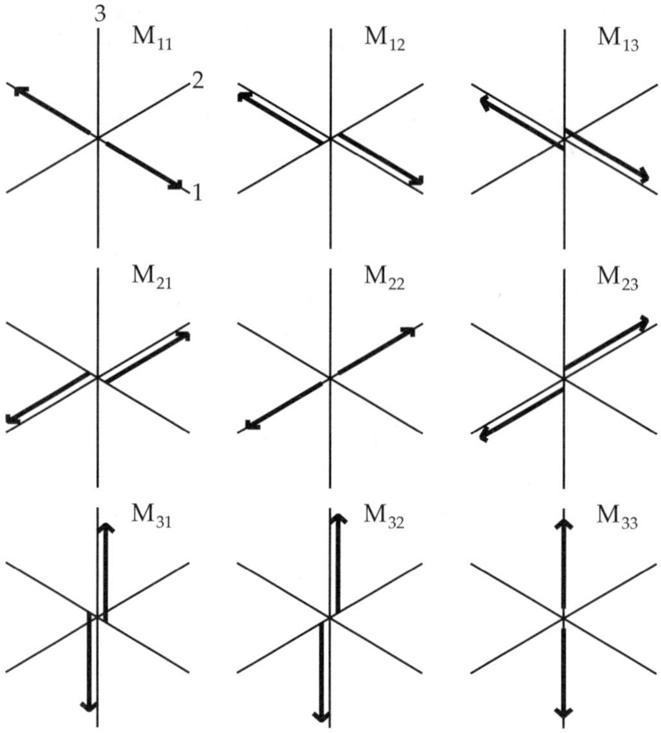

Figure 4.1. The representation of the elements of the moment tensor as weights for a set of dipoles and couples.

4.1.1 An explosion

A small explosive source can be simulated with an isotropic moment tensor

$$M_{ij} = M_o(t)\delta_{ij} \tag{4.1.3}$$

where the equivalent force system consists of three perpendicular dipoles of equal strength. The P wave radiation pattern is spherical and there is no direct S radiation.

4.1.2 Displacement on a fault

A simple model of an earthquake involves slip on a fault separating two blocks of material so that the displacement is forced to lie in the plane connecting the two blocks. The fault slip can be regarded as built up from a combination of simple models of fault slip depending on the relative motion of the block above the fault (hanging wall) and the block below the fault (foot wall). These basic faults are illustrated in figure 4.2. When the slip on a fault has both dip-slip and strike-slip components of comparable amplitude it is called an oblique slip fault.

As demonstrated in Chapter 11 the equivalent force system for fault slip can be described in terms of vectors representing the fault surface and the slip which are

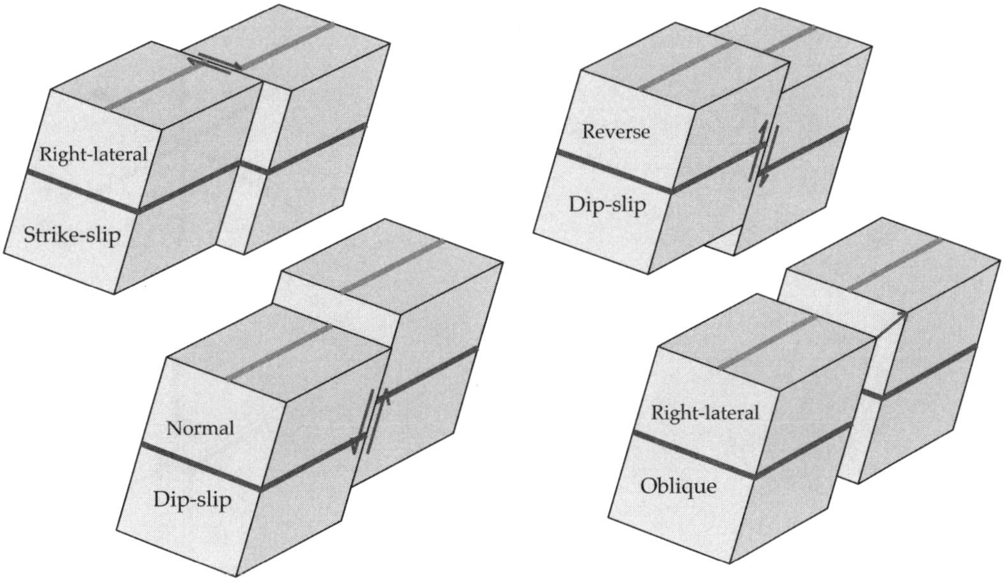

Figure 4.2. Simple models of fault slippage.

constrained by the geometry of faulting to be orthogonal. Consider a fault surface with normal **n** with a shear displacement in the direction $\boldsymbol{\nu}$ perpendicular to **n** (i.e., $\boldsymbol{\nu} \cdot \mathbf{n} = 0$. The equivalent moment tensor in the frequency domain is given by

$$M_{ij} = A\mu[\bar{u}(\omega)]\{n_i \nu_j + \nu_i n_j\}, \tag{4.1.4}$$

where A is the area of slip and $\bar{u}(\omega)$ is the averaged displacement at frequency ω. Because **n** and $\boldsymbol{\nu}$ are perpendicular the isotropic part of the moment tensor vanishes

$$\sum_{i=1}^{3} M_{ii} = 0. \tag{4.1.5}$$

The expression for the moment tensor components is symmetric in **n** and $\boldsymbol{\nu}$, so that the radiation pattern does not allow one to distinguish between the fault plane with normal **n** and the perpendicular auxiliary plane with normal $\boldsymbol{\nu}$.

In a coordinate frame with the 3-axis along **n** and the slip $\boldsymbol{\nu}$ in the direction of the 1-axis, the moment tensor has only 2 non-zero components,

$$M_{13} = M_{31} = M_o(\omega), \quad \text{all other } M_{ij} = 0, \tag{4.1.6}$$

which corresponds directly to the double couple case [cf. figure 4.3(b)]. The far-field P wave radiation in spherical polar coordinates (azimuth ϕ, colatitude θ) is

$$\sin 2\theta \cos\phi \tag{4.1.7}$$

and the amplitude of the far-field S radiation is

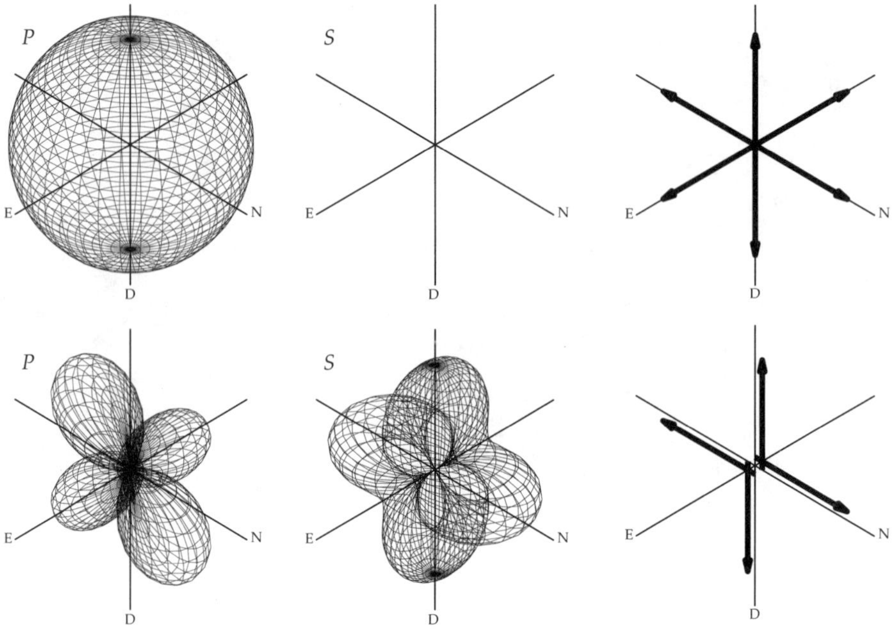

Figure 4.3. Equivalent force systems and *P*, *S* wave radiation patterns for (a) an explosive source, and (b) a 31/13 double couple.

$$(\cos^2 2\theta \cos^2 \phi + \cos^2 \theta \sin^2 \phi)^{1/2}. \tag{4.1.8}$$

The nodal planes for *P* wave radiation are the 12 and 23 planes. The characteristic four-lobed radiation pattern for *P* is used to determine the faulting characteristics of earthquakes.

4.2 Description of slip on a fault

We will represent displacement on a general fault by the equivalent double-couple source described in terms of the normals $\mathbf{n}$, $\boldsymbol{\nu}$ to the fault plane and the auxiliary plane. In order to describe the radiation from the source we have to establish a reference coordinate system; following Aki & Richards (1980) and Bullen & Bolt (1985) we choose a right-handed coordinate system with:

$$\begin{array}{ll} x \text{ direction to the North,} & \text{unit vector } \hat{\mathbf{N}}, \\ y \text{ direction to the East,} & \text{unit vector } \hat{\mathbf{E}}, \\ z \text{ direction downwards (towards nadir),} & \text{unit vector } \hat{\mathbf{D}}, \end{array} \tag{4.2.1}$$

The configuration of the fault system can be described by three angles (figure 4.4):

ϕ_S the strike angle, measured clockwise from North to the surface trace of the fault plane;

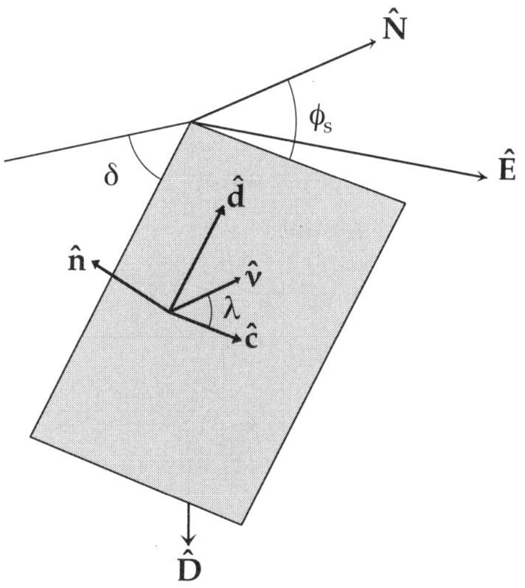

Figure 4.4. Configuration of the fault plane defined by the strike angle ϕ_S and the dip angle δ, with slip defined by the rake angle λ. The orientations of the normal to the fault plane $\mathbf{n}$ and the slip vector $\boldsymbol{\nu}$ are shown, as well as the auxiliary vectors $\mathbf{c}$, $\mathbf{d}$.

δ the dip angle, measured down from the horizontal at right angles to the strike direction;

λ the rake angle, the angle between the strike direction and the slip vector.

These are the angles used in the "Preliminary Determination of Epicenters" by the US National Earthquake Information Centre (NEIC) , which are later reprinted in the Bulletin of the International Seismological Centre (ISC).

A number of other conventions exist for describing the orientation of the slip relative to the strike. The angle in the vertical plane is known as the plunge

$$\sin p = \sin\lambda \sin\delta,$$

and the angle in horizontal plane as the trend

$$\sin t = \sin\lambda \cos\delta / (1 - \sin^2\lambda \sin^2\delta)^{1/2}.$$

Jarosch & Aboodi (1970) provide a good description of conversions between the different source conventions.

In the North, East, Down coordinate frame, the normal to the fault

$$\mathbf{n} = -\hat{\mathbf{N}} \sin\delta \sin\phi_S + \hat{\mathbf{E}} \sin\delta \cos\phi_S - \hat{\mathbf{D}} \cos\delta. \tag{4.2.2}$$

The strike of the fault lies in the direction

$$\mathbf{n} \wedge \hat{\mathbf{D}} = \mathbf{c} \sin\delta, \tag{4.2.3}$$

where the strike vector

$$\mathbf{c} = -\hat{\mathbf{N}} \cos\phi_S + \hat{\mathbf{E}} \sin\phi_S. \tag{4.2.4}$$

The vector which lies along the intersection of the fault plane and a vertical plane orthogonal to the strike is

$$\mathbf{d} = \mathbf{n} \wedge \mathbf{c} = \hat{\mathbf{N}} \cos\delta \sin\phi_S - \hat{\mathbf{E}} \cos\delta \cos\phi_S - \hat{\mathbf{D}} \sin\delta. \tag{4.2.5}$$

The vectors **c** and **d** provide an orthogonal basis on the fault plane so that the unit slip vector (orthogonal to the auxiliary plane)

$$\boldsymbol{\nu} = \mathbf{c} \cos\lambda + \mathbf{d} \sin\lambda, \tag{4.2.6}$$

in terms of the rake λ. Thus in the original basis

$$\begin{aligned} \boldsymbol{\nu} = {} & \hat{\mathbf{N}}(\cos\lambda \cos\phi_S + \sin\lambda \cos\delta \sin\phi_S) \\ & + \hat{\mathbf{E}}(\cos\lambda \sin\phi_S - \sin\lambda \cos\delta \cos\phi_S) - \hat{\mathbf{D}} \sin\lambda \sin\delta. \end{aligned} \tag{4.2.7}$$

With the identifications $x \equiv$ North, $y \equiv$ East, $z \equiv$ Down, we can now evaluate the six independent components of the moment tensor for a shear dislocation as

$$\begin{aligned} M_{xx} &= -M_0(\sin\delta \cos\lambda \sin 2\phi_S + \sin 2\delta \sin\lambda \sin^2\phi_S), \\ M_{xy} &= M_{yx} = M_0(\sin\delta \cos\lambda \cos 2\phi_S + \sin 2\delta \sin\lambda \sin\phi_S \cos\phi_S), \\ M_{yy} &= M_0(\sin\delta \cos\lambda \sin 2\phi_S - \sin 2\delta \sin\lambda \cos^2\phi_S), \\ M_{xz} &= M_{zx} = -M_0(\cos\delta \cos\lambda \cos\phi_S + \cos 2\delta \sin\lambda \sin\phi_S), \\ M_{yz} &= M_{zy} = -M_0(\cos\delta \cos\lambda \sin\phi_S - \cos 2\delta \sin\lambda \cos\phi_S), \\ M_{zz} &= M_0 \sin 2\delta \sin\lambda, \end{aligned} \tag{4.2.8}$$

where M_0 is the scalar moment.

Care needs to be taken to check the convention used for reference axes. The convention we have adopted is common in work with body waves. However, many solutions for moment tensors are derived from long-period seismograms using normal-mode techniques for which an alternative convention is prevalent based on a spherical coordinate basis ($r \equiv$ Up, $\theta \equiv$ South, $\phi \equiv$ East). The centroid moment tensor solutions of the Harvard group (see, e.g., Dziewonski et al, 1983) are given in this form and conversion involves some sign changes

$$\begin{aligned} & M_{xx} = M_{\theta\theta}, \ M_{xy} = -M_{\theta\phi}, M_{xz} = M_{r\theta}, \\ & M_{yy} = M_{\phi\phi}, M_{yz} = -M_{r\phi}, M_{zz} = M_{rr}. \end{aligned} \tag{4.2.9}$$

4.3 Body-wave radiation patterns

We consider an arbitrary moment tensor source in a uniform isotropic medium and can then find the radiation pattern from the source by looking at the distribution of wave amplitudes on a unit sphere surrounding the source point **h**. A knowledge of the radiation patterns for P and S waves will summarise the behaviour of the source as seen in the far field. For distant observation points we need to take into account the propagation processes after the waves leave the source region, including for example interactions with the free surface, which will modify the way in which the source

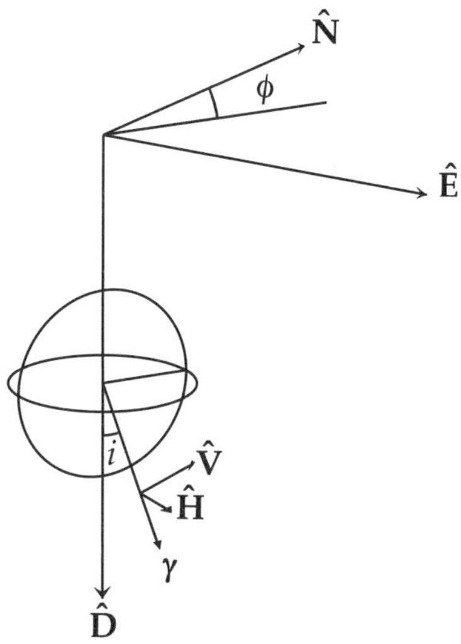

Figure 4.5. The relationship of the take-off angle i and azimuth ϕ to the focal sphere surrounding a source at $\mathbf{h}$. The orientation of the unit vectors in the P direction $\boldsymbol{\gamma}$, the SV direction $\hat{\mathbf{V}}$ and the SH direction $\hat{\mathbf{H}}$ are also indicated.

radiation is reflected in the character of the recorded seismograms. Such effects need to be considered when we attempt to recover source properties from observations.

For distant stations the directly arriving energy will leave the source downwards, and so attention is usually focussed on the lower hemisphere of the focal sphere. At closer ranges with local networks some far-field stations will be placed so that the energy that reaches them leaves the source travelling upwards; then the upper hemisphere of the focal sphere will be of interest.

The position on the focal sphere can be parameterised by the orientation of a radial unit vector $\boldsymbol{\gamma}$ described by the take-off angle measured up from the vertical (i) and the azimuth (ϕ) measured clockwise from north (see figure 4.5). P waves will be radiated along the direction $\boldsymbol{\gamma}$ which, in terms of the (North, East, Down) coordinate system, is given by

$$\boldsymbol{\gamma} = \hat{\mathbf{N}} \sin i \cos \phi + \hat{\mathbf{E}} \sin i \sin \phi + \hat{\mathbf{D}} \cos i. \tag{4.3.1}$$

In order to describe S wave radiation we introduce two further vectors orthogonal to $\boldsymbol{\gamma}$: $\hat{\mathbf{V}}$ lying in a vertical plane and $\hat{\mathbf{H}}$ confined to a horizontal plane.

$$\begin{aligned} \hat{\mathbf{V}} &= \hat{\mathbf{N}} \cos i \cos \phi + \hat{\mathbf{E}} \cos i \sin \phi - \hat{\mathbf{D}} \sin i, \\ \hat{\mathbf{H}} &= -\hat{\mathbf{N}} \sin \phi + \hat{\mathbf{E}} \cos \phi. \end{aligned} \tag{4.3.2}$$

This set of vectors allows us to polarise the S waves into SV and SH components at the source.

From the expression for the far-field radiation from a moment tensor source (4.1.1), the P wave contribution to the far-field displacement at a point on the focal sphere is given by

$$4\pi\rho_S\alpha_S^3\mathbf{u}^P(\boldsymbol{\gamma},t) = \boldsymbol{\gamma}\,[\gamma_p\gamma_q\partial_t M_{pq}(t-1/\alpha_S)]. \tag{4.3.3}$$

If all the components of the moment tensor have the same time dependence we can extract the radiation pattern for P waves in the form

$$\begin{aligned}\mathcal{F}^P &= \gamma_p\gamma_q M_{pq}\\ &= \sin^2 i\,[M_{xx}\cos^2\phi + M_{xy}\sin 2\phi + M_{yy}\sin^2\phi - M_{zz}]\\ &\quad +2\sin i\cos i\,[M_{xz}\cos\phi + M_{yz}\sin\phi] + M_{zz}.\end{aligned} \tag{4.3.4}$$

If we wish to combine the radiation pattern results with either ray theory or a plane wave decomposition of the wavefield, it can often be more useful to work in terms of the horizontal slowness $p = \sin i/\alpha_S$, or spherical slowness $\wp = r_S\sin i/\alpha_{Sr}$ for spherical stratification with a source at radius r_S.

For shear waves we need to separate the SV and SH polarisations. In a vertical plane the far-field SV contribution takes the form

$$4\pi\rho_S\beta_S^3\mathbf{u}^{SV}(\boldsymbol{\gamma},t) = \hat{\mathbf{V}}\,[V_p\gamma_q\partial_t M_{pq}(t-1/\beta_S)]. \tag{4.3.5}$$

The SV radiation pattern

$$\begin{aligned}\mathcal{F}^{SV} &= V_p\gamma_q M_{pq}\\ &= \sin i\cos i\,[M_{xx}\cos^2\phi + M_{xy}\sin 2\phi + M_{yy}\sin^2\phi - M_{zz}]\\ &\quad +\cos 2i\,[M_{xz}\cos\phi + M_{yz}\sin\phi].\end{aligned} \tag{4.3.6}$$

We note that the angular dependence of both the P and the SV radiation patterns depend on the same combinations of moment-tensor components. In the horizontal plane the SH contribution has a similar form

$$4\pi\rho_S\beta_S^3\mathbf{u}^{SH}(\boldsymbol{\gamma},t) = \hat{\mathbf{H}}\,[H_p\gamma_q\partial_t M_{pq}(t-1/\beta_S)] \tag{4.3.7}$$

The SH radiation pattern has a rather different dependence on the take-off angle i and the moment tensor components than the P or SV patterns:

$$\begin{aligned}\mathcal{F}^{SH} &= H_p\gamma_q M_{pq}\\ &= \sin i\,[(M_{yy}-M_{xx})\sin\phi\cos\phi + M_{xy}\cos 2\phi]\\ &\quad +\cos i\,[M_{yz}\cos\phi - M_{xz}\sin\phi].\end{aligned} \tag{4.3.8}$$

The analytic forms for the radiation patterns do not readily convey the geometrical character of the behaviour. For dislocations, the conventional display is to use a "beachball" with a projection of the lower focal hemisphere on which the dilatational segments of the P radiation are shaded. This representation gains its utility because of the form of the nodal surfaces which are just the projections of the fault plane and the auxiliary plane. However, even in this simple case, the variation of the radiation across the focal sphere is not adequately represented. For more general moment-tensor sources, the shapes of the nodal surfaces are useful for judging pictorially the extent of departures from double-couple behaviour, but are not very helpful in allowing an

Table 4.1 Parameters of the Xinjiang event used in figures 4.6 and 4.7

Date:	1997 iv 06
Origin time:	04:36:35
Latitude:	39.45°N
Longitude:	77.00°E
Depth:	31 km (ISC), 15 km (HRVD)

Centroid moment-tensor-solution

Scale: 10^{17} N m

$M_{xx} = M_{\theta\theta} = 2.09$	$M_{xy} = -M_{\theta\phi} = -6.73$	$M_{xz} = M_{r\theta} = 1.12$
$M_{yy} = M_{\phi\phi} = 3.99$	$M_{yz} = -M_{r\phi} = 5.95$	$M_{zz} = M_{rr} = -6.08$

Principal axes

T	10.8,	plunge 14°,	azimuth 126°
B	-0.5,	plunge 34°,	azimuth 27°
P	-10.3,	plunge 53°,	azimuth 235°

Best double couple: $M_0 = 1.1 \times 10^{18}$ N m.

NP1:	strike 253°,	dip 43°,	rake -36°
NP2:	strike 10°,	dip 67°,	rake -127°

assessment of the radiation in a particular direction. Frohlich (1996) provides a good account of the properties of the nodal surfaces for a general moment tensor and analytic expressions from which the nodes for *P*, *SV* and *SH* radiation can be plotted.

An alternative approach to displaying the character of the radiation patterns (Kennett, 1988b) is to display a projection of one focal hemisphere (usually the lower) using a regular grid of symbols, scaled to represent the local amplitude and sense of motion for a given wave type. Such displays can be readily generated from the radiation pattern formulae above for *P*, *SV* and *SH* waves for either a simple dislocation model or a more general moment tensor.

The example in figure 4.6 shows the radiation patterns for an event in Xinjiang, China in 1997 whose parameters are specified in Table 4.1. The size of the symbols indicates the amplitude of the radiation, and for *P* the nature of the symbols indicates the sign of the displacement. For *SV* and *SH* waves the direction of the movement is indicated by arrows. The diagram has been produced using a 10° grid in both azimuth and take-off angle i and is plotted using an equal area projection, which is commonly used in source studies. For a plot with radius R, and take-off angle i a symbol is placed at a distance $R \sin \frac{1}{2} i$ and azimuth ϕ from the origin. For a stereographic projection the mapping point would be placed at $R \tan \frac{1}{2} i$ at the same azimuth. Note that the most distant stations have the smallest take-off angles and so plot close to the origin, whilst stations at closer epicentral distances plot further away from the origin.

In order to get a suitable pattern of symbol sizes it is convenient to normalise the moment-tensor components by the modulus of the largest principal value, so that the sizes of the terms are comparable to the angular dependencies of the dislocation model (4.2.8). The principal values are frequently published along with moment-tensor

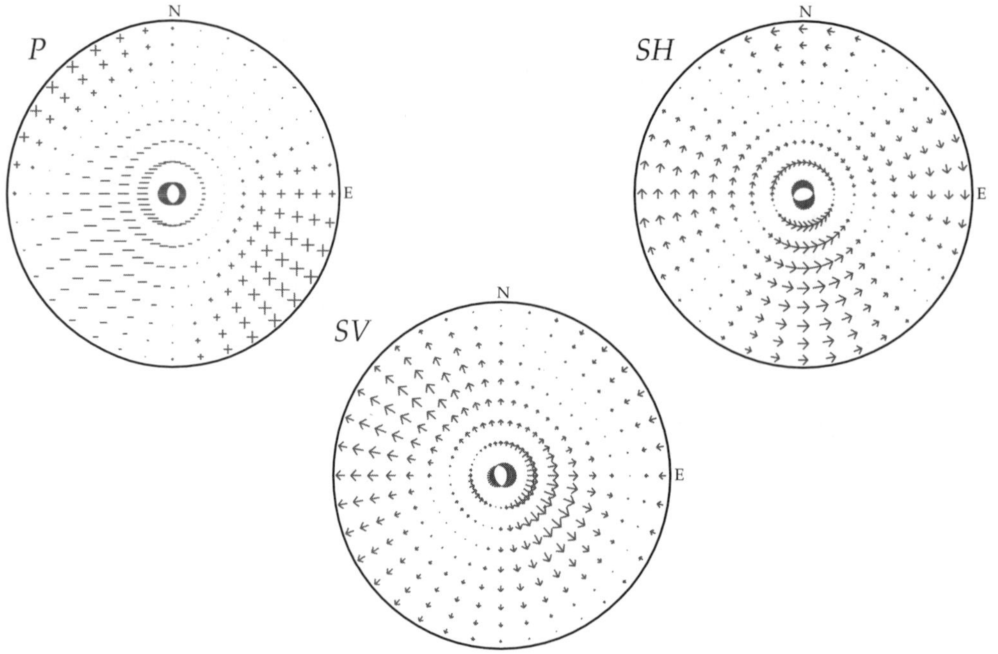

Figure 4.6. The radiation patterns for *P*, *SV* and *SH* for a Xinjiang event [Table 4.1]. The size and configuration of the symbols indicate the sense of motion (for *P* waves, pluses indicate compression and minuses dilatation). The symbols are plotted on a 10° grid in both take-off angle i and azimuth ϕ.

solutions, but can readily be found from the tensor components (see the appendix to Chapter 11).

One of the features of the style of display shown in figure 4.6 is that the significance of nodal and near-nodal regions is clearly displayed. The event shown in figure 4.6 has sufficient departure from a simple double-couple for there to be no crossing nodal surfaces, but the differences show up in the regions of low amplitude. Such near-nodal regions are up to 20-30° wide and can be seen to have a major influence on the radiation patterns. When the radiated amplitude is low, the presence of noise can mask the true behaviour and so the assignment of polarity can be very difficult.

For a general moment-tensor model the radiation patterns do not have the four-lobed symmetries of the dislocation model, and for surface reflected phases which leave the source in a upwards direction it can be worthwhile to draw an upper focal-hemisphere projection (with points plotted at $R\sin(\frac{1}{2}\pi - \frac{1}{2}i)$ to avoid excessive distortion).

The most stable estimates of source mechanism are derived from waveform studies at long-period because they avoid the influence of heterogeneity in the Earth and minor high-frequency variations in source character. Published moment tensors are derived either by using normal mode analysis (Dziewonski & Woodhouse, 1983) or using long-period body waves (Sipkin, 1994).

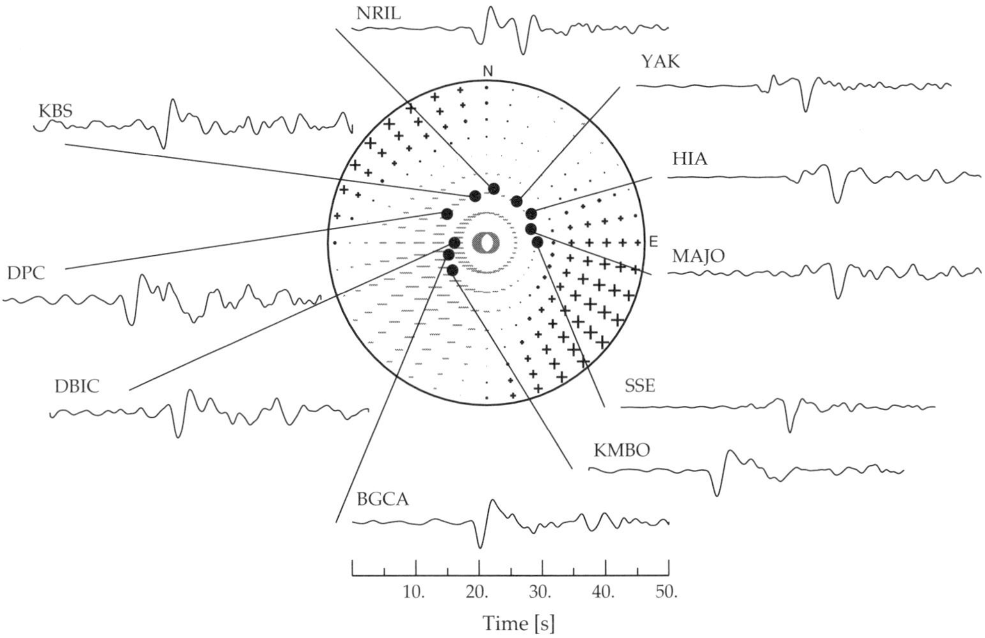

Figure 4.7. Filtered GSN records for the Xinjiang event in relation to their position on the focal sphere.

Figure 4.7 displays seismograms for the event in Xinjiang linked to the point of departure from the source in a projection of the lower hemisphere of the focal sphere. Stations close to the node such as HIA, MAJO, SSE show much reduced amplitude of first motion.

5

Seismic Phases

The onset of a seismogram comprising body waves is marked by the arrival of distinct bursts of energy which we associate with different classes of propagation path through the Earth. In order to understand the nature of these arrivals we need to be able to follow the way in which they travel through the Earth and the interactions with earth structure which dictate their character.

In this chapter we will consider the propagation of some of the major seismic phases in a spherical earth model and look at different aspects of their behaviour via ray paths, wavefronts and travel time relations. We will then examine the way in which the notation for the seismic phases can be linked to the nature of wave propagation through the globe.

5.1 Description of seismic phases

5.1.1 Notation for seismic phases

The character of a number of the major seismic phases propagating through the globe are indicated in figure 5.1 through the use of selected ray paths. The upper panel shows the waves which leave the source as *P* and the lower panel the waves which leave the source as *S*.

The convention for denoting the different types of arrivals is that a leg in the mantle is denoted by *P* or *S* depending on wave type, a compressional leg in the outer core by *K* (from the German Kern) and in the inner core by *I*. A shear wave leg in the inner core would be denoted *J*. Reflected waves from major interfaces are indicated by lower case letters: *m* for the Mohorovičić discontinuity (Moho), *c* at the core-mantle boundary and *i* at the inner core - outer core boundary. Waves leaving upward from the source are also indicated by lower case so that *pP* represents a surface reflection for *P* near the source and *sP* a reflection with conversion above the source. Such 'depth phases' can be very distinct for deep events, and their separation in time from the main phase provides a useful measure of depth.

The phase code for a path is built up by combining the different elements. Thus a *P* wave which is returned from depth, and then reflected at the surface with a further

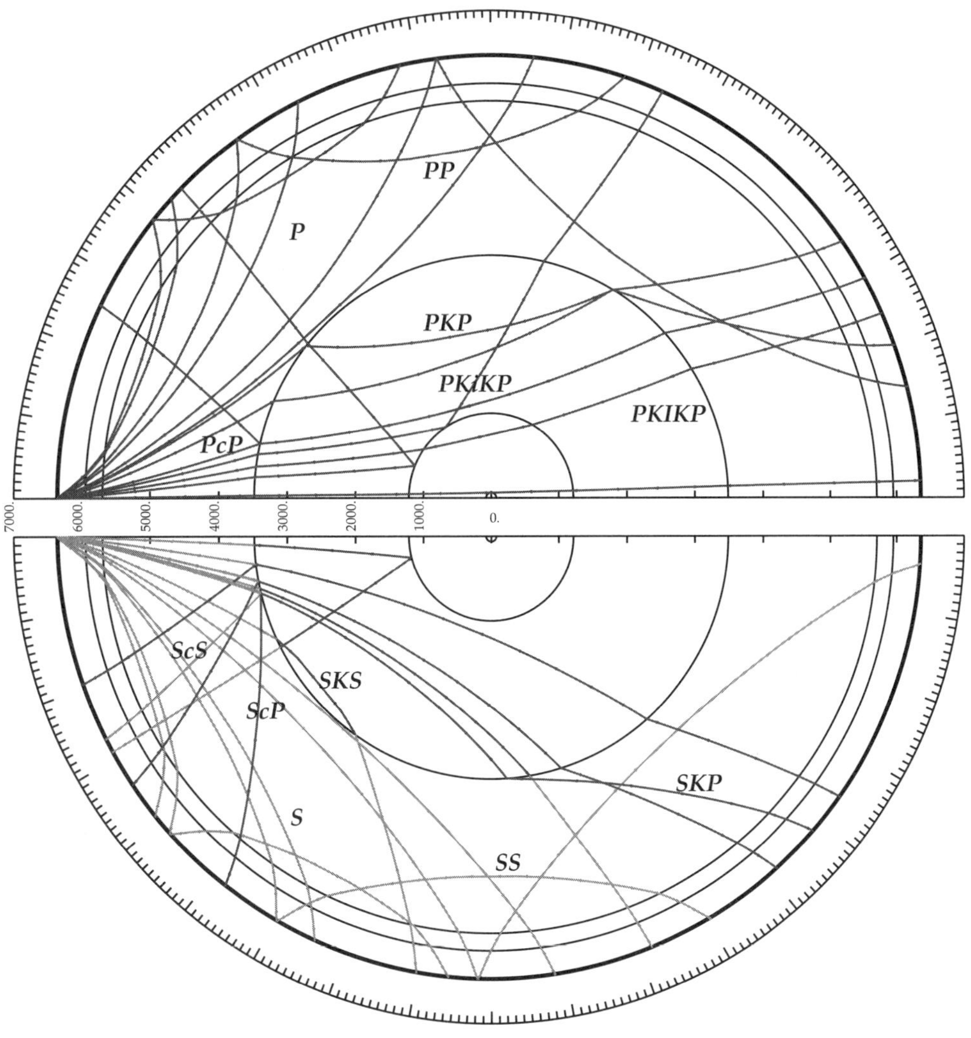

Figure 5.1. The main seismic phases in the Earth.

mantle leg will be represented by *PP*. If there is a conversion between wavetypes we have *PS*. An *S* wave reflected at the core-mantle boundary is indicated by *ScS*, and with conversion as well we get *ScP* a phase which can be quite clear for distances around 40° from the source.

PKIKP is a *P* wave which has travelled through the mantle and both the inner and outer cores, whereas the phase *PKiKP* is reflected back from the surface of the inner core.

The nature of the ray paths and wavefronts for a surface source are displayed in figure 5.2 for the major *P* and *S* phases. The wavefronts are indicated by placing time

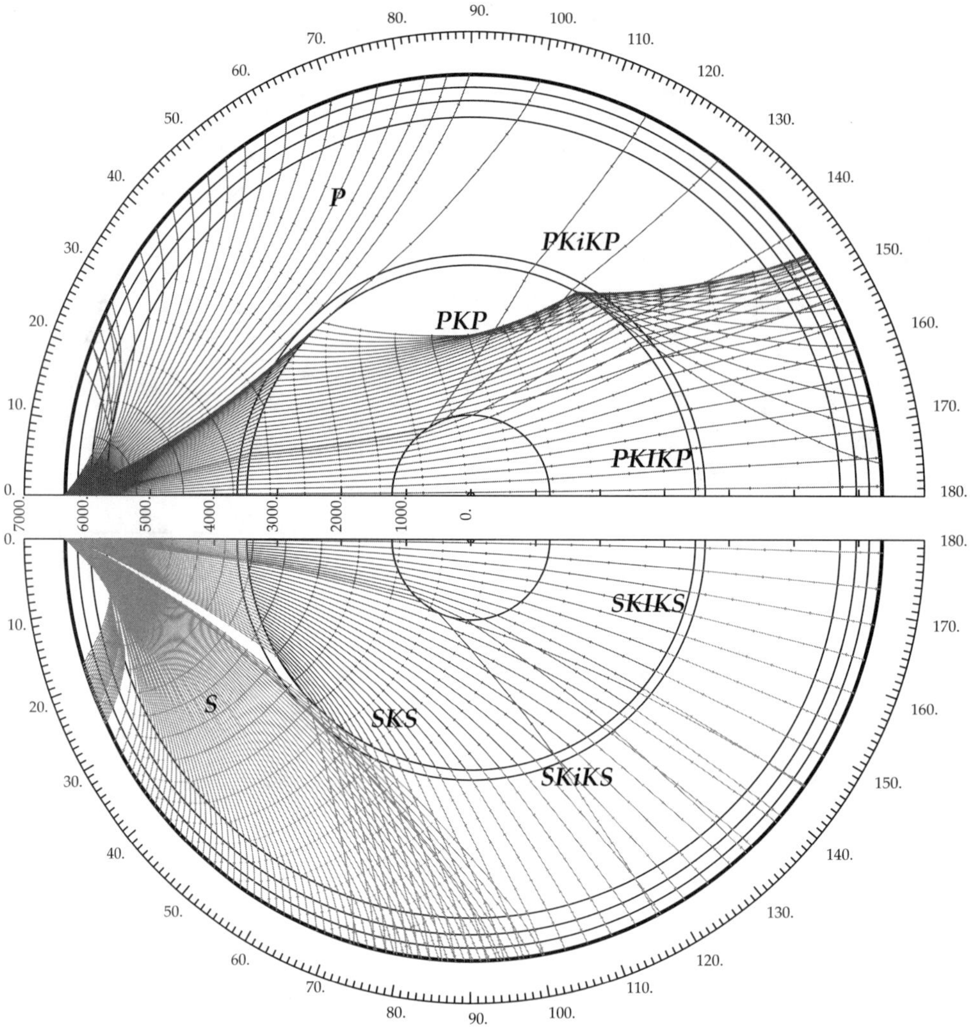

Figure 5.2. Ray paths for major *P* and *S* phases for the AK135 model of seismic wavespeeds.

ticks on the raypaths at 1 minute intervals. S wave legs are indicated by using grey tone.

For *P* waves [figure 5.2 - upper panel] there is a relatively complex pattern of propagation in the core. Because there is a drop in *P* wavespeed on entering the core, Snell's law requires the transmitted wave to be deflected to a much steeper angle to the normal to the boundary. The consequence is that a pronounced caustic is generated in the outer core with a concentration of *PKP* arrivals near 144°. Refraction through the inner core (*PKIKP*) produces arrivals which help to fill in the gap between direct *P* out to 100° and the *PKP* caustic. The identification of these arrivals lead to the discovery of the inner core by Inge Lehmann. Note that the presence of the *PKP* caustic means that the *P* waves do not sample the upper part of the core.

For *S* waves [figure 5.2- lower panel] the pattern of propagation in the mantle is similar to that for *P*. However, since the *P* wavespeed in the outer core is slightly higher than the *S* wavespeed at the base of the core, it is possible for an *SKS* wave to get ahead of the direct *S* wave at the same angular distance. The *SKS* arrivals sample the whole of the core and do not have the caustic patterns seen for *PKP*.

5.1.2 Seismic ray properties in a spherical Earth

In the previous section we have illustrated the behaviour of the rays for major seismic phases. To understand the character of these rays and their associated travel time behaviour, it is useful to look a little more closely at the commonly occurring features in spherically stratified models.

A detailed discussion of rays in spherical stratification is presented in Chapter 10 of Part II with consideration of amplitude relations and illustrations of the interaction of slowness, distance and travel time for different classes of velocity model.

A ray path in a wavespeed model which depends only on radius r is required to follow a continuous version of Snell's law. In terms of the inclination i of the ray to a radial vector, the ray parameter

$$\wp = \frac{r \sin i}{v(r)}, \tag{5.1.1}$$

is constant along the path, $v(r)$ is the current seismic wavespeed (either *P* or *S*). The ray parameter $\wp$ characterises the trajectory of the ray from the source to a receiver at a point on the surface of the Earth.

Consider a ray descending from a source at the surface ($0 < i < \pi/2$), into a medium in which $v(r)/r$ grows with depth. From the relation $\sin i = \wp v/r$, $\sin i$ must increase and so therefore must the angle i until $i = \pi/2$ at which point $\wp = r/v(r)$. This represents the 'turning point', the lowest point of the ray trajectory. After this i continues to grow but v/r decreases. The ray climbs back to the surface $\pi/2 < i < \pi$ with symmetry about the turning point.

The propagation time T for a ray between a source and the various stations is a function of the epicentral distance Δ and the depth of the source h. The relationship $T(\Delta)$ is referred to as the travel-time curve for the particular phase.

Two neighbouring rays leaving from the same source S (figure 5.3) will emerge at the surface at two nearby points R_1 and R_2. Let Δ_1 and Δ_2 be the epicentral distances to R_1 and R_2 i.e. the angles subtended at the centre by the two arcs SR_1 and SR_2 then the angular distance

$$\widehat{R_1R_2} = r(\Delta_1 - \Delta_2) \tag{5.1.2}$$

The difference in the travel times is related to the incremental distance in the propagation path (HR_2)

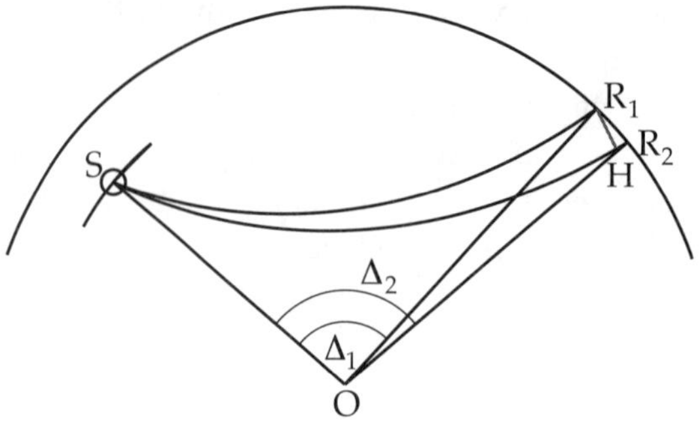

Figure 5.3. Definition of seismic ray parameter $\wp$.

$$T_2 - T_1 = \frac{HR_2}{v_0} = \frac{r(\Delta_1 - \Delta_2)\sin i_0}{v_0} \tag{5.1.3}$$

where v_0 is the velocity at the surface. In the limit as the rays tend to coalesce

$$\frac{r \sin i_0}{v_0} = \frac{\mathrm{d}T}{\mathrm{d}\Delta} = \wp \tag{5.1.4}$$

Thus the ray parameter $\wp$ is equal to the slope of the travel time curve at the point (T, Δ) and so has the dimensions of slowness, e.g., s/°.

5.1.2.1 Reflection and Refraction

When a seismic ray hits the surface of the Earth or an internal surface of discontinuity in $v(r)$ it is reflected. The angle of incidence is the same as the angle of reflection if there is no change in wavetype. If, however we have a converted reflection from a P ray making an angle i_1 to the local vertical to an S ray making an angle j_1 then from Snell's law

$$\frac{\sin i_1}{\alpha_1} = \frac{\sin j_1}{\beta_1}, \tag{5.1.5}$$

where α_1 and β_1 are the wavespeeds of the two rays. The same equation holds for the reflection of a S wave into a P wave but then we need $\sin j_1 < \beta_1/\alpha_1$ for a propagating P wave.

We have a comparable set of relations in transmission at an internal interface. An incoming wave has the potential to produce four waves, reflection and transmission in the same wave type and reflection and transmission with conversion. The angles of the rays are connected by

$$\frac{\sin i_1}{\alpha_1} = \frac{\sin j_1}{\beta_1} = \frac{\sin i_2}{\alpha_2} = \frac{\sin j_2}{\beta_2}, \tag{5.1.6}$$

where α_2 and β_2 are the wavespeeds on the transmission side of the boundary and i_2, j_2 are the corresponding angles.

As shown in Section 9.4, in the high frequency ray approximation the reflection and transmission coefficients to be applied to ray amplitudes are those for a plane boundary with the same contrast in seismic wavespeed

5.1.2.2 The influence of structure

The principal complications in the behaviour of seismic rays in a spherically stratified model come from the presence of discontinuities in $v(r)/r$. We will discuss briefly the influence of three special cases.

(a) sharp increase in $v(r)$

Figure 5.4(a) shows the velocity distribution in terms of $v(r)/r$ and the corresponding travel-time relations.

Whilst the turning point of the ray penetrates deeper and deeper up to C′ the travel-time curve progresses regularly up the point C and then turns back to D. The branch CD corresponds to reflection from the discontinuity surface C′D′. It is concave upwards, like all retrograde branches. The amplitude is normally large at the point D.

Rays impinging at angles of incidence less than the critical angle i_c ($\sin i_c = V_1/V_2$) are partly reflected and partly refracted into the lower medium. The branch DF corresponds to the refracted waves. Note that at the turning point of the ray $\wp = r/v(r)$ and thus the slope at D is less than that at C.

Examples of this situation occur for *P* and *S* wave interaction with the upper mantle discontinuities and at the core-mantle boundary for the *SKS* phase.

(b) sharp decrease in $v(r)$

In this case the significant feature is a geometrical shadow zone between Δ_1 and Δ_2 [figure 5.4(b)]. If at some deeper level (A′) $v(r)/r$ becomes greater than its value at the level X′, the ray emerges again and the travel-time curve is split.

The rate of growth of $v(r)/r$ determines whether or not there should be a retrograde branch. In all cases those rays which descend below a certain level B′ will emerge at distances greater than Δ_2; the branch BC is prograde and is concave downwards. The epicentral distance has a minimum at the point B which represents a caustic.

Such low velocity zones can occur in the upper mantle, for example beneath the fast oceanic lithosphere and we recall that the outer core shows a sharp decrease in *P* wavespeed.

(c) a sharp decrease in velocity followed by an increase

This case is of importance in the Earth for *P* waves, with the decrease occurring in the transition from mantle to core and the increase on passing from the core to the inner core [figure 5.4(c)]. There is still a shadow zone between Δ_1 and Δ_2. The branches AB and BC have a similar character to case (b). At the point C we are back to the conditions of example (a): the branch CD is reflected and DF corresponds to refracted waves in

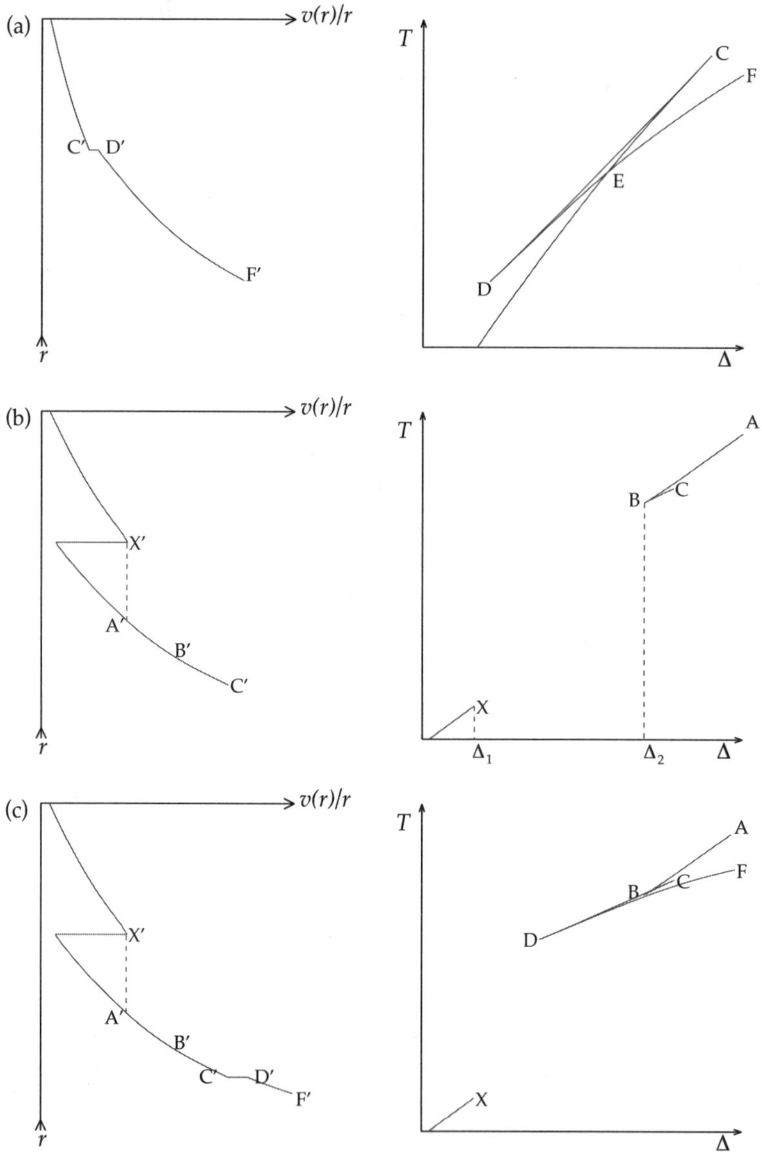

Figure 5.4. Influence of seismic structure (a) a velocity discontinuity, (b) a sharp decrease in wavespeed, (c) a combination of a sharp decrease in wavespeed followed by an increase.

the medium below C′D′. The loop ABCDF represents the full travel-time curve and the point B corresponds to a caustic

The main example for this case is the *PKP* phase which propagates through the core. The increase in velocity CD is then at the boundary between the inner and outer core. The branch *PKP*DF is also designated *PKIKP* because the refracted wave below CD propagates in the inner core.

5.2 Rays and seismic phases

5.2.1 Crustal and upper mantle propagation

The pattern of propagation of seismic rays can provide valuable insights into the character of the seismic wavefield. As an example we show in figure 5.5 the ray field for *P* waves and the corresponding travel-time behaviour for a crustal and upper mantle model. The wavespeed model is derived from the analysis of long-range refraction profiles in France (Hirn et al., 1973), in which a large number of recording stations were deployed along a 900 km line extending across the centre of France from a shot-point in the sea off the coast of Brittany. A prominent feature of the results was a clear *en echelon* pattern of energy return from the mantle with a succession of distinct arrivals with range, but with a similar apparent velocity. The combination of a low velocity zone and regions of increased velocity gradient in the mantle is designed to simulate the observations, and we see three distinct sets of mantle return in figure 5.5.

An advantage of a ray approach is that it is easy to select parts of the wavefield by the choice of the range of ray parameter. In figure 5.5 we have chosen to cover the reflections from the base of the crust (*PmP*) and propagation in the upper mantle (*Pn*). The rays leave the surface source at equal angular intervals and so the density of rays

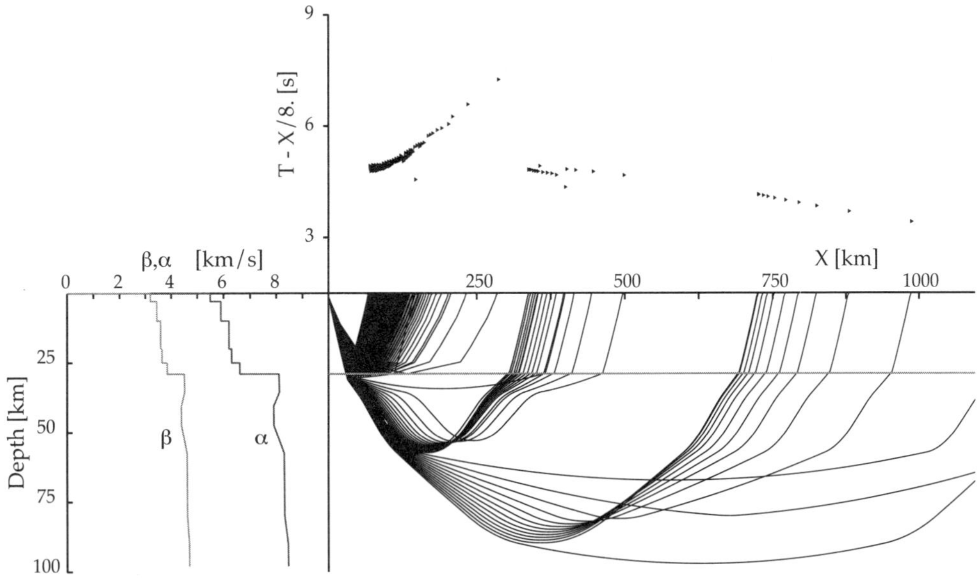

Figure 5.5. Ray and travel-time behaviour for a crustal and upper mantle model derived from long-range refraction profiles across France.

returning to the surface provides a measure of the variation of the amplitudes of the arrivals with range. Note the low density of ray emergence points associated with the refracted waves from the nearly uniform zone at the top of the mantle, compared with the stronger returns from the regions of increased gradient near 50 and 80 km depth. Figure 5.5 provides a good illustration of the influence of a low velocity zone, with the rays bent towards the vertical as they enter into the region of reduced velocity and then refracted back by the velocity gradient beneath to produce a caustic from crossing rays which reaches the surface at 300 km range. The deeper gradient zone near 80 km depth produces a triplication in the travel times but only a portion close to the caustic near 700 km is visible in figure 5.5, because the refracted waves from above and below the region of increased gradient reach the surface beyond the span of the figure.

5.2.2 Major phases from an intermediate depth event

We now consider the global pattern of propagation from an intermediate depth source (260 km deep) in the AK135 reference model. Because the source lies away from the surface we need to allow for both the direct downward radiation from the source and also the upward radiation which will be reflected back by the free surface and then propagate to distant stations. Thus each *P* phase is accompanied by surface reflected phases, e.g. the core reflection *PcP* is followed by the reflected phase *pPcP* and then by *sPcP* with conversion from *S* to *P* in the surface reflection. Similarly *ScS* is followed by *pScS* and *sScS*. As a result each of the travel-time branches is tripled, as can be seen in figure 5.6 which displays the travel-time distance relations for the major phases from the intermediate depth source.

The travel-time curves of figure 5.6 give a useful rendering of the behaviour of the seismic phases for comparison with observations, but do not provide any ties to the velocity model itself. The link to velocity structure can be made by employing a parametric representation in terms of spherical slowness $\wp$.

Although we could employ the travel time $T(\wp)$ directly, it is convenient to introduce the auxiliary variable

$$\tau(\wp) = T(\wp) - \wp\Delta(\wp), \tag{5.2.1}$$

which depends on both the travel time and the epicentral distance $\Delta(\wp)$, note that $\partial\tau/\partial\wp = -\Delta(\wp)$. Because the slowness $\wp$ represents the gradient of the travel-time curve for a particular phase, $\tau(\wp)$ represents the trajectory of the intercept of the tangent to the travel-time curve as a function of slowness $\wp$. The mapping to $\tau(\wp)$ has the useful property of unravelling triplications in the travel-time curves so that the interrelations between branches can be seen more clearly. In the presence of a low velocity zone $\tau(\wp)$ is discontinuous at the slowness $\wp_*$ corresponding to entry into the zone. $\tau(\wp)$ also plays an important role in generalised ray methods for seismogram synthesis.

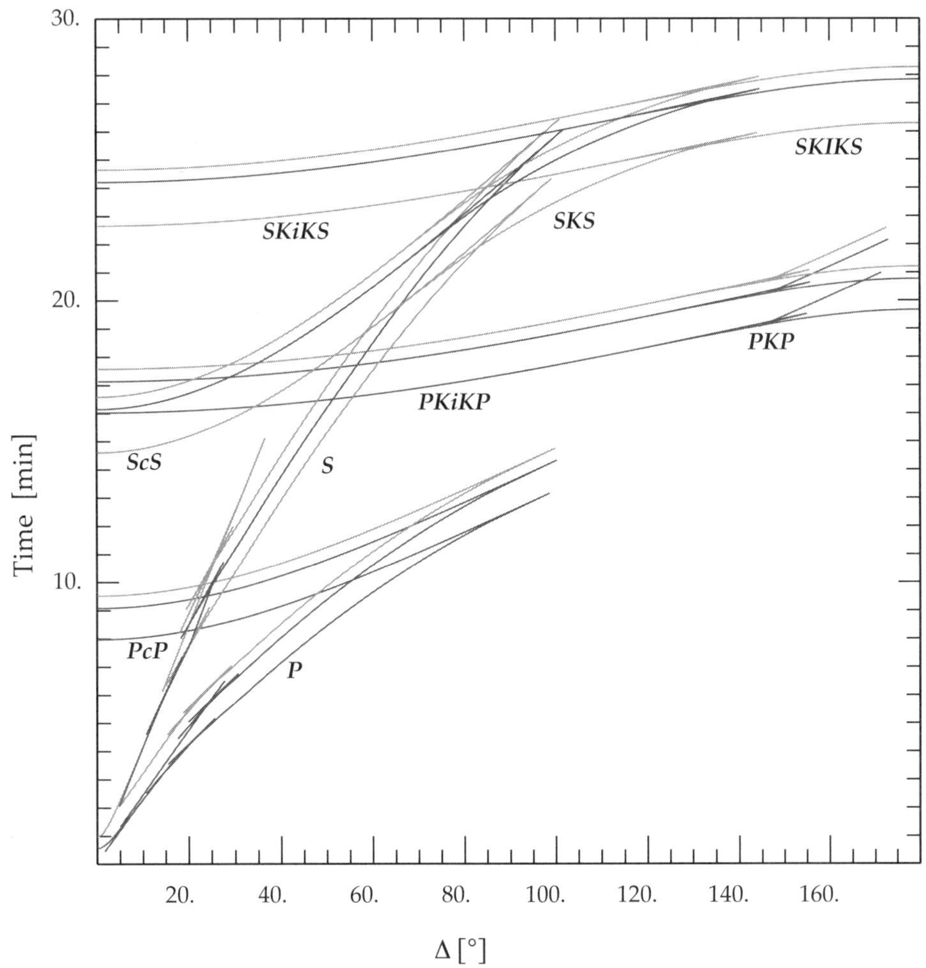

Figure 5.6. Travel-time curves for the major phases from an event at 260 km depth.

We map the velocity model into the slowness domain by the Snell's law relation (5.1.1); at the turning point r_p for the slowness $\wp$,

$$\wp = r_p / v(r_p) = \eta(r_p). \tag{5.2.2}$$

Thus a plot of $\eta(r)$ against radius r provides a direct mapping of the wavespeed profile into the slowness domain.

The different aspects of the phase behaviour for the AK135 model for this intermediate depth event are shown in figures 5.6-5.8. Figure 5.6 displays the travel-time curves for the major phases and the $\tau(\wp)$ relation for the same phases is shown in figure 5.7; note that the $T(\Delta)$ and $\tau(\wp)$ values correspond for zero range and slowness. The epicentral distance/slowness relationship $\Delta(\wp)$ is illustrated in

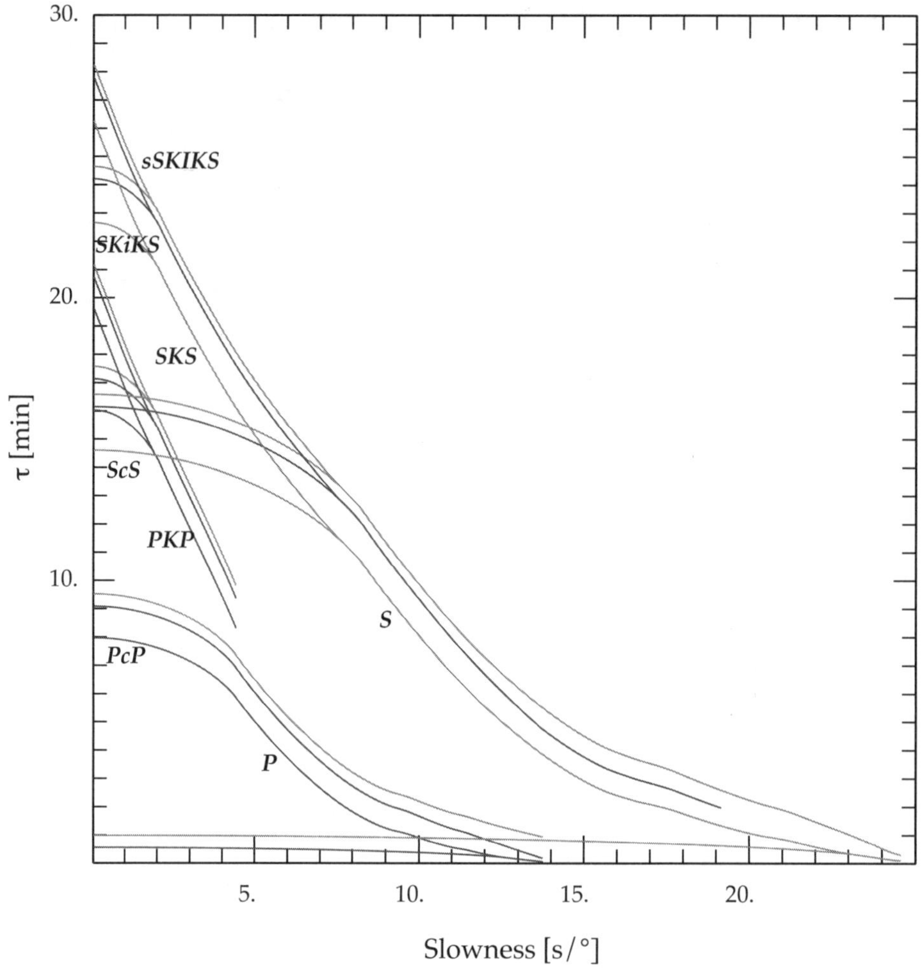

Figure 5.7. $\tau(\wp)$ behaviour for the major phases from an event at 260 km depth.

figure 5.8 and can be contrasted with the slowness behaviour for the AK135 model in figure 5.9.

The use of different domains enables us to look more closely at the character of the phase behaviour. For example, the *P* wave refracted in the the mantle and the *PcP* wave reflected from the core-mantle boundary have very similar raypaths for near grazing incidence (see also figure 5.10), and so the travel time curves converge close to 90° epicentral distance. In the $\tau(\wp)$ relation there is a smooth continuation of the *P* phase into *PcP*; the transition is marked by the discontinuity in $\tau(\wp)$ between *P* and *PKP*. The core phase *PKP* has the behaviour described in section 5.1.2.2c and links to the *PKiKP* reflections from the inner core boundary. The distinction between reflections and the

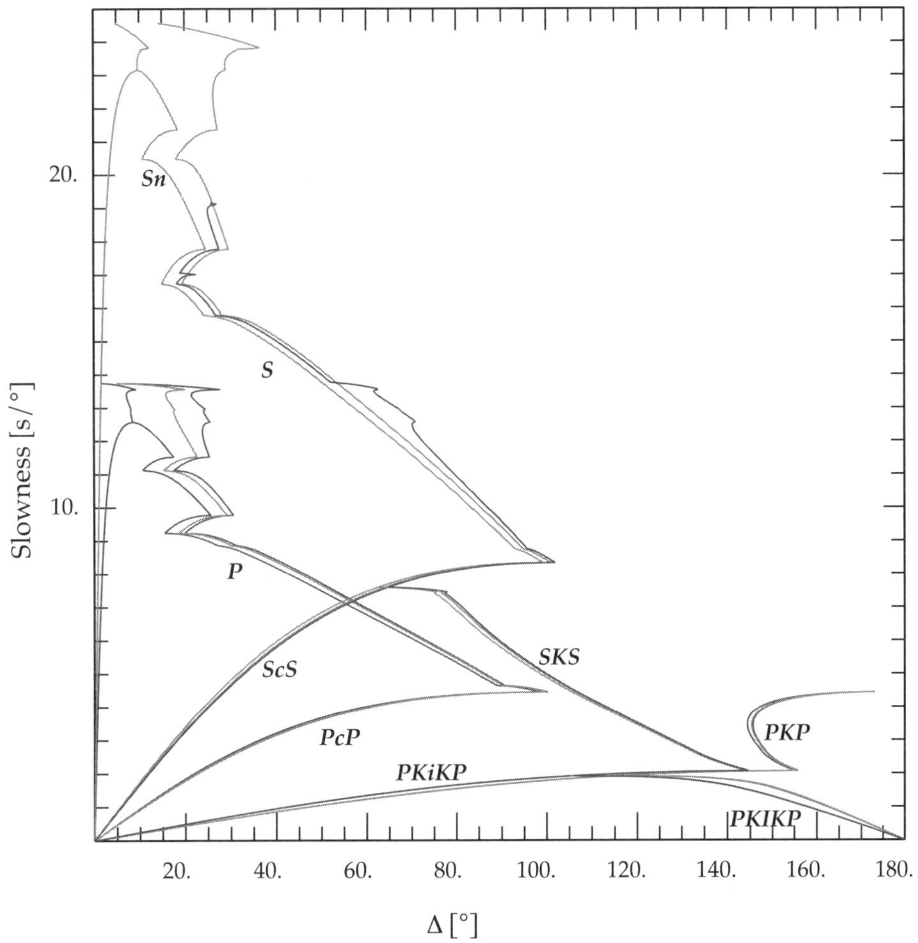

Figure 5.8. Epicentral distance Δ as a function of slowness ℘ for the major phases from an event at 260 km depth.

PKIKP phase penetrating the inner core can be clearly seen in both the $\tau(\wp)$ and $\Delta(\wp)$ displays (figures 5.7, 5.8).

The details of triplications, such as that associated with *S* and *SKS*, are most readily appreciated in the $\Delta(\wp)$ display (figure 5.8) and can then be tied to the velocity model itself through the slowness display (figure 5.9). The presence of the low velocity zone for *P* waves in the outer core is very clear in figure 5.9 through the temporary increase in the *P* slowness below the core mantle boundary (radius < 3480 km). The *P* refraction leg for *SKS* in the outer core arises from the slight decrease from the *S* slowness at the core-mantle boundary. The $\tau(\wp)$ relation for *S* and *SKS* is continuous and the reflected *ScS* phase joins the *S* slowness for the core-mantle boundary.

The slowness model also enables us to appreciate the influence of the velocity

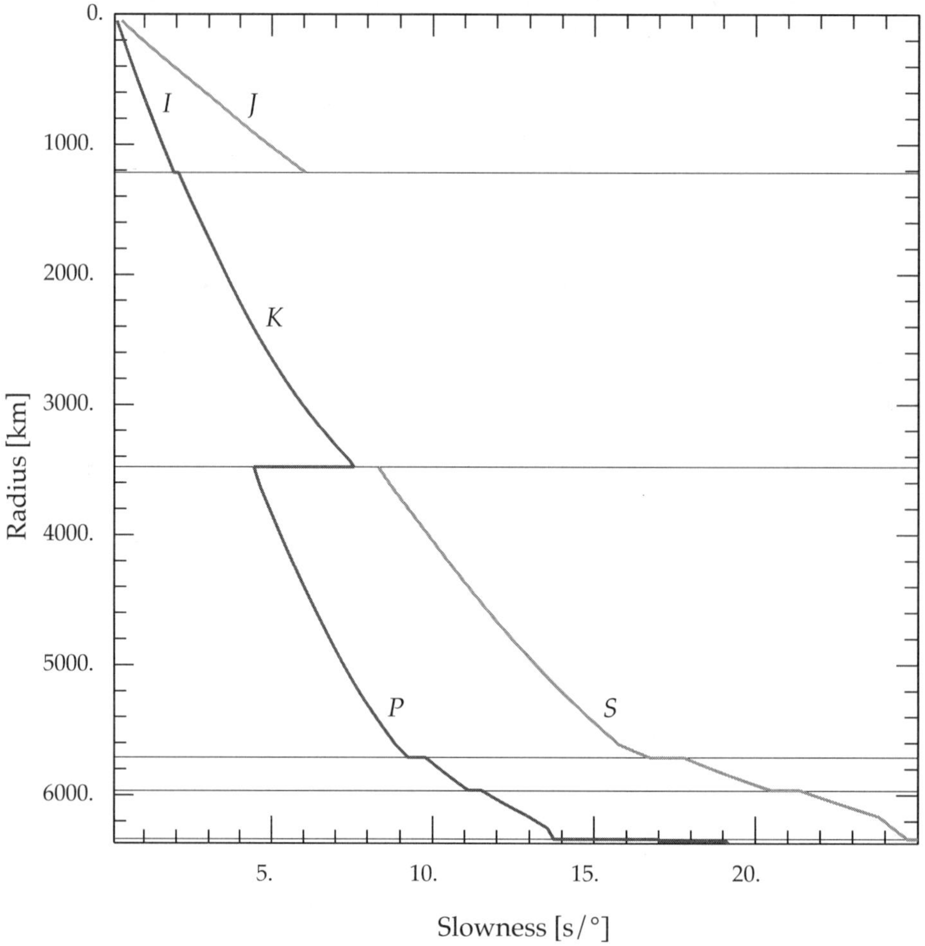

Figure 5.9. Slowness mapping of the *P* and *S* wavespeed profiles for the AK135 model. The major discontinuities are indicated by horizontal lines.

discontinuities at 410 km and 660 km depth, which produce the overlapping triplications in the travel-time curves near 20°. The triplications are most easily followed in the $\Delta(\wp)$ plot. For *S* waves the slownesses in the transition zone overlap with the *P* wave slownesses in the crust, which gives rise to quite complex behaviour for surface reflection with both *S* and *P* legs possible in the shallower part of the crust.

The ray paths associated with this intermediate depth event are illustrated in figure 5.10 for both *P* and *S* phases. In each case we have separated the mantle propagation (*P*, *S*) to the left, from the phases which interact with the core-mantle boundary (*PcP*, *ScS*) or propagate through the core itself to the right. The apparent duplication of rays arise from the presence of the surface reflected phases, which also make it difficult to

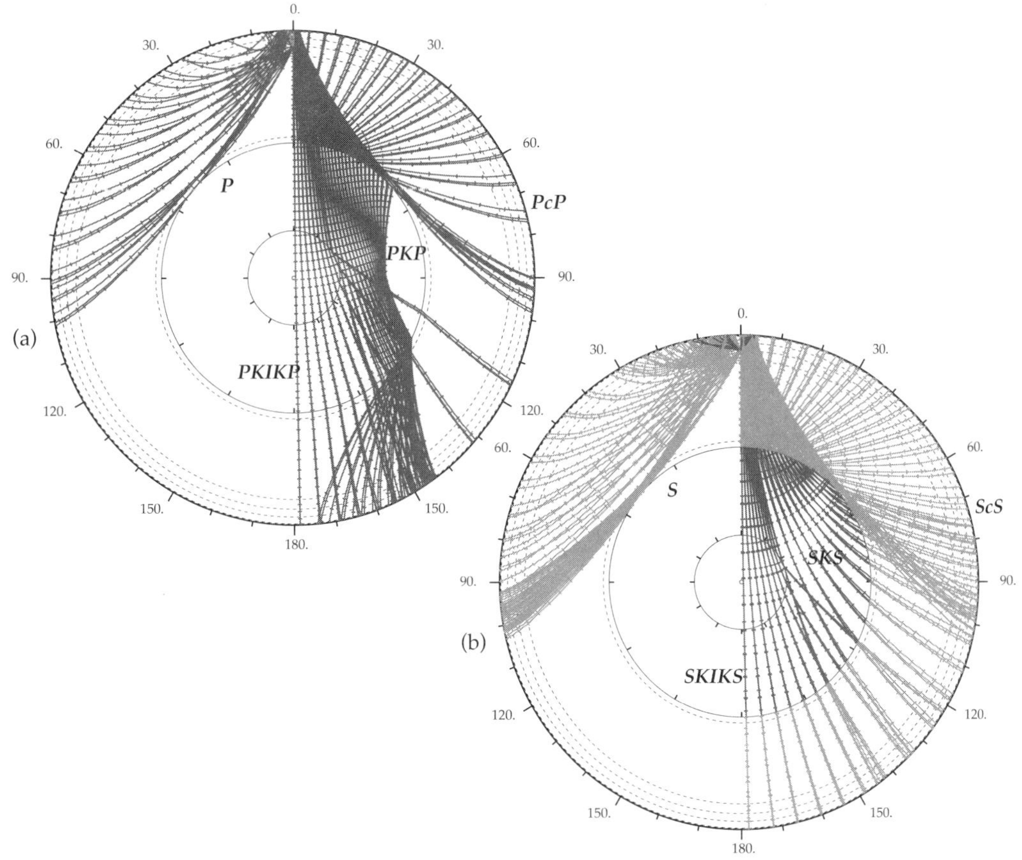

Figure 5.10. Ray paths for the major phases from an event at 260 km depth (a) *P* phases, (b) *S* phases.

follow the different classes of wavefronts in figure 5.10, even though tick marks have been placed at 60 s intervals along each of the rays.

A selected set of wavefronts for the phases included in figures 5.6-5.8 are displayed in figure 5.11. For *P* waves [figure 5.11(a)] we have chosen to plot the wavefronts at 480 s (8 min) and 960 s (16 min) after source initiation. The upper set of wavefronts correspond to the earlier time (480 s) and show refraction through the mantle accompanied by reflection from the core. The *P* wave in the core has advanced in part into the inner core, and the core refractions associated with the surface reflections *pP*, *sP* are just beginning to separate from the mantle legs. At the later time (960 s) we see the complexities of the *PKP*/*PKIKP* wavefronts imposed by the ray caustic in the outer core.

For *S* waves we show the wavefronts at 600 s (10 min) and 1200 s (20 min) after source initiation [figure 5.11(b)]. The time frames have been chosen so that propagation in and through the core is at a similar stage to the P waves in figure

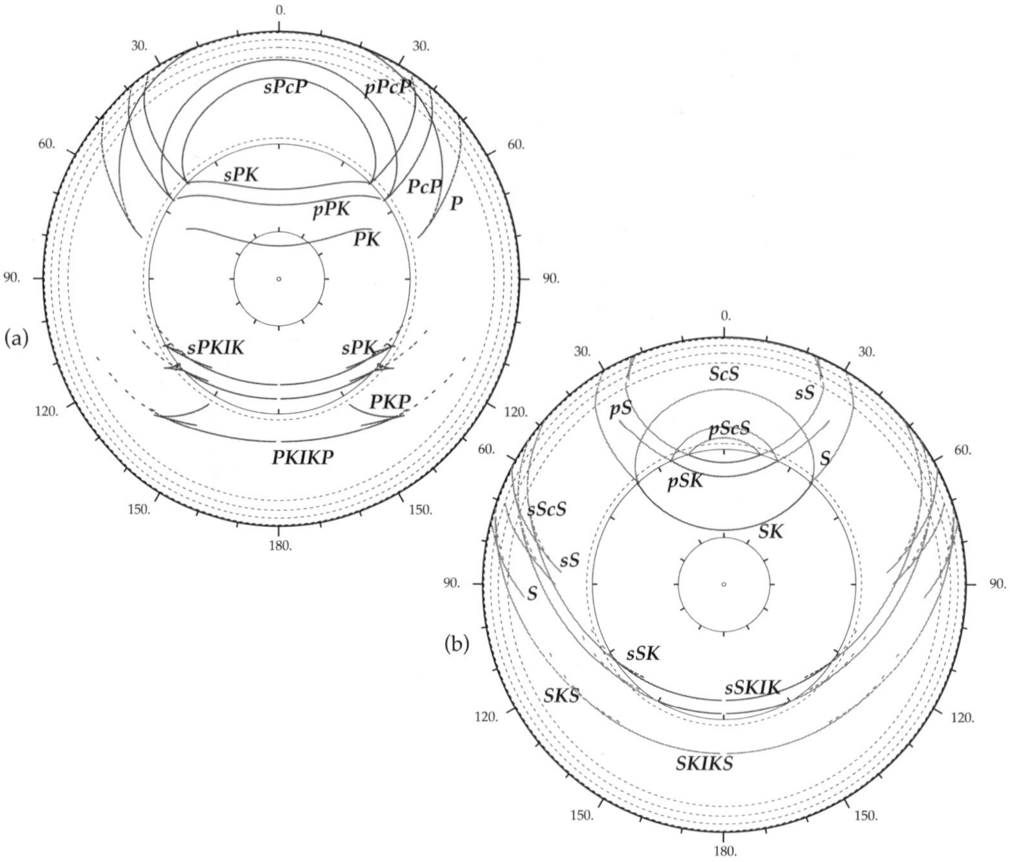

Figure 5.11. Wavefronts for the major phases from an event at 260 km depth (a) *P* phases at 480 s and 960 s, (b) *S* phases at 600 s and 1200 s.

5.11(a). At 600 s, the *SKS* wave in the core is just getting ahead of *S* in the mantle and the set of core reflections can be seen very clearly. At 1200 s, *SKS* has passed through the core and we can see the triplications in the *S* arrivals, associated with the faster *P* leg in the core, developing in the near-surface wavefronts near 70°. The core reflections (*ScS*) produce the trailing wavefronts from the *sS*, *pS* phases.

The use of wavefronts provides us with a snapshot of the wave patterns in the Earth and provides a convenient link to numerical calculations. In figure 5.12 we show an image at 600 s from a pseudospectral calculation of the wavefield in spherical stratification (Furumura, T. et al., 1998) for the same source depth of 260 km in the AK135 model. To avoid numerical complications, the waves are not allowed to penetrate the inner core. The calculation is carried out for the displacement field but to aid comparison with the other displays, *P* and *SV* contributions have been extracted as the divergence and curl of the wavefield. The *P* component is shown in dark grey and the *S* component in light grey. The time of the frame is the same as the upper part

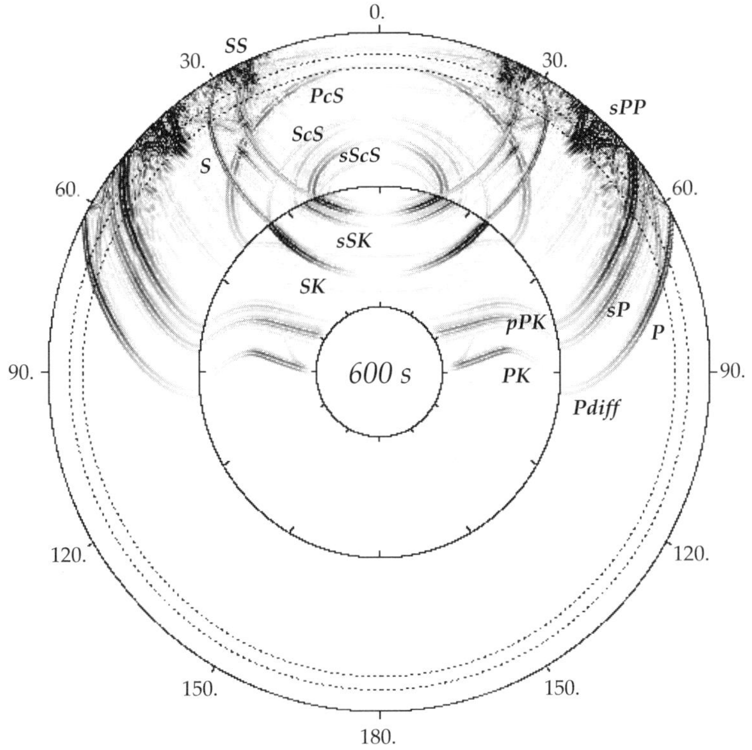

Figure 5.12. A snapshot of the seismic wavefield from a source at 260 km deep propagating in the AK135 model at 600 s after source initiation. The *P* wave contribution is shown in dark gray and the *SV* waves in light gray. The discontinuities at 410 km and 660 km are indicated by dotted lines and the core boundaries with solid lines.

of the *S* wavefront diagram (figure 5.11b) and we see a close correspondence between the ray response and the numerical simulation. The pseudo-spectral results include the full response of the model so that there are additional *S* phases as well as those arising from *P* wave propagation. However, we can see that the ray picture for the limited range of phases is able to capture much of the behaviour.

5.2.3 Relation to observed arrivals

The travel-time and ray displays in the previous section are the result of calculations with the AK135 reference model. This model was constructed to provide a good fit to a wide range of seismic phases (Kennett et al, 1995). But, to what extent does such a reference model actually represent observed times?

Kennett & Engdahl (1991) have assembled a set of 104 events (83 earthquakes, 21 explosions) for which the hypocentre is well controlled, and these events have a rich set of later phase readings (57 655 phases in all). In figure 5.13 we display the travel times for the AK135 model superimposed on the reported phases, which have been

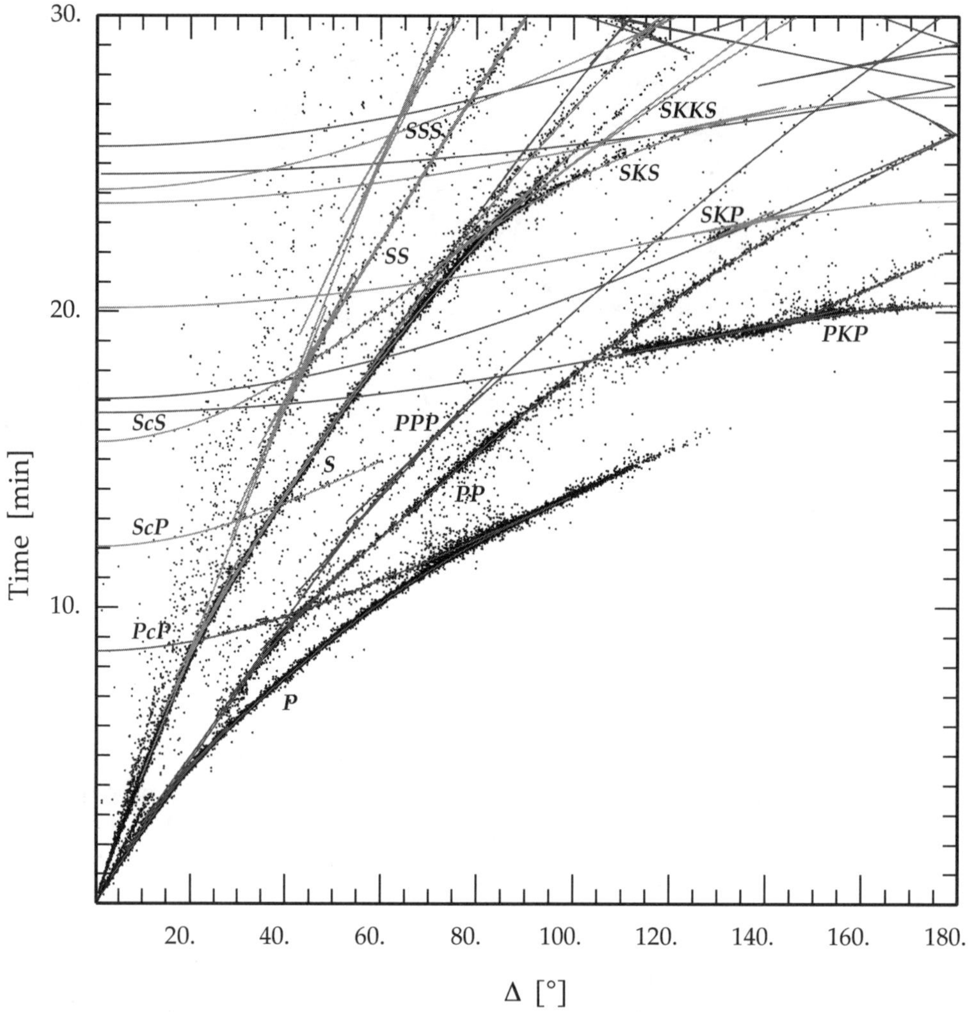

Figure 5.13. Display of AK135 travel times superimposed on the times of phases for the test events corrected to surface focus.

corrected to surface focus. The agreement between the observed and calculated times is very good for the full set of phases.

In figures 5.14-5.16 we show more detail of the comparisons using displays in terms of reduced time $T - \wp_r\Delta$, where the reduction slowness $\wp_r$ is chosen so that the behaviour of a group of phases may be more readily followed.

Figure 5.14 shows the results for both P and S times out to 35°, covering the distance span of energy return from the upper mantle and transition zone. We expect strong regional variations in structure in the upper mantle and this is reflected in the scatter of the observed travel times. Nevertheless the correspondence between the

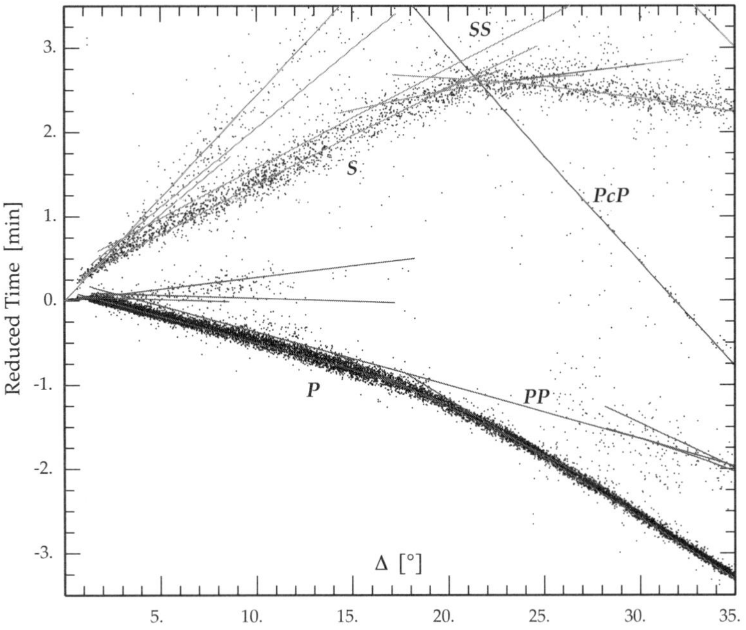

Figure 5.14. Reduced time display (reduction slowness 17.5 s/°) for upper mantle phases from the test event data, corrected to surface focus, compared with the AK135 times.

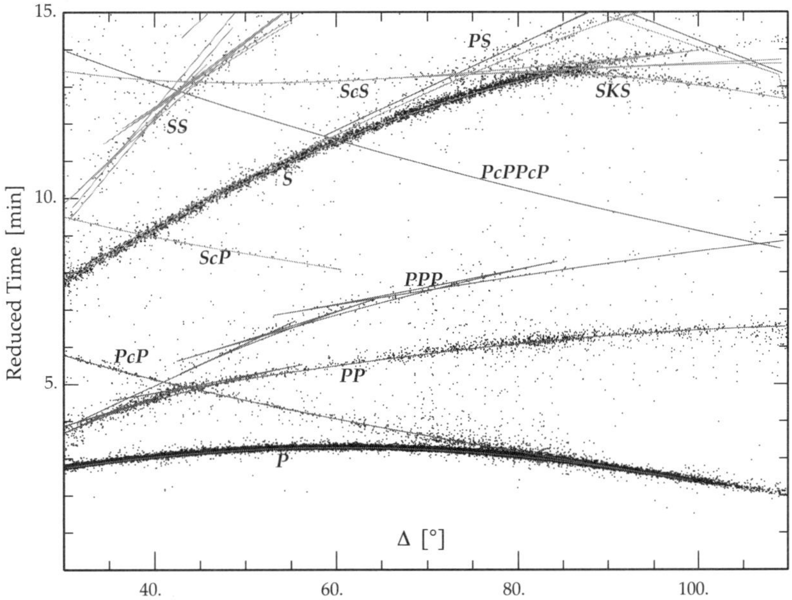

Figure 5.15. Reduced time display (reduction slowness 6.85 s/°) for teleseismic P and S and associated phases from the test event data, corrected to surface focus, compared with the AK135 times.

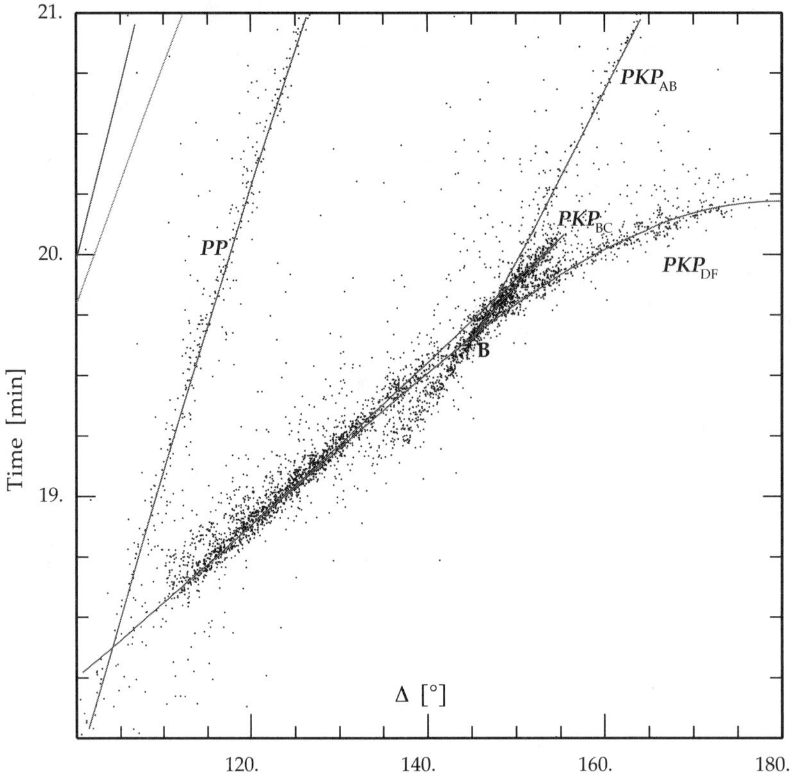

Figure 5.16. Time display of *PKP* phases from the test event data, corrected to surface focus, compared with AK135 times.

triplications for the upper mantle branches in the AK135 model and the times from the test events is good for both *P* and *S* waves. The deviations from the predictions of the reference model carry information about the variations in mantle structure which can be exploited using delay-time tomography (see, e.g., Inoue et al, 1990).

Teleseismic *P* and *S* wave times for the test events are compared to the AK135 predictions in figure 5.15, the agreement is good for *P* and *S* as well as a broad span of other phases including *PcP*, *PP*, *ScP*, *ScS* and the core refraction *SKS*. Observations of later *S* phases are rather sparse but still agree well with the reference model.

As a final comparison we show in figure 5.16 the behaviour for the *PKP* core phase. The DF branch (*PKIKP*) is clear in the test data set over the interval from 110° to 130° and beyond 145°. We can see the full expected structure of the travel time curves [figure 5.4(c)] in the observations with well constrained AB and BC branches. The predicted B caustic agrees well with the test data, but we can see a splatter of observations extending from the B caustic towards shorter distances. These arrivals are now recognised to arise from scattering at the base of the mantle and are not predicted by the ray theory we have been using. In figure 5.16 we also see the *PP* phase that agrees well with the AK135 predictions even though the number of observations is limited.

5.3 Propagation processes for seismic phases

The use of seismic rays as in section 5.2 provides an effective means of examining the behaviour of phase propagation, but needs to be supplemented when we want to characterise the amplitude of particular arrivals. In order to provide a full description of a phase we need to look at the full set of processes from generation at the source, transmission through the Earth, and recording at a seismic station.

5.3.1 Description of the source

It is common to employ a simplified description of a seismic source, typically a point moment-rate tensor, designed to provide a good approximation to far-field radiation (see Chapters 4 and 11). Such a description is effective if the immediate environment of the source can be regarded as uniform; there are complications when a source lies at a discontinuity in seismic properties. In chapter 4 we have presented the radiation characteristics for a point moment tensor source in terms of the dependence of *P*, *SV* and *SH* waves on take-off angle and hence the horizontal slowness for the particular wavetype.

We can therefore use the radiation pattern on a small focal sphere surrounding the point source as the initiation conditions for phase propagation. Normally we will be most interested in the lower focal hemisphere, but for close distances, and when we include surface reflections, the take-off will be from the upper hemisphere.

For shallow sources in the crust there will be little separation between the primary radiation (*P* or *S*) and the phases reflected at the surface (*pP*, *sP* or *pS*, *sS*) For distant stations it can then be more appropriate to think of an effective source which comprises the superposition of *P*, *pP*, and *sP* or *S*, *pS* and *sS* as an interference packet whose character changes with the depth of the source. Further complications can occur when the source lies in an oceanic region: in addition to the *pP* phase reflected at the sea bed because of the contrast between the properties of the rock and the sea water

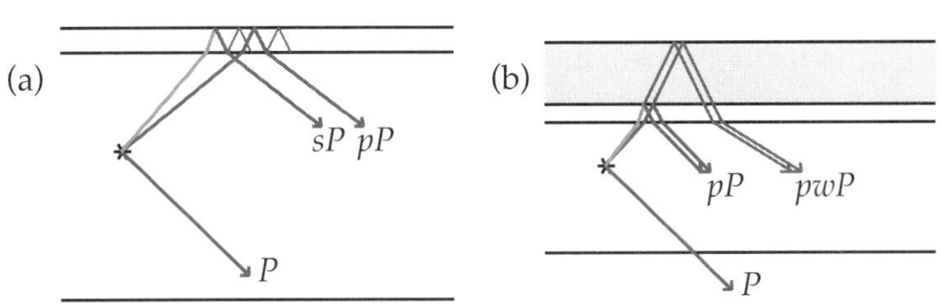

Figure 5.17. Near source processes for a crustal source (a) continental source, (b) oceanic source.

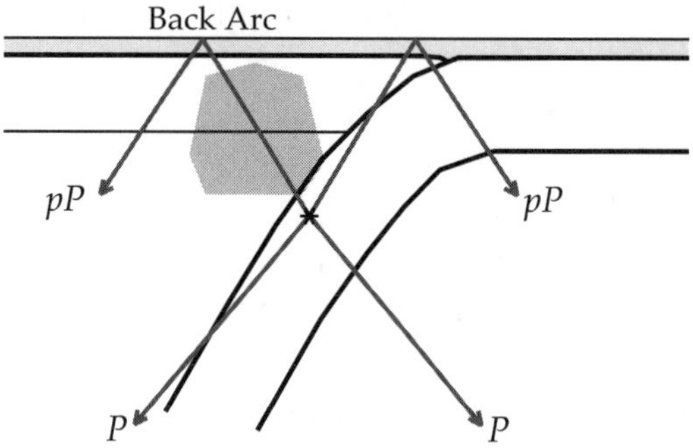

Figure 5.18. Differing environments for surface reflections from a deep source on the back-arc and fore-arc sides of a subduction zone.

there will be an additional arrival *pwP* reflected at the ocean surface (see figure 5.17. The *pwP* phase is often of significant amplitude and for ocean depths near trenches in excess of 6 km arrives several seconds after the surface reflection *pP*.

Internal reflections in the crust can be significant for sources within the crust, particularly where there is a very sharp Moho. Reverberations in shallow sediments are also likely to accompany the surface-reflected phases as indicated in figure 5.17(a). The wavetrain leaving the base of the crust to propagate to a distant station can thus have acquired quite a complex character.

For intermediate depth sources the time differentials between the direct phase and the surface reflections are larger. For a 100 km source the time difference between *pP* and *P* is about 25 s and between *sP* and *P* is 35 s. Thus it is only for long period waves that there will be any significant interference associated with the depth phases.

For deep events the surface reflections are well separated from the main arrivals and have distinct propagation paths. Generally the *P* paths separate rapidly from their subduction zone environments and have relatively simple character. The surface reflected phases travel up to the surface in the surrounds of the subduction zone and their interaction with structure can depend markedly on the geometry of reflection. A reflected phase interacting with a back-arc region will encounter a near surface zone of lowered velocities and marked attenuation, whereas the comparable phase reaching on the fore-arc side will propagate through a higher velocity environment with much less attenuation. The different environments will change the character of the surface reflected pulse as indicated schematically in figure 5.18.

The character of the seismic record changes markedly with increasing source depth as illustrated in figure 5.19 for unfiltered *P* wave arrivals from sources in the Tonga-Kermadec subduction zone recorded at a common portable station (KA3) in northern Australia at a similar epicentral distances. The mantle and receiver paths for the different events are very close, since the projection of the ray path for the deep source back to the surface is at just beyond 52°. The two shallow events are of similar magnitude (mb 5.7) and the deep event is somewhat larger (mb 6.1). The records from

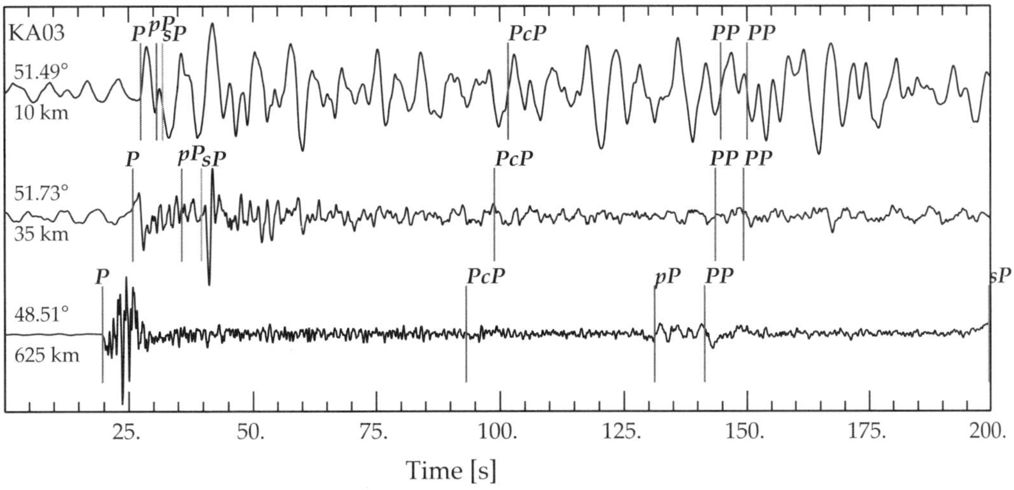

Figure 5.19. Unfiltered vertical component *P* seismograms for events at different depths in the Tonga-Kermadec subduction zone as recorded at similar epicentral distances at a single broadband station in northern Australia. Note the simplification of the character of the arrivals with increasing depth.

the shallowest source have a complex low frequency character with interference of the *P*, *pP* and *sP* arrivals and an extended *P* coda. The event at 35 km depth is by contrast simpler with a clear separation of the *sP* arrivals from the *P* and *sP* train. The deepest event has a complex set of high and low frequency arrivals following *P*, whereas the *pP* phase is relatively simple.

5.3.2 Passage through the Earth

For the main propagation component we can use ray theory as a guide, with wave passage within and across the different zones of the Earth and the imposition of reflection and transmission at the zone boundaries. At typical seismic frequencies, the zone that influences the behaviour of a phase is not confined to the ray itself. Frequently multiple propagation effects need to be included to provide an adequate description of the group of energy we would associated with an observed seismic phase. We can, however, still represent the behaviour of this part of the wavefield via the interaction of a set of causal processes represented through products in the frequency domain, i.e., convolutions in time.

We can set the background by considering the phases *P*, *PcP* and *ScP* travelling from a shallow source to a common receiver as illustrated on figure 5.20, and have used the geometry appropriate to the records shown in figure 5.19. We can use the ray path for these phases as a guide for the sequence of operations which represent the propagation processes. We divide up the Earth by the major discontinuities (indicated in fine lines in figure 5.20) and then characterise the nature of the interaction of seismic waves

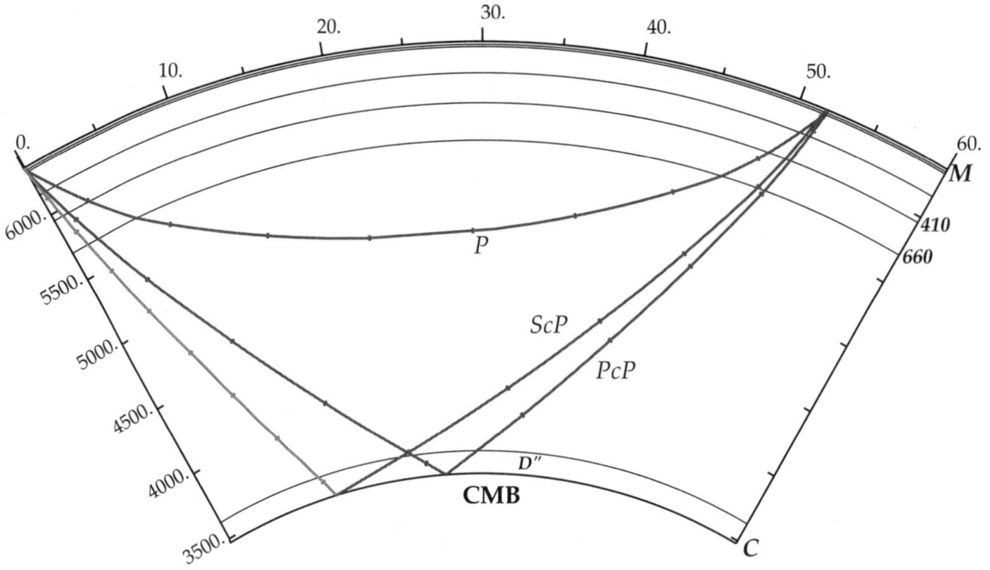

Figure 5.20. Propagation of *P*, *PcP* and *ScP* phases from a shallow source to a receiver at an epicentral distance of 52° in relation to the major divisions of Earth structure.

with the regions between discontinuities in terms of transmission through the region, reflection from wavespeed gradients or reflection from the bounding discontinuities.

The *P* waves emerging through the Moho at the base of the crust, (*M* in figure 5.20), will include the interference of the direct wave and its surface reflections. In order to reach the station at 52° epicentral distance, the *P* waves are turned back by the wavespeed gradients in the lower mantle. Passage through the upper mantle and transition zone is dominated by transmission, both on entry to and exit from the lower mantle. There will be only weak interactions with the upper mantle discontinuities to generate minor phases accompanying *P*.

For the *PcP* phase at the same epicentral distance, the *P* waves emerge more steeply from the crust and are transmitted right through the mantle to be reflected from the core-mantle boundary (CMB, *C* in figure 5.20). The strong contrast in properties between the solid mantle and the fluid core beneath with lowered *P* wavespeed produces substantial reflected energy. The reflected *P* waves are then transmitted again through the entire mantle on their way to the receiver. Because the propagation path is much longer and the boundary reflection is less efficient than the complete reflection from the wavespeed gradient, the amplitude of *PcP* will be significantly smaller than that for *P* even for an explosive source with no angular dependence of radiation.

The *ScP* wave starts as an *SV* wave and is transmitted through the mantle at a small inclination to the vertical. On arrival at the core-mantle boundary, conversion from *SV* to *P* occurs in reflection from the solid-fluid interface. The *P* wavespeed at the

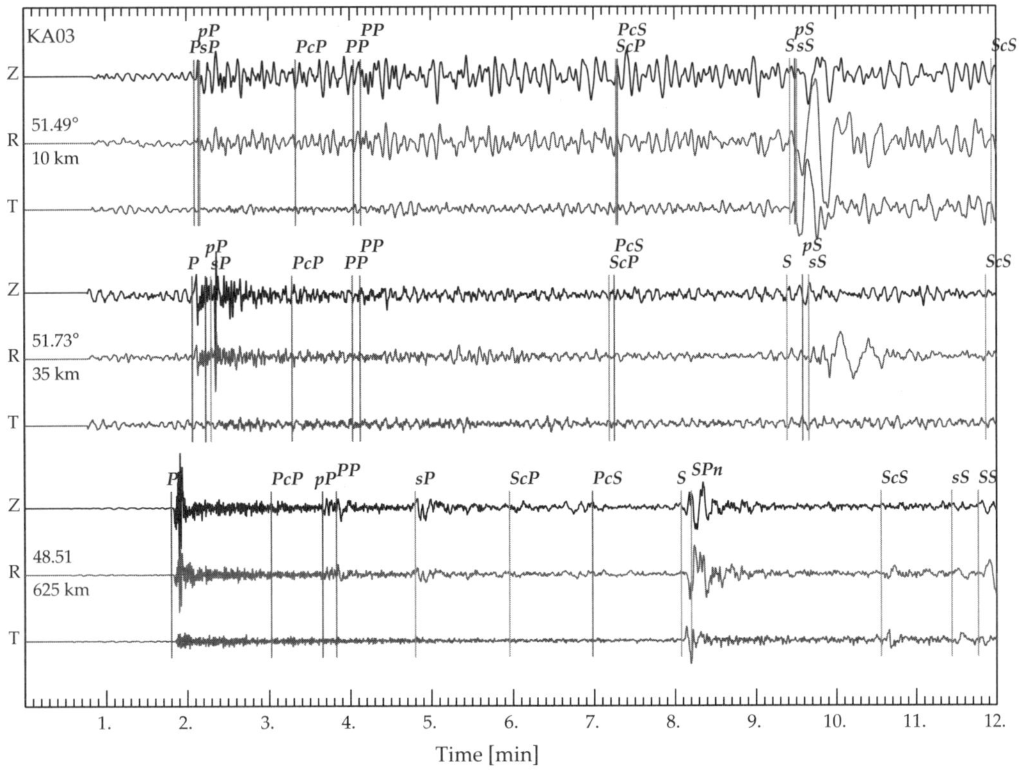

Figure 5.21. Unfiltered three-component seismograms for the events illustrated in figure 5.19, showing the improved clarity of *S* for deeper events as the *P* wave coda is diminished. The traces for each event have a common normalisation.

core-mantle boundary (13.69 km/s) is much larger than the *S* wavespeed (7.30 km/s) and so from Snell's law the resulting *P* wave leaves the core-mantle boundary at a large inclination to the vertical. The *P* wave is then transmitted through the mantle towards the surface. The inclination of the *P* wave to the vertical cannot exceed 90° and so there is limit on the reflection process for the phase, leading to a cut-off for the *ScP* phase just beyond 62°.

A description of a phase should follow the different stages of the propagation in sequence so that it is faithful to the propagation processes in the Earth. Thus reflection should follow downward transmission and be in its turn followed by upward transmission. The distinctions are clear when phase conversion is involved but should be borne in mind even for a single wavetype.

A phase such as *P* arrives against the background of seismic noise and so can readily stand out both by amplitude and by change in frequency content (see figure 5.19). The core reflections *PcP*, *ScP* have a longer transmission path and so have lost more energy both to wavefront spreading and attenuation; the reflection process is also much less efficient than the return from a wavespeed gradient. The resulting arrivals are therefore generally much smaller than *P* and can be masked by the *P* coda. This can be seen clearly in the 3-component displays of figure 5.21 for the same events as used in figure 5.19.

The family of *S* arrivals follow similar types of paths to their *P* counterparts but now appear mixed in with the late *P* coda, with often a similar frequency content to microseismic noise. The advent of digital broadband recording has markedly improved the utility of *S* wave information and good *S* arrivals at reasonably high frequencies can often be seen well into the teleseismic regime (beyond 30°) as in figures 1.2, 5.21.

The influence of the *P* coda on *S* is illustrated in figure 5.21 where we show the *P* and *S* portions of the record for the events previously used in figure 5.19. For the shallowest event the *S* wave has to overcome a long and complex *P* coda, but because of a favourable radiation pattern *S* is very clear on both vertical (Z) and radial (R) components. In an ideal earth there would be no *P* wave energy on the transverse (T) component, and as can be seen in figure 5.21 the rotation of the records has been fairly successful in suppressing the *P* coda so that *SH* arrivals emerge from a simpler background. For the deepest event the *P* coda diminishes quite quickly so that *S* is distinct.

5.3.3 Processes near the receiver

As seismic waves impinge on the structure beneath the receiver they acquire characteristics associated with reverberation and conversion near the receiver. A common scenario is to have strong multiple reflections in surface sedimentary layering. If the contrast in properties between sediment and basement is large then conversion between *P* and *S* waves can occur in transmission at oblique angles. Conversion can also occur at the Moho or deeper discontinuities for distant events and multiple reflections between the surface and the base of the crust can add to the complexity of the receiver-induced effects.

For an incident *P* wave the conversions and reverberations follow the onset of *P* and can produce a significant coda (see e.g figure 5.19). The technique of receiver function analysis seeks to exploit the character of this coda to extract information on crustal and upper mantle structure. For such a structural study the influence of the source needs to be minimised and this is commonly achieved by deconvolving the radial component of motion with the vertical component to produce a trace in which conversion will be most prominent. The simplicity of the main arrival associated with deeper events means that the *P* coda induced by receiver effects is particularly clear (see figure 5.19).

For an incident *S* wave the process of conversion produces *P* waves which then travel faster than *S* waves. The converted phases are therefore to be sought as precursors to the *S* wave arrival, whilst *S* wave reverberations will form a coda to *S*. The presence of precursory arrivals contributes to the difficulty of recognising the onset of S against the background of the coda of the P phase.

The near-surface regime is a zone of significant heterogeneity, and as a result in addition to the signals produced by local layering we often find significant scattered energy. For the deepest event in figure 5.21 a burst of high frequency energy occurs on the transverse component starting about 5 s after P at the time that an S conversion from the Moho should reach the surface. The longer period component may well arise from the deviation of the structure from a stratified isotropic medium; the high frequency energy is due to local scattering of S.

A different class of near-receiver effect can also be seen in the S records for the deepest event in figure 5.21. Closely following S there is predicted to be the phase SPn in which the mantle S arrival couples to a P wave travelling at the top of the upper mantle. This arrival is only expected for SV waves and thus appears only on the vertical (Z) and radial (R) components with a more complex wavetrain for S than the SH pulse on the transverse (T) component.

5.3.4 Composite representation of seismic phases

We can represent each of the steps influencing the nature of a phase by frequency dependent operators. The near-source contribution $C_S(\omega)$ includes the influence of the radiation pattern and directivity of the source, together with the interference of the associated surface reflections and any associated reverberation in the source structure. The main propagation contribution $\mathcal{P}(\Delta, \omega)$ represents the effect of passage through the Earth including both amplitude and phase influences and will depend on the nature of the particular phase. The near-receiver operator $C_R(\omega)$ includes the amplification at the surface from the interaction of the incident and reflected waves, together with the processes of conversion and reverberations in the shallow structure.

The source effects, propagation and receiver effects can be linked into a composite representation for the surface ground motion

$$\mathbf{u}_0(\Delta, \omega) = C_R(\omega)\mathcal{P}(\Delta, \omega)C_S(\omega)\mathrm{i}\omega M(\omega), \tag{5.3.1}$$

where we have introduced the far-field source-time function as seen through the appropriate instrument through $\mathrm{i}\omega M(\omega)$.

For the teleseismic regime (epicentral distances from 28° - 90°), where passage through the upper layers is at relatively steep angles, a useful high-frequency representation of the mantle component for P and S is through a term representing the reflection process multiplied by an attenuation operator

$$\mathcal{P}(\Delta, \omega) = G(\Delta, \omega) \exp\left[\mathrm{i}\omega T(\wp_{\mathrm{ray}}) - \mathrm{i}\tfrac{1}{2}\pi\right] \mathcal{Q}(\omega). \tag{5.3.2}$$

$G(\Delta, \omega)$ represents the amplitude loss due to spread of the wavefronts on passage through the mantle. $T(\wp_{\mathrm{ray}})$ is the travel time from source to receiver predicted by ray theory for slowness $\wp_{\mathrm{ray}}$. The additional phase term in the complex exponential arises from complete reflection from the wavespeed gradients. The source and receiver components also depend on the slowness of the phase through the take-off angle at the source and the incidence angle at the free surface.

The effect of attenuation in the mantle $\mathcal{Q}(\omega)$ for each wavetype can be included via an empirical operator in terms of t^*, the integral of the loss-factor along the time path. For P waves,

$$t^*_\alpha = \int ds[Q_\alpha \alpha]^{-1}, \tag{5.3.3}$$

and is commonly taken to be a constant with a value around 1. For S waves t^*_β is rather higher, typically 3-4. The attenuation operator is approximately

$$\mathcal{Q}(\omega) \approx \exp\left[-i\omega t^* \frac{1}{\pi} \ln\left(\frac{\omega}{2\pi}\right)\right] \exp\left[-\tfrac{1}{2}\omega t^*\right], \tag{5.3.4}$$

including an allowance for the velocity dispersion required for causal attenuation.

For a phase such as *PcP* the propagation term needs to be modified to allow for the reflection process at the core-mantle boundary and so an additional term R_C is inserted into the propagation term and the phase shift associated with complete reflection removed,

$$\mathcal{P}_{PcP}(\Delta, \omega) = G(\Delta, \omega) R_C(\omega) \exp\left[i\omega T(\wp_{\mathrm{ray}})\right] \mathcal{Q}(\omega). \tag{5.3.5}$$

For reflection from a single interface R_C would be frequency independent, but the structure near the core-mantle boundary is both complex and heterogeneous and so a full representation may need to be frequency-dependent, e.g., from modelling the reflection of P from a layered structure.

6
Building a Seismogram

6.1 Seismogram analysis

In the previous chapter we have discussed the way in which we can build up the representation of the nature of a single seismic phase. A seismogram comprises many such distinct phases, but later in the record we encounter the large amplitude surface wavetrain which has been guided along the surface of the Earth, rather than penetrating into interior. The separation of a seismogram into body wave and surface wave components is a convenient way of characterising the appearance of a seismogram and has had a considerable influence on the development of seismic wave theory.

The advent of broad-band recording blurs the separation of seismic signals into body waves and surface waves. Both are present on the same records and indeed we see features that cannot be readily assigned to either class. The portion of the seismogram which lies between the surface multiple *SS* and the low-frequency fundamental-mode surface waves is a debatable territory in which the features of the seismic record can be ascribed to either the interference of progressively higher order *S* wave multiples (a ray picture), or alternatively as the superposition of the higher modes of surface waves. Both approaches can provide a satisfactory result and have in common the feature that the interactions of many wave processes need to be considered.

In this chapter we will look at the way in which the main seismic phases contribute to the body wavefield and then give attention to the character of surface waves. The treatment will be largely descriptive and is supplemented by a more thorough mathematical treatment in Chapters 14–16 of Part II.

The character of the seismograms recorded at the surface changes with epicentral distance as the time separation between different components of the wavefield increases and the attenuation of seismic waves progressively reduces the significance of high frequencies.

The changes in the character of the wavefield out to a range of 40° are illustrated in figure 6.1 with a calculation for the full wavefield in a spherical model using the pseudospectral method (Furumura, M. et al., 1999). The seismogram section in figure 6.1 is for the vertical component of motion from a double-couple source at 60 km

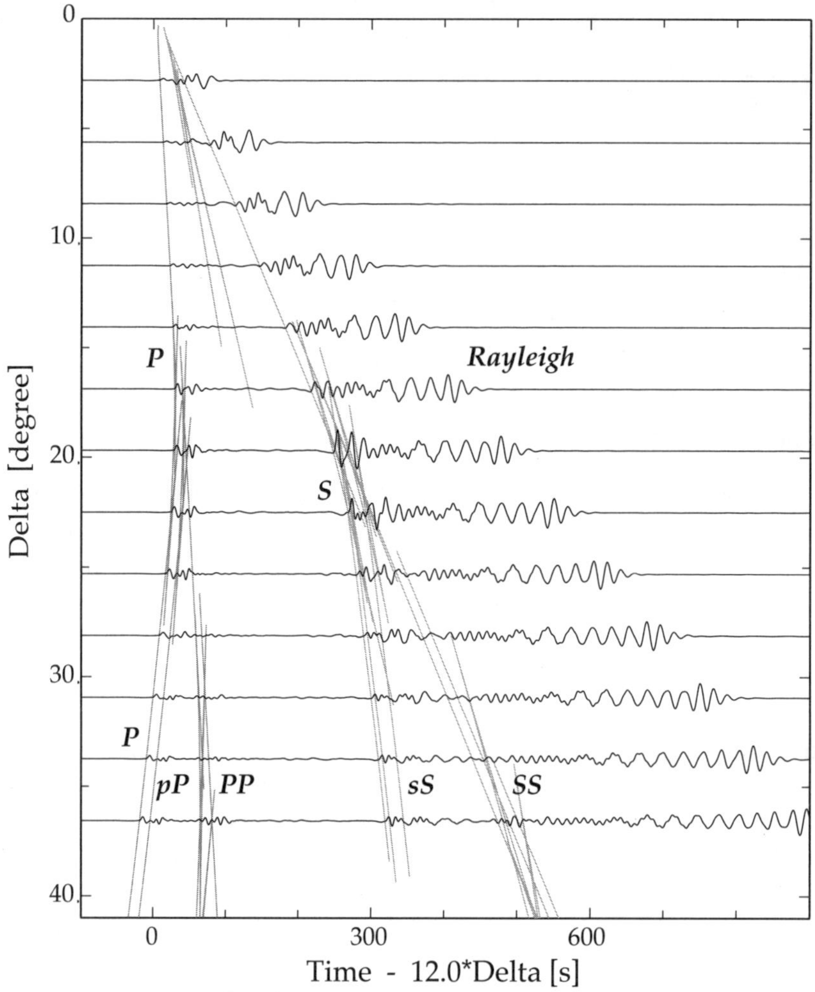

Figure 6.1. Record section for vertical component synthetic seismograms, calculated using the pseudospectral method, for a source at 60 km depth in the AK135 model illustrating the evolution of the wavefield with range.

depth in the AK135 model, and so represents the *P-SV* wavefield with a full wavetrain from the onset of *P* to the fundamental mode Rayleigh wave. The record section is presented with a reduction slowness of 12 s/°, corresponding to the slowness of a *P* wave at 18°. In this way we can nearly align the *P* arrivals and so reduce the time span needed for the presentation of the records.

The *P* arrivals form a distinct group, but the corresponding *S* phases are closely linked to the following dispersive Rayleigh waves. At the largest distances we can see the clear separation of the *P* and *S* waves and their associated depth phases, *pP*, *sS*, from the surface multiples *PP*, *SS*. At 40° the multiples with two surface reflections *PPP*,

SSS do not have a clear identity (compare figure 6.6). The larger amplitudes of *P* and *S* near 20° arise from the interaction of the various branches associated with reflection and refraction from the 410 km and 660 km discontinuities. The interference patterns occur again at approximately twice the range for *PP*, *SS* and we can see the beginning of the pattern in the travel time curves displayed in figure 6.1.

The relatively high frequency arrivals following *SS* represent the combination of many surface multiples of *S* in which the individual contributions lose their identity. This portion of the seismogram can also be described as the superposition of higher modes of Rayleigh waves. The ray representation in terms of *S* surface multiples and the Rayleigh modes are equivalent (see Section 16.2) and either can be used depending on circumstances. In general the ray description becomes more economical at high frequencies, whilst only a few modes provide a good representation at low frequencies.

The distinctive arrival at the end of each seismogram is the contribution from the fundamental mode Rayleigh wave. The lowest frequency components overlap with the higher modes. The frequency reduces progressively along the trace and terminates rather abruptly with an arrival with a frequency close to 0.05 Hz (period 20s), which corresponds to a minimum in the group velocity for the fundamental Rayleigh mode (see section 6.3).

6.2 Higher frequency aspects of the wavefield - body waves

The example of the evolution of the seismic wavefield used in the last section was based on numerical simulation, and we now turn to observed records to illustrate the appearance of the wavefield and the relation of the different contributions to the wavetrains.

We use a set of seismograms from a common receiver, SA04, in Queensland, Australia, which is a portable broad-band station on the north Australian craton, with generally low noise. This station lies within 50° of a number of major zones of seismicity and so it is possible to build up record sections with relatively uniform distance separation between the seismic traces. The configuration of the station and the sources used in later figures is shown in figure 6.2, together with the great circle paths. With the exception of event 1, the paths include a significant oceanic component. The propagation path for event 1 in New Guinea is purely continental, since the Australian craton extends beneath the shallow sea between Australia and New Guinea. Each of the sources is shallow with depths between 14 km and 29 km, and therefore excites a complex wavefield.

In figures 6.3-6.5 we display record sections for these six events for the 3-components of motion corrected to a common source depth of 24 km. Figure 6.3 shows the vertical component. The horizontal components have been rotated along the great-circle paths shown in figure 6.2. The radial component directed towards the station SA04 is displayed in figure 6.4 and the tangential component (perpendicular to the great-circle) is shown in figure 6.5. In each case the records are unfiltered ground

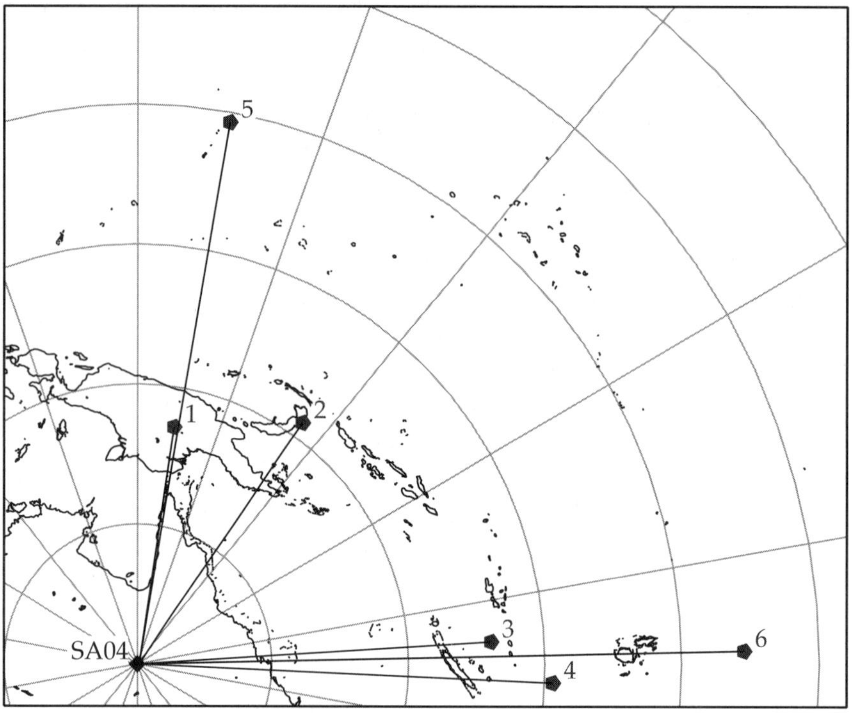

Figure 6.2. Event location and great-circle propagation paths for the six events recorded at the portable broad-band station SA04 in Queensland, Australia for which seismograms are displayed in figures 6.3-6.5, and 6.7-6.9. Azimuthal equidistant projection centred on the station location, with a 10° grid in range.

velocity derived from a Güralp CMG-3ESP seismometer with a 24 bit digitiser. The combined response of seismometer and recording system is flat to ground velocity from 0.03 Hz to 8.0 Hz so that the high frequency components of the wavefield are well represented. The traces are plotted in terms of time after the origin time of the events and are normalised with a common scale factor applied to all components derived from the maximum energy across the three components. The differences in the character of the sources for the different events and the varying position of the station relative to the radiation patterns means that the significance of the *P* and *S* body waves and the later surface waves varies from event to event. In addition to the seismograms for the six events, we display the travel-time curves for the AK135 model with a source depth of 24 km. These travel time curves are intended to provide a guide to the identification of the arrivals rather than a precise match to the records. The major phases are indicated with annotation close to 35°; the core phase *PKiKP*, which has almost constant arrival time and is expected to have small amplitude, is annotated near 15°.

The rotation of the horizontal components has been quite successful. There is very

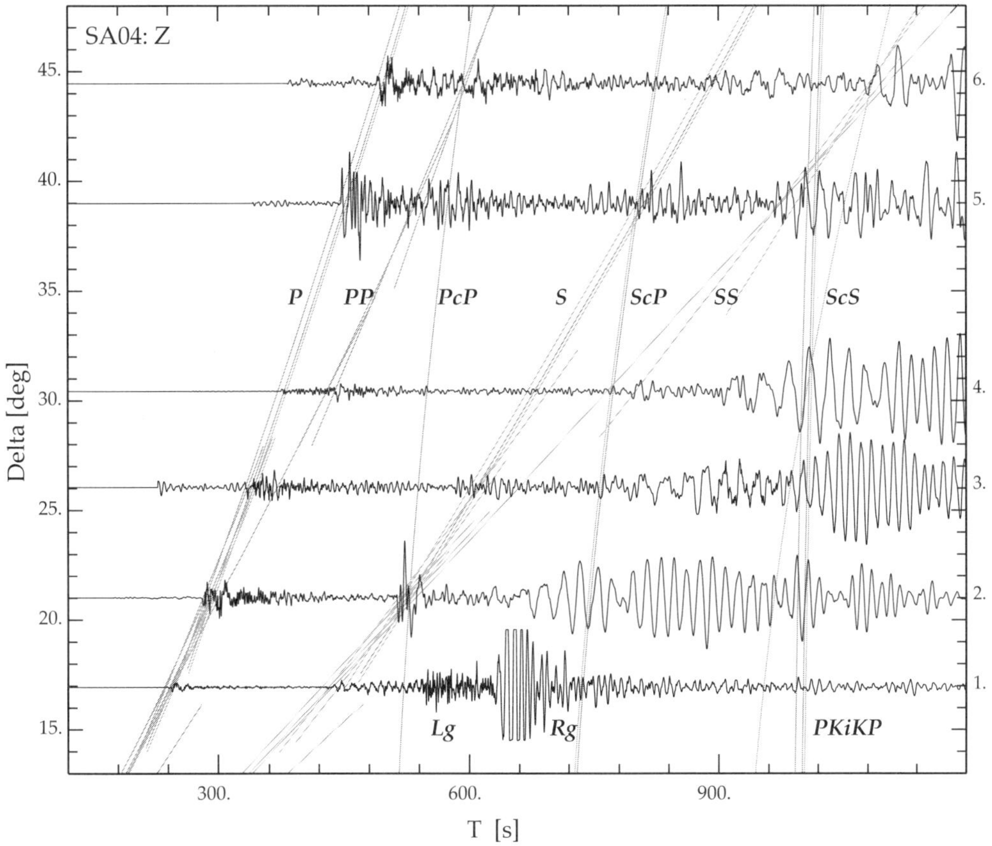

Figure 6.3. Record section of unfiltered vertical component records from six events recorded at station SA04.

little *P* energy on the tangential component records in figure 6.5 at the onset of *P*. However, as is common, a high-frequency scattered *P* component builds up after a short while. The slow decay of this *P* coda is characteristic of a cratonic station.

The seismograms for event 1 in New Guinea, in figures 6.3 - 6.5, are dominated by guided waves in the crust; the high frequency *Lg* phase arrives just after the expected time for *PcP* followed by surface waves with lower frequency. The surface waves are much higher frequency than for the other events and are extremely strong on the vertical and radial components. The distribution of energy in the *Lg* wave group differs between *SH* and *P-SV* waves. *Lg* is very energetic in the tangential (*SH*) component and there is more amplitude in the early part of the phase group. As we shall see in volume II, *Lg* has its complex character because of the superposition of *S* wave multiples in the crust and the total reflection of *SH* waves at the free surface tends to enhance the *SH* component relative to *P-SV*. The mantle arrivals for event 1 are distinct and accompanied by an extended high-frequency coda, characteristic of propagation

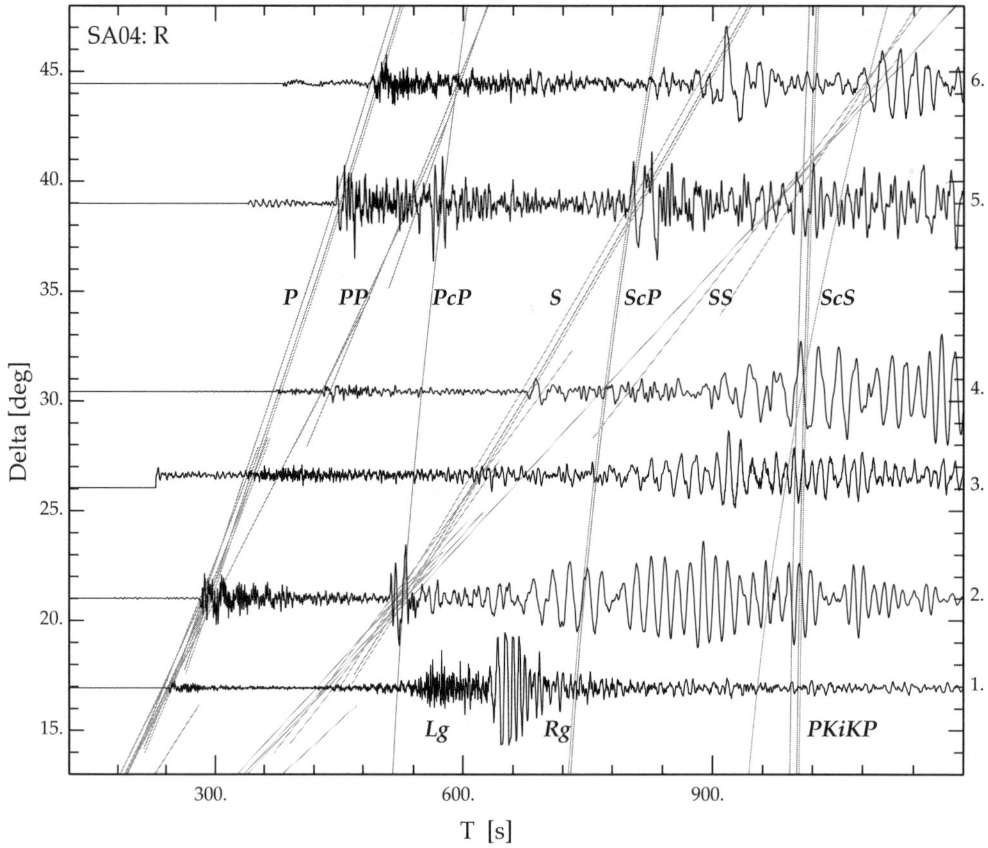

Figure 6.4. Record section of unfiltered radial component records from six events recorded at station SA04.

in the cratonic lithosphere of Australia (Kaiho & Kennett, 2000). Their visibility is suppressed in the record sections because of the very large amplitude later arrivals.

Events 2 and 3 occur in the distance range where the arrivals of *P* and *S* are complicated by the interference of the multiple branches associated with reflection and refraction at the upper mantle discontinuities (410 and 660 km), and compounded by the superposition of the sets of phases for the depth phases. The multiple propagation processes arriving at the same range tend to enhance the amplitude and give complex composite pulse. The presence of a number of distinct arrivals is clearly seen in the *S* pulse for event 2.

For epicentral distances beyond 30° the *P* and *S* waves turn in the lower mantle and the propagation processes are simpler. The records can still be complex since they carry the signature of the source time function and its interference with the depth phases (see section 6.4). The differential attenuation between *P* and *S* becomes very

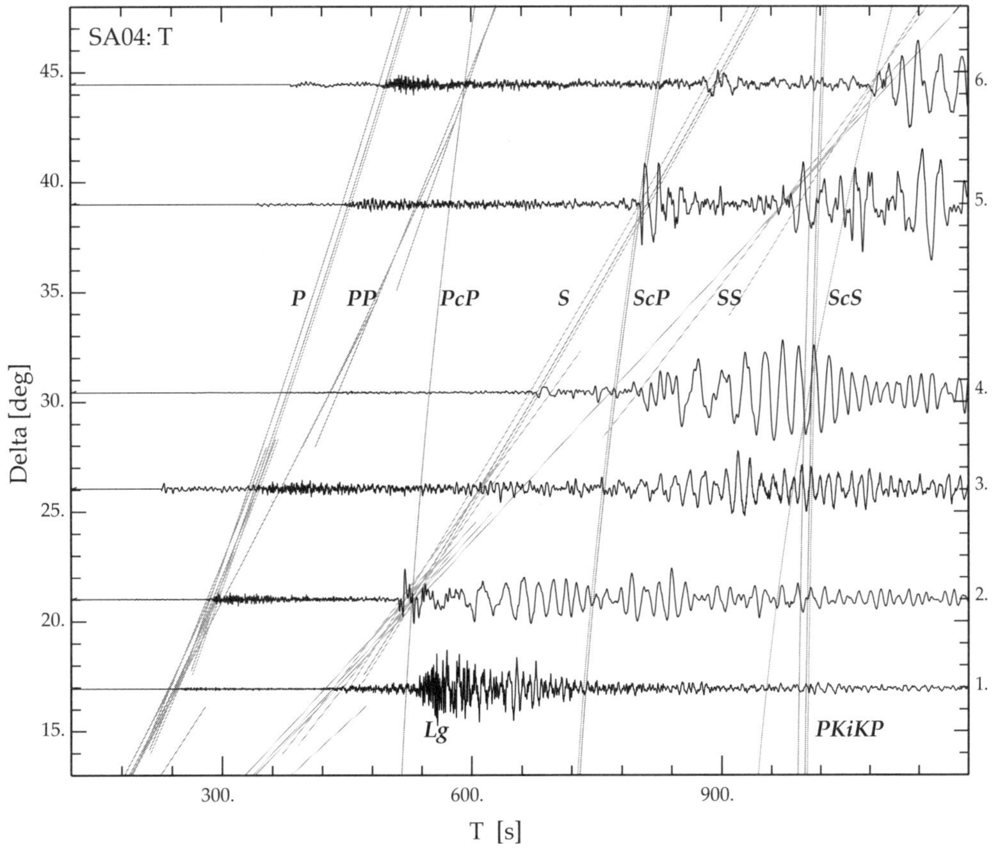

Figure 6.5. Record section of unfiltered tangential component records from six events recorded at station SA04.

evident as the epicentral distance increases, through the much lower frequencies of the *S* arrivals.

The surface multiples *PP* and *SS* begin to emerge from the *P* and *S* wave packets around 20°. Initially they form a lower frequency extension of the *P* or *S* arrival but at greater distances emerge as distinct pulses. The double bounce path has a turning point closer to the surface than for the direct wave. Thus *PP* and *SS* spend longer in the zone of higher attenuation near the surface with a consequent loss of high frequencies. This effect can be seen quite clearly for the horizontal components for event 5. The multiple branches associated with upper mantle structure become significant for *PP* and *SS* near 40° and tend to enhance the amplitude of these surface-reflected phases (event 5). The relative amplitude of *S* and *SS* depends on the position of their take-off angles relative to the radiation pattern and can vary markedly between *SV* and *SH* waves as can be seen for events 5 and 6.

The expected propagation processes for *S* waves from a shallow source for an

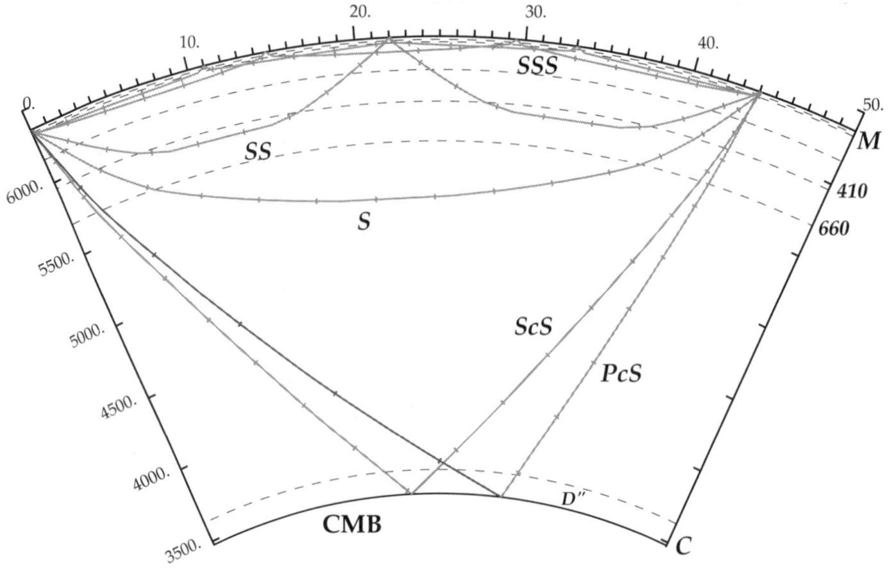

Figure 6.6. Propagation of different classes of *S* wave phases from a shallow source to a receiver at an epicentral distance of 44°, showing the clearly separated paths of *S*, *SS* and the core phases *ScS*, *PcS*, whilst the higher order surface multiples *SSS*, *SSSS*, crowd close to the surface and build up the higher mode surface wave train.

epicentral distance of 44° (corresponding to event 6) are illustrated in figure 6.6. The *S* wave penetrates well into the lower mantle, whereas *SS* is expected to turn in the transition zone. The higher surface multiples *SSS*, *SSSS*, are confined to propagation paths in the upper part of the mantle and are expected to arrive close together in time so that, as in the theoretical simulation in figure 6.1, they do not have a separate identity.

The superposition of such multiple *S* arrivals form the higher frequency parts of the extended wave train in the *S* coda, which is clearly seen in figures 6.3-6.5. These are followed by lower frequency waves which are best described as contributions from a few modes of surface waves.

6.3 Lower frequency aspects of the wavefield - Surface waves

In the unfiltered record sections of figures 6.3-6.5, the pulse-like body waves are accompanied by extended wavetrains with lower frequencies. These surface waves have been guided along the surface of the Earth and have displacements confined to the upper mantle.

We have introduced the idea of surface waves with a simple illustration in Chapter 3. The constructive interference concepts that we used in that discussion have quite general validity and are presented for a general Earth model in Section 16.3. Surface

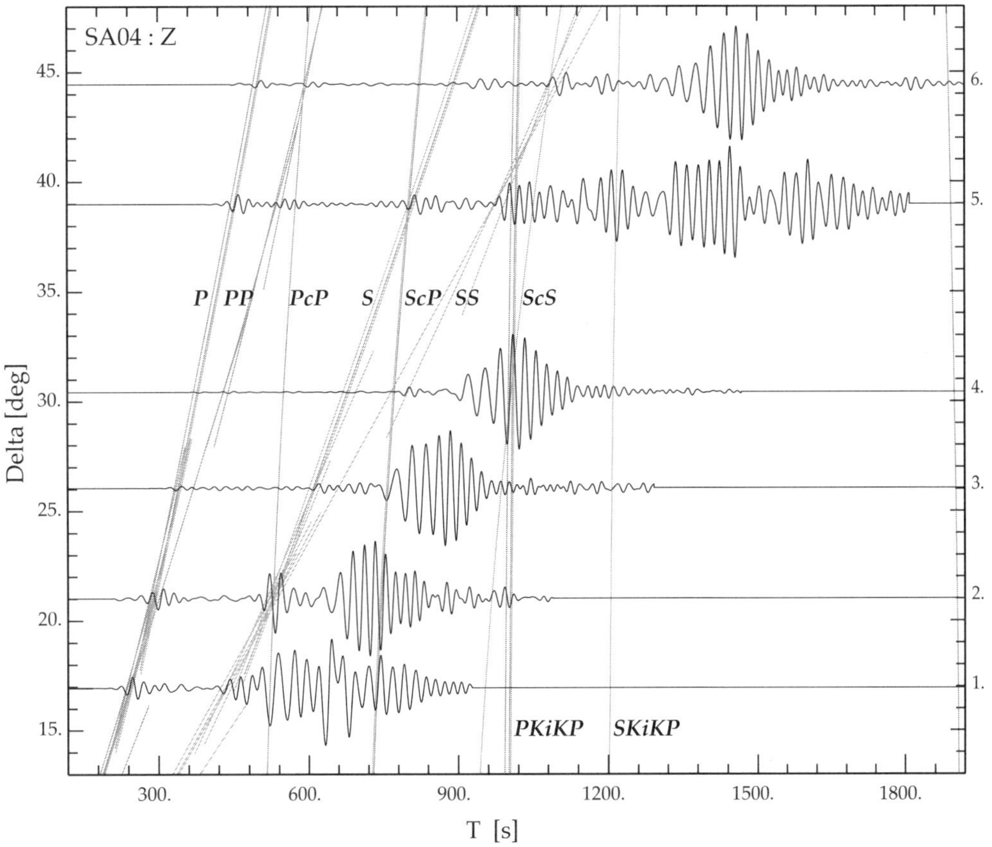

Figure 6.7. Record section of vertical component records, bandpass filtered from 0.01 to 0.05 Hz, from six events recorded at station SA04.

waves are dispersive; the propagation speed depends on frequency in a way that depends on the nature of the structure over which they are propagating.

Body waves penetrate deeper into the Earth and their characteristics are dominated by the vertical gradient of seismic velocities. As a result their arrival times can be predicted well by the results for a 1-D reference model (as we see using AK135 in figures 6.3-6.5). The deviations from the expected times provide information on the integrated effect of 3-D structure along the path. With a very large number of crossing paths, the patterns of travel time residuals can be exploited in a tomographic inversion to build up images of 3-D structure within the earth (see Part V in Volume II).

The surface waves exist because of the trapping of energy between the surface and the increase of wavespeed, but their properties reflect the closer balance between vertical and horizontal gradients in seismic parameters near the surface. The character of surface wave dispersion varies markedly between continental and oceanic paths. But, with the aid of a suitable reference model, the dominant dispersion behaviour

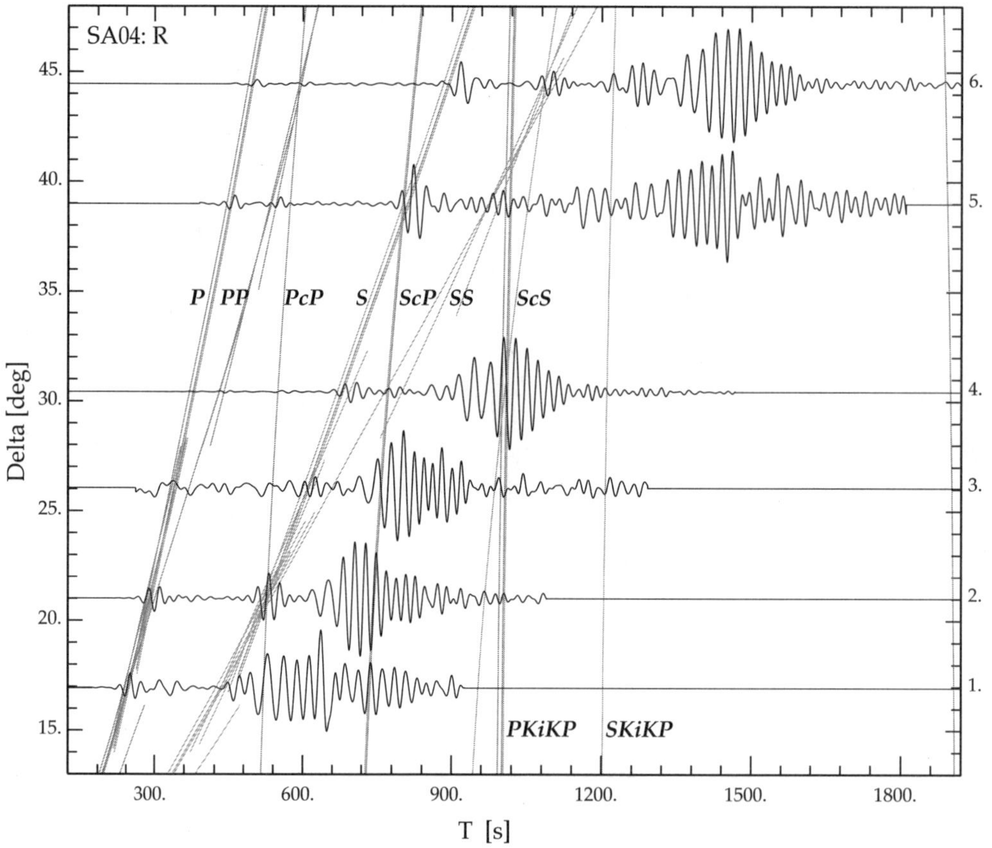

Figure 6.8. Record section of radial component records, bandpass filtered from 0.01 to 0.05 Hz, from six events recorded at station SA04.

can be predicted and the residual phase dispersion provides information on the structure encountered along the path. By waveform inversion, or direct measurement of dispersion, the geographic pattern of phase or group velocity can be extracted as a function of frequency. A subsequent inversion allows the reconstruction of 3-D shear structure (see Part V).

For horizontally polarised *S* waves in a stratified structure, we have a sequence of Love modes whose properties are determined by the requirement that successive multiple *SH* reflections from the structure are in phase with each other. The *P-SV* system is more complex because of the linking of the *P* and *SV* waves through the free-surface boundary condition. The higher modes of Rayleigh waves can be explained by the constructive interference of *SV* waves with allowances for conversion to *P*. In addition the fundamental Rayleigh mode couples evanescent *P* waves to *SV* waves, and can have a phase velocity lower than the *S* wave at the surface through a propagation process which links evanescent *P* and *SV* wave fields.

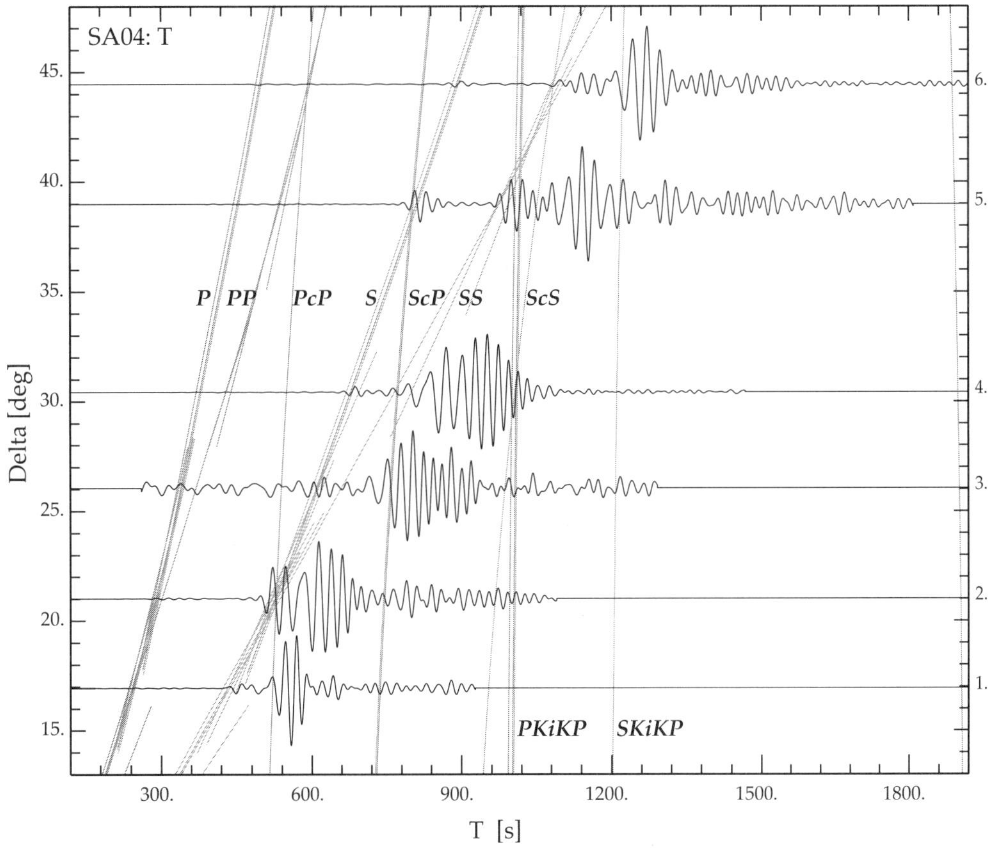

Figure 6.9. Record section of tangential component records, bandpass filtered from 0.01 to 0.05 Hz, from six events recorded at station SA04.

The displacements in the surface waves are concentrated near the surface where the variations in structure are strongest, and their observed character bears the imprint of the paths the waves have followed. In figures 6.3-6.5, in the wavetrain following *S* for events 2-4 we see low frequency arrivals followed by a complex of somewhat higher frequency waves. These elongated trains of waves are characteristic of paths with a significant oceanic component and arise because there is a large variation in group velocity over a small frequency interval (see figure 6.12). The *S* waves in the solid material link into *P* waves in the ocean and the influence of the oceanic layer becomes greater as the frequency increases.

Although surface waves are often evident in unfiltered records they become the most prominent feature of the wavefield at low frequencies. In figures 6.7-6.9 we show record sections for the same six events but have now filtered the three components of motion with a bandpass filter (4-pole Butterworth) with corner frequencies at 0.01 Hz and 0.05 Hz. Over this frequency window the dominant contribution from the shallow sources comes from the fundamental modes of the surface waves, although long period *S* body waves are also evident.

The prominent dispersive wave train on the vertical and radial components (figures 6.7, 6.8), grading from rather long period towards a more uniform oscillation, is the fundamental Rayleigh mode. The oceanic paths (2-4,6) show a very consistent character that is rather different from the continental path from event 1, and the mixed oceanic-continent path from event 5 to the north. The high frequency Rayleigh wave which is so strong in the unfiltered record for event 1 (figure 6.4) is accompanied by longer period waves with deeper penetration into the earth.

The differences between the *P-SV* and *SH* wave systems are very clear when we turn our attention to the tangential component records (figure 6.9). The character of the dispersion of the wave pulses for the fundamental Love mode is different, and the wave group arrives somewhat earlier than for Rayleigh waves just after the expected arrival time for *SS*. The fundamental Love waves can be viewed as representing the interference of the low frequency components of all the higher order surface multiples, *SSS* etc. The group velocities of the fundamental and higher modes of Love waves are very similar, especially for oceanic paths (see figures 6.11, 6.12). As a result the Love wave group has mixed character. The differences in the shape of the wavetrain for the different events reflects the way that the character of the source affects the excitation of the fundamental and higher modes.

For the Rayleigh waves, the long period multiple *S* waves are mostly represented by higher modes, and these are beginning to separate from the fundamental mode at the larger epicentral distances in figure 6.8. For events 5 and 6, following the *SS* arrival we see a longer period group of waves before the larger amplitude burst associated with the fundamental mode. This wave group, arising mostly from the first higher Rayleigh mode, has a similar group velocity to the fundamental Love mode.

Following the main surface wave arrivals we see smaller amplitude waves that represent waves which have been scattered or refracted by three-dimensional heterogeneity in the vicinity of their path between source and receiver. Frequently, a detailed study of the arrivals of surface waves on long paths indicates that there are slight polarisation anomalies, the Rayleigh and Love waves appear to arrive along azimuths a few degrees away from that expected for the great circle. The polarisation anomalies carry information on the wavespeed gradients which can be exploited in inversion for 3-D structure (see, e.g., Laske & Masters, 1996; Yoshizawa et al., 1999).

In areas of strong heterogeneity there is the possibility of surface wave energy arriving by multiple paths with different apparent group velocities. The complexity of the surface wavetrains in the unfiltered records of figures 6.3-6.5 is likely to be partially influenced by such effects. However, such contributions from paths away from the neighbourhood of the great-circle path are generally small at long periods. Only for event 5 which has a complex oceanic-continental path crossing the island of New Guinea (figure 6.2) do we see an extended train of low frequency waves.

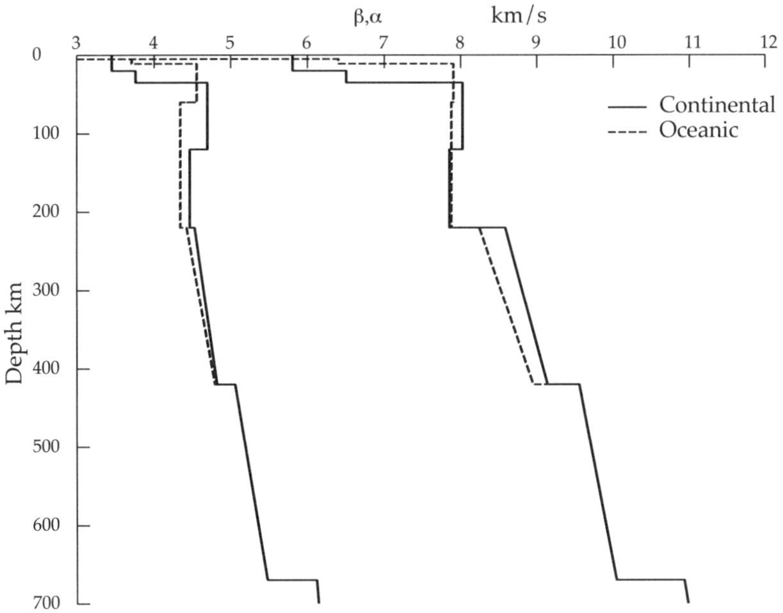

Figure 6.10. Wavespeed models for continental and oceanic regions used for surface wave dispersion calculations.

We illustrate the dependence of the properties of surface waves on structure by calculations of dispersion and mode shapes with depth for continental and oceanic models. We use the upper mantle from the models PEM-C and PEM-O constructed by Dziewonski, Hales & Lapwood (1975) to represent averaged continental and oceanic structures (figure 6.10). The difference in wavespeed structure in these two models extend to the 410 km discontinuity but are most marked at shallow depth, because of the major differences between the thickness of the oceanic and continental crusts.

In the oceanic model, the seafloor acts as a perfect reflector for *SH* waves so that there is no coupling of Love waves into the seawater. For Rayleigh waves we have the possibility of conversion between *SV* waves in the crust and *P* in the water, and the reverberations of *P* waves in the ocean have a significant influence on the dispersion of the Rayleigh modes at high frequencies.

For the continental model the free surface is in common to both Love wave and Rayleigh waves. However, the presence of *P* waves in the crust with lower wavespeeds than are encountered in the *S* distribution at depth means that there will be coupling between crustal *P* propagation and *SV* waves, which will influence the detailed character of the Rayleigh wave dispersion. The fundamental mode Rayleigh wave intrinsically links evanescent *P* waves with shallow propagation of *SV* waves.

The dispersion behaviour for the continental and oceanic models is shown in figures 6.11, 6.12 for both Love and Rayleigh waves. The variation of both phase velocity (solid lines) and group velocity (chain dotted lines) is displayed as a function of period.

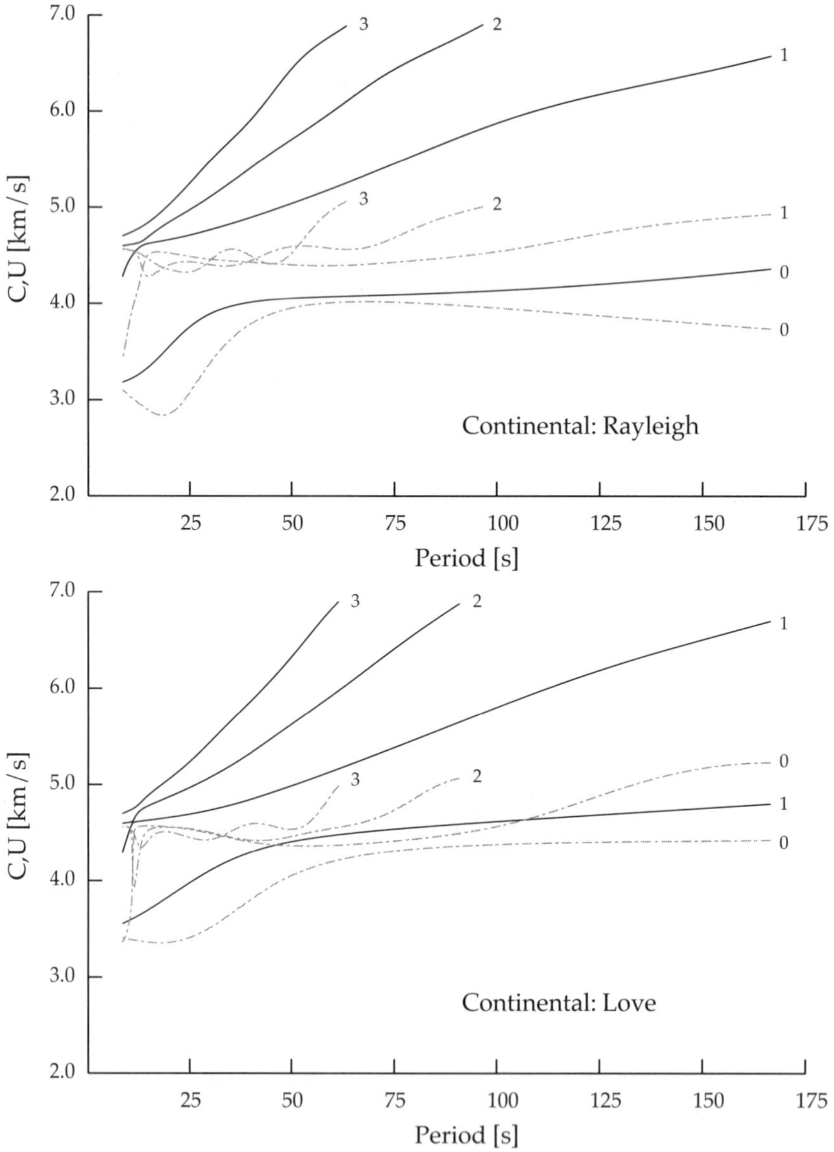

Figure 6.11. Rayleigh and Love wave dispersion for the fundamental and three higher modes of the continental model as a function of period. The phase velocity is shown by solid lines and the group velocity by chain dotted lines in tone.

For the continental model PEM-C, the fundamental mode of Rayleigh waves has a distinct minimum in group velocity near 20 s period (0.05 Hz). The minimum means that both higher and lower frequency waves travel with higher group velocities and arrive earlier. This gives rise to the characteristic appearance of an Airy phase, the wavetrain commences with long periods and then acquires small high frequency

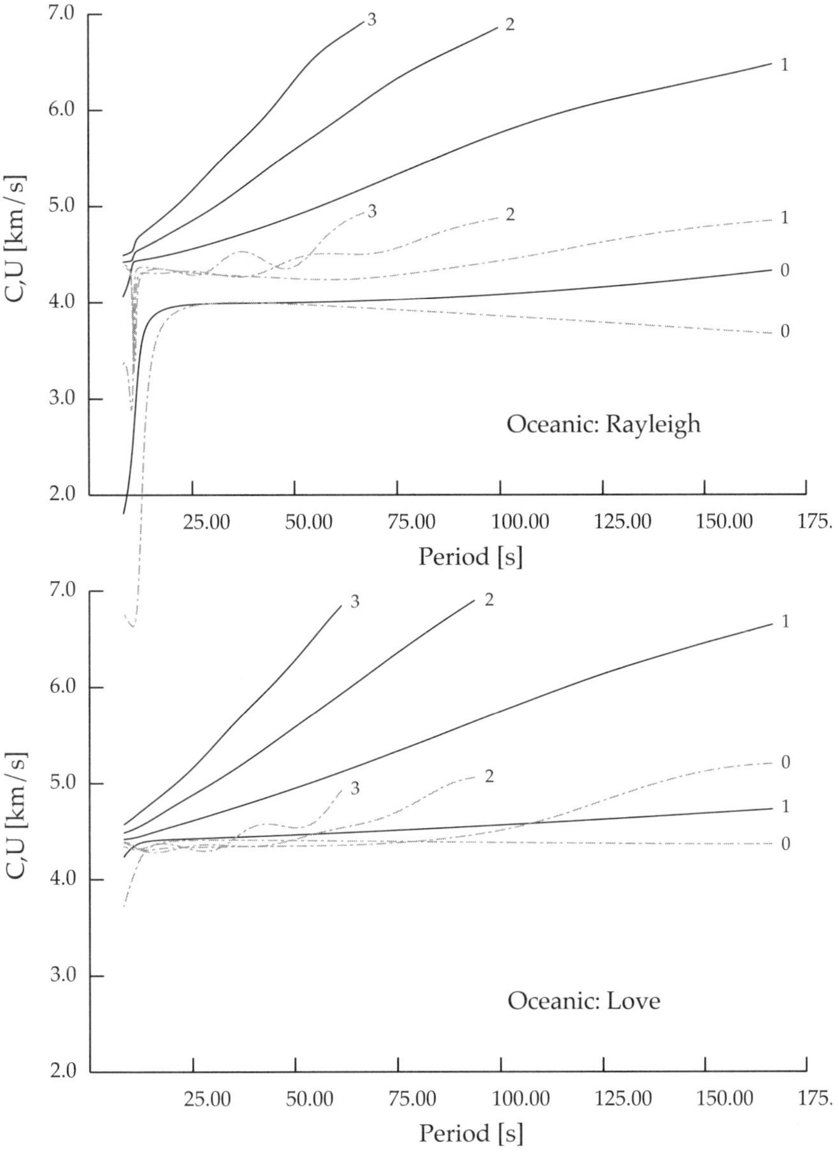

Figure 6.12. Rayleigh and Love wave dispersion for the fundamental and three higher modes of the continental model as a function of period. The phase velocity is shown by solid lines and the group velocity by chain dotted lines in tone.

waves riding on top, and the wavetrain comes to an abrupt halt at the frequency corresponding to the group velocity minimum. The crowding of different frequencies near the minimum tends to enhance the amplitude at the tail of the fundamental mode wavetrain. Event 5 with a mixed continental-oceanic path displays this Airy phase phenomenon for the fundamental mode and this is best seen on the radial component

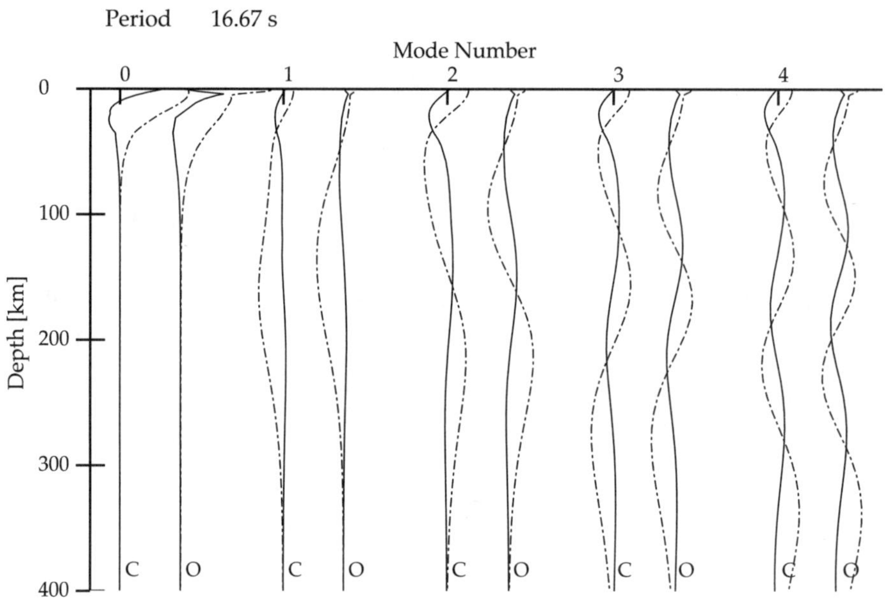

Figure 6.13. Mode shapes for the fundamental and four higher modes of the Rayleigh waves in the continental and oceanic models at a period of 16.67 s (0.06 Hz).

(figure 6.8). The corresponding group velocity minimum for the fundamental modes of Love waves is much shallower and does not produce any distinctive feature of the seismogram. The first higher Love mode has a similar group velocity to the fundamental mode for periods between 60 and 100 s (0.01–0.016 Hz) and so the two waves combine within a single wave group. For periods less than 50 s, the higher modes of both Rayleigh and Love waves have a tangled skein of group velocities near 4.45 km/s with both maxima and minima associated with different modes.

For the continental model, the fundamental modes separate from the higher modes with lower group velocities. This separation is maintained for Rayleigh waves, but there is no difference in the group velocities of the fundamental and higher Love modes for the oceanic model PEM-O over the period range 20 s–100 s (0.01–0.05 Hz). This is consistent with the character of the concentrated Love pulse seen for the oceanic events 2,3,4,6 in figure 6.9. There is a rapid transition in the dispersion character of the fundamental Rayleigh mode for an oceanic structure at short period, once the oceanic layer itself becomes significant; the group and phase velocities vary rapidly over a narrow period range. As a result a broad spread in arrival times occurs for a small change in period producing elongated wave trains with a period near 10 s, as we have noted in our discussion of figures 6.3, 6.4.

The higher mode Rayleigh waves for the oceanic structure have similar group velocities to those for the continental model. We would therefore expect the onset of the higher mode group for the paths for the six events to station SA04 to have a group velocity around 4.45 km/s, and so arrive just after 1000 s at 40° and this is indeed seen for the Rayleigh waves in figures 6.7, 6.8.

Each of the surface wave modes represents a combination of phase slowness and frequency (phase velocity and period) for which it is possible to simultaneously satisfy the surface boundary conditions of vanishing traction and the requirement that the wavefield decays at depth. This is the basis of the constructive interference condition for the *S* waves and means that each mode has a distinctive pattern of variation in depth. Successive higher modes include more oscillation with depth as we have seen in the simple example in Section 3.2. The shape of these modal *eigenfunctions* is influenced by the nature of the structures across which they propagate. We illustrate the behaviour for Rayleigh modes in the continental and oceanic structures in figure 6.13. The differences in the mode shapes are most marked in the crustal and uppermost mantle zone where the wavespeed models differ (figure 6.10). At fixed period the phase velocity steadily increases with mode number and the higher modes penetrate to greater depth since the decay in the wavefield only sets in once the material velocities are higher than the depth corresponding to the surface wavespeed, allowing for the effects of sphericity.

6.4 Modelling of teleseismic arrivals

The waveforms of *P* and *S* at teleseismic distances contain information on the nature of the source and the structures encountered along the propagation path. For shallow sources, complexity can be introduced by reverberations near source and receiver, but fortunately the contribution from the mantle is generally simple. The depth of the source plays an important role in determining the appearance of the *P* and *S* waveforms because the depth phases such as *pP*, *sS* arrive soon after the main arrival and interfere with it to produce complex pulses. The shape of the interference packet changes with depth and this information gives a means of estimating the source depth by comparing the observations with suitable synthetic seismograms (see, e.g., Langston & Helmberger, 1975). With a spread of stations around the source, reasonable sampling of the radiation pattern by the direct waves and the depth phases can be achieved and can be employed to infer the mechanism of the event. For short-period observations, the best constraints on source characteristics are achieved by inverting for source-mechanism and depth simultaneously (see, e.g., Goldstein & Dodge, 1999; Marson-Pidgeon & Kennett, 2000b).

As an illustration of the way in which modelling of teleseismic waveforms can be employed, we consider an event near Fiji which was recorded on an array of portable broad-band seismic stations in north-western Australia (Marson-Pidgeon & Kennett, 2000a). The epicentral distances range from 48.4° to 53.6° and this group of stations gives us an opportunity to look at the consistency of teleseismic arrivals between nearby stations. The event (1997 viii 8) has a moment magnitude of 6.6; the records

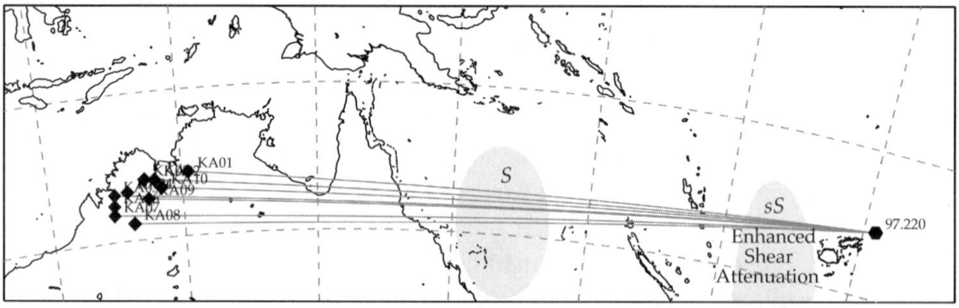

Figure 6.14. Map of the location of the Fiji event and the mantle ray paths relative to the portable stations. Likely areas of high attenuation for S are indicated by shading.

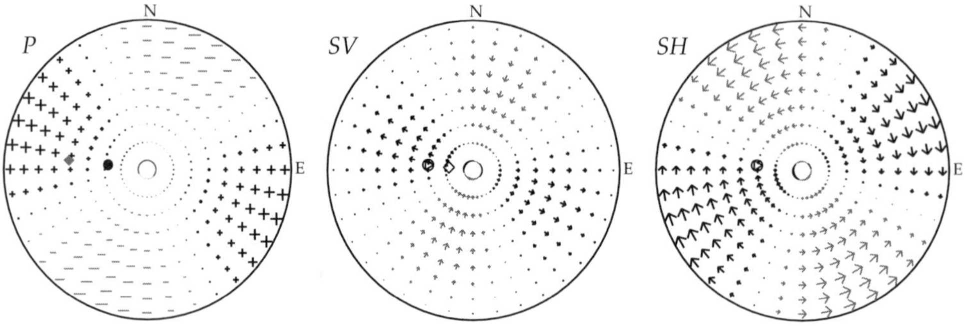

Figure 6.15. Radiation patterns for *P*, *SV* and *SH* waves for the Fiji event (1997 viii 08). The symbols indicate the average take-off angle and azimuth to the group of stations for various phases; solid circle for *P*, solid triangle for *pP*, and solid diamond for *pS*; open circle for *S*, open triangles for *sS*, and open diamond for *sP*.

have a good signal-to-noise ratio without any need for filtering, and include a noticeable long-period component.

We use the source mechanism and source depth from the Harvard CMT catalog and generate theoretical seismograms using the procedure discussed in Section 16.1. The source mechanism is displayed in figure 6.15 in terms of the radiation patterns for *P*, *SV* and *SH* waves. The average take-off angle and azimuth to the group of stations is indicated for the direct *P* and *S* waves and the depth phases. A source depth estimate of 17 km is taken from the Harvard CMT catalog, rather than the 10 km depth from the NEIC Preliminary Determination of Epicenters (PDE), because the fit to the observed seismograms was much improved. The source function for the event was taken as a simple trapezoid whose parameters are adjusted by trial-and-error to achieve the best fit to the character of all the records. The empirical time parameters obtianed for the pulse are 1.25 s for the up ramp, 2.5 s for the constant portion, and 1.25 s for the down ramp.

A comparison of the observed and synthetic seismograms for the vertical and radial components of the *P* wave is shown in figure 6.16. Figure 6.17 displays

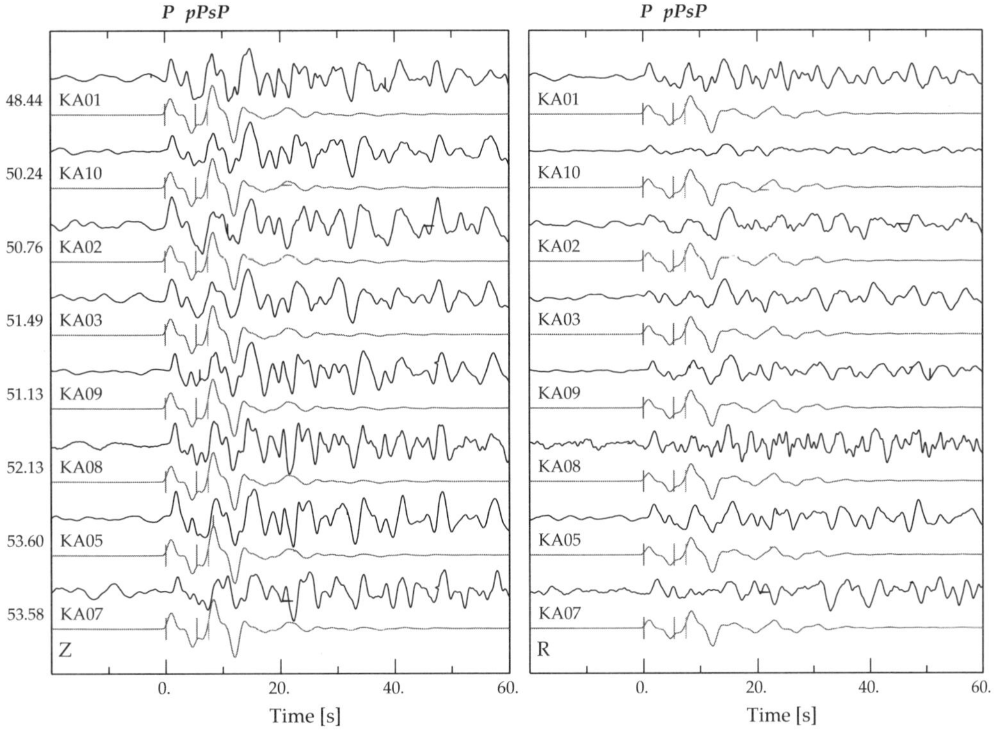

Figure 6.16. Comparison between observed and synthetic *P* wave seismograms for the Fiji event, for the vertical (left) and radial (right) components. The black traces are the observed seismograms, and the grey traces are the synthetic seismograms. The traces are aligned on the expected *P* arrival for the AK135 model, and the travel times for the depth phases are superimposed on the synthetic seismograms.

the 3-component seismograms for both observed and synthetic *S* waves for two of the stations. The synthetics are calculated using the AK135 velocity model and the corresponding attenuation profile of Montagner & Kennett (1996). The synthetic seismograms include all reverberations in the stratified structures on the source and receiver sides above 210 km, and deeper structure is included with a 'full-wave' approximation with an allowance for a phase shift of $-\pi/2$ in total reflection in the mantle (see Section 16.1).

The modelling achieves a good match for the onset of the *P* wavetrain (including the phases *P*, *pP* and *sP*). The source depth of 17 km provides a good fit to the relative arrival times of the direct *P* wave and its surface reflections (figure 6.16). We note that a better fit is obtained for the vertical components than the radial components. The *P* wave amplitude is strongest on the vertical component due to the near-vertical incidence at 50°. The differences between the actual and assumed crustal structure are likely to have more influence on the radial component, because conversions are more significant and these are exploited in receiver function studies. The use of the

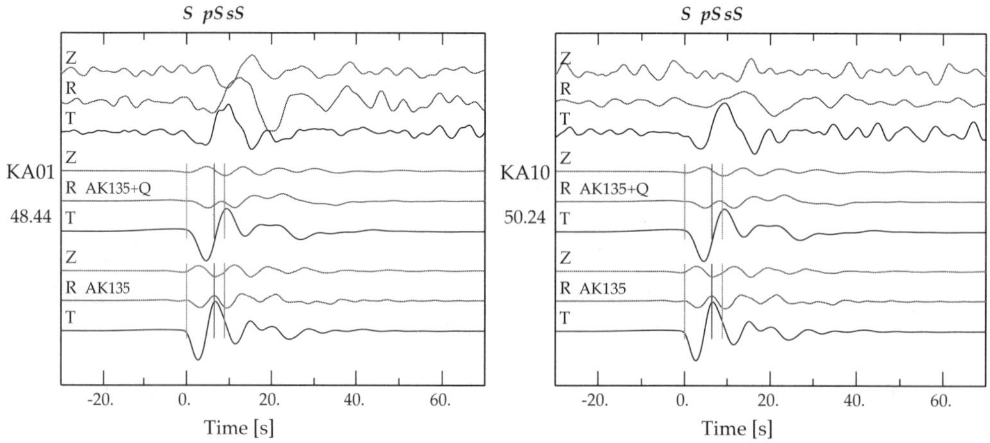

Figure 6.17. Comparison between observed and synthetic *S* wave seismograms for the Fiji event, at stations KA01 and KA10. The top three traces are the observed seismograms, the middle three traces are synthetic seismograms calculated for the AK135 model with modified *S* attenuation, and the lower three traces are synthetic seismograms calculated for the AK135 model. The traces are aligned on the expected *S* arrival for the AK135 model, and the travel times for the depth phases are superimposed on the synthetic seismograms.

AK135 attenuation model with our empirical source function provides a good fit to the observed *P* wave seismograms. The *P* arrivals at the stations KA09, KA08, KA05 and KA07 are a little delayed compared with the arrivals at the other stations; these stations lie in a major fold belt and so the differences are likely to represent structural variations in the crust.

The initial *P* phases have been successfully modelled, but following these is set of arrivals which are moderately consistent set of arrivals across the stations. The late arriving energy is fairly coherent over the epicentral distance range of around 570 km, and so is unlikely to be due to effects of receiver-side structure. It is therefore likely that the arrivals come from source-side effects for the oceanic subduction zone which were not considered in the modelling, which employs simple stratification.

In contrast to the *P* wave results, there are significant differences between the observed and synthetic seismograms for the onset of the *S* arrivals. The *S* waves show broad pulses and the *sS* contribution appears to be even longer period than *S* (figure 6.17). The results for two stations (KA01, KA10) are shown; similar behavior is found for the other stations. The lower set of synthetics in figure 6.17 were calculated using the *ak135* model, with the attenuation profile of Montagner & Kennett (1996), and the pulse length is clearly too short. However, by introducing increased *S* attenuation an improved match to the observations can be achieved as shown in the middle set of traces in figure 6.17. The modified attenuation profile has a 44 per cent decrease in Q_β and yields much broader *S* pulses.

The enhanced attenuation for *S* arises from the propagation path. As the waves leave the source they cross the Lau Basin (figure 6.14) which is known to be an area of back-arc spreading, with potential for partial-melting, which is expected to cause significant attenuation for *S*. The double passage of *sS* through a highly attenuating structure would also account for its broader pulse shape. There are also indications of low Q_β beneath the Coral Sea from other studies. The two areas of likely increased attenuation for *S* are shown by shading in figure 6.14.

The relative amplitudes of the tangential and radial components differ somewhat between the observed and theoretical seismograms. The amplitude of the radial component can be increased by modifying the receiver structure but we should note that the take-off points for *S* and *sS* are close to the node in the *S* radiation patterns, and so the synthetics will be sensitive to even small errors in the source mechanism.

There appears to be a small time offset between the observed radial and tangential components in figure 6.17; the time separation is probably due to shear wave splitting induced by anisotropy along the ray path and is not modelled in the synthetic seismograms.

Part II

Seismic Wave Propagation: General

7

Stress and Strain

The characteristic feature of a solid medium is resistance to shear which leads to the existence of both compressional and shear waves. The description of seismic wave phenomena has therefore to combine the representation of deformation within the solid material with the prevailing force system. Hooke's law provides the link between the geometrical description of deformation though the strain tensor and the stress tensor representing the forces. The linear relation between stress and strain is specified through a fourth rank tensor of elastic moduli, and material symmetries reduce the number of independent moduli needed to describe the behaviour.

The material in this chapter provides a brief summary of the properties of stress and strain, with the object of establishing the results which we will need for describing seismic wave propagation. We will use a cartesian representation for the displacements etc. and employ suffix notation for vector and tensor components.

7.1 Continuum representation

We adopt a viewpoint in which the details of the microscopic structure of the medium through which seismic waves propagate is ignored. The material is supposed to comprise a continuum of which every subdivision possesses the macroscopic properties. This description is reasonable because even at the MHz frequencies used in some laboratory studies the seismic wavelengths are of the order of 1 mm.

We can therefore assign field variables for velocity $\mathbf{v}(\mathbf{x}, t)$, and density $\rho(\mathbf{x}, t)$ as a function of the current position $\mathbf{x}$ of a material element. Alternatively, we may refer to such field variables via the position $\boldsymbol{\xi}$ of the material element at some reference time t_0.

A convenient description of the current configuration of the material is provided by the displacement $\mathbf{u}$ between the current and reference positions

$$\mathbf{u} = \mathbf{x} - \boldsymbol{\xi}, \tag{7.1.1}$$

The deformation properties of the material depend on the differential displacement between nearby points. Consider therefore a line element of length $\mathrm{d}\mathbf{x}$ at $\mathbf{x}$ which was originally of length $d\boldsymbol{\xi}$ at $\boldsymbol{\xi}$. We can compare the lengths before and after deformation. The squared length of the element

$$|d\mathbf{x}|^2 = dx_i dx_i = (d\xi_i + du_i)(d\xi_i + du_i) = (d\xi_i + \frac{\partial u_i}{\partial \xi_j} d\xi_j)(d\xi_i + \frac{\partial u_i}{\partial \xi_k} d\xi_k), \tag{7.1.2}$$

where we have employed the convention of summation over repeated suffices. We can then recast the relation in terms of the differences from the original length $|d\boldsymbol{\xi}|^2 = d\xi_i d\xi_i$, so that

$$dx_i dx_i = d\xi_i d\xi_i + \left(\frac{\partial u_i}{\partial \xi_j} + \frac{\partial u_j}{\partial \xi_i}\right) d\xi_i d\xi_j + O\left(\frac{\partial \mathbf{u}}{\partial \boldsymbol{\xi}}\right)^2. \tag{7.1.3}$$

We can extract the symmetric part of the displacement derivatives which describes the change of length as the strain tensor

$$e_{ij} = \frac{1}{2}\left(\frac{\partial u_i}{\partial \xi_j} + \frac{\partial u_j}{\partial \xi_i}\right), \tag{7.1.4}$$

and then the relation between the squared lengths of the line elements before and after deformation is

$$dx_i dx_i = d\xi_i d\xi_i + 2e_{ij} d\xi_i d\xi_j + O(e^2). \tag{7.1.5}$$

The material properties of earth materials are essentially linear for strains up to 10^{-5} and in this range we can neglect higher order contributions to the deformation. Strains higher than 10^{-5} are normally only encountered in the immediate vicinity of earthquakes or underground explosions.

In the small strain limit the fractional change in volume due to deformation is just the trace of the strain tensor

$$e_{ii} = \frac{\partial u_1}{\partial \xi_1} + \frac{\partial u_2}{\partial \xi_2} + \frac{\partial u_3}{\partial \xi_3}. \tag{7.1.6}$$

7.2 Stress

The deformation of the material is accompanied by the presence of an internal force field. Consider a surface element dS at $\mathbf{x}$ with normal $\mathbf{n}$, the associated traction $\boldsymbol{\tau}(\mathbf{n})$ represents the force per unit area due to the side on which $\mathbf{n}$ points (figure 7.1).

In general, the force will depend on the orientation of the surface element dS. We introduce traction vectors $\boldsymbol{\sigma}_i$, for which the normal points in the direction of the ith coordinate axis. The cartesian components of these vectors

$$\sigma_{ij}(\mathbf{x}) = (\boldsymbol{\sigma}_i)_j(\mathbf{x}), \tag{7.2.1}$$

constitute the stress tensor at the point $\mathbf{x}$.

By considering the conservation of linear and angular momentum applied to a small element of volume, we can demonstrate that

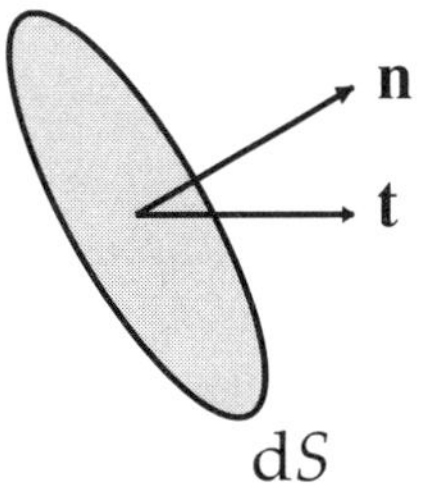

Figure 7.1. The configuration of the traction vector $\boldsymbol{\tau}(\mathbf{n})$.

$$\tau_i(\mathbf{n}) = \sigma_{ij} n_j, \tag{7.2.2}$$

and that the stress tensor is symmetric (in the absence of external couples).

7.3 Equation of motion

Consider a volume V with surface S and look at the balance of linear momentum in the presence of an externally imposed force field $\mathbf{g}$. We require the sum of the surface force field, derived from the tractions $\boldsymbol{\tau}$, and the external forces to balance the mass acceleration in the volume i.e.

$$\int_S \mathrm{d}S\, \tau_i + \int_V \mathrm{d}V\, \rho g_i = \int_V \mathrm{d}V\, \rho f_i, \tag{7.3.1}$$

where $\mathbf{f}$ is the local acceleration field.

In the linear approximation the acceleration $\mathbf{f}$ can be represented as the second time derivative of the displacement field $\mathbf{u}$ at $\mathbf{x}$:

$$f_i(\mathbf{x}) = \frac{\partial^2}{\partial t^2} u_i(\mathbf{x}). \tag{7.3.2}$$

We can rewrite the traction integral in terms of the stress tensor σ_{ij} and then apply the divergence theorem to convert the surface integral into a volume integral

$$\int_S \mathrm{d}S\, \tau_i = \int_S \mathrm{d}S\, \sigma_{ij} n_j = \int_V \mathrm{d}V \frac{\partial}{\partial x_j} \sigma_{ij}, \tag{7.3.3}$$

and so from (7.3.1) and (7.3.2) we obtain

$$\int_V \mathrm{d}V \left[\frac{\partial}{\partial x_j} \sigma_{ij} + \rho g_i \right] = \int_V \mathrm{d}V \frac{\partial^2}{\partial t^2} u_i. \tag{7.3.4}$$

We now recognise that the volume V is arbitrary so that we can establish a local equation of motion

$$\frac{\partial}{\partial x_j} \sigma_{ij} + \rho g_i = \frac{\partial^2}{\partial t^2} u_i. \tag{7.3.5}$$

7.4 Relating stress and strain

In order to relate the geometrical description of the deformation provided by the strain tensor e_{ij} to the summary of the force system provided by the stress tensor σ_{ij}, we need to introduce a description of the physical properties of the continuum.

For a linear elastic medium, the components of the stress tensor σ_{ij} are assumed to be linearly related to the strain tensor components e_{kl}, through a generalisation of Hooke's law to tensorial behaviour

$$\sigma_{ij} = c_{ijkl} e_{kl}, \tag{7.4.1}$$

The fourth rank tensor of elastic moduli c_{ijkl} has the symmetry properties

$$c_{ijkl} = c_{jikl} = c_{ijlk}, \tag{7.4.2}$$

from the symmetry of σ_{ij} and e_{kl}, and also

$$c_{ijkl} = c_{klij}. \tag{7.4.3}$$

This last relation can be derived from thermodynamic arguments for isentropic disturbances based on the existence of an internal energy function as a function of strain. The time scale for thermal conduction is so long compared with the period of the waves that heat does not have a chance to escape during the passage of an elastic wave past any particular point.

The tensor c_{ijkl} has 81 components, the symmetries (7.4.2) reduce this to 36 independent components and the application of (7.4.3) leaves only 21 independent components for a fully anisotropic medium. Most minerals exhibit some form of anisotropy, but the polycrystalline aggregates constituting most rocks are often close to isotropic in their macroscopic properties (see also Section 8.2).

With the specific assumption of isotropy, we can write c_{ijkl} in the form

$$c_{ijkl} = \lambda \delta_{ij} \delta_{kl} + \mu(\delta_{ik}\delta_{jl} + \delta_{il}\delta_{kj}), \tag{7.4.4}$$

Here μ is the shear modulus and the Lamé modulus λ is related to the bulk modulus κ by $\lambda = \kappa - \frac{2}{3}\mu$.

The assumption of perfect elasticity does not allow for the observed dissipation of seismic wave energy in the course of propagation. A suitable model which can account for such anelastic dissipation is to work with a linear viscoelastic material in which the stress depends on the prior history of the local strain

$$\sigma_{ij}(t) = c^o_{ijkl} e_{kl}(t) + \int_{t_0}^{t} \mathrm{d}s\, C_{ijkl}(t-s) e_{kl}(s). \tag{7.4.5}$$

where c^o_{ijkl} are the instantaneous elastic moduli and the C_{ijkl} describe the dependence on past deformation. Now take a Fourier transform with respect to time, so that

$$\bar{v}(\omega) = \int_{-\infty}^{\infty} \mathrm{d}t\, v(t) \mathrm{e}^{\mathrm{i}\omega t}. \tag{7.4.6}$$

The transform of the convolution with the relaxation function will then reduce to a product of transforms so that, for frequency ω, the constitutive equation for linear viscoelasticity can be written as

$$\bar{\sigma}_{ij}(\omega) = [c^{o}_{ijkl} + \bar{C}_{ijkl}]\bar{e}_{kl}(\omega). \tag{7.4.7}$$

Thus the behaviour of the stress at each frequency for the linear viscoelastic medium mirrors the perfectly elastic case but with a complex set of elastic moduli. The imaginary part of the elastic moduli leads to the dissipation of seismic waves since now we have disturbances which are out of phase with the original strain field. We will discuss the attenuation of seismic waves in more detail in Section 8.3.

7.5 The effect of prestress

When we describe the propagation of seismic waves within the Earth we have to recognise that the small deformations associated with the passage of the waves are superimposed on the ambient state, which is dominated by the self-gravitation of the Earth. In this equilibrium state the gradient of the stress tensor σ^0_{ij} matches the gravitational accelerations derived from a potential ψ^0

$$\frac{\partial \sigma^0_{ij}}{\partial x_i} + \rho^0 \frac{\partial \psi^0}{\partial x_j} = 0. \tag{7.5.1}$$

where ρ^0 is the equilibrium density. This ambient stress field will be predominantly hydrostatic, and if the isostatic compensation of shallow variations in properties was complete at some level, there would be no deviatoric component at greater depths. The style of compensation is not important, so that an Airy or Pratt mechanism or some combination could prevail.

Except in the immediate vicinity of a seismic source the strain levels associated with seismic waves are small. We therefore suppose that the incremental changes of displacement (**u**) and stress (σ_{ij}) from the equilibrium state induced by the waves behave as for an elastic medium, and so from (7.3.5) these quantities satisfy the equation of motion

$$\frac{\partial \sigma_{ij}}{\partial x_i} + \rho f_j = \rho \frac{\partial^2 u_j}{\partial t^2}. \tag{7.5.2}$$

The body force term f_j includes the effect of self-gravitation and in particular the perturbation in the gravitational potential consequent on the displacement.

In order to express (7.5.2) in terms of the displacement alone we need a constitutive relation between the stress and strain increments away from the reference state. The usual assumption is that this incremental relation can be represented via linear elasticity and so we model σ_{ij} by a stress field τ_{ij} generated from the displacement **u**,

$$\tau_{ij} = c_{ijkl}\,\partial_l u_k, \tag{7.5.3}$$

where we have used the abbreviated notation $\partial_l = \partial/\partial x_l$.

If however there is a significant level of stress in the reference state the relation (7.5.3) would not be appropriate and a more suitable form (Dahlen, 1972) is provided by

$$\tau_{ij} = d_{ijkl}\partial_k u_l - u_k \partial_k \sigma^0_{ij}. \tag{7.5.4}$$

The second term arises because it is most convenient to adopt a Lagrangian viewpoint for the deformation of the solid material. A comprehensive discussion of the relevant continuum mechanics in a seismological framework is provided by Dahlen & Tromp (1998).

The constants d_{ijkl} depend on the initial stress

$$d_{ijkl} = c_{ijkl} + \tfrac{1}{2}\left(\delta_{ij}\sigma^0_{kl} - \delta_{kl}\sigma^0_{ij} + \delta_{il}\sigma^0_{jk} - \delta_{jk}\sigma^0_{il} + \delta_{jl}\sigma^0_{ik} - \delta_{ik}\sigma^0_{jl}\right), \tag{7.5.5}$$

and the tensor c_{ijkl} possesses the symmetries (7.4.2), (7.4.3). For a hydrostatic initial stress state d_{ijkl} reduces to c_{ijkl}. The slight influence of the second term in (7.5.4), is frequently neglected. The dominant variation of the stress tensor σ^0_{ij} is normally with depth and, for the hydrostatic component, the gradient is about 40 Pa/m in the Earth's mantle. In the mantle the individual moduli c_{ijkl} have values of the order of 10^{10} Pa. For a disturbance with a frequency near 0.05 Hz and thus a wavelength of about 200 km, the term $c_{ijkl}\partial_k u_l$ will be about 10^4 times the correction $u_k\partial_k\sigma^0_{ij}$. The ratio between the terms will increase with increasing frequency. For teleseismic studies the correction is therefore negligible.

Deviatoric components of the initial stress are likely to be most significant in the outer portions of the Earth, where spatial variability of the elastic constants is also important. The initial stress state may therefore have significant spatial variation on scales comparable to seismic wavelengths, and so the correction $u_k\partial_k\sigma^0_{ij}$ in (7.5.4) will be of greater significance than at depth.

7.6 Representation of sound waves in a fluid

We have so far considered the deformation of a solid but the same continuum viewpoint can be used for a fluid. The small incremental displacement associated with the passage of wave will induce slight variations in the pressure in the fluid.

The pressure within the fluid is purely isotropic and so the stress tensor is just diagonal

$$\sigma_{ij} = -(p_o + p')\delta_{ij}, \tag{7.6.1}$$

where p_o is the hydrostatic pressure and p' arises from the influence of the deformation. The incremental equation of motion for the small wave disturbances is then

$$-\frac{\partial p'}{\partial x_i} = \rho\frac{\partial^2 u_i}{\partial t^2}. \tag{7.6.2}$$

When we neglect the effects of viscosity, the incremental pressure and the local volume change are related by

$$p' = -\kappa\Delta = -\kappa\frac{\partial u_k}{\partial x_k}, \tag{7.6.3}$$

in terms of the bulk modulus κ, since the volume change is the dilatation

$$\Delta = \partial u_k/\partial x_k.$$

The displacement field thus satisfies

$$\kappa\frac{\partial^2 u_k}{\partial x_k \partial x_i} = \rho\frac{\partial^2 u_i}{\partial t^2}. \tag{7.6.4}$$

By working in terms of the dilatation Δ, (7.6.4) can be cast in the form of a wave equation for sound waves in the fluid

$$\rho\frac{\partial^2 \Delta}{\partial t^2} - \kappa\frac{\partial^2 \Delta}{\partial x_i \partial x_i} = 0 \tag{7.6.5}$$

with a wavespeed

$$c = (\kappa/\rho)^{1/2}. \tag{7.6.6}$$

Such sound waves are purely longitudinal.

8

Seismic Waves I - Plane Waves

In this chapter we examine the nature of seismic waves in unbounded media by using plane wave solutions. We start by establishing the governing equations for elastic wave propagation and then show the way in which three wavetypes emerge for a general anisotropic medium. For the special case of an isotropic medium, the wavespeeds of the two solutions corresponding to shear wave propagation are identical but it is convenient to distinguish *SV* and *SH* via their polarisation.

Natural materials are intrinsically anisotropic and it is only in the aggregate that the seismic properties approximate isotropy. We examine the particular case of hexagonal symmetry, where the properties are symmetric about a single axis (often termed *transverse isotropy*), for which the algebraic details are not too formidable.

The Earth's crust and mantle are dominated by silicate mineral assemblages which are not purely elastic in behaviour. Seismic energy is dissipated in passage through a variety of anelastic effects that can be represented through a linear visco-elastic component in the stress-strain relation. At fixed frequency, the behaviour is similar to the elastic case but with complex wavespeeds. The imaginary part of the wavespeed represents the dissipative component, and this has also an accompanying small correction to the real part which introduces slight dispersion to the seismic wavespeeds. The presence of heterogeneity gives another mode of transfer of seismic energy from the primary propagation path and can also be described via an imaginary attenuative term in the wavespeed, but this scattering component has no associated dispersion.

8.1 The elastodynamic equation

In the previous chapter we have established the equation of motion for elastic disturbances

$$\frac{\partial \tau_{ij}}{\partial x_j} + \rho g_i = \rho \frac{\partial^2 u_i}{\partial t^2}. \tag{8.1.1}$$

and the constitutive relation (Hooke's Law)

$$\tau_{ij} = c_{ijkl}e_{kl} = c_{ijkl}\frac{\partial u_k}{\partial x_l}. \tag{8.1.2}$$

where we have exploited the $k \leftrightarrow l$ symmetry of the elastic modulus tensor c_{ijkl} to express the strain in terms of the displacement derivatives.

These two equations (8.1.1), (8.1.2) can now be combined to give a single equation for the dynamic behaviour of the displacement $\mathbf{u}$ in an elastic deformation

$$\frac{\partial}{\partial x_j}\left(c_{ijkl}\frac{\partial u_k}{\partial x_l}\right) + \rho g_i = \rho\frac{\partial^2 u_i}{\partial t^2}. \tag{8.1.3}$$

After Fourier transformation with respect to time, the elastodynamic equation takes the form

$$\frac{\partial}{\partial x_j}\left(c_{ijkl}\frac{\partial u_k}{\partial x_l}\right) + \rho g_i + \rho\omega^2 u_i = 0. \tag{8.1.4}$$

We can now introduce a linear viscoelastic model by incorporating complex moduli at each frequency.

We normally consider deformation away from the hydrostatically pre-stressed configuration of the Earth. In this case $\mathbf{g}$ will represent the incremental effects of self-gravitation in the Earth which are only significant for very long-period waves (frequencies less than 0.01 Hz).

8.2 Plane waves

For a region with uniform properties, in the absence of body forces, the elastodynamic equation reduces to

$$c_{ijkl}\frac{\partial^2 u_k}{\partial x_l \partial x_j} = \rho\frac{\partial^2 u_i}{\partial t^2}. \tag{8.2.1}$$

Equation (8.2.1) controls the spatial and temporal development of the displacement field and admits solutions in the form of travelling waves. Consider then a plane wave travelling in an anisotropic medium

$$\begin{aligned} u_i &= U_i \exp[\mathrm{i}\omega p \mathbf{n}.\mathbf{x} - \mathrm{i}\omega t], \\ &= U_i \exp[\mathrm{i}\omega p n_k x_k - \mathrm{i}\omega t], \end{aligned} \tag{8.2.2}$$

$\mathbf{n}$ represents the direction of travel of the phase fronts and p the apparent slowness (inverse of wave velocity) in that direction. On substituting this plane wave form into (8.2.1) we obtain

$$-\omega^2 p^2 c_{ijkl} n_j n_l U_k + \omega^2 \rho U_i = 0, \tag{8.2.3}$$

which constitutes an eigenvalue problem for the slowness p for with waves travelling in the direction $\mathbf{n}$

$$\left[p^2 c_{ijkl} n_j n_l - \rho\delta_{ik}\right] U_k = 0. \tag{8.2.4}$$

For each direction of propagation $\mathbf{n}$ we can define a 3×3 Christoffel matrix g_{ik}

$$g_{ik}(\mathbf{n}) = c_{ijkl}n_j n_l, \tag{8.2.5}$$

which is symmetric and real for a perfectly elastic medium. In terms of g_{ik} we can rewrite (8.2.4) as

$$\left[g_{ik}(\mathbf{n}) - \rho p^{-2}(\mathbf{n})\delta_{ik}\right] U_k(\mathbf{n}) = 0, \tag{8.2.6}$$

which can be recognised as an eigenvalue problem for the matrix $\mathbf{g}$ for which the three roots for $p^2(\mathbf{n})$ are to be determined from

$$\det\left[g_{ik}(\mathbf{n}) - \rho p^{-2}(\mathbf{n})\delta_{ik}\right] = 0. \tag{8.2.7}$$

The slownesses $p^{(m)}(\mathbf{n})$ will be associated with polarisations of the displacement specified by orthogonal eigenvectors $\mathbf{U}^{(m)}(\mathbf{n})$. The corresponding phase velocity $c^{(m)}$ is the reciprocal of $p^{(m)}$. When there is degeneracy with two $p^{(m)}$ equal, as for shear waves in an isotropic medium, the eigenvectors for these slownesses will be orthogonal to the eigenvector for the third slowness and two eigenvectors can be chosen to be mutually orthogonal.

In a general anisotropic medium, the slownesses vary with direction and can give quite complex slowness surfaces: an illustrative example is presented in Section 8.2.2 below.

8.2.1 Isotropic media

A constitutive relation such as (8.1.2) expresses the macroscopic characteristics of the material within the Earth. On a fine scale we will have a relatively chaotic assemblage of crystal grains with anisotropic elastic moduli. However, the overall properties of a cube with the dimensions of a typical seismic wavelength (a few kilometres in the mantle) will generally be nearly isotropic. In consequence the elastic constant tensor may often be approximated by the isotropic form, in terms of only the bulk modulus κ and shear modulus μ,

$$c_{ijkl} = (\kappa - \tfrac{2}{3}\mu)\delta_{ij}\delta_{kl} + \mu(\delta_{ik}\delta_{jl} + \delta_{il}\delta_{jk}). \tag{8.2.8}$$

For such an isotropic medium, the gradient of the stress tensor takes the form

$$\frac{\partial}{\partial x_j}\tau_{ij} = \frac{\partial}{\partial x_j}\left[\lambda\delta_{ij}\frac{\partial u_k}{\partial x_k} + \mu\left(\frac{\partial u_i}{\partial x_j} + \frac{\partial u_j}{\partial x_i}\right)\right], \tag{8.2.9}$$

where the Lamé modulus $\lambda = \kappa - \frac{2}{3}\mu$. The elastodynamic equation can thus be written as

$$\frac{\partial}{\partial x_i}\left[\lambda\frac{\partial u_k}{\partial x_k}\right] + \frac{\partial}{\partial x_j}\left[\mu\left(\frac{\partial u_i}{\partial x_j} + \frac{\partial u_j}{\partial x_i}\right)\right] = \rho\frac{\partial^2}{\partial t^2}u_i. \tag{8.2.10}$$

When we substitute a plane wave displacement of the form (8.2.2) in (8.2.10), the eigenvalue equation for slowness in an isotropic medium takes the form

$$[p^2(\lambda+\mu)n_i n_k + p^2\mu\delta_{ik} - \rho\delta_{ik}]v_k = 0. \tag{8.2.11}$$

8.2.1.1 P waves

Consider a displacement field oriented along the propagation direction $\mathbf{u}_P = c\mathbf{n}$, then the slowness equation is

$$[(\lambda+2\mu)p^2 - \rho]n_i = 0, \tag{8.2.12}$$

so that we have a wave disturbance with slowness $a = [\rho/(\lambda+2\mu)]^{1/2}$, and associated phase velocity $\alpha = [(\lambda+2\mu)/\rho]^{1/2}$. This longitudinal wave solution is called the P wave

$$\mathbf{u}_P = A_P\mathbf{n}\exp[i\omega(a\mathbf{n}.\mathbf{x} - t)]. \tag{8.2.13}$$

8.2.1.2 S waves

There is an alternative type of elastic wave motion in which the displacement is transverse to the direction of motion. Consider a displacement field oriented along a vector $\mathbf{s}$ perpendicular to the propagation direction $\mathbf{n}$,

$$\mathbf{u}_S = c\mathbf{s} \quad \text{with} \quad \mathbf{s}.\mathbf{n} = 0, \tag{8.2.14}$$

for which the slowness equation reduces to

$$[\mu p^2 - \rho]s_i = 0, \tag{8.2.15}$$

so that we have a wave disturbance with slowness $b = [\rho/\mu]^{1/2}$, and associated phase velocity $\beta = [\mu/\rho]^{1/2}$. This form of solution holds for any direction orthogonal to the direction of motion, i.e., we have a degenerate eigenvalue problem from which we can choose two orthogonal S wave vectors. It is conventional to choose one vector in the vertical plane (denoted *SV*) and the other purely horizontal (denoted *SH*). This choice simplifies the analysis of wave propagation in horizontally stratified media, and so we represent the S wave field as

$$\mathbf{u}_S = [B_V\mathbf{s}_V + B_H\mathbf{s}_H]\exp[i\omega(b\mathbf{n}.\mathbf{x} - t)], \tag{8.2.16}$$

$\mathbf{s}_V$ lies in the vertical plane through $\mathbf{n}$ and $\mathbf{s}_H$ in the horizontal plane $[\mathbf{s}_V.\mathbf{n} = \mathbf{s}_H.\mathbf{n} = 0]$.

In terms of the bulk modulus κ and the shear modulus μ, the P wavespeed

$$\alpha = [(\kappa + \tfrac{4}{3}\mu)/\rho]^{1/2}, \tag{8.2.17}$$

and so the purely dilatational P wave motion depends on the shear modulus.

8.2.2 Anisotropic media

Although we would expect that most rocks would exhibit some degree of anisotropy, the approximation of isotropy is frequently effective. However, there is increasing direct evidence for the presence of significant influence from anisotropy in a variety of circumstances (see, e.g., Babuška & Cara, 1991). For near-surface rocks, patterns of cracking and jointing can give rise to anisotropic variation in wavespeed.

Significant large-scale anisotropy has been established in a number of cases in the uppermost mantle; for example, at the top of the upper mantle under the oceans there is about five per cent anisotropy in *P* wavespeed (Raitt, 1969), and similar behaviour is sometimes seen for mantle *Pn* in continental areas (Bamford, 1977). Such anisotropy is likely to arise when there is preferential alignment of crystal grains, associated with some prevailing tectonic stress.

Studies of the phase *SKS* have revealed clear indications of shear waves with different polarisation having slightly different propagation speeds (Silver, 1996; Vinnik et al., 1992). The presence of such shear-wave splitting can be recognised by the presence of a tangential component to *SKS* which would not be expected for an isotropic medium. The observations are commonly interpreted in terms of anisotropic material at the top of the mantle, which may be associated with past deformation which has imposed a distinct fabric on the rocks or be a consequence of contemporary deformation.

In the period before a major earthquake, significant prestrain can be built up in the epicentral region, which will be relieved by the earthquake itself. The presence of such a strain modifies the local constitutive relation as in (7.5.4), where σ_{ij}^0 is to be taken now as the stress associated with the prestrain (Walton, 1974), and this will give rise to apparent anisotropy for propagation through the region. The presence of non-hydrostatic stress will have a significant effect on the crack distribution in the crust. At low ambient stress and low pore pressure within the rocks, systems of open cracks may be differentially closed. At high ambient stress, systems of closed cracks may be opened if the pore pressure and non-hydrostatic stress are large enough, and new cracks may also be formed. Such *dilatancy* effects lead to aligned crack systems over a fair size area, and this will give apparent anisotropy to seismic wave propagation through the region. The anisotropy is evidenced by differences in the times of passage of shear waves of different polarisation (shear wave splitting or seismic birefringence). Theoretical models have been developed for the deformation to be expected before fracturing based on fluid migration along pressure gradients between neighbouring grain boundary cracks and pores with a low-aspect ratio at different orientations to the stress field (Zatsepin & Crampin, 1997). These models predict the effects of the build-up of stress before earthquakes on shear wave splitting (Crampin & Zatsepin, 1997) and have been applied to observations in Iceland to forecast occurence of an M=5 earthquake (Crampin, Volti and Stefánsson, 1999).

Transverse anisotropy, where the vertical and horizontal wavespeeds differ, can be simulated by very fine bedding in sedimentary sequences below the scale of seismic disturbances. There is now extensive evidence from both surface recording and exploration logs to suggest this effect can be important for compressional wave propagation in many situations in reflection seismology.

Transverse isotropy has also been postulated for the outer part of the upper mantle above 250 km depth, in an attempt to reconcile the observed dispersion of Love and Rayleigh waves at moderate periods (Dziewonski & Anderson, 1981). Here the differences in horizontal and vertical wavespeed are needed principally for shear waves.

The generalised Hooke's law for anisotropic media

$$\tau_{ij} = c_{ijkl} e_{kl}, \tag{8.2.18}$$

can be rewritten in a compressed matrix notation

$$\tau_p = c_{pq} e_q, \quad p, q = 1, 6, \tag{8.2.19}$$

where now

$$c_{ijkl} \equiv c_{pq}, \quad i, j, k, l = 1, 3, \quad p, q = 1, 6, \tag{8.2.20}$$

with the correspondence

$$\begin{array}{ccccccc} ij/kl & 11 & 22 & 33 & 23{,}32 & 31{,}13 & 12{,}21 \\ p/q & 1 & 2 & 3 & 4 & 5 & 6 \end{array} \tag{8.2.21}$$

This notation due to Love (1927) enables the structure of the anisotropic moduli to be visualised by writing out the 21 elements of the upper triangular part of the matrix c_{pq} (the lower triangular part can be recovered from the symmetry properties of c_{pq}).

For example, for a medium with hexagonal symmetry with the symmetry axis vertical (i.e. along the 3-axis) there are 5 independent moduli. This situation corresponds to "transverse isotropy" with the properties perpendicular to the symmetry axis equal, but different from those along the symmetry axis itself. In Love's notation

$$c_{11} = A, \quad c_{22} = A, \quad c_{33} = C, \quad c_{44} = L, \quad c_{55} = L, \quad c_{66} = N,$$
$$c_{12} = H = A - 2N, \quad c_{13} = F, \quad c_{23} = F. \tag{8.2.22}$$

The upper triangle of the matrix c_{pq} can be represented as

$$\begin{array}{cccccc} \mathrm{A} & \mathrm{H} & \mathrm{F} & 0 & 0 & 0 \\ & \mathrm{A} & \mathrm{F} & 0 & 0 & 0 \\ & & \mathrm{C} & 0 & 0 & 0 \\ & & & \mathrm{L} & 0 & 0 \\ & & & & \mathrm{L} & 0 \\ & & & & & \mathrm{N} \end{array} \tag{8.2.23}$$

For the special case of isotropy

$$A = C = \lambda + 2\mu, \quad F = H = \lambda, \quad L = N = \mu. \tag{8.2.24}$$

For a transversely isotropic medium, the phase velocity c and polarisation state $\mathbf{U}$ (U_1, U_2, U_3) for propagation in the direction $\mathbf{n}$ (n_1, n_2, n_3) are to be determined from (8.2.4) or (8.2.6).

These equations can be rewritten explicitly as

$$\begin{aligned}
&[An_1^2 + Nn_2^2 + Ln_3^2 - \rho c^2]U_1 + [(A-N)n_1n_2]U_2 + [(F+L)n_1n_3]U_3 = 0,\\
&[(A-N)n_1n_2]U_1 + [Nn_1^2 + An_2^2 + Ln_3^2 - \rho c^2]U_2 + [(F+L)n_2n_3]U_3 = 0,\\
&[(F+L)n_1n_3]U_1 + [(F+L)n_2n_3]U_2 + [L(n_1^2+n_2^2) + Cn_3^2 - \rho c^2]U_3 = 0.
\end{aligned} \tag{8.2.25}$$

A convenient representation of the properties of the transverse isotropic medium is to work with the wavespeeds

$$\begin{aligned}
\alpha_h &= (c_{11}/\rho)^{1/2} = (A/\rho)^{1/2}, \quad \alpha_v = (c_{33}/\rho)^{1/2} = (C/\rho)^{1/2},\\
\beta_h &= (c_{66}/\rho)^{1/2} = (N/\rho)^{1/2}, \quad \beta_v = (c_{44}/\rho)^{1/2} = (L/\rho)^{1/2},
\end{aligned} \tag{8.2.26}$$

where the subscripts h, v indicate horizontal and vertical propagation. The additional parameter needed for a full description of the behaviour can be represented in a number of ways: one suitable choice is

$$\gamma = (c_{13}/\rho)^{1/2} = (F/\rho)^{1/2}. \tag{8.2.27}$$

An alternative adopted by Takeuchi & Saito (1972) is

$$\eta = F/(A-2L), \tag{8.2.28}$$

which is unity for an isotropic medium.

None of the vertical or horizontal velocities involve F. This means that neither recordings of direct waves nor near-vertical geometries such as recording directly over a source, as e.g in borehole surveys with a source near the borehole, are sufficient to extract the full properties of a transversely isotropic medium.

With a vertical symmetry axis, the wave propagation velocities are invariant for notations about the 3-axis but vary with the angle of inclination θ of the normal to the plane wavefront. In the transversely isotropic medium the phase velocities of *SH* and *SV* waves are distinct.

For *SH* waves there is no vertical component and

$$c_H(\theta) = [\beta_v^2 \cos^2\theta + \beta_h^2 \sin^2\theta]^{1/2} \tag{8.2.29}$$

so that along the 3 symmetry-axis $c_H(0) = \beta_v$, and orthogonal to the symmetry axis $c_H(\frac{\pi}{2}) = \beta_h$.

Those waves polarised in a vertical plane depend on all the elastic parameters and can be classified into quasi-*P*-waves and quasi-*SV* waves by their affinities in the special case of isotropy.

For the *qSV* waves

$$\begin{aligned}
c_S(\theta) = \Big[&\alpha_v^2 \cos^2\theta + \alpha_h^2 \sin^2\theta + \beta_v^2\\
&-[(\alpha_h^2 \sin^2\theta - \alpha_v^2\cos^2\theta + \beta_v^2\cos 2\theta)^2 + (\gamma^2+\beta_v^2)^2\sin^2 2\theta]^{1/2}\Big]^{1/2}/\sqrt{2},
\end{aligned} \tag{8.2.30}$$

and so the phase velocity is equal along the symmetry axis and orthogonal to it, $c_S(0) = c_S(\frac{\pi}{2}) = \beta_v$, but varies with angle θ for intermediate inclinations.

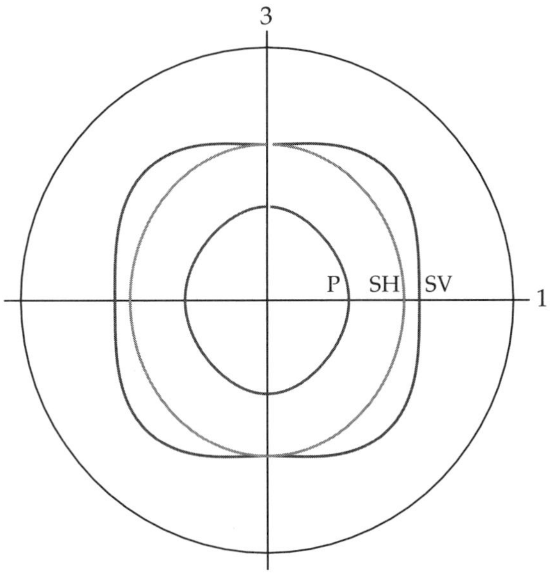

Figure 8.1. The configuration of slowness surfaces for a transversely isotropic material with faster horizontal velocities.

For the qP waves

$$c_P(\theta) = \Big[\alpha_v^2 \cos^2\theta + \alpha_h^2 \sin^2\theta + \beta_v^2 \qquad (8.2.31)$$

$$+[(\alpha_h^2 \sin^2\theta - \alpha_v^2 \cos^2\theta + \beta_v^2 \cos 2\theta)^2 + (\gamma^2 + \beta_v^2)^2 \sin^2 2\theta]^{1/2}\Big]^{1/2} /\sqrt{2},$$

so that the phase velocity along the symmetry axis $c_P(0) = \alpha_h$ and orthogonal to the symmetry axis $c_P(\frac{\pi}{2}) = \alpha_v$.

The behaviour of the phase slownesses $p_H(\theta)$, $p_S(\theta)$, $p_P(\theta)$ as a function of the inclination θ to the vertical are illustrated in figure 8.1 for the set of parameters

$$\alpha_h = 4000\text{m/s}, \quad \alpha_v = 3600\text{m/s},$$

$$\beta_h = 2400\text{m/s}, \quad \beta_v = 2160\text{m/s}, \quad \eta = 1.1,$$

representing an anisotropic shale. We draw the slowness surface rather than the equivalent phase velocity surfaces, because we will see later that the process of reflection and refraction can be directly related to the properties of the slowness surface. An important consequence of even this limited form of anisotropy is that the degeneracy in shear wavespeeds for isotropy is broken, and the horizontally and vertically polarised shear waves have different wave velocities.

Because the phase velocities vary with angle, the direction of energy propagation deviates from the inclination θ to the vertical by an angle

$$\phi = \tan^{-1}\left(\frac{1}{c}\frac{\mathrm{d}c}{\mathrm{d}\theta}\right) \qquad (8.2.32)$$

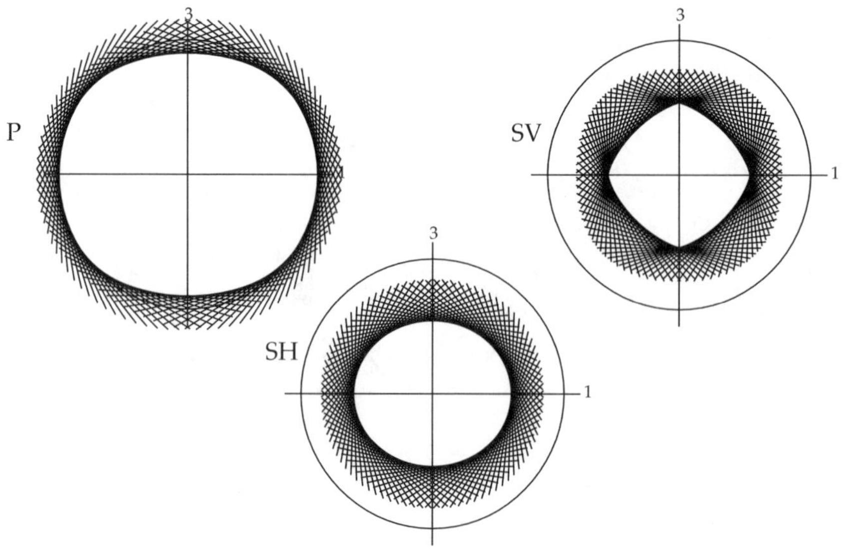

Figure 8.2. Wave surfaces for the transversely isotropic material used in figure 8.1, formed from the envelope of plane wavefronts.

for phase velocity c, and the group velocity

$$v_g = c / \cos\phi. \tag{8.2.33}$$

The complications introduced by the variations in phase velocity with inclination can be illustrated by drawing the wave surfaces corresponding to the locus of energy emitted from a central source (figure 8.2). The simple spherical wavefronts in an isotropic medium are replaced by near elliptical wavefronts for *qP* and *SH* waves. The wave surface for *qSV* is more complex and can develop cusps for some choices of the elastic parameters.

A convenient construction for the wave surfaces is to build them up as an envelope of plane wavefronts travelling in slightly different directions (see, e.g., Keith & Crampin, 1977). We envisage a collection of plane wavefronts which all pass through the origin at $t = 0$ and a unit time later the normals to the plane wavefront will be of length $|\mathbf{p}|^{-1}$. The envelope defines the group velocity surface as a function of inclination θ. The plane wave envelopes for *qP*, *qS* and *SH* are illustrated in figure 8.2 for the same set of parameters as in figure 8.1.

For general anisotropy there is no axis of symmetry and so there can be a very complicated dependence of wave slownesses on the direction of propagation as illustrated by Garmany (1989). The polarisation properties of the different wavetypes can deviate significantly from the simple longitudinal and transverse polarisations for *P* and *S* in an isotropic medium.

The transverse isotropy model has often been employed in exploration seismology for sedimentary materials. The vertical symmetry axis is taken to arise from the effect of compaction of originally unconsolidated sediments under hydrostatic load. The actual pattern of anisotropy in shales will be more complex but the differences between propagation velocities in the horizontal plane are generally somewhat smaller than the difference between the vertical and horizontal propagation velocities.

Another situation in which transverse isotropy provides a good summary of the behaviour of a material arises when there is very fine cyclic bedding in depth at a scale smaller than the wavelengths of the incident energy. The interaction of the wavefield with the fine layering will give different averaging properties in the vertical and horizontal directions and so the effective propagation speeds will differ even though the laminate is composed of thin isotropic layers.

The macroscopic properties of material with oriented sets of fractures can also be represented by an anisotropic relation which is strongly dependent on the properties of the cracks and any entrapped fluid (see, e.g., Crampin, 1981; Hudson, 1991)

8.3 Attenuation

The mineral assemblages in the crust and mantle have a very complex rheology. Over geological time scales they can sustain flow, and in the fold belts of mountain systems we can see considerable deformation without fracture. However, on the relatively short time scales appropriate to seismic wave propagation (0.01 s – 1000 s) we cannot expect to see the influence of the long-term rheology, and the behaviour will be nearly elastic. The small incremental strains associated with seismic disturbances suggest that departures from the constitutive relations (7.4.1) should obey some linear law as sketched in (7.4.7).

Anelastic processes will lead to the dissipation of seismic energy as a wave propagates through the Earth. Among phenomena which are likely to be of importance are crystal defects, dislocation motion, grain boundary processes and some thermoelastic effects (see, e.g., Jackson & Anderson, 1970; Karato, 1998). Dislocation motion is very likely to be important for attenuation in coarse-grain materials (e.g., Jackson et al, 1992) but grain-boundary processes may become more significant for fine-grained materials (e.g., Tan et al, 1997). Karato (1998) discusses the way in which the behaviour of dislocations in silicates can account for seismic wave attenuation and long-term steady-state creep. In his model micro-creep from the migration of geometrical kinks in dislocations is dominant for the short time scales appropriate to seismic waves, and continuous deformation involving successive nucleation and migration of kinks will be active on long time scales. The two classes of process have very different time dependence which would preclude the type of extrapolation made by Jeffreys (1958) in which transient creep behaviour inferred from seismic attenuation is transferred to geological time scales.

Anelastic behaviour may be included in our constitutive laws by introducing the assumption that the stress at a point depends on the time history of strain, so that the material has a 'memory' (Boltzman, 1876). The theory of such linear viscoelasticity has been reviewed by Hudson (1980). For anisotropic media, the appropriate modification of the constitutive law is, cf. (7.4.5),

$$\tau_{ij}(t) = c^{o}_{ijkl}\partial_l u_k(t) + \int_{t_0}^{t} \mathrm{d}s\, C_{ijkl}(t-s)\partial_l u_k(s). \tag{8.3.1}$$

The equivalent relation for isotropy is

$$\begin{aligned}\tau_{ij}(t) &= \lambda_0\partial_k u_k(t) + \mu_0(\partial_i u_j(t) + \partial_j u_i(t))\\ &\quad + \int_0^t \mathrm{d}s \left[\dot{R}_\lambda(t-s)\delta_{ij}\partial_k u_k(s) + \dot{R}_\mu(t-s)\left(\partial_i u_j(s) + \partial_j u_i(s)\right)\right].\end{aligned} \tag{8.3.2}$$

Here $\lambda_0 = \kappa_0 - \frac{2}{3}\mu_0$ and μ_0 are the instantaneous elastic moduli which define the local wavespeeds and R_λ, R_μ are the relaxation functions specifying the dependence on the previous strain states.

We can work with the frequency dependence of the stress and strain by taking the Fourier transform of (8.3.2) with respect to time. At frequency ω the stress and displacement are related by

$$\bar{\tau}_{ij}(\omega) = [\lambda_0 + \lambda_1(\omega)]\delta_{ij}\partial_k\bar{u}_k(\omega) + [\mu_0 + \mu_1(\omega)](\partial_i\bar{u}_j(\omega) + \partial_j\bar{u}_i(\omega)), \tag{8.3.3}$$

where λ_1 and μ_1 are the transforms of the relaxation terms

$$\lambda_1(\omega) = \int_0^\infty \mathrm{d}t\dot{R}_\lambda(t)\mathrm{e}^{\mathrm{i}\omega t}, \qquad \mu_1(\omega) = \int_0^\infty \mathrm{d}t\dot{R}_\mu(t)\mathrm{e}^{\mathrm{i}\omega t}. \tag{8.3.4}$$

So that, if we are indeed in the regime of linear departures from elastic behaviour, the stress-strain relation at frequency ω is formally equivalent to an elastic medium but now with complex moduli, as noted in Section 7.4.

A convenient measure of the rate of energy dissipation at frequency ω is provided by the loss factor $Q^{-1}(\omega)$, which may be defined as the ratio of the energy loss in a cycle $\Delta E(\omega)$ to the 'elastic' energy stored in the oscillation E_0:

$$Q^{-1}(\omega) = -\Delta E(\omega)/(2\pi E_0(\omega)). \tag{8.3.5}$$

E_0 is the sum of the strain and kinetic energy calculated with just the instantaneous elastic moduli, and the energy dissipation ΔE arises from the the imaginary part of the elastic moduli.

For purely dilatational disturbances

$$Q_\kappa^{-1}(\omega) = -\mathrm{Im}\,\{\kappa_1(\omega)\}/\kappa_0, \tag{8.3.6}$$

and for purely deviatoric effects

$$Q_\mu^{-1}(\omega) = -\mathrm{Im}\,\{\mu_1(\omega)\}/\mu_0. \tag{8.3.7}$$

For the Earth it appears that loss in pure dilatation is much less significant than loss in shear, and so $Q_\kappa^{-1} \ll Q_\mu^{-1}$.

Since the relaxation contributions to (8.3.2) depend only on the past history of the strain, $\dot{R}_\mu(t)$ vanishes for $t < 0$, so that the transform $\mu_1(\omega)$ must be analytic in the upper half plane ($\mathrm{Im}\,\omega \geq 0$). In consequence the real and imaginary parts of $\mu_1(\omega)$ are required to be the Hilbert transforms of each other (see, for example, Titchmarsh 1937). Thus the real part of μ_1 is given by

$$\mathrm{Re}\{\mu_1(\omega)\} = \frac{1}{\pi}\mathrm{P}\int_{-\infty}^{\infty} \mathrm{d}\omega' \frac{\mathrm{Im}\{\mu_1(\omega')\}}{\omega' - \omega}, \tag{8.3.8}$$

where P denotes the Cauchy principal value. Thus it is not possible to have dissipative effects on seismic disturbances without some frequency dependent modification of the elastic moduli. This property is associated with any causal dissipative mechanism and the analogous result to (8.3.8) in electromagnetic work is known as the Kramers-Krönig relations.

Using the definition (8.3.7) of Q_μ^{-1} we can rewrite the relation (8.3.8) in a way which shows the dependence of $\mathrm{Re}\{\mu_1(\omega)\}$ on the behaviour of the loss factor with frequency,

$$\mathrm{Re}\{\mu_1(\omega)\} = -\frac{2\mu_0}{\pi}\mathrm{P}\int_{0}^{\infty} \mathrm{d}\omega' \frac{\omega' Q_\mu^{-1}(\omega')}{\omega'^2 - \omega^2}. \tag{8.3.9}$$

Observational information for the loss factor $Q_\mu^{-1}(\omega)$ will only cover a limited range of frequencies and is not sufficient to determine the dispersive component. The detailed form of $\mathrm{Re}\{\mu_1(\omega)\}$ depends on the extrapolation of $Q_\mu^{-1}(\omega)$ to both high and low frequencies.

The distribution of the loss factor Q_μ^{-1} in the Earth is still imperfectly known, because of the difficulties in isolating all the factors which effect the amplitude of a recorded seismic wave. Most studies of attenuation suggest a moderate loss factor in the crust ($Q_\mu^{-1} \sim 0.004$) with an increase in the uppermost mantle ($Q_\mu^{-1} \sim 0.01$) and then a decrease to crustal values, or lower, in the mantle below 1000 km. Over the frequency band 0.001-1 Hz the intrinsic loss factor Q_μ^{-1} appears to be essentially constant. However, in order for there to be a physically realisable loss mechanism, Q_μ^{-1} must depend on frequency outside this band. A number of different forms for the frequency dependence have been suggested (Azimi et al., 1968; Kanamori & Anderson, 1977; Jeffreys, 1958) but provided Q_μ^{-1} is not too large ($Q_\mu^{-1} < 0.01$) these lead to the approximate relation

$$\mathrm{Re}\{\mu_1(\omega)\} = \frac{2\mu_0}{\pi}\ln(\omega a) Q_\mu^{-1}, \tag{8.3.10}$$

in terms of some time constant a.

A similar development may be made for the complex bulk modulus $\kappa_0 + \kappa_1(\omega)$ in terms of the loss factor Q_κ^{-1}.

For a locally uniform region, at a frequency ω, substitution of the stress-strain relation (8.3.3) into the equations of motion shows that, as in a perfectly elastic medium, two sets of plane waves exist. The S waves have a complex wavespeed $\bar{\beta}$ given by

$$\bar{\beta}^2(\omega) = [\mu_0 + \mu_1(\omega)]/\rho, \tag{8.3.11}$$

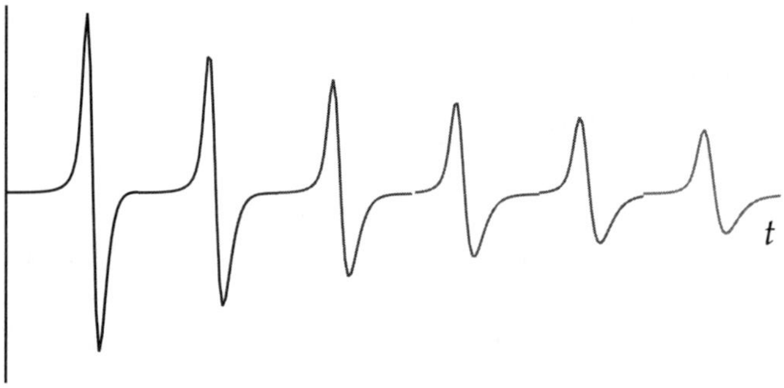

Figure 8.3. Pulses at progressively later times, after passage through a medium with causal Q_μ^{-1} and associated velocity dispersion. The pulses broaden and reduce in amplitude.

influenced only by shear relaxation processes. In terms of the wavespeed $\beta_0 = (\mu_0/\rho)^{1/2}$ calculated for the instantaneous modulus, (8.3.11) may be rewritten as

$$\bar{\beta}^2(\omega) = \beta_0^2\left(1 + \frac{\mathrm{Re}\{\mu_1(\omega)\}}{\mu_0} - i\,\mathrm{sgn}(\omega)Q_\mu^{-1}(\omega)\right), \tag{8.3.12}$$

where we have used the definition of Q_μ^{-1} from (8.3.7). Even if Q_μ^{-1} is frequency independent in the seismic band, our previous discussion shows that $\bar{\beta}$ will have weak frequency dispersion through $\mathrm{Re}\{\mu_1(\omega)\}$.

For a small loss factor ($Q_\mu^{-1} \ll 1$), using (8.3.10), the ratio of the complex velocities at two different frequencies ω_1 and ω_2 will be approximately

$$\frac{\bar{\beta}(\omega_1)}{\bar{\beta}(\omega_2)} = 1 + \frac{1}{\pi}Q_\mu^{-1}\ln\left(\frac{\omega_1}{\omega_2}\right) - i\,\mathrm{sgn}(\omega)\tfrac{1}{2}Q_\mu^{-1}. \tag{8.3.13}$$

The problem of the unknown constant a can therefore be overcome by specifying a reference frequency (most commonly 1 Hz) and then

$$\bar{\beta}(\omega) \approx \beta_1\left[1 + \frac{1}{\pi}Q_\mu^{-1}\ln\left(\frac{\omega}{2\pi}\right) - i\,\mathrm{sgn}(\omega)\tfrac{1}{2}Q_\mu^{-1}\right], \tag{8.3.14}$$

where β_1 is the velocity at 1 Hz. The presence of the frequency dependent terms in (8.3.12), (8.3.14) arise from the requirement that all dissipative processes will be causal, so that no seismic energy arrives with a wavespeed faster than β_0, the wavespeed for the reference elastic medium.

A pulse travelling through the lossy medium suffers a steady amplitude reduction with a lengthening of the tail of the pulse as propagation proceeds as illustrated in figure 8.3. The onset of the pulse remains fairly sharp when velocity dispersion is included.

When $Q_\mu(\omega)$ has some significant frequency dependence we will still obtain a similar structure to (8.3.14) although the nature of the frequency dependence $\mathrm{Re}\{\bar{\beta}(\omega)\}$ will vary. Following the suggestion of Jeffreys (1958), Smith & Dahlen (1981) have shown that a weak frequency variation in loss factor, $Q_\mu^{-1} \propto \omega^{-\gamma}$ with $\gamma \approx 0.1$, will fit the observed period (435.2 days) and damping of the Chandler wobble, as well as the results in the seismic band. The value of γ is dependent on the reference loss model and is primarily influenced by the properties of the lower mantle. Lundquist & Cormier (1980) have suggested that the loss factor in the upper mantle may vary significantly for frequencies between 1 and 10 Hz, and relate this to relaxation time scales for absorption processes. A number of studies have studied power law frequency dependence,

$$Q_\mu^{-1} = Q_{\mu\,\mathrm{ref}}^{-1}\left(\frac{\omega}{\omega_{\mathrm{ref}}}\right)^{-\gamma} \quad \text{with} \quad 0 < \gamma < 1, \tag{8.3.15}$$

for frequencies higher than 1 Hz; and in this case we expect dispersion of the form (Brennan & Smylie, 1981)

$$\frac{\mu(\omega)}{\mu(\omega_{\mathrm{ref}})} = 1 - \tan[(1-\gamma)\frac{\pi}{2}][Q^{-1}(\omega) - Q^{-1}(\omega_{\mathrm{ref}})]. \tag{8.3.16}$$

There is strong variability in the Q^{-1} values and the apparent frequency dependence at both crustal and mantle levels (Mitchell, 1995; Mitchell & Cong, 1998). Regions of current or recent tectonic activity have higher levels of attenuation than other regions and old stable regions generally have low loss factor Q^{-1}, particularly in the mantle. In most cases the frequency exponent γ is larger when attenuation and, thus, the loss factor Q^{-1} are larger. This suggests that there is a common controlling factor which is most likely thermal activation of the attenuation mechanisms (Jackson, 2000).

For shallow propagation at high frequencies (10-60 Hz), O'Brien & Lucas (1971) have shown that the constant Q^{-1} model gives a good explanation of observed amplitude loss in prospecting situations.

For P waves we have to take into account both the anelastic effects in pure dilatation and in shear. The complex wavespeed $\bar{\alpha}$ is given by

$$\begin{aligned}\bar{\alpha}^2(\omega) &= [\kappa_0 + \tfrac{4}{3}\mu_0 + \kappa_1(\omega) + \tfrac{4}{3}\mu_1(\omega)]/\rho, \\ &= \alpha_0\{1 + A(\omega) + \mathrm{i}\,\mathrm{sgn}(\omega) Q_A^{-1}(\omega)\},\end{aligned} \tag{8.3.17}$$

where the wavespeed $\alpha_0 = [(\kappa_0 + \frac{4}{3}\mu_0)/\rho]^{1/2}$ is calculated using the instantaneous modulus. The loss factor Q_A^{-1} for P waves introduced in (8.3.17) is

$$Q_A^{-1} = -\frac{\mathrm{Im}\{\kappa_1 + \frac{4}{3}\mu_1\}}{(\kappa_0 + \frac{4}{3}\mu_0)}. \tag{8.3.18}$$

Normally we expect that loss in dilatation is very small compared with that in shear so that $Q_\kappa^{-1} \ll Q_\mu^{-1}$, and then

$$Q_A^{-1} \approx \frac{4\beta_0^2}{3\alpha_0^2} Q_\mu^{-1}, \tag{8.3.19}$$

as suggested by Anderson, Ben-Menahem & Archambeau (1965). In general the dispersive correction to the wave speed $A(\omega)$ would be expected to have a rather complex form. However, when loss in compression is small compared to that in shear, as in (8.3.19), we will have a similar form to (8.3.10)

$$A(\omega) = \frac{2}{\pi}(\kappa_0 + \tfrac{4}{3}\mu_0)Q_A^{-1}\ln(\omega a). \tag{8.3.20}$$

The complex wavespeed may therefore be represented in terms of the wavespeed at 1 Hz (α_1),

$$\bar{\alpha}(\omega) = \alpha_1\left[1 + \frac{1}{\pi}Q_A^{-1}\ln\left(\frac{\omega}{2\pi}\right) - \mathrm{isgn}(\omega)\tfrac{1}{2}Q_A^{-1}\right]. \tag{8.3.21}$$

In the frequency domain, calculations with these complex velocities turn out to be little more complicated than in the perfectly elastic case.

In addition to the dissipation of elastic energy by anelastic processes, the apparent amplitude of a seismic wave can be diminished by scattering which redistributes the elastic energy. As we have noted above, our choice of elastic moduli defines a reference medium whose properties smooth over local irregularities in the properties of the material. The fluctuations in the actual material will lead to scattering of the seismic energy out of the primary wave, which will be cumulative along the propagation path. The apparent velocity of transmission of the scattered energy will vary from that in the reference medium. Since locally the material may be faster or slower than the reference wavespeed, the effect of scattering is to give a pulse shape which is broadened and diminished in amplitude relative to that in the reference, with an emergent onset before the expected travel time for the reference medium.

The effects of scattering can be described by a loss factor ${}_sQ^{-1}(\omega)$. Unlike the loss factor for anelastic processes, ${}_sQ^{-1}$ is not a strictly a measure of energy loss per cycle but, rather, a measure of energy redistribution. ${}_sQ^{-1}$ depends strongly on frequency and is very path dependent, since it depends on the particular heterogeneity spectrum encountered by a wavefield propagating through the Earth. The scattering loss ${}_sQ^{-1}$ is usually modeled with stochastic operators, or randomisation coefficients. As the wavelength diminishes, the effect of local irregularities becomes more pronounced and so ${}_sQ^{-1}$ tends to increase until the wavelength is of the same order as the scale of variation of the heterogeneity.

This scattering mechanism of seismic wave attenuation is important in the lithosphere where heterogenieity occurs on a wide variety of scales. There are also considerable regional variations, with earthquake zones showing the most significant effects (Aki, 1981; Sato & Fehler, 1998).

For each wave type the overall rate of seismic attenuation Q^{-1}, which is the quantity which would be derived from observations, will be the sum of the loss factors from intrinsic anelasticity and scattering. Thus for S waves

$$Q_\beta^{-1}(\omega) = Q_\mu^{-1}(\omega) + {}_sQ_\beta^{-1}(\omega). \tag{8.3.22}$$

For P waves,

$$Q_\alpha^{-1}(\omega) = Q_A^{-1}(\omega) + {}_sQ_\alpha^{-1}(\omega); \tag{8.3.23}$$

since the scattering component here is arising from a totally distinct mechanism to the dissipation there is no reason to suppose that ${}_sQ_\alpha^{-1}$, ${}_sQ_\beta^{-1}$ are related in a similar way to (8.3.19).

The influence of scattering can give fairly strong frequency dependence to Q_β^{-1}, Q_α^{-1}, however we have to be wary about how such dependence is interpreted. In particular it is not necessarily appropriate to apply wavespeed dispersion corrections such as (8.3.9), (8.3.16). It is only for the anelastic portion Q_μ^{-1}, Q_A^{-1} that we have dispersive wavespeed terms. The scattering contribution ${}_sQ_\beta^{-1}$, ${}_sQ_\alpha^{-1}$ does not have the same restriction to a local 'memory' effect and there is no consequent dispersion.

The preceding discussion has been presented in terms of isotropic media so that the dominant behaviour for the two wavetypes can be analysed. The constitutive relation for linear viscoelasticity in an anisotropic medium (8.3.1) includes anisotropy in the relaxation functions C_{ijkl} that specify the dependence on the prior strain state. In consequence, the loss factor Q^{-1} has to be anisotropic. The current state of measurements of Q^{-1} is such that this anisotropic effect has not been found to be important, but it may contribute to the scatter in the results for Q^{-1}.

9

Seismic Waves II - Wavefronts and Rays

The plane wave solutions for seismic waves introduced in the previous chapter are particularly suitable for regions that have locally homogeneous properties. Here we develop an approach based on wavefronts and rays that can be more readily extended to heterogeneous media.

We consider the nature of the seismic wavefield near a wavefront and look at the way in which wavefront surfaces evolve in passage through a medium. We will see that we arrive at a *local* condition for the three seismic wavespeeds in a general anisotropic medium that matches the plane wave representation for a material with the same properties. We are also able to follow the pattern of energy transmission associated with the wavefront via the properties of ray trajectories. This leads to a representations of wave amplitudes in terms of the divergence of ray tubes. The ray equations simplify somewhat for isotropic media. When a wavefront impinges on a surface at which material properties change, we have to restart the ray tracing process and energy is redistributed between reflected and transmitted waves in a way which can be described by using local plane waves.

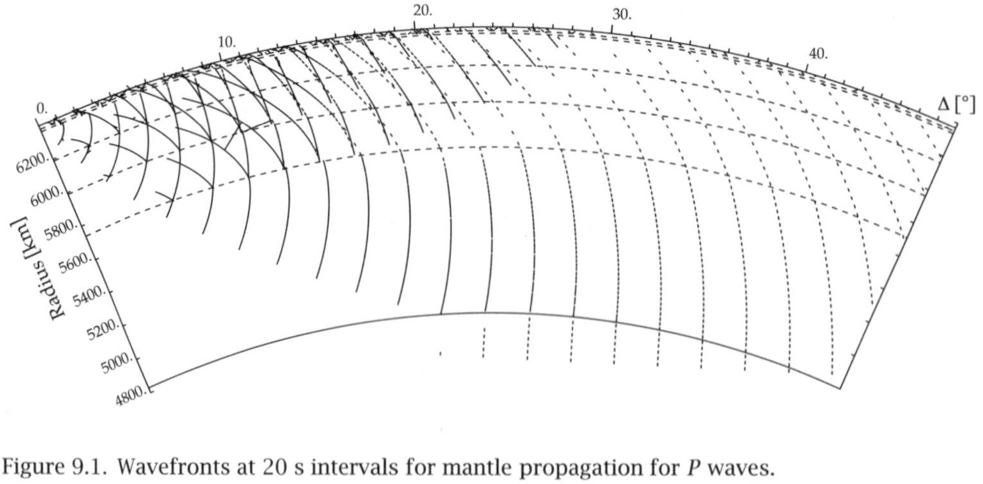

Figure 9.1. Wavefronts at 20 s intervals for mantle propagation for *P* waves.

9.1 Wavefronts and ray series

When we look at the propagation of seismic waves away from a source we are interested in the behaviour of *wavefronts*, i.e., surfaces of discontinuity in the displacement or its derivatives. At such wavefronts (see figure 9.1) the solution of the elastodynamic equations cannot be represented in simple analytic terms and it is convenient to look for solutions in the form of an expansion in a sequence of terms with increasing smoothness.

Consider a wavefront expanding from the source whose spatial variation at time t after initiation can be represented in terms of a phase function $\vartheta(\mathbf{x})$ so that

$$t = \vartheta(\mathbf{x}). \tag{9.1.1}$$

As shown in the appendix to this chapter, we can represent the displacement field near the wavefront as

$$\mathbf{u}(\mathbf{x}, t) = \sum_{r=0}^{\infty} \mathbf{U}^{[r]}(\mathbf{x}) f_r(t - \vartheta(\mathbf{x})), \tag{9.1.2}$$

an expansion in terms of 'generalised progressive waves' with $f_r' = f_{r-1}$, so that each successive term is smoother than its predecessors. As demonstrated by Babich (1961) this *ray series* will converge if $\vartheta(\mathbf{x})$ is an analytic function.

Consider a situation in which we have no disturbance ahead of the wavefront but a non-zero displacement following, then we could describe the situation with a leading term f_0 which is a generalised function

$$f_0(s) = \frac{s_+^\lambda}{\Gamma(1+\lambda)}, \qquad s_+^\lambda = \begin{cases} s^\lambda, & s > 0, \\ 0, & s < 0. \end{cases} \tag{9.1.3}$$

Note that $s < 0$ corresponds to $t < \vartheta(\mathbf{x})$, i.e. ahead of the wavefront. The successive f_r have the form

$$f_r(s) = \frac{s_+^{\lambda+r}}{\Gamma(1+\lambda+r)}, \qquad r \geq 0, \tag{9.1.4}$$

with progressively milder singularities at the wavefront. This process is illustrated in fig 9.2 for f_0 in the form of a Heaviside step function, i.e., $\lambda = 0$. Each of the f_r then has a discontinuity in the rth derivative across the wavefront.

$$f_0(t) = H(t), \quad f_1(t) = tH(t), \quad f_2(t) = \tfrac{1}{2}t^2H(t), \quad f_3(t) = \tfrac{1}{6}t^3H(t), \ldots$$

This set of $f_r(t)$ illustrates well the nature of the discontinuities at the wavefront but we note that the higher order terms grow significantly away from the wavefront $t = 0$.

For a high frequency pulse such that

$$f_0(s) = \int_{\omega_0}^{\infty} \mathrm{d}s\, \bar{f}_0(\omega) \mathrm{e}^{-i\omega s}, \tag{9.1.5}$$

for some large ω_0, we obtain an asymptotic representation

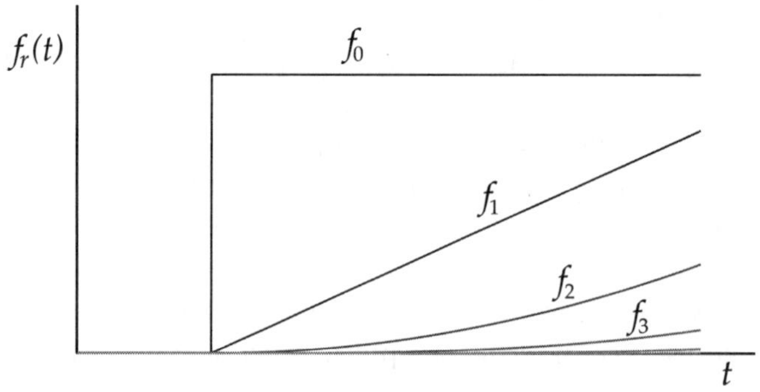

Figure 9.2. Progressive terms in a ray series expansion showing the milder singularities at higher order.

$$\mathbf{u}(\mathbf{x}, t) \sim \sum_{r=0}^{\infty} \mathbf{U}^{[r]}(\mathbf{x}) f_r(t - \vartheta(\mathbf{x})), \tag{9.1.6}$$

where the successive terms f_r have the form

$$f_r(s) = \int_{\omega_0}^{\infty} \mathrm{d}s \, \frac{\bar{f}_0(\omega)}{(\mathrm{i}\omega)^r} \mathrm{e}^{-\mathrm{i}\omega s}, \tag{9.1.7}$$

so that they have less high frequency content. Once again we require the wavefront function $\vartheta(\mathbf{x})$ to be analytic for the asymptotic representation (9.1.6) to be valid. This assumption will fail in the vicinity of a caustic where the wavefront is folded.

9.2 Ray theory for seismic waves

Consider an heterogeneous medium with a state specified by displacement $\mathbf{u}$ and incremental stress tensor τ_{ij}. The elastic modulus tensor c_{ijkl} will be a function of position so that the local stress-strain conditions take the form

$$\tau_{ij}(\mathbf{x}) = c_{ijkl}(\mathbf{x}) \partial_l u_k(\mathbf{x}). \tag{9.2.1}$$

The elastodynamic equation is then

$$\mathcal{L}_i(\mathbf{u}) = \partial_j \left(c_{ijkl}(\mathbf{x}) \partial_l u_k(\mathbf{x}) \right) = \rho(\mathbf{x}) \partial_{tt} u_i(\mathbf{x}), \tag{9.2.2}$$

in terms of the differential operator $\boldsymbol{\mathcal{L}}$.

We will look for solutions of the elastodynamic equation (9.2.2) in terms of a ray expansion (9.1.2)

$$u_i(\mathbf{x}, t) = \sum_{r=0}^{\infty} U_i^{[r]}(\mathbf{x}) f_r(t - \vartheta(\mathbf{x})), \tag{9.2.3}$$

where $f_r' = f_{r-1}$, so that the successive time functions are the integral of the preceding function. The acceleration term is then

$$\rho(\mathbf{x})\partial_{tt}u_i(\mathbf{x},t) = \rho(\mathbf{x})\sum_{r=0}^{\infty} U^{[r]}(\mathbf{x})f_r''(t-\vartheta(\mathbf{x})), \tag{9.2.4}$$

and the stress-gradient contributions are

$$\begin{aligned}\partial_j\left(c_{ijkl}\partial_l\left[U_k^{[r]}f_r(t-\vartheta)\right]\right) &= \partial_j\left(c_{ijkl}\partial_l U_k^{[r]}f_r(t-\vartheta)\right) - \partial_j\left(c_{ijkl}U_k^{[r]}\partial_l\vartheta f_r'(t-\vartheta)\right)\\ &= \partial_j\left(c_{ijkl}\partial_l U_k^{[r]}\right)f_r(t-\vartheta)\\ &\quad -\left[\partial_j\left(c_{ijkl}U_k^{[r]}\partial_l\vartheta\right) + c_{ijkl}\partial_l U_k^{[r]}\partial_j\vartheta\right]f_r'(t-\vartheta)\\ &\quad +c_{ijkl}U_k^{[r]}\partial_l\vartheta\partial_j\vartheta f_r''(t-\vartheta).\end{aligned} \tag{9.2.5}$$

On substituting the ray series into the elastodynamic equation we find that

$$\sum_{r=0}^{\infty}\left[\mathcal{L}_i(\mathbf{U}^{[r]})f_r(t-\vartheta) + \mathcal{M}_i(\mathbf{U}^{[r]})f_r'(t-\vartheta) + \mathcal{N}_i(\mathbf{U}^{[r]})f_r''(t-\vartheta)\right] = 0, \tag{9.2.6}$$

where we have introduced the terms

$$\mathcal{M}_i(\mathbf{U}^{[r]}) = -\left[\partial_j\left(c_{ijkl}U_k^{[r]}\partial_l\vartheta\right) + c_{ijkl}\partial_l U_k^{[r]}\partial_j\vartheta\right], \tag{9.2.7}$$

and

$$\mathcal{N}_i(\mathbf{U}^{[r]}) = \left[c_{ijkl}\partial_l\vartheta\partial_j\vartheta - \rho\delta_{ik}\right]U_k^{[r]}. \tag{9.2.8}$$

When we exploit the relationship between the successive $f_r(t)$ we can rewrite (9.2.6) as

$$\sum_{r=0}^{\infty}\left[\mathcal{L}_i(\mathbf{U}^{[r-2]}) + \mathcal{M}_i(\mathbf{U}^{[r-1]}) + \mathcal{N}_i(\mathbf{U}^{[r]})\right]f_r(t-\vartheta) = 0, \tag{9.2.9}$$

and then equating the coefficients of successive f_r to zero we require

$$\mathcal{L}_i(\mathbf{U}^{[r-2]}) + \mathcal{M}_i(\mathbf{U}^{[r-1]}) + \mathcal{N}_i(\mathbf{U}^{[r]}) = 0, \tag{9.2.10}$$

with

$$\mathbf{U}^{[-2]} = \mathbf{U}^{[-1]} = 0, \quad \mathbf{U}^{[0]} \neq 0. \tag{9.2.11}$$

For $r = 0$ (9.2.10) reduces to $\mathcal{N}_i(\mathbf{U}^{[0]}) = 0$ so that we require

$$\left[c_{ijkl}\partial_l\vartheta\partial_j\vartheta - \rho\delta_{ik}\right]U_k^{[0]} = 0, \tag{9.2.12}$$

Since $\mathbf{U}^{[0]}$ is non-zero we require

$$\det\left[c_{ijkl}\partial_l\vartheta\partial_j\vartheta - \rho\delta_{ik}\right] = 0, \tag{9.2.13}$$

which will have three solutions for the gradient of the wavefront function $\nabla\vartheta$ associated with the three wavetypes in the anisotropic medium. We can recognise the similarity with our earlier discussion of plane waves by working in terms of the local direction $\mathbf{n}(\mathbf{x})$ and magnitude $p(\mathbf{x})$ of the gradient of ϑ,

$$\nabla\vartheta(\mathbf{x}) = p(\mathbf{x})\mathbf{n}(\mathbf{x}) \tag{9.2.14}$$

and then (9.2.12) becomes

$$\left[c_{ijkl}(\mathbf{x})p^2(\mathbf{x})n_l(\mathbf{x})n_j(\mathbf{x}) - \rho(\mathbf{x})\delta_{ik}\right] U_k^{[0]} = 0, \tag{9.2.15}$$

which is a local version of the eigenvalue equation for plane-wave slowness (8.2.4). Locally therefore the evolution of the three possible wavefront functions $\vartheta^{(m)}$ will correspond to the propagation of the different types of plane waves whose properties are determined by the elastic parameters $c_{ijkl}(\mathbf{x})$, $\rho(\mathbf{x})$ at a point $\mathbf{x}$.

9.2.1 Wavefronts and rays

Introducing a local Christoffel matrix $g_{ik}(\mathbf{n})$, cf. (8.2.5),

$$g_{ik}(\mathbf{n}) = c_{ijkl}n_j n_l, \tag{9.2.16}$$

we can rewrite (9.2.15) in the form

$$\left[g_{ik}(\mathbf{n}) - \rho p^{-2}(\mathbf{n})\delta_{ik}\right] U_k^{[0]}(\mathbf{n}) = 0, \tag{9.2.17}$$

and the three solutions for the slowness $p^{(m)}(\mathbf{n})$ are then to be found from

$$\det\left[g_{ik}(\mathbf{n}) - \rho p^{-2}(\mathbf{n})\delta_{ik}\right] = 0. \tag{9.2.18}$$

There are therefore three different wavefront surfaces $\vartheta^{(m)}(\mathbf{x})$ associated with the slownesses $p^{(m)}(\mathbf{n})$ with displacement polarisations characterised by corresponding orthogonal eigenvectors $\mathbf{V}^{(m)}(\mathbf{n})$ satisfying (9.2.17). The local phase velocities $c^{(m)}(\mathbf{n})$ for the three wavetypes are the reciprocals of the slownesses $p^{(m)}(\mathbf{n})$, i.e., $c^{(m)}(\mathbf{n}) = [p^{(m)}(\mathbf{n})]^{-1}$.

For each of the three wavefront surfaces the slowness vector

$$\mathbf{p}^{(m)}(\mathbf{x}) = \nabla\vartheta^{(m)}(\mathbf{x}) = p^{(m)}(\mathbf{x})\mathbf{n}(\mathbf{x}), \tag{9.2.19}$$

lies normal to the wavefront at the point $\mathbf{x}$. However, the corresponding ray direction, along which energy is transported, will not in general lie along $\mathbf{p}^{(m)}(\mathbf{x})$, but instead lie normal to the slowness surfaces defined by (9.2.15), (9.2.18).

As shown by Musgrave (1970, §7.3), the direction $\mathbf{m}$ of the normal to a slowness surface is given by

$$m_j \propto c_{ijkl}V_i^{(m)}V_k^{(m)}p^{(m)}n_l, \tag{9.2.20}$$

and, further, the energy flux associated with a plane wave specified by the slowness vector $\mathbf{p}^{(m)}$ is

$$\mathcal{F}_j \propto \omega^2 c_{ijkl}V_i^{(m)}V_k^{(m)}p^{(m)}n_l, \tag{9.2.21}$$

so that the direction of energy transport and the normal to the slowness surface $\mathbf{m}$ coincide.

The group velocity vector $\mathbf{v}^{(m)}$ for the mth wavetype lies along $\mathbf{m}$ and, following Helbig (1958) can be expressed in terms of the phase velocity $c^{(m)}$ as

$$v_j^{(m)} = c^{(m)}n_j + \frac{\partial c^{(m)}}{\partial n_j}, \tag{9.2.22}$$

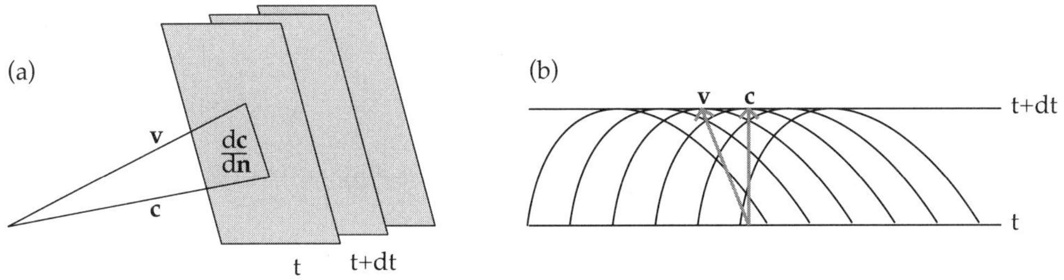

Figure 9.3. Deviation of phase and group velocity vectors in anisotropic media. The phase velocity vector **c** is perpendicular to the phase surface (wavefront) while the group velocity vector **v** tracks the direction of energy propagation.

where the second term is orthogonal to the phase velocity vector $c^{(m)}\mathbf{n}$ (see figure 9.3). As the wave surfaces $\vartheta^{(m)}$ evolve the energy associated with each wavetype propagates along ray trajectories determined by the directions of the the normals to the slowness surface $\mathbf{m}(\mathbf{x})$ in the heterogeneous medium. The components of the group velocity are given by

$$v_j^{(m)} = \frac{1}{\rho} c_{ijkl} V_i^{(m)} V_k^{(m)} p^{(m)} n_l. \tag{9.2.23}$$

The calculation of the ray paths in heterogeneous anisotropic media is a complex procedure, but can be cast in the form of a coupled set of first order equations for the evolution of the position **x** and slowness **p** with time θ along the ray path. Červený (1972) writes the slowness equation in the form

$$(\Gamma_{ik} - \delta_{ik})\, U_k^{[0]} = 0, \quad \text{with} \quad \Gamma_{ik} = \frac{p^2}{\rho} g_{ik} = \frac{1}{\rho} c_{ijkl} p_k p_l, \tag{9.2.24}$$

and determines wavefronts for the mth wavetype from the condition that the mth eigenvalue of $\boldsymbol{\Gamma}$ is unity,

$$G^{(m)}(\mathbf{x}, \mathbf{p}) = 1; \tag{9.2.25}$$

and also the property that the $G^{(m)}$ are homogeneous function of second order p_i

$$p_i \frac{\partial G^{(m)}}{\partial p_i} = 2G^{(m)}. \tag{9.2.26}$$

The development of the rays in $(\mathbf{p}, \mathbf{x})$ phase space can then be found from the Hamiltonian equations

$$\frac{\mathrm{d}x_i}{\mathrm{d}\theta} = \tfrac{1}{2} \frac{\partial G^{(m)}}{\partial p_i}, \qquad \frac{\mathrm{d}p_i}{\mathrm{d}\theta} = -\tfrac{1}{2} \frac{\partial G^{(m)}}{\partial x_i}. \tag{9.2.27}$$

The derivatives of $G^{(m)}$ can be found by implicit differentiation without having to construct an explicit expression for $G^{(m)}$ and Červený (1972) has provided convenient forms for the ray tracing equations in terms of the elements Γ_{ij} which are suitable for direct numerical implementation.

9.2.2 Amplitude relations

The wavefront for each wavetype $\vartheta^{(m)}$ is specified by

$$\left[c_{ijkl}(\mathbf{x})\partial_l\vartheta^{(m)}\partial_j\vartheta^{(m)} - \rho(\mathbf{x})\delta_{ik}\right]U_k^{[0]} = 0 \tag{9.2.28}$$

and the leading term in the ray-series $\mathbf{U}^{[0]}$ has to be aligned with the displacement eigenvector $\mathbf{V}^{(m)}$ for that wavetype.

We consider the development of the ray system from an initial surface $t_0 = \vartheta^{(m)}(\mathbf{x})$ and set up a cylindrical coordinate system in this surface. Each ray is specified by a coordinate pair (q_1, q_2) and each point on the ray by the three parameters (q_1, q_2, θ) where θ represents time along the ray path. The entire region 'illuminated' by the rays is then covered by a three-dimensional system of coordinates in which the ray direction $\nabla\theta$ is perpendicular to the (q_1, q_2) plane.

Consider the position vector $\mathbf{x}(q_1, q_2)$, the surface element of the wavefront within a tube of rays defined by the parameter limits q_1, $q_1 + \mathrm{d}q_1$; q_2, $q_1 + \mathrm{d}q_2$ is given by

$$\mathrm{d}\sigma = J\mathrm{d}q_1\mathrm{d}q_2 = \left|\frac{\mathrm{d}\mathbf{x}}{\mathrm{d}q_1} \wedge \frac{\mathrm{d}\mathbf{x}}{\mathrm{d}q_2}\right| \mathrm{d}q_1\mathrm{d}q_2. \tag{9.2.29}$$

Note that in anisotropic media the wavefront is not orthogonal to the rays so that the surface element $\mathrm{d}\sigma$ is not a perpendicular cross section of the ray tube. To consider the development of this ray tube to a neighbouring wavefront surface $\theta + \mathrm{d}\theta$ we can use the divergence theorem,

$$\int_V \mathrm{d}^3\mathbf{x}\,\nabla\cdot\mathbf{f} = \int_S \mathrm{d}\mathbf{S}\cdot\mathbf{f}, \tag{9.2.30}$$

to examine the behavior of a vector function $\mathbf{f}(\mathbf{x})$ directed everywhere along $\nabla\theta$,

$$\nabla\cdot\mathbf{f}\,J\mathrm{d}q_1\mathrm{d}q_2\mathrm{d}s = \left[fJ\mathrm{d}q_1\mathrm{d}q_2\right]_\theta^{\theta+\delta\theta} = \frac{\partial}{\partial\theta}(fJ)\mathrm{d}q_1\mathrm{d}q_2, \tag{9.2.31}$$

in terms of a path element $\mathrm{d}s$. Thus,

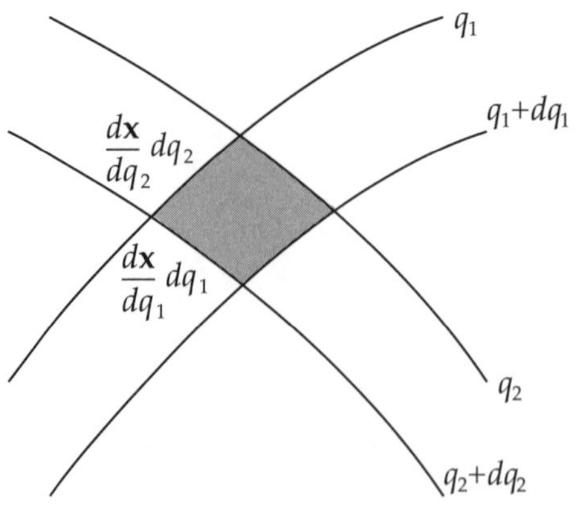

Figure 9.4. An element of a wavefront surface within a ray-tube.

$$\nabla \cdot \mathbf{f} = \frac{\mathrm{d}\theta}{\mathrm{d}s} \frac{1}{J} \frac{\partial}{\partial \theta}(fJ) = \frac{1}{cJ} \frac{\partial}{\partial \theta}(fJ), \tag{9.2.32}$$

where c is the velocity of advance along the ray path.

Consider now the way in which the amplitude of the leading term for the mth wavetype varies with position. From (9.2.10) for $r = 1$ we have

$$\mathcal{M}_i(\mathbf{U}^{[0]}) + \mathcal{N}_i(\mathbf{U}^{[1]}) = 0, \tag{9.2.33}$$

and if we now take the dot product with the eigenvector $\mathbf{V}^{(m)}$

$$\mathcal{M}_i(\mathbf{U}^{[0]})V_i^{(m)} + \mathcal{N}_i(\mathbf{U}^{[1]})V_i^{(m)} = 0. \tag{9.2.34}$$

Reintroducing the explicit form of $\mathcal{N}_i(\mathbf{U}^{[1]})$ we obtain

$$\mathcal{N}_i(\mathbf{U}^{[1]})V_i^{(m)} = \left[c_{ijkl}(\mathbf{x})\partial_l\vartheta^{(m)}\partial_j\vartheta^{(m)} - \rho(\mathbf{x})\delta_{ik}\right] U_k^{[1]}V_i^{(m)} = 0 \tag{9.2.35}$$

where we have used the symmetry of the elastic modulus tensor and the properties of the eigenvector $\mathbf{V}^{(m)}$.

Now, since $\mathbf{U}^{[0]}$ lies along $\mathbf{V}^{(m)}$ we can write the displacement as $\mathbf{U}^{[0]} = \phi^{[0]}\mathbf{V}^{(m)}$ so that (9.2.35) becomes

$$\partial_j \left(c_{ijkl}p^{(m)}n_l\phi^{[0]}V_k^{(m)}\right) V_i^{(m)} + c_{ijkl}\partial_l \left(\phi^{[0]}V_k^{(m)}\right) p^{(m)}n_lV_i^{(m)} = 0, \tag{9.2.36}$$

where we have set $\partial_l\vartheta^{(m)} = p^{(m)}n_l$ as in (9.2.14). We can rearrange (9.2.36) as a differential equation for $\phi^{[0]}$:

$$2\partial_j\phi^{[0]}c_{ijkl}p^{(m)}n_lV_k^{(m)}V_i^{(m)} + \phi^{[0]}\partial_j \left(c_{ijkl}p^{(m)}n_lV_k^{(m)}V_i^{(m)}\right) = 0. \tag{9.2.37}$$

From (9.2.23) we can recognise the combination $c_{ijkl}p^{(m)}n_lV_k^{(m)}V_i^{(m)}$ as ρv_j^m in terms of the components of the group velocity $\mathbf{v}^{(m)}$. Thus (9.2.37) can be rewritten as

$$\partial_j\phi^{[0]}v_j^{(m)} + \frac{\phi^{[0]}}{\rho}\partial_j(\rho v_j^{(m)}) = \partial_\theta\phi^{[0]} + \frac{\phi^{[0]}}{\rho}\partial_j(\rho v_j^{(m)}) = 0, \tag{9.2.38}$$

where θ represents time along the ray path.

The group velocity is directed along the ray path, and so we can exploit the representation (9.2.32) above in terms of the ray coordinates (q_1, q_2, θ) and express the spatial derivatives $\partial_j(\rho v_j^{(m)})$ in the form

$$\frac{\partial}{\partial x_j}(\rho v_j^{(m)}) = \frac{1}{c^{(m)}J}\frac{\mathrm{d}}{\mathrm{d}\theta}(\rho c^{(m)}J). \tag{9.2.39}$$

where $c^{(m)} = [p^{(m)}]^{-1}$. The differential equation for $\phi^{[0]}$ can now be expressed as

$$\frac{\mathrm{d}\phi^{[0]}}{\mathrm{d}\theta} + \frac{\phi^{[0]}}{2\rho c^{(m)}J}\frac{\mathrm{d}}{\mathrm{d}\theta}(\rho c^{(m)}J) = 0. \tag{9.2.40}$$

The evolution of the amplitude along the ray with time θ is given by

$$\phi^{[0]}(\theta) = \phi^{[0]}(t_0)\left[\frac{(\rho c^{(m)}J)_{t_0}}{(\rho c^{(m)}J)_\theta}\right]^{1/2}. \tag{9.2.41}$$

The calculation of the amplitude depends on the computation of J and Červený (1972) provides a set of first order equations for the evolution of the derivatives $\partial x_i/\partial q_k$, $\partial p_i/\partial q_k$ from which J may be constructed.

A comparable development can be made for the part of the higher order ray series term $\mathbf{U}^{[r]}$ aligned with $\mathbf{U}^{[0]}$, which is termed the *principal component.* In general there will also be an *additional component* which is orthogonal to $\mathbf{U}^{[0]}$. We can illustrate this by considering $\mathbf{U}^{[1]}$; we write

$$\mathbf{U}^{[1]} = \mathbf{U}_0^{[1]} + \mathbf{U}_\perp^{[1]}, \tag{9.2.42}$$

where $\mathbf{U}_\perp^{[1]}$ lies perpendicular to $\mathbf{U}^{[0]}$. From (9.2.33) we find

$$\mathcal{N}_i(\mathbf{U}_\perp^{[1]}) + \mathcal{M}_i(\mathbf{U}^{[0]}) = 0, \tag{9.2.43}$$

since $\mathcal{N}_i(\mathbf{U}_0^{[1]}) = 0$ because $\mathbf{U}_0^{[1]}$ is parallel to the eigenvector $\mathbf{U}^{[0]}$. The term $\mathcal{M}_i(\mathbf{U}^{[0]})$ depends on the spatial derivatives of the elastic moduli c_{ijkl} and $\mathbf{U}^{[0]}$ (9.2.7), and so if there is any component of these derivatives perpendicular to $\mathbf{U}^{[0]}$ then $\mathbf{U}^{[1]}$ will have a non-zero additional component $\mathbf{U}_\perp^{[1]}$.

9.3 Rays in isotropic media

For an isotropic medium the wavefront equation (9.2.11) has the specific form

$$(\lambda + \mu)(\mathbf{U}^{[0]} \cdot \nabla\vartheta) + [\mu(\nabla\vartheta)^2 - \rho]\mathbf{U}^{[0]} = 0, \tag{9.3.1}$$

taking the scalar and vector products of (9.3.1) with the gradient of the wavefront function, $\nabla\vartheta$, we obtain

$$\left[(\lambda + 2\mu)(\nabla\vartheta)^2 - \rho\right]\left(\mathbf{U}^{[0]} \cdot \nabla\vartheta\right) = 0, \tag{9.3.2}$$

$$\left[\mu(\nabla\vartheta)^2 - \rho\right]\left(\mathbf{U}^{[0]} \wedge \nabla\vartheta\right) = 0. \tag{9.3.3}$$

Thus as we would expect we have two possible solutions
(a) for P waves:

$$(\nabla\vartheta)^2 = \frac{\rho}{\lambda + 2\mu}, \quad \text{with} \quad \mathbf{U}^{[0]} \wedge \nabla\vartheta = 0, \tag{9.3.4}$$

(b) for S waves

$$(\nabla\vartheta)^2 = \frac{\rho}{\mu}, \quad \text{with} \quad \mathbf{U}^{[0]} \cdot \nabla\vartheta = 0. \tag{9.3.5}$$

These represent the *eikonal* equations for compressional and shear waves. As in section 8.2.1 we introduce the P and S wavespeeds α, β and associated slownesses a,b

$$\alpha = \frac{1}{a} = \left[\frac{\lambda + 2\mu}{\rho}\right]^{1/2}, \qquad \beta = \frac{1}{b} = \left[\frac{\mu}{\rho}\right]^{1/2}. \tag{9.3.6}$$

The two independent solutions of (9.3.1),

$$(\nabla\vartheta)^2 = a^2, \qquad (\nabla\vartheta)^2 = b^2, \tag{9.3.7}$$

define two distinct sets of wavefronts and corresponding rays specified by the direction of $\nabla\vartheta$, since for isotropic propagation the phase and group velocity vectors coincide.

For inhomogeneous media the elastodynamic equation cannot generally be separated into two wave equations characterising the propagation of equivoluminal (S) and irrotational (P) waves because of coupling produced by gradients in material properties. However, for high frequency disturbances, there exist two independent wavefronts provided λ, μ, ρ and their derivatives are continuous. The faster wavefront propagates with the local P wavespeed (α) and the other propagates with the local S wavespeed (β); between these two wavefronts the disturbance is normally not irrotational.

For the isotropic case, the normals to the wavefront and slowness surfaces coincide. The ray trajectories thus comprise lines which are everywhere orthogonal to ϑ = const and so lie along $\nabla\vartheta$. In terms of the distance s along the raypath, the P rays satisfy

$$\frac{\mathrm{d}\mathbf{x}(s)}{\mathrm{d}s} = \alpha\nabla\vartheta, \qquad \mathrm{d}s^2 = (\mathrm{d}\mathbf{x})^2, \tag{9.3.8}$$

and also

$$\frac{\mathrm{d}\vartheta}{\mathrm{d}s} = \nabla\vartheta\cdot\frac{\mathrm{d}\mathbf{x}(s)}{\mathrm{d}s} = a = \frac{1}{\alpha},$$

so that the wavefront function

$$\vartheta(\mathbf{x}) = \vartheta_0 + \int_{\mathbf{x}_0}^{\mathbf{x}} \mathrm{d}s\, a, \tag{9.3.9}$$

where the integration is taken along the ray. Further,

$$\frac{\mathrm{d}}{\mathrm{d}s}(\nabla\vartheta) = \nabla a,$$

and so from (9.3.8) we obtain the vector differential equation determining the raypath

$$\frac{\mathrm{d}}{\mathrm{d}s}\left(a\frac{\mathrm{d}\vartheta}{\mathrm{d}s}\right) = \nabla a. \tag{9.3.10}$$

An alternative form as a set of first order equations is more commonly used for numerical solution (see Červený, Pšenčík & Molotkov, 1977). In terms of the slowness vector $\mathbf{p} = \nabla\vartheta$ and the time θ along the ray

$$\frac{\partial x_i}{\partial\theta} = \frac{p_i}{\alpha^2}, \qquad \frac{\partial p_i}{\partial\theta} = -\frac{1}{\alpha}\frac{\partial\alpha}{\partial x_i}. \tag{9.3.11}$$

The amplitude of the leading term in the ray series is given by (9.2.41)

$$U^{[0]}(\theta) = U^{[0]}(\theta_0)\left[\frac{\rho(\theta_0)\alpha(\theta_0)J(\theta_0)}{\rho(\theta)\alpha(\theta)J(\theta)}\right]^{1/2}. \tag{9.3.12}$$

connecting two points on the same ray. For this leading term the particle velocity along the ray

$$\partial_t u = U^{[0]} f_0'(t - \vartheta(\mathbf{x})),$$

and the corresponding stress

$$\tau_{ss} = (\lambda + 2\mu)\left[-\frac{U^{[0]}}{\alpha}f_0'(t - \vartheta(\mathbf{x}))\right] + \text{terms in } f_0.$$

Thus the energy flux in the ray tube $(q_1, q + \mathrm{d}q_1)$, $(q_2, q + \mathrm{d}q_2)$, is given by

$$J\mathrm{d}q_1\mathrm{d}q_2\rho\alpha(U^{[0]})^2[f_0'(t - \vartheta(\mathbf{x}))^2], \tag{9.3.13}$$

and the transport equation (9.3.12) may be interpreted as the conservation of energy along a ray tube.

For S waves, the displacement in the leading term of the ray series $\mathbf{U}^{[0]}$ is directed perpendicular to the ray along which propagation occurs with wavespeed β, slowness b. The ray equations and transport equations are entirely analogous to (9.3.10), (9.3.11) and (9.3.12) with the change of wavespeed.

For these shear waves

$$\mathcal{N}(\mathbf{U}^{[r]}) = (\lambda + \mu)(\mathbf{U}^{[r]} \cdot \nabla\vartheta)\nabla\vartheta, \tag{9.3.14}$$

and thus $\mathcal{N}(\mathbf{U}^{[1]})$ is directed along the ray. From the transport equation (9.2.33)

$$\mathcal{M}(\mathbf{U}^{[0]}) + \mathcal{N}(\mathbf{U}^{[1]}) = 0, \tag{9.3.15}$$

so that the component of $\mathcal{M}(\mathbf{U}^{[0]})$ transverse to the ray must vanish. In explicit terms

$$-\mu\left[\mathbf{U}^{[0]}\nabla^2\vartheta + \frac{2}{\beta^2}\left(\frac{\partial\mathbf{U}^{[0]}}{\partial\vartheta}\right)_\perp\right] - (\nabla\mu \cdot \nabla\vartheta)\mathbf{U}^{[0]} = 0, \tag{9.3.16}$$

and thus $\mathbf{U}^{[0]}$, $(\partial\mathbf{U}^{[0]}/\partial\vartheta)_\perp$ are collinear.

In order to examine the polarisation of $\mathbf{U}^{[0]}$ we need to use a coordinate system connected to the ray. We use the ray direction $\mathbf{s}$, the normal to the raypath $\mathbf{n}$ and the binormal $\boldsymbol{\nu} = \mathbf{s} \wedge \mathbf{n}$. Since $\mathbf{U}^{[0]}$ lies perpendicular to $\mathbf{s}$ we can write

$$\mathbf{U}^{[0]} = U_n\mathbf{n} + U_\nu\boldsymbol{\nu}, \tag{9.3.17}$$

and then

$$\frac{\partial\mathbf{U}^{[0]}}{\partial\theta} = \beta\frac{\partial\mathbf{U}^{[0]}}{\partial s} = \frac{\partial U_n}{\partial\theta}\mathbf{n} + \frac{\partial U_\nu}{\partial\theta}\boldsymbol{\nu} + \beta U_n\frac{\partial\mathbf{n}}{\partial s} + \beta U_\nu\frac{\partial\boldsymbol{\nu}}{\partial s}. \tag{9.3.18}$$

From the standard Frénet formulae

$$\frac{\partial\mathbf{n}}{\partial s} = T\boldsymbol{\nu} - K\mathbf{s}, \qquad \frac{\partial\mathbf{n}}{\partial s} = -T\mathbf{n}, \tag{9.3.19}$$

where K is the local radius of curvature and T the radius of torsion. Thus

$$\frac{\partial\mathbf{U}^{[0]}}{\partial\theta} = \left(\frac{\partial U_n}{\partial\theta} - \beta U_\nu T\right)\mathbf{n} + \left(\frac{\partial U_\nu}{\partial\theta} + \beta U_n T\right)\boldsymbol{\nu} - K\beta U_n\mathbf{s}. \tag{9.3.20}$$

Substituting (9.3.19) into (9.3.15) and separating the $\mathbf{n}$ and $\boldsymbol{\nu}$ components, we have

$$\begin{aligned} 2\frac{\partial U_n}{\partial\vartheta} - 2T\beta U_\nu + \left(\beta^2\nabla^2 + \frac{1}{\mu}\frac{\partial\mu}{\partial\vartheta}\right)U_n = 0,\\ 2\frac{\partial U_\nu}{\partial\vartheta} + 2T\beta U_n + \left(\beta^2\nabla^2 + \frac{1}{\mu}\frac{\partial\mu}{\partial\vartheta}\right)U_\nu = 0. \end{aligned} \tag{9.3.21}$$

By eliminating the term in brackets from (9.3.20) we obtain

$$U_n^2 \frac{\partial}{\partial \vartheta}\left(\frac{U_\nu}{U_n}\right) + T\beta (U^{[0]})^2 = 0, \tag{9.3.22}$$

where $(U^{[0]})^2 = U_n^2 + U_\nu^2$. When the angle between $\mathbf{U}^{[0]}$ and $\mathbf{n}$ is φ,

$$U_n = U^{[0]} \cos \varphi, \qquad U_\nu = U^{[0]} \sin \varphi,$$

and then

$$\frac{\partial \varphi}{\partial \theta} = -T\beta, \qquad \frac{\partial \varphi}{\partial s} = -T. \tag{9.3.23}$$

But, from the Frenét formulae (9.3.19)

$$\frac{\partial \boldsymbol{\nu}}{\partial s} = -T\mathbf{n},$$

so the system $(\mathbf{n}, \boldsymbol{\nu})$ rotates about the ray with angular velocity T, and the vector $\mathbf{U}^{[0]}$ in this system rotates with the same velocity but in the *opposite* direction. As a result the vector $\mathbf{U}^{[0]}$ does not turn about the ray, and in smoothly varying media the polarisation of the shear wave is preserved.

9.4 Boundary conditions for rays

The representation of a solution of the seismic wave equations by a ray series is only possible when the elastic moduli $c_{ijkl}(\mathbf{x})$ and the phase function $\vartheta(\mathbf{x})$ are analytic functions of $\mathbf{x}$.

In deriving the expressions for the $U^{[r]}$ we perform successive differentations, so $U^{[r]}$ depends on the rth derivative of the elastic constants. Thus, if a ray crosses a surface on which the rth derivative is discontinuous, we have to introduce 'interface' conditions at the surface to determine the solution.

In particular, if the elastic constants are themselves discontinuous on a surface Σ we require boundary conditions appropriate to a junction of elastic media. We will consider isotropic media and assume welded contact across the interface so that

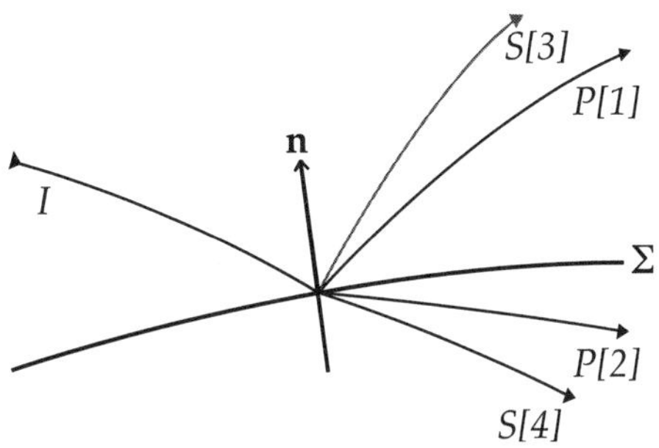

Figure 9.5. Ray configuration at a material interface.

displacement and traction are continuous. We define a local normal to the surface $\mathbf{n}$ directed into the medium from which the ray is incident.

We suppose that the incident wave can be represented in the form of a ray series. This incident wave will generate reflected and transmitted P and S waves, for each of which we will take a ray series representation:

$$\mathbf{u}_l(\mathbf{x},t) = \sum_{r=0}^{\infty} \mathbf{U}_l^{[r]}(\mathbf{x}) f_r(t - \vartheta_l(\mathbf{x})), \tag{9.4.1}$$

where

$l = 0$ – incident wave,
$l = 1, 2$ – reflected, refracted P,
$l = 3, 4$ – reflected, refracted S,

We require the phase functions for the reflected and refracted waves to coincide with the phase function for the incident wave ϑ_0 on the surface Σ,

$$\vartheta_0(\mathbf{x}) = \vartheta_l(\mathbf{x}) \quad \text{on } \Sigma. \tag{9.4.2}$$

Each phase function satisfies an eikonal equation of the form

$$(\nabla\vartheta_l)^2 = 1/v_l^2 = \mathbf{p}_l \cdot \mathbf{p}_l, \tag{9.4.3}$$

in terms of the slowness vector $\mathbf{p}_l = \nabla\vartheta_l$. The continuity conditions at the discontinuity are that

$$\mathbf{u}_0 + \mathbf{u}_1 + \mathbf{u}_3 = \mathbf{u}_2 + \mathbf{u}_4, \tag{9.4.4}$$

$$\tau_{ij}(\mathbf{u}_0 + \mathbf{u}_1 + \mathbf{u}_3)n_j = \tau_{ij}(\mathbf{u}_2 + \mathbf{u}_4)n_j. \tag{9.4.5}$$

From 9.4.3 we require that the derivatives lying in the surface Σ must be the same for ϑ_0 and each of the reflected and refracted waves, thus

$$\mathbf{n} \wedge \mathbf{p}_l = \mathbf{n} \wedge \mathbf{p}_0, \quad \text{on } \Sigma \text{ for } l = 1, 2, 3, 4. \tag{9.4.6}$$

Thus the incident wave, reflected and refracted waves and the normal to the surface have to be locally coplanar and

$$\frac{\sin i_0}{v_0} = \frac{\sin i_l}{v_l}, \tag{9.4.7}$$

which is simply a local Snell's law relation. If $v_l \sin i_0 / v_0 > 1$, no real i_l exists and we get a complex $\nabla\vartheta_l$ and hence an evanescent wave.

We now equate the coefficients of f_r at the interface in the equations (9.4.5). The displacement equation gives

$$\sum_{l=1}^{4} (-1)^l U_l^{[r]} = U_0^{[r]}. \tag{9.4.8}$$

The traction for an individual wave may be expressed in the form

$$\mathbf{t}_n(U^{[r]} f_r) = \mathbf{H}(\mathbf{U}^{[r]}) f_r - \mathbf{G}(\mathbf{U}^{[r]}) f_{r-1}, \tag{9.4.9}$$

where

$$\mathbf{H}(\mathbf{U}^{[r]}) = \lambda(\nabla\cdot\mathbf{U}^{[r]})\mathbf{n} + 2\mu\left(\frac{\partial\mathbf{U}^{[r]}}{\partial n} + \tfrac{1}{2}\mathbf{n}\wedge(\nabla\wedge\mathbf{U}^{[r]}\right), \tag{9.4.10}$$

$$\mathbf{G}(\mathbf{U}^{[r]}) = \lambda(\mathbf{U}^{[r]}\cdot\nabla\vartheta)\mathbf{n} + \mu\left(\mathbf{n}\cdot\nabla\vartheta)\mathbf{U}^{[r]} + \mu(\mathbf{n}\cdot\mathbf{U}^{[r]}\right)\nabla\vartheta. \tag{9.4.11}$$

Thus, equating the traction terms at the interface, we have

$$\sum_{l=1}^{4}(-1)^l\mathbf{G}(\mathbf{U}_l^{[r]}) = \mathbf{G}(\mathbf{U}_0^{[r]}) + \sum_{l=1}^{4}(-1)^l\mathbf{H}(\mathbf{U}_l^{[r-1]}) - \mathbf{H}(\mathbf{U}_0^{[r-1]}) \tag{9.4.12}$$

For the leading terms there is a considerable simplification

$$\sum_{l=1}^{4}(-1)^l\mathbf{U}_l^{[0]} = \mathbf{U}_0^{[0]}, \qquad \sum_{l=1}^{4}(-1)^l\mathbf{G}(\mathbf{U}_l^{[0]}) = \mathbf{G}(\mathbf{U}_0^{[0]}). \tag{9.4.13}$$

For P waves

$$\mathbf{G}(\mathbf{U}^{[0]}) = \frac{1}{\alpha}U^{[0]}[\lambda\mathbf{n} + 2\mu(\mathbf{n}\cdot\mathbf{s})\mathbf{s}],$$

where $\mathbf{s}$ is the unit vector along the ray $\nabla\vartheta$. For S waves

$$\mathbf{G}(\mathbf{U}^{[0]}) = \frac{1}{\beta}U^{[0]}[(\mathbf{n}\cdot\boldsymbol{\nu})\mathbf{s} + (\mathbf{s}\cdot\mathbf{n})\boldsymbol{\nu}],$$

and $\boldsymbol{\nu}$ is the unit vector perpendicular to the ray in the common plane. The relations specified by (9.4.13) correspond to those for a plane harmonic wave on a plane boundary (see section 13.2), and thus the leading coefficients can be found by considering this problem.

For $r > 0$ we have both principal and additional components so that writing

$$\mathbf{U}^{[r]} = {}_{\rm p}\mathbf{U}^{[r]} + {}_{\rm a}\mathbf{U}^{[r]}, \tag{9.4.14}$$

we obtain

$$\begin{aligned}\sum_{l=1}^{4}(-1)^l{}_{\rm p}\mathbf{U}_l^{[r]} &= \mathbf{U}_0^{[0]} - \sum_{l=1}^{4}(-1)^l{}_{\rm a}\mathbf{U}_l^{[r]},\\ \sum_{l=1}^{4}(-1)^l\mathbf{G}({}_{\rm p}\mathbf{U}_l^{[r]}) &= \mathbf{G}(\mathbf{U}_0^{[r]}) - \mathbf{H}(\mathbf{U}_0^{[r-1]})\\ &\quad - \sum_{l=1}^{4}(-1)^l[\mathbf{G}({}_{\rm a}\mathbf{U}_l^{[r]}) - \mathbf{H}({}_{\rm a}\mathbf{U}_l^{[r-1]})].\end{aligned} \tag{9.4.15}$$

There are thus six equations in six unknown components yielding a set of 'generalised' coefficients. For $r = 0$, the SH waves separate, but for higher values of r all the P and S waves are coupled through the previous $\mathbf{U}^{[r-1]}$. To determine the values of the $\mathbf{U}_l^{[r]}$ it is not sufficient to solve the system for index $r-1$, since the operator $\mathbf{H}$ requires derivatives of $\mathbf{U}_l^{[r-1]}$ normal to the surface Σ but the equations only give $\mathbf{U}_l^{[r-1]}$ on Σ. Thus after solving (9.4.15) we need to construct the vectors $\mathbf{U}_l^{[r-1]}$ in the vicinity of the boundary and these will be affected by the shape of the boundary and of the wavefront.

Appendix: Behaviour near a wavefront

Each displacement component in the vicinity of a wavefront at $t = \vartheta(\mathbf{x})$ can be represented in the form

$$u(\mathbf{x}, t) = G(\mathbf{x}, t - \vartheta(\mathbf{x})) + H(\mathbf{x}, t), \tag{9.a.1}$$

where $H(\mathbf{x}, t)$ is analytic in the neighbourhood of the wavefront and $G(\mathbf{x}, t-\vartheta(\mathbf{x}))$ is singular at $t = \vartheta(\mathbf{x})$ and may be written in the form,

$$G(\mathbf{x}, t') = g_0(\mathbf{x}, t') + \int_{-\infty}^{t'} \mathrm{d}s\, g_0(\mathbf{x}, s) h(\mathbf{x}, t'-s), \tag{9.a.2}$$

where h is analytic and $g_0 \to 0$ as $t \to \infty$. The leading term contains the singularities since integration is a smoothing process.

Since $h(\mathbf{x}, t')$ is analytic, we expand about the wavefront $t' = 0$ to obtain

$$h(\mathbf{x}, s) = a_1(\mathbf{x}) + a_2(\mathbf{x})s + a_3(\mathbf{x})s^2 + \ldots + a_{n-1}s^{n-2} + R_n(\mathbf{x}, s), \tag{9.a.3}$$

and so substituting back into (9.a.2) we obtain

$$G(\mathbf{x}, t') = \sum_{r=0}^{n-1} a_r g_r(\mathbf{x}, t') + \int_{-\infty}^{t'} \mathrm{d}s\, R_n(\mathbf{x}, t'-s) g_0(\mathbf{x}, s), \tag{9.a.4}$$

with

$$a_0(\mathbf{x}) = 1; \qquad g_r(\mathbf{x}, t') = \int_{-\infty}^{t'} \mathrm{d}s\, g_{r-1}(\mathbf{x}, s).$$

If in addition we assume that the major singularity has the same character all along the wavefront so that

$$g_0(\mathbf{x}, t') = A_0(\mathbf{x}) f_0(t), \tag{9.a.5}$$

then we can simplify the representation of the singular term

$$G(\mathbf{x}, t') = \sum_{r=0}^{n-1} A_r(\mathbf{x}) f_r(t') + \int_{-\infty}^{t'} \mathrm{d}s\, R_n(\mathbf{x}, t'-s) A_0(\mathbf{x}) f_0(s), \tag{9.a.6}$$

with

$$A_r(\mathbf{x}) = A_0 a_r; \qquad f_r'(t) = f_{r-1}(t).$$

Provided that the wavefront function $\vartheta(\mathbf{x})$ is analytic, this *ray series* will converge in the neighbourhood of the wavefront (Babich, 1961). We can therefore express the displacement associated with the wavefront as

$$u(\mathbf{x}, t) = \sum_{r=0}^{\infty} U^{[r]}(\mathbf{x}) f_r(t - \vartheta(\mathbf{x})), \tag{9.a.7}$$

an expansion in terms of 'generalised progressive waves' with $f_r' = f_{r-1}$.

Consider the Fourier transform of $G(\mathbf{x}, t - \vartheta(\mathbf{x}))$

$$\begin{aligned}\bar{G}(\mathbf{x}, \omega) &= \int_{-\infty}^{\infty} \mathrm{d}t\, G(\mathbf{x}, t - \vartheta) \mathrm{e}^{\mathrm{i}\omega t} = \mathrm{e}^{\mathrm{i}\omega\vartheta} \int_{-\infty}^{\infty} \mathrm{d}s\, G(\mathbf{x}, s) \mathrm{e}^{\mathrm{i}\omega s}, \\ &= \mathrm{e}^{\mathrm{i}\omega\vartheta} \bar{g}_0(\mathbf{x}, \omega)[1 + \bar{h}(\mathbf{x}, \omega)].\end{aligned} \tag{9.a.8}$$

The function $h(\mathbf{x}, t)$ is only defined for $t > 0$ and is analytic there, we set $h(\mathbf{x}, t) = 0$ for $t < 0$, and then $\bar{h}(\mathbf{x}, \omega)$ exists for $\mathrm{Im}\omega > 0$. From (9.a.3) we have the asymptotic representation for large $\mathrm{i}\omega$:

$$\bar{h}(\mathbf{x}, \omega) \sim \frac{a_1(\mathbf{x})}{(i\omega)^1} + \frac{a_2(\mathbf{x})}{(i\omega)^2} + \dots . \tag{9.a.9}$$

Then making the assumption that the major singularity has the same character along the wavefront, as in (9.a.5), we find

$$\bar{G}(\mathbf{x}, \omega) \sim e^{i\omega\vartheta} A_0(\mathbf{x}) \bar{f}_0(\omega) \sum_{r=0}^{\infty} \frac{A_r(\mathbf{x})}{(i\omega)^r}. \tag{9.a.10}$$

On Fourier transformation back to time we obtain the ray series form as an asymptotic representation (9.1.6).

10

Rays in Stratification

Although there is convincing evidence of pervasive three-dimensional heterogeneity within the Earth, the dominant pattern of variation is with depth (radius). We can therefore achieve a good understanding of the main characteristics of seismic wave propagation by looking at models which vary only with depth and then later include the influence of heterogeneity.

The ray-tracing equations introduced in Section 9.3 for isotropic media take particular simple forms when the wavespeed distribution depends only on one coordinate. In this chapter we will consider two important special cases; firstly, horizontal stratification where the wavespeed depends only on the depth coordinate (z), and secondly spherical stratification with a wavespeed distribution $v(r)$ in terms of radius r.

10.1 Rays in horizontal stratification

10.1.1 Vertical wavespeed variation

We will generate forms appropriate to a P wave but equivalent results hold for S waves with simply a replacement of the wavespeed and slowness distributions (i.e., replace α, a by β, b)

If the wavespeed (and slowness) distribution is only a function of z, $a = a(z)$ and the ray-tracing equations (9.3.10) simplify considerably since

$$\frac{\mathrm{d}}{\mathrm{d}s}\left(a\frac{\mathrm{d}y}{\mathrm{d}s}\right) = \frac{\mathrm{d}}{\mathrm{d}s}\left(a\frac{\mathrm{d}x}{\mathrm{d}s}\right) = 0. \tag{10.1.1}$$

Hence, since $a\mathrm{d}y/\mathrm{d}s = \text{const}$, if $\mathrm{d}y/\mathrm{d}s = 0$ initially it will remain zero so that $\mathrm{d}y/\mathrm{d}s = 0$ all along the ray. A ray that starts in the (x, z) plane will remain in that plane.

For such a ray we define the inclination to the vertical via the angle $i(z)$, (see figure 10.1), such that

$$\sin i(z) = \mathrm{d}x/\mathrm{d}s, \tag{10.1.2}$$

and then the continuous analogue of Snell's law is that

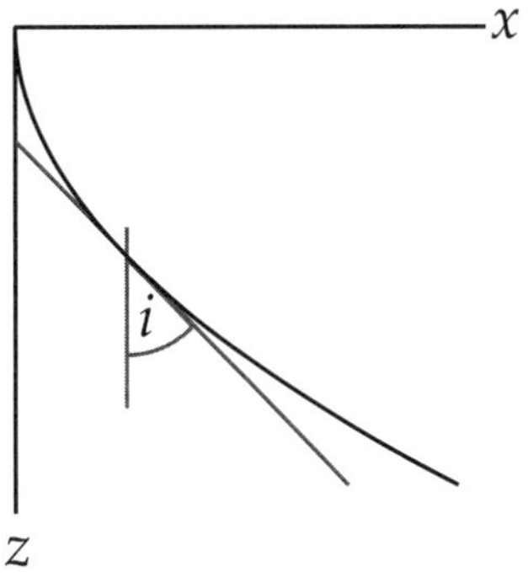

Figure 10.1. Configuration of a ray in the (x, z) plane.

$$\frac{\sin i(z)}{\alpha(z)} = a(z)\sin i(z) = p, \tag{10.1.3}$$

where p is a constant along the ray, known as the *ray parameter* for the ray.

10.1.2 Travel times

For a medium where the wavespeed v for a particular wavetype is a piecewise continuous function of depth z alone, the ray parameter p is constant along a ray. In a small interval $\mathrm{d}s$ along a ray,

$$\mathrm{d}x = \mathrm{d}s \sin i, \quad \mathrm{d}z = \mathrm{d}s \cos i, \quad \mathrm{d}x = \mathrm{d}z \tan i, \quad \mathrm{d}t = \mathrm{d}s/v. \tag{10.1.4}$$

Thus, over a depth interval (z_a, z_b), the increment in horizontal distance

$$\Delta x = \left| \int_{z_a}^{z_b} \mathrm{d}z \tan i(z) \right| = \left| \int_{z_a}^{z_b} \mathrm{d}z\, p \left[\frac{1}{v^2} - p^2 \right]^{-1/2} \right|, \tag{10.1.5}$$

and the increment in travel time

$$\Delta t = \left| \int_{z_a}^{z_b} \mathrm{d}z \frac{1}{v^2} \left[\frac{1}{v^2} - p^2 \right]^{-1/2} \right|. \tag{10.1.6}$$

The vertical component of the slowness vector

$$q(z) = \frac{\cos i}{v} = \left[\frac{1}{v^2} - p^2 \right]^{1/2}, \tag{10.1.7}$$

and as we shall see later this vertical slowness plays an important role in the description of wave behaviour in stratified media.

The normal tendency is for the seismic wavespeeds to increase with depth and so the inclination of the ray $i(z) = \sin^{-1}(pv)$ also increases. At a depth Z_p such that $p = 1/v(Z_p)$ the inclination i will reach 90°, this will produce a turning point for the ray which will then head back towards the surface. For the direct ray from a surface source the total distance and travel time will be

$$X(p) = 2\int_0^{Z_p} \mathrm{d}z\, p\left[\frac{1}{v^2} - p^2\right]^{-1/2} = 2\int_0^{Z_p} \mathrm{d}z\,\frac{p}{q(z)}, \tag{10.1.8}$$

$$T(p) = 2\int_0^{Z_p} \mathrm{d}z\,\frac{1}{v^2}\left[\frac{1}{v^2} - p^2\right]^{-1/2} = 2\int_0^{Z_p} \mathrm{d}z\,\frac{1}{v^2 q(z)}, \tag{10.1.9}$$

with an implied summation over the zones with different functional dependence on z. More complicated ray paths can be built up by summation of appropriate segments.

A useful auxiliary parameter is $\tau(p)$, the integral of the vertical slowness,

$$\tau(p) = T(p) - pX(p) = 2\int_0^{Z_p} \mathrm{d}z\left[\frac{1}{v^2} - p^2\right]^{1/2} = 2\int_0^{Z_p} \mathrm{d}z\, q(z). \tag{10.1.10}$$

The function $\tau(p)$ has an alternative interpretation as the intercept on the time axis of the tangent to the travel time curve at $X(p)$, $T(p)$ with slope

$$p = \mathrm{d}T/\mathrm{d}X, \tag{10.1.11}$$

a relation which can be derived from (10.1.10), or via the geometry of successive wavefronts (cf. the discussion on spherical stratification in Section 10.2.2).

10.2 Rays in spherical stratification

10.2.1 Radial wavespeed variation

When the wavespeed (and slowness) depend solely on the radial distance r from a fixed point O, $a = a(r)$ and then

$$\frac{\mathrm{d}}{\mathrm{d}s}(a\hat{\mathbf{t}}) = \nabla a = \frac{\mathrm{d}a}{\mathrm{d}r}\hat{\mathbf{e}}_r, \tag{10.2.1}$$

where $\hat{\mathbf{t}}$ is the unit tangent along ray, and $\hat{\mathbf{e}}_r$ the unit vector in the radial direction (see figure 10.2).

From (10.2.1)

$$\frac{\mathrm{d}\hat{\mathbf{t}}}{\mathrm{d}s} = \frac{1}{a}\frac{\mathrm{d}a}{\mathrm{d}r}\hat{\mathbf{e}}_r - \frac{1}{a}\frac{\mathrm{d}a}{\mathrm{d}s}\hat{\mathbf{t}}, \tag{10.2.2}$$

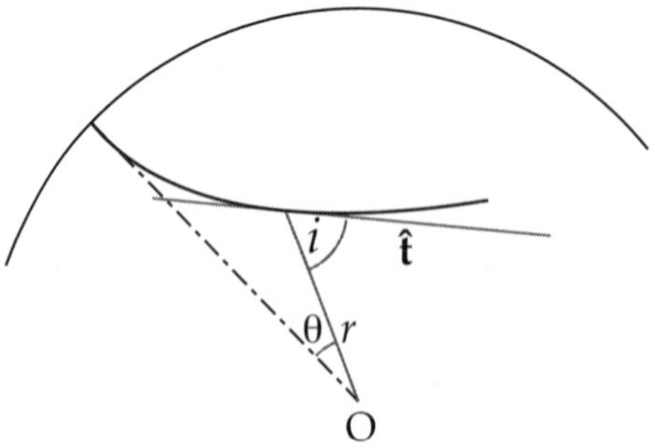

Figure 10.2. Configuration of a ray in a radial velocity variation.

and so the ray again lies in a plane and can be described in terms of (r, θ) coordinates. The tangent to the ray path can be written in terms of these coordinates as

$$\hat{\mathbf{t}} = -\hat{\mathbf{e}}_r \cos i + \hat{\mathbf{e}}_\theta \sin i,$$

so that

$$\frac{\mathrm{d}}{\mathrm{d}s}(a\hat{\mathbf{t}}) = \left[\frac{\mathrm{d}}{\mathrm{d}s}(a \sin i) - a \cos i \frac{\mathrm{d}\theta}{\mathrm{d}s}\right] \hat{\mathbf{e}}_\theta - \left[\frac{\mathrm{d}}{\mathrm{d}s}(a \cos i) + a \sin i \frac{\mathrm{d}\theta}{\mathrm{d}s}\right] \hat{\mathbf{e}}_r,$$

with $r\mathrm{d}\theta/\mathrm{d}s = \sin i$, $\mathrm{d}r/\mathrm{d}s = -\cos i$. However, from the original equation (9.4.5) there can be no θ-component, so

$$0 = \frac{\mathrm{d}}{\mathrm{d}s}(a \sin i) - \frac{a}{r} \cos i \sin i = \frac{\mathrm{d}}{\mathrm{d}s}(a \sin i) + \frac{a}{r} \sin i \frac{\mathrm{d}r}{\mathrm{d}s}.$$

We therefore require

$$\frac{\mathrm{d}}{\mathrm{d}s}[a(r) r \sin i(r)] = 0, \tag{10.2.3}$$

which we can alternatively express as

$$a(r) r \sin i(r) = \frac{r}{\alpha(r)} \sin i(r) = \wp, \tag{10.2.4}$$

where, again, $\wp$ is a constant along the ray. The invariance of the spherical ray parameter $\wp$ is again a manifestation of Snell's law.

10.2.2 Travel times

Consider a medium in which the seismic wavespeeds α, β are piecewise continuous functions of the radius r. For any ray in such a medium the ray parameter $\wp$ will be constant along the ray (10.2.4). For a ray associated with wavespeed v it is useful to introduce the auxiliary variable $\eta(r) = r/v(r)$ and then (10.2.4) takes the form

$$\wp = \frac{r}{v(r)} \sin i(r) = \eta(r) \sin i(r). \tag{10.2.5}$$

In a small interval $\mathrm{d}s$ along the ray

$$\mathrm{d}r = \mathrm{d}s \cos i, \qquad r\mathrm{d}\theta = \mathrm{d}s \sin i = \wp/\eta. \tag{10.2.6}$$

Over a zone between radii r_a and r_b the angular increment along the ray path is

$$\Delta\theta = \int_{r_a}^{r_b} \mathrm{d}r \, \frac{\wp}{r[\eta^2 - \wp^2]^{1/2}}, \tag{10.2.7}$$

The travel time increment is the integral $\int \mathrm{d}s/v$ along this segment,

$$\Delta T = \left| \int_{r_a}^{r_b} \mathrm{d}r \, \frac{\tan i}{v} \right| = \left| \int_{r_a}^{r_b} \mathrm{d}r \, \frac{\eta^2}{r[\eta^2 - \wp^2]^{1/2}} \right|. \tag{10.2.8}$$

As the ray descends into the Earth the seismic wavespeeds increase in value and so the inclination i becomes larger for fixed $\wp$. When the inclination i reaches 90° the ray will reach its lowest point at which

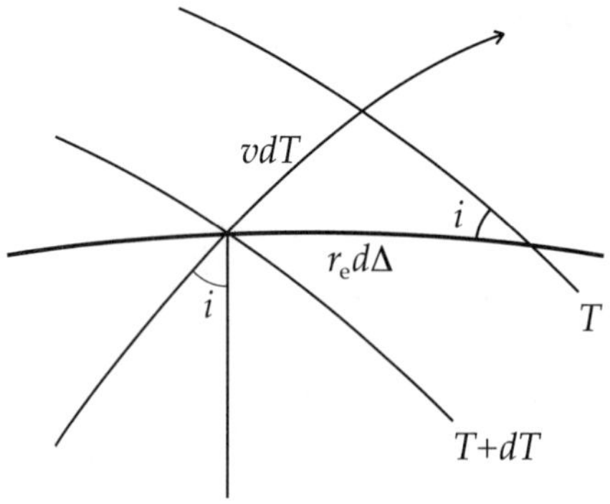

Figure 10.3. The passage of wavefronts across the surface.

$$\wp = \eta_p = r_p / v_p, \tag{10.2.9}$$

subsequently it will switch direction and climb upwards towards the surface.

When both the source and the receiver lie at the Earth's surface r_e the total angular distance (epicentral distance)

$$\Delta(\wp) = 2 \int_{r_p}^{r_e} \mathrm{d}r \, \frac{\wp}{r[\eta^2 - \wp^2]^{1/2}}, \tag{10.2.10}$$

and the total travel-time

$$T(\wp) = 2 \int_{r_p}^{r_e} \mathrm{d}r \, \frac{\eta^2}{r[\eta^2 - \wp^2]^{1/2}}. \tag{10.2.11}$$

It is convenient to introduce the quantity $\tau(\wp)$

$$\tau(\wp) = T - \wp\Delta = \int_{r_p}^{r_e} \mathrm{d}r \, \frac{1}{r}[\eta^2 - \wp^2]^{1/2}. \tag{10.2.12}$$

From (10.2.12) we can recognise that the slope of the travel time curve $T(\Delta)$ is the ray parameter $\wp$

$$\frac{\mathrm{d}T}{\mathrm{d}\Delta} = \wp. \tag{10.2.13}$$

The function $\tau(\wp)$ has therefore a simple geometrical interpretation: the tangent to the travel time curve at the point with coordinates $\Delta(\wp)$, $T(\wp)$ is characterised by slope $\wp$ and a time intercept $\tau(\wp)$.

The relation (10.2.13) can also be seen from the geometry of wavefronts (figure 10.3). Consider the passage of wavefronts over the Earth's surface

$$\sin i = v \frac{\delta T}{r_e \delta\Delta}, \tag{10.2.14}$$

and so using (10.2.5)

$$\frac{\mathrm{d}T}{\mathrm{d}\Delta} = \wp. \tag{10.2.15}$$

The auxiliary variable $\tau(\wp)$ can also be regarded as the integral of the radial waveslowness,

$$\bar{q}(r,\wp) = \left(\frac{1}{v^2} - \frac{\wp^2}{r^2}\right)^{1/2}, \tag{10.2.16}$$

over radius

$$\tau(\wp) = \int_{r_p}^{r_e} \mathrm{d}r\, \frac{1}{r}[\eta^2 - \wp^2]^{1/2} = \int_{r_p}^{r_e} \mathrm{d}r\, \bar{q}(r,\wp). \tag{10.2.17}$$

In terms of this radial slowness we can recast the expressions for the epicentral distance and travel time as

$$\Delta(\wp) = 2\int_{r_p}^{r_e} \mathrm{d}r\, \frac{\wp}{r^2\bar{q}(r,\wp)}, \quad T(\wp) = 2\int_{r_p}^{r_e} \mathrm{d}r\, \frac{1}{v^2\bar{q}(r,\wp)}. \tag{10.2.18}$$

When the source lies at a depth z_s (radius $r_s = r_e - z_s$) and the receiver at a depth z_r (radius $r_r = r_e - z_r$), the symmetric forms of (10.2.10), (10.2.11) about the turning point need to be replaced by

$$\Delta(\wp) = \int_{r_p}^{r_s} \mathrm{d}r\, \frac{\wp}{r[\eta^2 - \wp^2]^{1/2}} + \int_{r_p}^{r_r} \mathrm{d}r\, \frac{\wp}{r[\eta^2 - \wp^2]^{1/2}}, \tag{10.2.19}$$

$$T(\wp) = \int_{r_p}^{r_s} \mathrm{d}r\, \frac{\eta^2}{r[\eta^2 - \wp^2]^{1/2}} + \int_{r_p}^{r_r} \mathrm{d}r\, \frac{\eta^2}{r[\eta^2 - \wp^2]^{1/2}}. \tag{10.2.20}$$

The expression for the travel time for a surface source and receiver can be recast as

$$T(\wp) = 2\int_{r_p}^{r_e} \mathrm{d}r\, \frac{\eta^2}{r[\eta^2 - \wp^2]^{1/2}} = 2\int_{r_p}^{r_e} \mathrm{d}r \left(\frac{\eta}{r\mathrm{d}_r\eta}\right) \frac{\mathrm{d}}{\mathrm{d}r}[\eta^2 - \wp^2]^{1/2}, \tag{10.2.21}$$

where for brevity we have written $\mathrm{d}_r\eta$ for $\mathrm{d}\eta/\mathrm{d}r$. Integrating by parts we obtain

$$T(\wp) = \xi_e[\eta_e^2 - \wp^2]^{1/2} - \int_{r_p}^{r_e} \mathrm{d}r\, \frac{\mathrm{d}\xi}{\mathrm{d}r}[\eta^2 - \wp^2]^{1/2}. \tag{10.2.22}$$

where we have followed Bullen (1963) by introducing

$$\zeta = \frac{\mathrm{d}\ln v}{\mathrm{d}\ln r} = \frac{r}{v}\frac{\mathrm{d}v}{\mathrm{d}r}, \qquad \xi = \frac{2}{1-\zeta} = 2\frac{\eta}{r\mathrm{d}_r\eta} = 2\frac{\mathrm{d}\ln r}{\mathrm{d}\ln \eta}. \tag{10.2.23}$$

Using the chain rule for differentiation and (10.2.12) $\mathrm{d}\Delta/\mathrm{d}\wp = \wp \mathrm{d}T/\mathrm{d}\wp$ and so

$$\frac{\mathrm{d}\Delta}{\mathrm{d}\wp}(\wp) = -\frac{\xi_e}{[\eta_e^2 - \wp^2]^{1/2}} + \int_{r_p}^{r_e} \mathrm{d}r\, \frac{\mathrm{d}\xi}{\mathrm{d}r}\frac{1}{[\eta^2 - \wp^2]^{1/2}}. \tag{10.2.24}$$

If the velocity is piecewise continuous and differentiable there will be additional contributions from the interfaces, cf (10.2.44).

Analytic integrals for the distance and time contributions can be made for a variety of forms of the radial dependence $v(r)$. A simple and convenient form is a linear gradient with radius (Azbel & Yanovskaya, 1972),

$$v(r) = u + cr. \tag{10.2.25}$$

Recalling (10.2.5) we can express the distance and time increments through a layer from r_a to r_b in the form

$$\delta T = \frac{1}{c}\left[\ln\left(\tan\tfrac{1}{2}i\right) + J(i)\right]_{i_a}^{i_b} \tag{10.2.26}$$

$$\delta\Delta = -\left[i + c\wp J(i)\right]_{i_a}^{i_b} \tag{10.2.27}$$

where

$$J(i) = \int \mathrm{d}i \frac{1}{[\sin i - c\wp]} \quad \text{and} \quad \sin i = \frac{\wp(u + cr)}{r}. \tag{10.2.28}$$

If r_a lies below the turning point radius r_p then i_a should be replaced by $\pi/2$.

The expression for $J(i)$ depends on the value of $|c\wp|$. Substituting $|c\wp| = \cosh\gamma$ or $|c\wp| = \cos\gamma$, we find

$$\begin{aligned} |c\wp| > 1, &\qquad J(i) = \frac{1}{\sinh\gamma}\tan^{-1}\left\{\frac{1 - c\wp\sin i}{\sinh\gamma\cos i}\right\}, \\ |c\wp| = 1, &\qquad J(i) = \frac{-\cos i}{1 + \sin i}, \\ |c\wp| < 1, &\qquad J(i) = -\frac{1}{2\sin\gamma}\ln\left\{\frac{1 + \sin(\gamma + i)}{1 - \sin(\gamma - i)}\right\}. \end{aligned} \tag{10.2.29}$$

10.2.3 Amplitudes

We use the conservation of energy along ray-tubes as a means of determining the amplitude behaviour due to geometrical spreading from the source. Consider a point source at S which radiates energy

$$E = E_S(i_S) 2\pi \sin i_S \mathrm{d}i_S, \tag{10.2.30}$$

into the angle range i_S to $i_S + \mathrm{d}i_S$. The rays corresponding to the limits of this angle interval will emerge at epicentral distances Δ and $\Delta + \mathrm{d}\Delta$. Thus if the energy density per unit area is σ, the original energy must be equated to

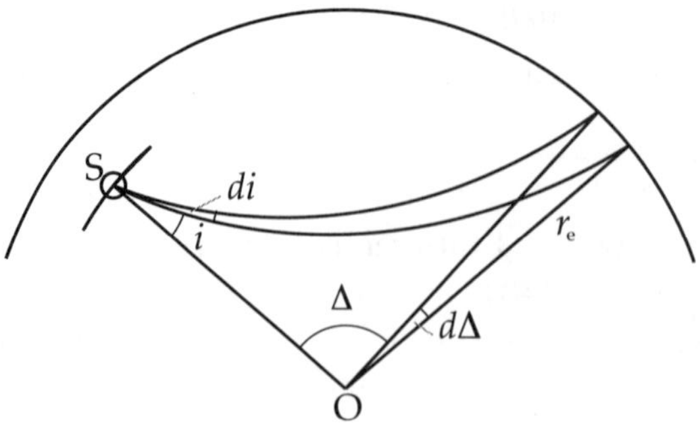

Figure 10.4. Geometry of a ray tube from a source in a sphere.

$$E = \sigma 2\pi r_e^2 \sin\Delta |d\Delta| \cos i_e, \tag{10.2.31}$$

and thus

$$\sigma = \frac{E_s(i_s)}{\sin\Delta} \frac{\sin i_s}{\cos i_e} \left| \frac{di_s}{d\Delta} \right|. \tag{10.2.32}$$

Now

$$\wp = \eta_s \sin i_s = \eta_e \sin i_e = dT/d\Delta,$$

and so

$$\frac{d\wp}{d\Delta} = \frac{d^2T}{d^2\Delta} = \eta_s \cos i_s \frac{di_s}{d\Delta}. \tag{10.2.33}$$

With this expression for the derivative of the ray parameter we can recast the expression for the energy density (10.2.32) in the form

$$\sigma = E_s(i_s) \frac{1}{\eta_s^2 \cos i_s \cos i_e} \frac{\wp}{\sin\Delta} \left| \frac{d\wp}{d\Delta} \right|. \tag{10.2.34}$$

The amplitude density depends on $\sigma^{1/2}$ and will thus be proportional to $[\wp(d\wp/d\Delta)/\sin\Delta]^{1/2}$. The true surface amplitude will need to include the effect of reflection at the free surface (see Chapter 13).

We recall from (9.3.12) that the amplitude of the leading term of the ray expansion at the surface receiver **R** is given by

$$U^{[0]}(\mathbf{R}) = U^{[0]}(\mathbf{S}) \left[\frac{\rho_s v_s J_s}{\rho_e v_e J_e} \right]^{1/2} = \left[\frac{\rho_s v_s d\sigma_s}{\rho_e v_e d\sigma_e} \right]^{1/2}. \tag{10.2.35}$$

From the argument we used to derive (10.2.32), taking $U^{[0]}(\mathbf{S})$ to be specified on a unit sphere about the source

$$\left(\frac{d\sigma_s}{d\sigma_s} \right)^{1/2} = \left(\frac{1}{\sin\Delta} \frac{\sin i_s}{\cos i_e} \left| \frac{di_s}{d\Delta} \right| \right)^{1/2}, \tag{10.2.36}$$

The amplitude at the receiver can then be expressed as

$$U^{[0]}(\mathbf{R}) = U^{[0]}(\mathbf{S}) g_s(i_s) \left(\frac{\rho_s v_s}{\rho_e v_e} \right)^{1/2} \left(\frac{1}{\eta_s^2 \cos i_s \cos i_e} \right)^{1/2} \left(\frac{\wp}{\sin\Delta} \left| \frac{d\wp}{d\Delta} \right| \right)^{1/2}, \tag{10.2.37}$$

where we have allowed for the radiation characteristics of the source through $g_s(i_s)$ and have again used (10.2.33). In this ray approximation the amplitude is strongly dependent on the curvature of the travel time curves through $|d\wp/d\Delta|$.

For a velocity model which includes discontinuities in the velocity model we have to take account of the partition of energy at the interface and the change in the area of the ray tube crossing the interface.

Across a discontinuity at radius r_j

$$\frac{d\sigma_j^{out}}{d\sigma_j^{in}} = \frac{\cos i_j^{out}}{\cos i_j^{in}}.$$

The amplitude at the receiver thus takes the form

$$U^{[0]}(\mathbf{R}) = U^{[0]}(\mathbf{S}) g(i_s) \left(\frac{\rho_s v_s}{\rho_e v_e}\right)^{1/2} \left(\frac{1}{\eta_s^2 \cos i_s \cos i_e}\right)^{1/2} \cdot \prod_j R_j \left(\frac{\cos i_j^{\text{out}}}{\cos i_j^{\text{in}}}\right)^{1/2} \left(\frac{\wp}{\sin\Delta}\left|\frac{d\wp}{d\Delta}\right|\right)^{1/2}, \tag{10.2.38}$$

where R_j is the appropriate energy-normalised reflection or transmission coefficient for the interaction of the ray with the interface at r_j. Note that for a symmetric path in a single wavetype the term associated with the changes in ray tube area at the interfaces

$$\prod_j \frac{\cos i_j^{\text{out}}}{\cos i_j^{\text{in}}} = 1,$$

since a ray tube will return to the surface with its original size.

10.2.4 Properties of travel time curves

We will examine the influence of the nature of the velocity distribution on the relation between travel time T, slowness $\wp$ and epicentral distance Δ as well as the character of $\tau(\wp)$. For simplicity we will consider the situation with source and receiver both at the surface. We illustrate the various cases with computations for simple velocity models with a common form of display for each case.

The behaviour seen for these simple models enables us to understand the properties for the various seismic phases in their passage through the Earth, and to provide a detailed description of their characteristics to complement the qualitative treatment in Chapter 5.

We will examine first the case of continuous refraction and then look at the way in which the properties of the travel times and slowness relations are affected by rapid changes in velocity, as well as discontinuities in velocity gradient and velocity. These complications in the velocity distribution lead to triplications in the travel time curves, and are directly relevant to models for the upper mantle transition zone.

At the core-mantle boundary the P wavespeed in the mantle is significantly higher than the P wavespeed in the core, but this again is higher than the S wavespeed in the mantle. The passage of P waves into the core in the *PKP* corresponds to the low velocity zone in Section 10.2.4.5 but is further complicated by the velocity increase at the inner core boundary which can be described by the treatment in Section 10.2.4.4. This case of a positive jump in velocity also corresponds to the situation when S waves convert to P at the core-mantle boundary for the *SKS* phase.

10.2.4.1 Smooth increase in velocity

For a smooth increase in velocity $v(r)$ as depth increases (and thus r decreases), the velocity parameter $\eta = v(r)/r > 0$. Because the slowness $\wp$ is constant the inclination of the ray i to the radial vector will increase with depth. In terms of the functions $\zeta(r)$, $\xi(r)$ introduced in (10.2.23), the condition $\zeta < 1$, or $\xi > 0$ (or $d\eta/dr > 0$), determines whether there can be a turning point a given level r. The travel time curve $T(\Delta)$ will be a single valued function of range if $d\xi/dr \le 0$.

In the smooth structure the rays are continuously refracted back to the surface. The travel-time behaviour is illustrated in figure 10.5 with a display of the velocity model and associated ray paths, the dependence of epicentral distance Δ on slowness $\wp$, the $\tau(\wp)$ relation and the reduced travel time $T - \wp_0\Delta$ as a function of Δ. The reduction slowness $\wp_0$ is chosen to track the travel time curve and so allow an expansion of the detail of the behaviour.

We also show the derivative $d\Delta/d\wp$ as a function of slowness $\wp$. The slight jitters in this plot arise from approximating the velocity distribution by a sequence of linear gradients of the form (10.2.25). This form of representation introduces a sequence of minor discontinuities in velocity gradient which lead to discontinuities in the local derivative (as discussed in Section 10.2.4.3).

The sequence of linear gradients is sufficient to give a very good approximation of the travel time behaviour expected for a smooth model, but is not adequate for derivatives. Careful numerical integration using, e.g., a Gauss-Legendre quadrature (with allowance for the square root singularity at the turning point) with a cubic spline interpolation of a tabulated model can provide a suitable accuracy for the derivative $d\Delta/d\wp$ (see, e.g., Chapman, 1971). The use of spline interpolation ensures continuity of the velocity and its first and second derivatives, so that the discontinuities are only introduced where they occur in the real model.

10.2.4.2 Rapid increase in velocity gradient

In the case of a gentle increase in velocity we have simple continuous refraction, but if the velocity begins to increase rapidly the refraction process can lead to crossing of rays with similar slowness and hence caustics, with a triplication in the travel-time curves.

We consider a continuous velocity profile and assume that, as before, $\xi_e > 0$. Then extrema can only occur in Δ if

$$\frac{d\Delta}{d\wp} = 0 \quad \text{for} \quad p \in (0, \eta_e), \tag{10.2.39}$$

and these extrema will lead to cusps in the travel time curve. We recall the expression (10.2.24),

$$\frac{d\Delta}{d\wp}(\wp) = -\frac{\xi_e}{[\eta_e^2 - \wp^2]^{1/2}} + \int_{r_p}^{r_e} dr\, \frac{d\xi}{dr} \frac{1}{[\eta^2 - \wp^2]^{1/2}}. \tag{10.2.40}$$

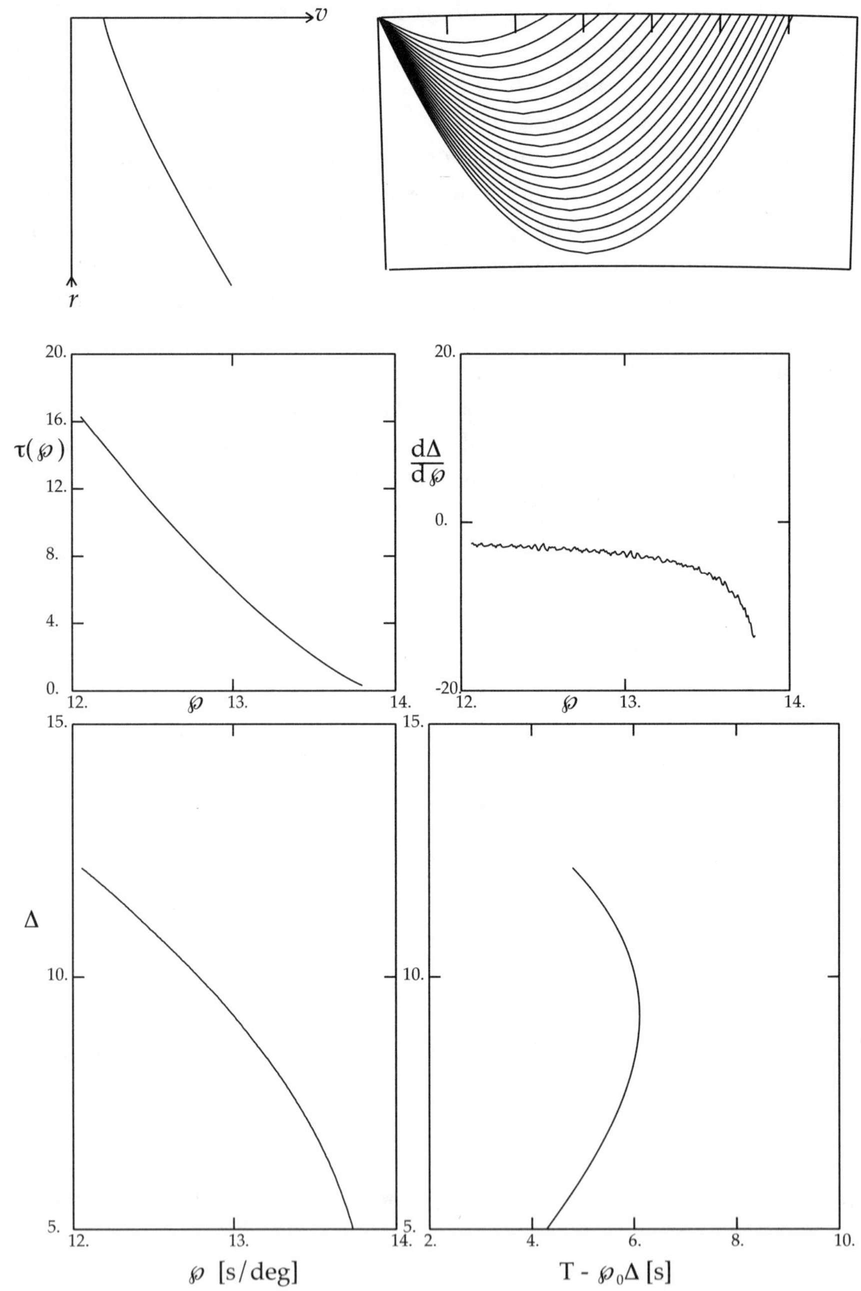

Figure 10.5. Velocity model, ray paths, travel time, slowness and epicentral distance behaviour for a smooth increase in velocity.

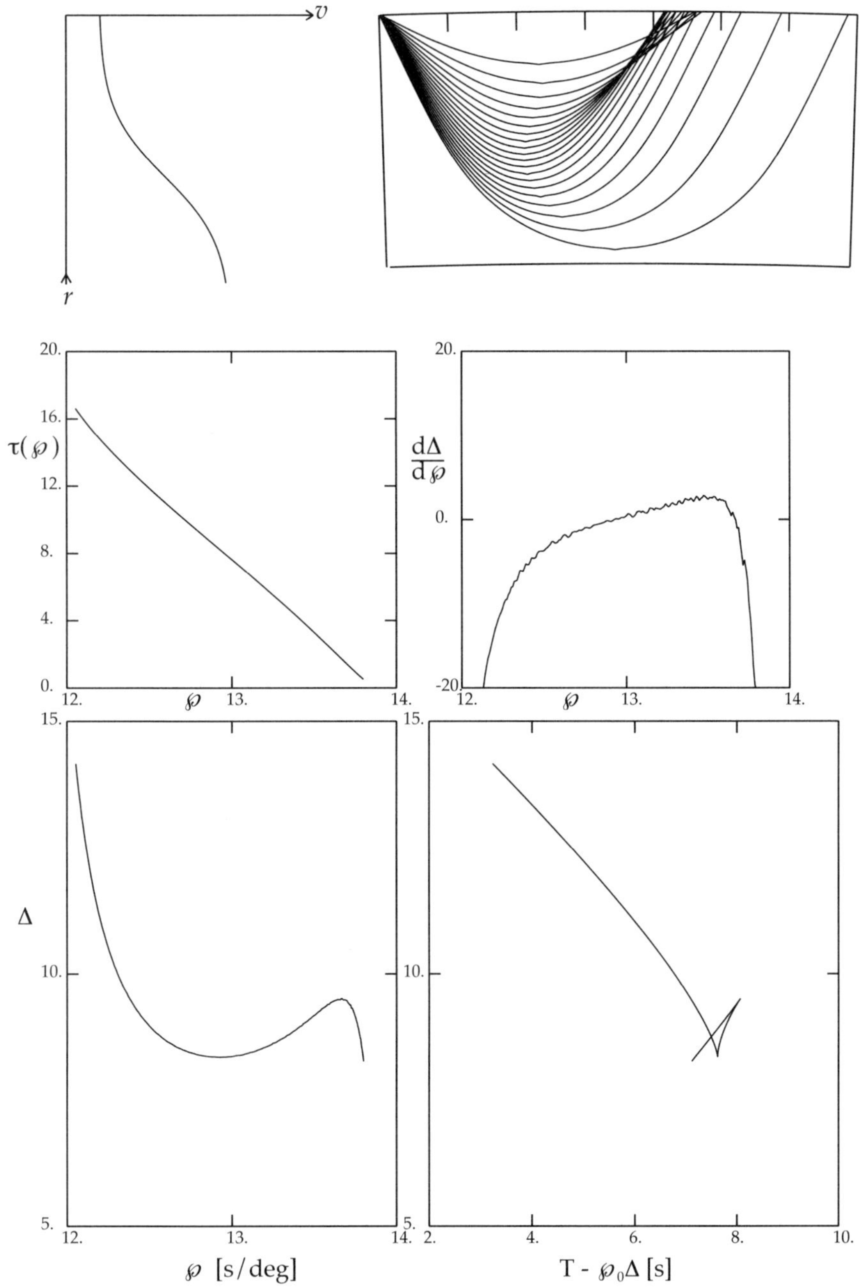

Figure 10.6. Velocity model, ray paths, travel time, slowness and epicentral distance behaviour for a rapid increase in velocity gradient.

From (10.2.40), $\mathrm{d}\Delta/\mathrm{d}\wp$ will be negative for $\wp$ close to η_e and can only become positive if the integral is positive, i.e., if $\mathrm{d}\xi/\mathrm{d}r > 0$ due to a rapid increase in velocity. Since the integral in (10.2.40) is zero for $\wp = \eta_e$ and tends in general towards ∞ as $r_p \to 0$, the travel time curve can have either zero or an even number of cusps.

The sense of progression of the travel time curve will change as it passes through a cusp, and so there is a range of distances Δ_1, Δ_2 for which we expect three arrivals associated with the same travel time curve. This class of triplication is illustrated in figure 10.6 and we see the direct association of the cusps in the travel-time curve with the caustics in the ray pattern. If however we reach a situation where $\mathrm{d}\Delta/\mathrm{d}\wp$ just touches the $\wp$ axis, then there is no triplication but the travel-time curve has a point of inflexion.

Near a simple zero of $\mathrm{d}\Delta/\mathrm{d}\wp$ we may write

$$\frac{\mathrm{d}\Delta}{\mathrm{d}\wp} = d_i(\wp - q_i) + \mathrm{O}(\wp - q_i)^2, \tag{10.2.41}$$

so that

$$\Delta(\wp) = \Delta_i + \tfrac{1}{2}d_i(\wp - q_i)^2, \quad T(\wp) = T_i + q(\Delta - \Delta_i) \pm \tfrac{2}{3}[2(\Delta - \Delta_i)^3/d_i]^{1/2}. \tag{10.2.42}$$

The zeroes of $\mathrm{d}\Delta/\mathrm{d}\wp$ correspond to the travel-times cusps, local extrema in $\Delta(\wp)$ and caustics, as can be seen in figure 10.6. The simple amplitude formula (10.2.38) would suggest that infinite geometrical amplitude would be expected. The amplitudes will indeed be large but as we noted in Chapter 9 the ray expansion becomes invalid at a caustic where the phase function ϑ is no longer analytic.

10.2.4.3 Discontinuity in velocity gradient

We now consider a situation where the velocity is continuous but the velocity gradient with depth has a discontinuous increase at a radius r_1. Then

$$|\xi(r_1+)| > |\xi(r_1-)|\,. \tag{10.2.43}$$

For slownesses $\wp$ greater than $\eta(r_1)$ (the slowness at the gradient discontinuity) we can use (10.2.40) directly. However for $\wp < \eta(r_1)$ we must introduce terms to allow for the discontinuity in ξ.

$$\begin{aligned}\frac{\mathrm{d}\Delta}{\mathrm{d}\wp}(\wp) = &-\frac{\xi_e}{[\eta_e^2 - \wp^2]^{1/2}} - \frac{[\xi_{1+} - \xi_{1-}]}{[\eta_1^2 - \wp^2]^{1/2}} \\ &+ \int_{r_p}^{r_1} \mathrm{d}r\, \frac{\mathrm{d}\xi}{\mathrm{d}r}\frac{1}{[\eta^2 - \wp^2]^{1/2}} + \int_{r_1}^{r_e} \mathrm{d}r\, \frac{\mathrm{d}\xi}{\mathrm{d}r}\frac{1}{[\eta^2 - \wp^2]^{1/2}}.\end{aligned} \tag{10.2.44}$$

Since ξ_{1+}, ξ_{1-} are both positive in the case of an increase in velocity gradient with depth, $\mathrm{d}\Delta/\mathrm{d}\wp$ is discontinuous at $\wp = \eta_1$ and jumps to a positive infinite values at η_1-.

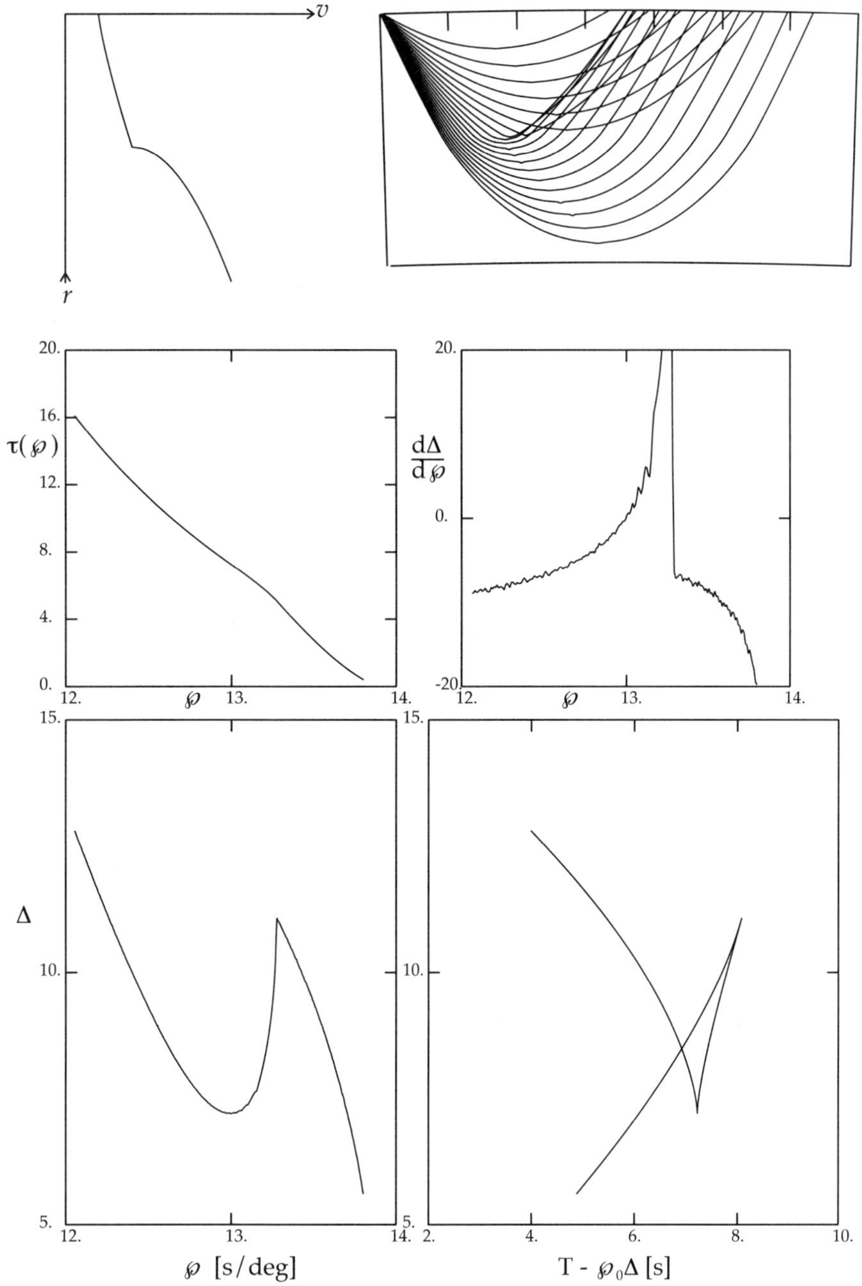

Figure 10.7. Velocity model, ray paths, travel time, slowness and epicentral distance behaviour for a discontinuity in velocity gradient.

Once again we have a triplication in the travel-time curve (figure 10.7, but now only one caustic. Now,

$$\frac{\mathrm{d}\tau(\wp)}{\mathrm{d}\wp} = -\Delta(\wp) \tag{10.2.45}$$

as can be seen by direct differentiation of (10.2.12). As a result the more pronounced triplication imposes more character on the $\tau(\wp)$ distribution.

The geometrical amplitude produced by the discontinuous velocity gradient is only infinite for the cusp nearest the epicentre At the other point $|\mathrm{d}\wp/\mathrm{d}\Delta|$ will be discontinuous, taking a finite value for $\wp = \eta_1+$ and being zero for $\wp = \eta_1-$. The behaviour is reflected in the shape of the epicentral distance as a function of slowness $\wp$ (figure 10.7). The smooth minimum corresponds to a break down of the geometric amplitude formula, whilst the angular maximum has a discontinuity in amplitude. Only the closer minimum corresponds to a caustic in the ray pattern.

In contrast, for a decrease in velocity gradient with depth $\mathrm{d}\Delta/\mathrm{d}\wp$ jumps to $-\infty$ at $\wp = \eta_1-$, and there will be no cusps or caustics.

10.2.4.4 Discontinuity in velocity

Consider the situation where there is a discontinuous increase in velocity at the radius r_1. Above r_1, the rays will be continuously refracted by the structure as before. However, the ray which arrives at grazing incidence at r_1 will be reflected back from the interface. For slowness $\eta_e > \wp > \eta(r_1 + 0)$ we have continuous refraction, with $\Delta(\wp)$, $T(\wp)$ and $\mathrm{d}\Delta/\mathrm{d}\wp$ given by

$$\begin{aligned}
\Delta(\wp) &= 2\int_{r_p}^{r_e} \mathrm{d}r\, \frac{\wp}{r[\eta^2 - \wp^2]^{1/2}}, \qquad T(\wp) = 2\int_{r_p}^{r_e} \mathrm{d}r\, \frac{\eta^2}{r[\eta^2 - \wp^2]^{1/2}}, \\
\frac{\mathrm{d}\Delta}{\mathrm{d}\wp}(\wp) &= -\frac{\xi_e}{[\eta_e^2 - \wp^2]^{1/2}} + \int_{r_p}^{r_e} \mathrm{d}r\, \frac{\mathrm{d}\xi}{\mathrm{d}r}\frac{1}{[\eta^2 - \wp^2]^{1/2}}.
\end{aligned} \tag{10.2.46}$$

where as before r_p is the turning radius.

For $\eta(r_1 + 0) > \wp > \eta(r_1 - 0)$, there is no possibility of a turning point below the interface and all the rays are reflected from the interface. We can employ similar forms to (10.2.46) with the interface level r_1 replacing r_p:

$$\begin{aligned}
\Delta(\wp) &= 2\int_{r_1}^{r_e} \mathrm{d}r\, \frac{\wp}{r[\eta^2 - \wp^2]^{1/2}}, \qquad T(\wp) = 2\int_{r_1}^{r_e} \mathrm{d}r\, \frac{\eta^2}{r[\eta^2 - \wp^2]^{1/2}}, \\
\frac{\mathrm{d}\Delta}{\mathrm{d}\wp}(\wp) &= -\frac{\xi_e}{[\eta_e^2 - \wp^2]^{1/2}} + \frac{\xi_{1+}}{[\eta_{1+}{}^2 - \wp^2]^{1/2}} + \int_{r_1}^{r_e} \mathrm{d}r\, \frac{\mathrm{d}\xi}{\mathrm{d}r}\frac{1}{[\eta^2 - \wp^2]^{1/2}}.
\end{aligned} \tag{10.2.47}$$

There is therefore a discontinuity in $\mathrm{d}\Delta/\mathrm{d}\wp$ across $\wp = \eta(r_1+)$.

For $\wp = \eta(r_1 - 0)$, the rays are refracted at grazing incidence along the underside of the interface, and for smaller slownesses $\wp < \eta(r_1-)$ we have the possibility of rays being either reflected from the interface or refracted into the lower medium (figure 10.8).

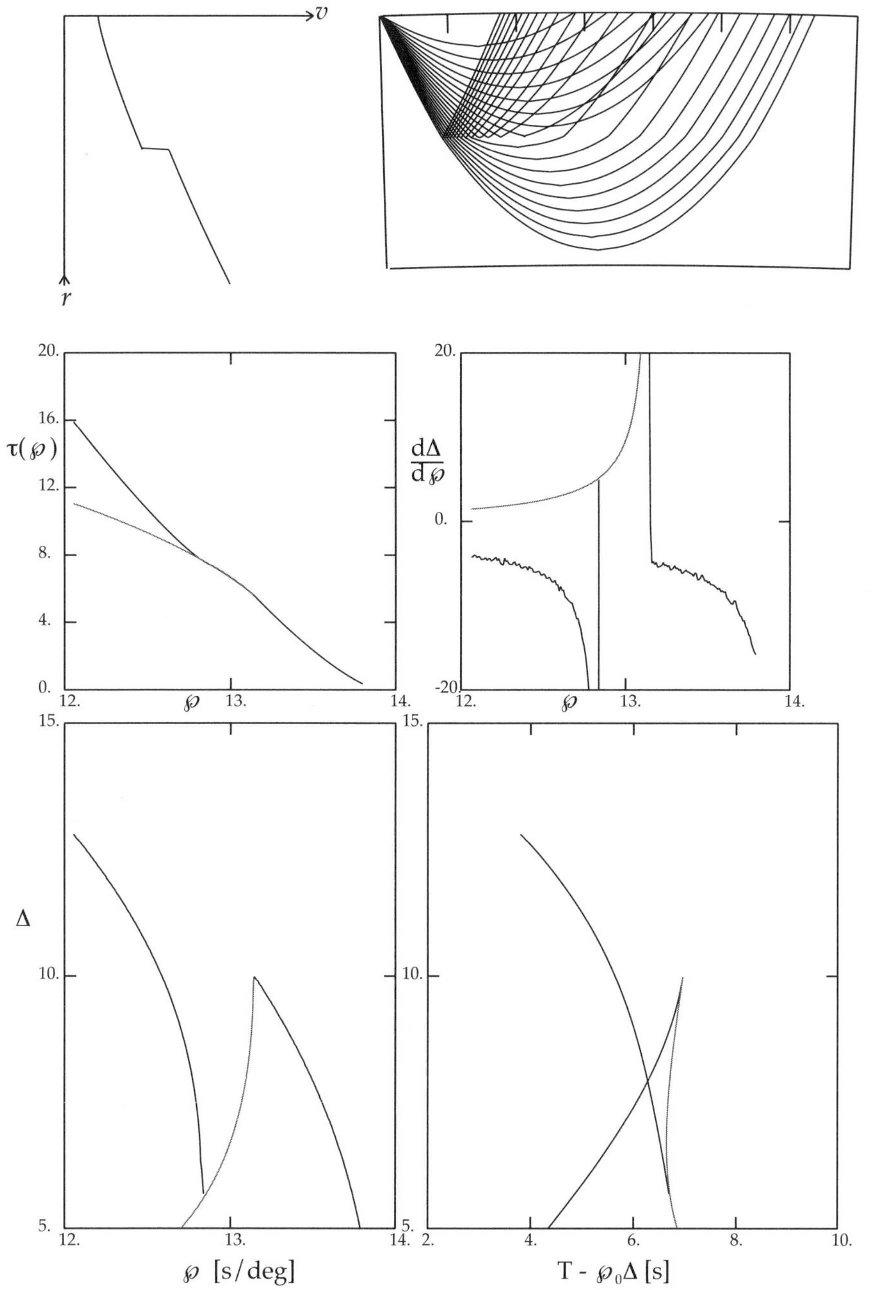

Figure 10.8. Velocity model, ray paths, travel time, slowness and epicentral distance behaviour for a discontinuity in velocity.

For the reflected waves in $\wp \le \eta(r_1-)$ we can use the expressions (10.2.47). However, for the waves refracted from below the interface at r_1 we must integrate once again from the turning radius r_p.

$$\Delta(\wp) = 2\int_{r_p}^{r_e} \mathrm{d}r \frac{\wp}{r[\eta^2 - \wp^2]^{1/2}}, \quad T(\wp) = 2\int_{r_p}^{r_e} \mathrm{d}r \frac{\eta^2}{r[\eta^2 - \wp^2]^{1/2}},$$
$$\frac{\mathrm{d}\Delta}{\mathrm{d}\wp}(\wp) = -\frac{\xi_e}{[\eta_e^2 - \wp^2]^{1/2}} + \frac{\xi_{1+}}{[\eta_{1+}{}^2 - \wp^2]^{1/2}} - \frac{\xi_{1-}}{[\eta_{1-}{}^2 - \wp^2]^{1/2}}$$
$$+ \int_{r_1}^{r_e} \mathrm{d}r \frac{\mathrm{d}\xi}{\mathrm{d}r} \frac{1}{[\eta^2 - \wp^2]^{1/2}} + \int_{r_p}^{r_1} \mathrm{d}r \frac{\mathrm{d}\xi}{\mathrm{d}r} \frac{1}{[\eta^2 - \wp^2]^{1/2}}. \quad (10.2.48)$$

There will be a further discontinuity in $\mathrm{d}\Delta/\mathrm{d}\wp$ across $\wp = \eta_1-$.

The two turning points in the $\Delta(\wp)$ relation are introduced by the transition from direct waves (refracted in the upper medium) to reflected waves, and from reflected waves to waves refracted in the lower medium. Both points correspond to discontinuous and finite curvature of the travel time curve. Unlike the case of discontinuities in velocity gradient we do not have any caustics.

In figure 10.8 we show the refracted branches in black and the reflected contributions in grey tone, so that their properties can be distinguished. The reflections for small slowness ('pre-critical') extend to short distances and show up very clearly as a bifurcation in the properties of $\tau(\wp)$.

10.2.4.5 A low-velocity zone

In all the preceding discussion we have assumed that the velocity increases with depth. Suppose now that at the level $r = r_1$ the velocity decreases, either abruptly or continuously and that the velocity distribution is such that it eventually increases with depth to the point where $\eta = r/v(r)$ is smaller than $\eta(r_1+)$.

As in our discussions of the effects of velocity gradient discontinuities and velocity discontinuities, there will be no effect in this ray approximation on rays with turning points above r_1, i.e. ray parameters in the range $\eta_e > \wp > \eta(r_1+)$.

The effect of a decrease in velocity below r_1 and thus an increase in η is to bend a ray towards, rather than away from the vertical. It is only when the ray encounters material for which $\eta < \eta(r_1+)$ that it will be bent again towards the horizontal and eventually present a turning point. The case is illustrated for a discontinuity in velocity gradient in figure 10.9, and we can see that the influence of the low velocity zone can extend over a considerable span in distance.

The travel time and epicentral distance have the usual form in terms of the ray parameter $\wp$,

$$\Delta(\wp) = 2\int_{r_p}^{r_e} \mathrm{d}r \frac{\wp}{r[\eta^2 - \wp^2]^{1/2}}, \quad T(\wp) = 2\int_{r_p}^{r_e} \mathrm{d}r \frac{\eta^2}{r[\eta^2 - \wp^2]^{1/2}}. \quad (10.2.49)$$

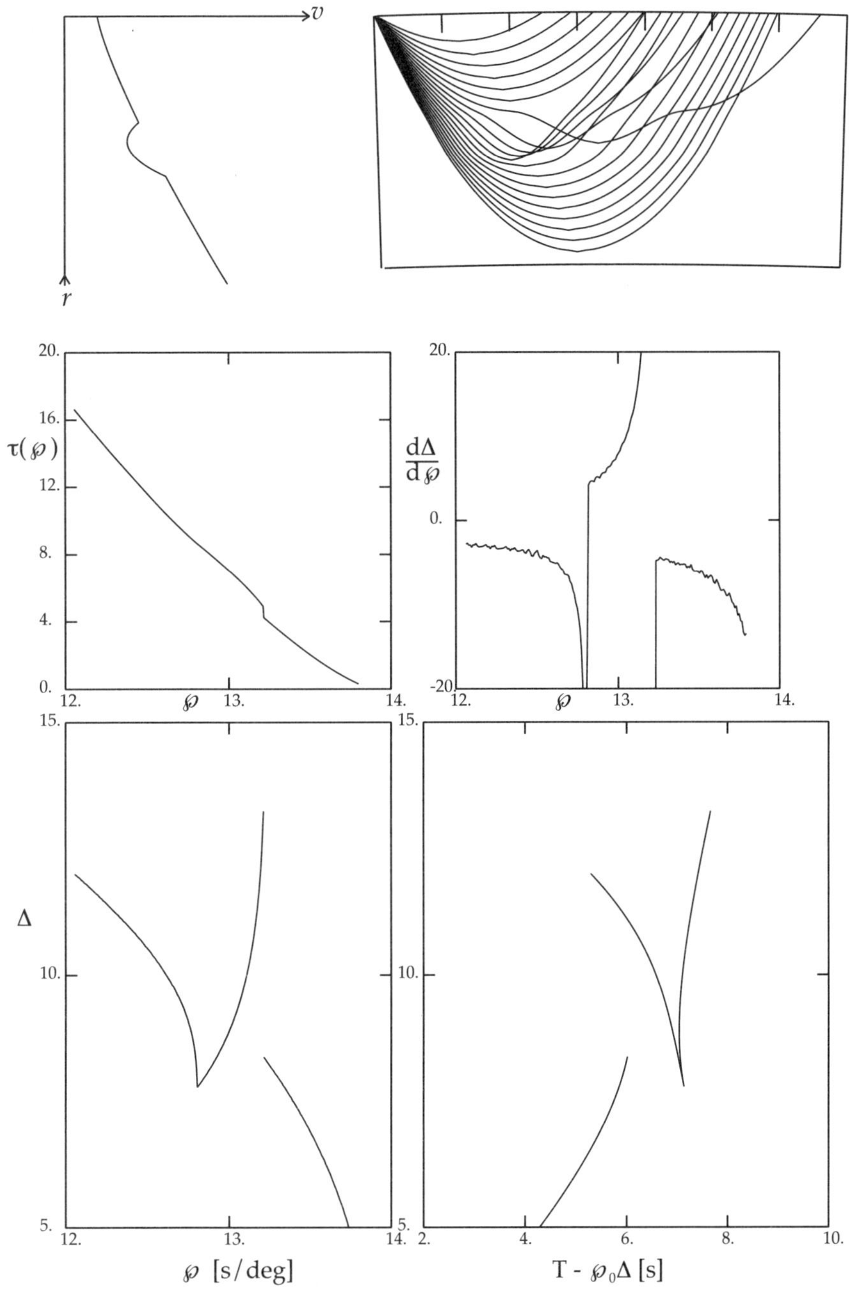

Figure 10.9. Velocity model, ray paths, travel time, slowness and epicentral distance behaviour for a low velocity zone.

but the 'low velocity zone' in the radius interval (r_2, r_1) for which $\eta(r) > \eta(r_1+)$ means that $\Delta(\wp)$, $T(\wp)$ are discontinuous across the ray parameter $\wp = \eta(r_1+)$. There is often a finite range of Δ, 'the shadow zone' in which geometrical ray theory predicts no arrivals.

Sometimes, as in figure 10.9, there is a slight overlap in Δ between the two distinct segments of the travel-time curves. In reality, diffraction of elastic waves will tend to fill in the shadow zone.

The general form for the behaviour of the derivative $d\Delta/d\wp$ in the case of a low velocity zone can be obtained from (10.2.48). For either a jump in velocity or velocity gradient $d\Delta/d\wp$ has a discontinuity at $\wp = \eta(r_1+)$ and there will therefore generally be a retrograde portion of the travel time curve with either a cusp or an angular point.

We may note that the effect of the low velocity zone has to been introduce a jump in the value of $\tau(\wp)$ at $\wp = \eta(r_1+)$ of magnitude

$$\sigma_1 = \int_{r_2}^{r_1} dr\, \frac{1}{r}[\eta^2 - \wp^2]^{1/2}. \tag{10.2.50}$$

This τ jump can be clearly seen in figure 10.9.

10.2.5 Near-distances

For a source at the surface in a spherically symmetric structure, the travel time T and the epicentral distance Δ are connected by

$$T = \wp\Delta + 2\int_{r_p}^{r_e} dr \left[\frac{1}{v^2} - \frac{\wp^2}{r^2}\right]^{1/2}, \tag{10.2.51}$$

for wavespeed $v(r)$ and ray parameter $\wp$. As in the previous section the suffix p denotes properties at the turning point of the ray. Thus $\wp = r_p/v_p$.

In terms of depth $z = r_e - r$, (10.2.51) can be rewritten as

$$T = \frac{r_p\Delta}{v_p} + 2\int_0^{z_p} dz \left[\frac{1}{v^2} - \left(\frac{r_p}{r}\right)^2 \frac{1}{v_p^2}\right]^{1/2}, \tag{10.2.52}$$

When propagation is only over a short distance, the rays will not penetrate to great depth and so $r_e - r_p$ is small compared with r_e. With this approximation we can rewrite (10.2.52) as

$$T \approx \frac{r_e\Delta}{v_p} + 2\int_0^{z_p} dz \left[\frac{1}{v^2} - \frac{1}{v_p^2}\right]^{1/2}. \tag{10.2.53}$$

We can recognise the distance of propagation along the surface $X = r_e\Delta$ and the ray parameter for a horizontally stratified medium $p = 1/v_p$ and so

$$T \approx pX + 2\int_0^{z_p} dz \left[\frac{1}{v^2} - p^2\right]^{1/2}. \tag{10.2.54}$$

which is just the form for horizontal stratification (cf 10.1.10).

It is possible to make an exact mapping of the ray properties from the sphere into horizontal stratification using a conformal transformation. This 'earth-flattening' transformation maps the wavespeed profile with radius r in a sphere into a new wavespeed distribution with depth z in a half space so that transit times from source to receiver are preserved. In the flattened model we take

$$z = r_e \ln(r_e/r), \tag{10.2.55}$$

and the wavespeeds after flattening α_f, β_f are

$$\alpha_f(z) = \alpha(r)(r_e/r), \qquad \beta_f(z) = \beta(r)(a/r), \tag{10.2.56}$$

where r_e is the radius of the Earth. The increased velocity gradients in the flattened model compensate for the crowding effect of sphericity as the radius diminishes.

The characteristic of a low velocity zone is that no rays have their turning point within the zone. It is interesting to note that for a velocity distribution with constant η, i.e., $v(r) = ur$ for some constant u, there are no turning points. Such a medium would correspond to a uniform layer in horizontal stratification under the earth flattening transformation.

10.2.6 Depth corrections

The travel-time and distance relations have a strong dependence on the depth of the source. On occasion it is useful to be able to make adjustments to the apparent source depth such as when new hypocentral information becomes available. In particular when data from a number of events are combined it is often convenient to work with a common source depth.

We can make suitable adjustments for direct phases such as P and S by projecting the ray path for a given slowness $\wp$ and wavetype to the new source depth. For example,

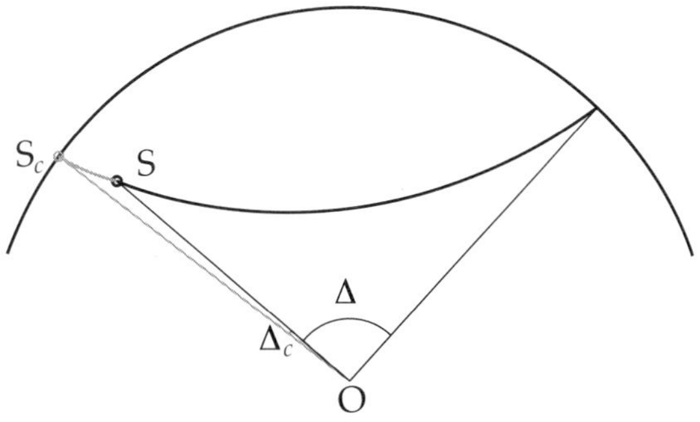

Figure 10.10. Schematic representation of the introduction of depth corrections. A source can be corrected back to surface focus S_c by projecting the ray path with the same slowness to include an extra leg shown in grey, with an addition to the epicentral distance of Δ_c and a consequent addition to the travel time (T_c).

figure 10.10 shows the extension of the raypath from the source S back to the surface to a new apparent source S_c. In the process we need to adjust the epicentral distance by Δ_c to compensate for the extra path from the surface shown in grey, and also to add a similar correction T_c to the travel time.

For a source at r_s and a corrected source depth r_c the adjustments in epicentral distance Δ_c and travel time for slowness $\wp$ take the form

$$\Delta_c(\wp) = 2\int_{r_s}^{r_c} \mathrm{d}r \, \frac{\wp}{r[\eta^2 - \wp^2]^{1/2}}, \qquad T_c(\wp) = 2\int_{r_s}^{r_c} \mathrm{d}r \, \frac{\eta^2}{r[\eta^2 - \wp^2]^{1/2}}. \tag{10.2.57}$$

Although it is possible to make comparable adjustments to the depth phases (e.g. *pP*, *sP*), the corrections for source depth are often applied directly to seismogram traces. Although the time and distance shift will compensate for the new source level, the differential times between the direct and surface reflected phases will not have been adjusted. Further, the depth corrections for *P* and *S* will differ and so a single correction cannot be applied to a whole seismogram.

11

Seismic Sources

As we have seen in Chapter 4, seismic sources are associated with the release of potential energy, strain energy for an earthquake, chemical or nuclear energy for an explosion, mechanical energy for surface impact. The energy radiated as seismic waves is only a small part of the total.

The deformation in the immediate vicinity of the source is very large and the stress is non-linearly related to the large strains and strain rates. It takes some distance from the source for the strains to be small enough that we can employ the linear theory for seismic waves developed in Chapters 7 and 8. Our observations will normally lie in the domain of linear stress-strain relations. Thus, when we try to reconstruct source characteristics from our observations we are not able to extrapolate the wavefield directly into the near-source region where non-linear effects are important. The best we can do is to provide a description of an equivalent source in terms of our linearised wave equations.

We introduce a moment tensor density to describe the differences between the model and actual stress fields. This volume representation can be simplified in many circumstances to provide a point equivalent source based on a multipole force system. The leading term is the point moment tensor source which is commonly used to characterise seismic events.

11.1 Equivalent forces and moment tensor densities

The basis of our development of seismic wave theory is that the behaviour of the material within the Earth is governed by linearised equations of motion for nearly elastic media. Sources need to be introduced wherever the actual stress distribution departs from that predicted from our linearised model.

The true, local, equation of motion in the continuum is

$$\rho \partial_{tt} u_j = \partial_i [\sigma_{ij} + \sigma^0_{ij}] + \rho \partial_j \psi, \tag{11.1.1}$$

in an Eulerian viewpoint, even in the non-linear regime. As in section 7.5, σ_{ij} represents the deviation of the local stress state from the initial stress state specified by σ^0_{ij}. ψ

is the gravitational potential. The total stress will be related to the displacement **u** via some constitutive equation, which in the non-linear zone may have a complex dependence on the history of the motion.

As a model of the incremental stress σ_{ij} we construct a field τ_{ij} based on simple assumptions about the nature of the constitutive relation between stress and strain and the variation of the elastic constants with position. In order to reconcile the displacements **u** derived from the linearised equation of motion based on this model stress τ_{ij} with those for the true situation (11.1.1), we have to modify the equation of motion by introducing an additional force distribution **e**. Thus our linearised equation becomes

$$\rho \partial_{tt} u_j = \partial_i \tau_{ij} + f_j + e_j, \qquad (11.1.2)$$

we include any self-gravitation effects in **f**. We shall refer to **e** as the 'equivalent force distribution' to the original source.

From our linearised viewpoint the *source region* will be that portion of the medium in which **e** is non-zero. When the displacements are small, the equivalent force system $\mathbf{e}(\mathbf{x}, t)$ will be equal to the gradient of the difference between the physical stress field $\boldsymbol{\sigma}$ and our postulated stress field $\boldsymbol{\tau}$

$$e_j = \partial_i \sigma_{ij} - \partial_i \tau_{ij}. \qquad (11.1.3)$$

If the source region extends to the earth's surface, we have to include additional surface tractions $\mathbf{e}_s$ to compensate for the difference between the tractions induced by $\boldsymbol{\tau}$ and $\boldsymbol{\sigma}$. Thus at the surface we take

$$e^s_j = n_i \tau_{ij} - n_i \sigma_{ij}, \qquad (11.1.4)$$

where **n** is the local outward normal.

For explosive sources, the principal source region will be a spherical zone within the 'elastic radius', which is somewhat arbitrarily taken as the range from the point of detonation at which the radial strain in the material drops to 10^{-6}. Within this volume our linearised equations will be inadequate and the equivalent forces **e** will be large. Outside the elastic radius we may expect to do a reasonable job of approximating the stress and so **e** will be small and tend to zero with increasing distance from the origin if our structural model is adequate. For small explosions, the equivalent forces will be confined to a very limited region which, compared with most seismic wavelengths, will approximate a point; we shall see later how we can represent such a source via a singular force distribution.

For an earthquake the situation is somewhat different. In slip on a pre-existing fault, the two faces of the fault will be little deformed during the motion and the major deformation will occur in any fault gouge present between the faces. When we use the stress estimated from the rock properties ($\boldsymbol{\tau}$) in place of that for the gouge material ($\boldsymbol{\sigma}$) we get a serious discrepancy and so a large equivalent force. This force **e** will be concentrated in a narrow zone along the line of the fault. For long seismic

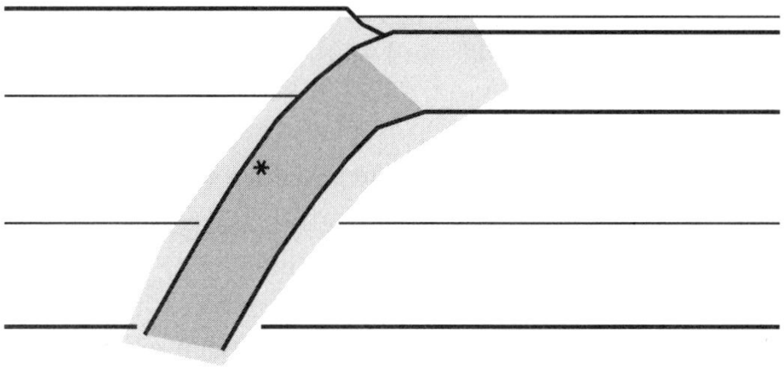

Figure 11.1. Intermediate depth event in a subduction zone; the region in which sources are introduced due to departures from the horizontally stratified reference model are indicated by shading.

wavelengths, the net effect will be equivalent to having a discontinuity across the fault with a singular force distribution concentrated on the fault plane.

Where we have clearly localised physical sources there is a direct association of the equivalent forces **e** with the true source position. However, from (11.1.3) we see that we also have to introduce some 'source' terms **e** whenever our model of the stress field $\boldsymbol{\tau}$ departs from the actual field. Such departures will occur whenever we have made the wrong assumptions about the character of the constitutive relation between stress and strain or about the spatial variation of the elastic parameters.

Suppose, for example, we have an *anisotropic* region and we use an *isotropic* stress-strain relation to find $\boldsymbol{\tau}$. In order to have the linearised equations (11.1.2) predict the correct displacements, we would need a force system **e**, (11.1.3), distributed throughout the region. This force depends on the actual displacement **u** and is therefore difficult to evaluate, but without it we get incorrect predictions of the displacement. We would expect the equivalent forces to be locally small, but their total effect may need to be significant if we are to produce the necessary modification of the displacement field.

Another case of particular importance is provided by the occurrence of many earthquakes in zones of localised heterogeneity (subduction zones). Commonly we locate events and synthesise their radiation using stratified reference models without taking full account of the influence of the lateral heterogeneity in these regions. Consider an intermediate depth event in a downgoing slab (figure 11.1). If we try to represent the radiation from such a source using the stratified model appropriate to the continental side, the model stress $\boldsymbol{\tau}$ will not match the actual physical stress in the neighbourhood of the source itself, and will also be in error in the downgoing slab and the upper part of the oceanic structure. The affected areas are indicated by shading in figure 11.1, with a greater density in those regions where the mismatch is likely to be significant. The equivalent sources **e** will depend on the actual displacement field and at low frequencies we would expect the principal contribution to be near the earthquake's hypocentre. The importance of the heterogeneity will increase with increasing frequency; for example at 0.05 Hz the shear wavelength will be comparable to the thickness of the downgoing slab and so the influence of the slab structure can begin to become important.

When an attempt is made to use observed teleseismic seismograms to invert for the source characteristics using a horizontally stratified reference model, the effective source will appear throughout the shaded region in figure 11.1 and so any localised source estimate is likely to be contaminated to some extent by the inadequacy of the reference model. The result is that the source mechanism estimates will be dependent on the form of stratified reference model which has been employed. The reference model used for centroid moment tensor determinations (Dziewonski et al, 1983) is the PREM model of Dziewonski & Anderson (1981). A large body of results have been built up using this approach based on long period observations (periods > 80 s), but when assessing published mechanisms we have to be aware of the possibility that some component may arise from unexpectedly large lateral heterogeneity.

For each equivalent force distribution $\mathbf{e}(\mathbf{x}, t)$ we follow Backus & Mulcahy (1976a,b) and introduce a moment tensor density $m_{ij}(\mathbf{x}, t)$ such that, in the Earth's interior

$$\partial_i m_{ij} = -e_j, \tag{11.1.5}$$

and if there are any surface traction effects,

$$n_i m_{ij} = e_j^s, \tag{11.1.6}$$

at the Earth's surface. For any situation in which forces are not imposed from outside the Earth (an *indigenous* source), there will be no net force or torque on the Earth. The total force and torque exerted by equivalent forces $\mathbf{e}$ for such a source must therefore vanish. In consequence the moment tensor density for an indigenous source is symmetric:

$$m_{ij} = m_{ji}, \tag{11.1.7}$$

reflecting also the symmetry of the stress tensor in the absence of external couples.

The moment tensor density is not unique, but all forms share the same equivalent forces and thus the same radiation. We see from (11.1.3) that a suitable choice for the moment tensor density is

$$m_{ij} = \Gamma_{ij} = \tau_{ij} - \sigma_{ij}, \tag{11.1.8}$$

the difference between the model stress and the actual physical stress. Backus & Mulcahy refer to Γ as the 'stress glut'. This choice of m_{ij} has the convenient property that it will vanish outside our source region.

11.2 The representation theorem

We would now like to relate our equivalent source descriptions to the seismic radiation which they produce. Since we wish to consider sources in dissipative media we will work in the frequency domain, and use complex moduli (7.4.7) in the constitutive relation connecting our model stress $\tau_{ij}(\mathbf{x}, \omega)$ to the displacement $\mathbf{u}(\mathbf{x}, \omega)$ so that

$$\tau_{ij}(\mathbf{x}, \omega) = c_{ijkl}(\mathbf{x}, \omega)\frac{\partial}{\partial x_k}u_l(\mathbf{x}, \omega). \tag{11.2.1}$$

We first establish a suitable form of the elastodynamic representation theorem (Burridge & Knopoff, 1964) relating the displacement at a point to the force distribution within a region and the displacements and tractions on the surface of that region.

The equation of motion in the presence of a body force $\mathbf{f}(\mathbf{x}, \omega)$ is

$$\partial_i\tau_{ij} + \rho\omega^2 u_j = -f_j. \tag{11.2.2}$$

and we will also need to introduce the Green's tensor $G_{jp}(\mathbf{x}, \boldsymbol{\xi}, \omega)$ for a concentrated source. We have a tensor character for $\mathbf{G}$ because the displacement at a point depends on the orientation of the force.

For a unit force in the pth direction at $\boldsymbol{\xi}$, with a Dirac delta function time dependence, the time transform of the displacement in the lth direction at $\mathbf{x}$ (subject to some boundary conditions) is then $G_{lp}(\mathbf{x}, \boldsymbol{\xi}, \omega)$. We also introduce the stress tensor $H_{ijp}(\mathbf{x}, \boldsymbol{\xi}, \omega)$ derived from the Green's tensor $G_{lp}(\mathbf{x}, \boldsymbol{\xi}, \omega)$,

$$H_{ijp}(\mathbf{x}, \boldsymbol{\xi}, \omega) = c_{ijkl}(\mathbf{x}, \omega)\frac{\partial}{\partial x_k}G_{lp}(\mathbf{x}, \boldsymbol{\xi}, \omega). \tag{11.2.3}$$

The Green's tensor then satisfies the equation of motion

$$\partial_i H_{ijp} + \rho\omega^2 G_{jp} = -\delta_{jp}\delta(\mathbf{x} - \boldsymbol{\xi}), \tag{11.2.4}$$

where we have written ∂_i for $\partial/\partial x_i$.

We construct the scalar product of (11.2.2) with $G_{jp}(\mathbf{x}, \boldsymbol{\xi}, \omega)$

$$G_{jp}(\mathbf{x}, \boldsymbol{\xi}, \omega)\partial_i\tau_{ij} + G_{jp}(\mathbf{x}, \boldsymbol{\xi}, \omega)\rho\omega^2 u_j = -G_{jp}(\mathbf{x}, \boldsymbol{\xi}, \omega)f_j, \tag{11.2.5}$$

and the scalar product of (11.2.4) with $u_j(\mathbf{x}, \omega)$,

$$u_j(\mathbf{x}, \omega)\partial_i H_{ijp} + u_j(\mathbf{x}, \omega)\rho\omega^2 G_{jp} = -u_j(\mathbf{x}, \omega)\delta_{jp}\delta(\mathbf{x} - \boldsymbol{\xi}). \tag{11.2.6}$$

On subtracting (11.2.6) from (11.2.5) the acceleration terms cancel, and then integrating over a volume V we find

$$\begin{aligned}\int_V \mathrm{d}^3\mathbf{x}\,[G_{jp}(\mathbf{x}, \boldsymbol{\xi}, \omega)\partial_i\tau_{ij}(\mathbf{x}, \omega) - u_j(\mathbf{x}, \omega)\partial_i H_{ijp}(\mathbf{x}, \boldsymbol{\xi}, \omega)]\\ = -\int_V \mathrm{d}^3\mathbf{x}\,G_{jp}(\mathbf{x}, \boldsymbol{\xi}, \omega)f_j(\mathbf{x}, \omega) + \int_V \mathrm{d}^3\mathbf{x}\,\delta_{jp}\delta(\mathbf{x} - \boldsymbol{\xi})u_j(\mathbf{x}, \omega).\end{aligned} \tag{11.2.7}$$

Using the properties of the delta function we can recast (11.2.7) in the form

$$\begin{aligned}\Theta(\boldsymbol{\xi})u_p(\boldsymbol{\xi}, \omega) = &\int_V \mathrm{d}^3\mathbf{x}\,G_{jp}(\mathbf{x}, \boldsymbol{\xi}, \omega)f_j(\mathbf{x}, \omega)\\ &+ \int_V \mathrm{d}^3\mathbf{x}\,[G_{jp}(\mathbf{x}, \boldsymbol{\xi}, \omega)\partial_i\tau_{ij}(\mathbf{x}, \omega) - u_j(\mathbf{x}, \omega)\partial_i H_{ijp}(\mathbf{x}, \boldsymbol{\xi}, \omega)],\end{aligned} \tag{11.2.8}$$

where

$$\Theta(\boldsymbol{\xi}) = \begin{cases} 1, & \boldsymbol{\xi} \in V, \\ 0, & \boldsymbol{\xi} \notin V. \end{cases} \tag{11.2.9}$$

If the source point $\boldsymbol{\xi}$ is excluded from the integration volume, the right hand side of (11.2.8) must vanish. The second integral in (11.2.8) can be converted into an integral over the surface ∂V of the region V, by use of the tensor divergence theorem, using the anisotropic constitutive equation (11.2.1) e.g.

$$\int_V \mathrm{d}^3\mathbf{x}\, G_{jp}(\mathbf{x},\boldsymbol{\xi},\omega)\partial_i \tau_{ij}(\mathbf{x},\omega) = \int_{\partial V} \mathrm{d}^2\mathbf{x}\, n_p G_{jp}(\mathbf{x},\boldsymbol{\xi},\omega)\tau_{ij}(\mathbf{x},\omega), \tag{11.2.10}$$

where $\mathbf{n}$ is the normal to the surface ∂V. We can therefore rewrite (11.2.8) in the form of a representation theorem for the displacement $\mathbf{u}(\mathbf{x},\omega)$,

$$\begin{aligned} \Theta(\mathbf{x})u_k(\mathbf{x},\omega) = & \int_V \mathrm{d}^3\boldsymbol{\xi}\, G_{qk}(\boldsymbol{\xi},\mathbf{x},\omega) f_q(\boldsymbol{\xi},\omega) \\ & + \int_{\partial V} \mathrm{d}^2\boldsymbol{\xi}\, n_p[G_{qk}(\boldsymbol{\xi},\mathbf{x},\omega)\tau_{pq}(\boldsymbol{\xi},\omega) - u_q(\boldsymbol{\xi},\omega)H_{pkq}(\boldsymbol{\xi},\mathbf{x},\omega)], \end{aligned} \tag{11.2.11}$$

where for subsequent convenience we have interchanged the roles of $\mathbf{x}$ and $\boldsymbol{\xi}$.

The representation theorem (11.2.11) applies to an arbitrary volume V and we may, for example, take V to be the whole Earth. Consider applying (11.2.11) directly to a Greens tensor $G_{kp}(\mathbf{x},\boldsymbol{\xi},\omega)$ with a homogeneous boundary condition on ∂V, such as vanishing traction for a free surface. The force distribution $f_q(\mathbf{x}) = \delta_{qp}\delta(\mathbf{x}-\boldsymbol{\xi})$ and so when $\boldsymbol{\xi}$ lies inside V

$$G_{kp}(\mathbf{x},\boldsymbol{\xi},\omega) = \int_V \mathrm{d}^3\boldsymbol{\eta}\, G_{qk}(\mathbf{x},\boldsymbol{\eta},\omega)\delta_{qp}\delta(\mathbf{x}-\boldsymbol{\xi}), \tag{11.2.12}$$

since the surface integral will vanish by the nature of the boundary conditions. Thus we obtain the reciprocity relation

$$G_{jp}(\mathbf{x},\boldsymbol{\xi},\omega) = G_{pj}(\boldsymbol{\xi},\mathbf{x},\omega). \tag{11.2.13}$$

This relation enables us to recast the representation for $\mathbf{u}(\mathbf{x},\omega)$ so that the Green's tensor elements correspond to a receiver at $\mathbf{x}$ and a source at $\boldsymbol{\xi}$, thus

$$\begin{aligned} \Theta(\mathbf{x})u_k(\mathbf{x},\omega) = & \int_V \mathrm{d}^3\boldsymbol{\xi}\, G_{kq}(\mathbf{x},\boldsymbol{\xi},\omega) f_q(\boldsymbol{\xi},\omega) \\ & + \int_{\partial V} \mathrm{d}^2\boldsymbol{\xi}\, [G_{kq}(\mathbf{x},\boldsymbol{\xi},\omega)t_q(\boldsymbol{\xi},\omega) - u_q(\boldsymbol{\xi},\omega)h_{kq}(\mathbf{x},\boldsymbol{\xi},\omega)], \end{aligned} \tag{11.2.14}$$

in terms of the traction components t_q and the traction elements h_{kq}, associated with the Green's tensor, on ∂V.

11.3 Source representation

Now we have established the representation theorem (11.2.14) we will use it to establish different classes of source descriptions. Equivalent forces $\mathbf{e}$ are introduced whenever the model stress tensor $\boldsymbol{\tau}$ departs from the actual physical stress and the contribution to the displacement field arising from these equivalent forces can be expressed as

$$u_k(\mathbf{x}, \omega) = \int_V \mathrm{d}^3\boldsymbol{\eta}\, G_{kq}(\mathbf{x}, \boldsymbol{\eta}, \omega) e_q(\boldsymbol{\eta}, \omega). \tag{11.3.1}$$

This radiation field may be alternatively expressed in terms of a moment tensor density. We express the equivalent force distribution in terms of the gradient of the moment tensor density by using the definitions (11.1.5), (11.1.6), so that

$$u_k(\mathbf{x}, \omega) = \int_V \mathrm{d}^3\boldsymbol{\eta}\, G_{kq}(\mathbf{x}, \boldsymbol{\eta}, \omega) \partial_p m_{pq}(\boldsymbol{\eta}, \omega), \tag{11.3.2}$$

and we then integrate by parts to yield

$$u_k(\mathbf{x}, \omega) = \int_V \mathrm{d}^3\boldsymbol{\eta}\, \partial_p G_{kq}(\mathbf{x}, \boldsymbol{\eta}, \omega) m_{pq}(\boldsymbol{\eta}, \omega). \tag{11.3.3}$$

The integration in (11.3.1) will be restricted to the source region Y and this restriction applies in (11.3.3) if the 'stress-glut' $\boldsymbol{\Gamma}$ is chosen as the moment tensor density.

11.3.1 Explosion sources

Suppose we have a numerical model which allows us to predict the full displacement and traction field generated by an explosion as a function of position. The calculation may be extended out into the linear regime, e.g., to the 'elastic radius' and the values for displacement $\mathbf{u}^s$ and traction $\mathbf{t}^s$ on a surface S surrounding the initiation point may then be used to determine the seismic radiation. The radiated displacement from this explosive source can then be found from

$$u_k(\mathbf{x}, \omega) = \int_S \mathrm{d}^2\boldsymbol{\xi}\, \{G_{kq}(\mathbf{x}, \boldsymbol{\xi}, \omega) t^s_q(\boldsymbol{\xi}, \omega) - u^s_q(\boldsymbol{\xi}, \omega) h_{kq}(\mathbf{x}, \boldsymbol{\xi}, \omega)\}. \tag{11.3.4}$$

For nuclear explosions such models based on finite element or finite difference calculations have reached a high level of sophistication. If we employ a Green's tensor which satisfies the condition of vanishing traction at the free surface, (11.3.4) will give a good representation of the radiated displacement up to the time of the return of surface reflections to the source region. For shallow explosions the values of $\mathbf{u}^s$ and $\mathbf{t}^s$ should include the effect of surface interactions, e.g. spalling, and then (11.3.4) can be used as a representation of the radiation induced by the explosion itself. Frequently, a major explosion will also initiate energy release by fault activation ('tectonic release') and this component will need to be modelled separately.

For a full description of the effective source associated with the detonation of the explosion, the Green's tensor to be used in (11.3.4) should correspond to propagation in the geological situation prevailing before the explosion, including any pre-existing stress fields. Even though prestress will be included in the calculation of $\mathbf{t}^s$, $\mathbf{u}^s$, it is commonly neglected away from the surface S and a relatively simple Green's function is used to estimate the radiation via (11.3.4).

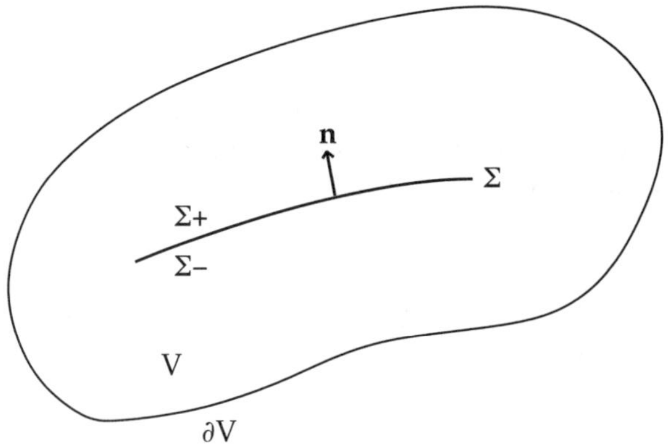

Figure 11.2. Surface of discontinuity in displacement and tractions Σ.

11.3.2 Earthquake faulting

A source representation such as (11.3.4) in terms of displacement and traction behaviour on a surface S is not restricted to explosions and could be used in association with some numerical modelling of the process of faulting in an earthquake.

Normally an earthquake is represented by some dynamic discontinuity in displacement across a fault surface Σ. This singular case can be derived from (11.3.4) by taking S to consist of the two surfaces Σ^+, Σ^- lying on either side of the fault surface Σ and joined at the termination of the dislocation (figure 11.2). When the Green's tensor corresponds to the original configuration in V, $\mathbf{G}_k(\mathbf{x}, \boldsymbol{\xi}, \omega)$ and its associated traction $\mathbf{h}_k(\mathbf{x}, \boldsymbol{\xi}, \omega)$ will be continuous across Σ. The two surface integrals over Σ^+, Σ^- can therefore be combined into a single integral

$$u_k(\mathbf{x}, \omega) = -\int_\Sigma \mathrm{d}^2\boldsymbol{\xi}\, \{G_{kq}(\mathbf{x}, \boldsymbol{\xi}, \omega)[t_q(\boldsymbol{\xi}, \omega)]_-^+ - [u_q(\boldsymbol{\xi}, \omega)]_-^+ h_{kq}(\mathbf{x}, \boldsymbol{\xi}, \omega)\}, \quad (11.3.5)$$

where $[t_q(\boldsymbol{\xi}, \omega)]_-^+$, $[u_q(\boldsymbol{\xi}, \omega)]_-^+$ represent the jumps in traction and displacement in passing from Σ^- to Σ^+. The normal $\mathbf{n}$ to Σ is taken to be directed from Σ^- to Σ^+. The equivalent source distribution will lie along the fault surface Σ. We allow for a general dislocation in displacement and also traction jumps in order to provide a flexible parameterisation of possible sources associated with a fault surface.

As we allow for seismic wave attenuation, the traction associated with the Green's tensor is given by

$$h_{kq}(\mathbf{x}, \boldsymbol{\xi}, \omega) = n_p(\boldsymbol{\xi}) c_{pqrs}(\boldsymbol{\xi}, \omega) \partial_r G_{ks}(\mathbf{x}, \boldsymbol{\xi}, \omega). \quad (11.3.6)$$

In (11.3.5) we are only concerned with the values of G_{kq} and h_{kq} on the fault surface Σ, so we introduce the Dirac delta function $\delta_\Sigma(\boldsymbol{\xi}, \boldsymbol{\eta})$ and its derivative localised on the surface Σ. The expression for the radiated seismic displacement (11.3.5) in terms of a surface integral over Σ can then be cast into the form

$$u_k(\mathbf{x},\omega) = -\int_V \mathrm{d}^3\boldsymbol{\eta}\, G_{kq}(\mathbf{x},\boldsymbol{\eta},\omega)\Big\{[t_q(\boldsymbol{\xi},\omega)]_-^+\delta_\Sigma(\boldsymbol{\xi},\boldsymbol{\eta}) \\ +[u_s(\boldsymbol{\xi},\omega)]_-^+ n_r(\boldsymbol{\xi})c_{pqrs}(\boldsymbol{\xi},\omega)\partial_p\delta_\Sigma(\boldsymbol{\xi},\boldsymbol{\eta})\Big\}, \quad (11.3.7)$$

where $-\partial_p\delta$ extracts the derivative of the function it acts upon and we have exploited the symmetry of the moduli $c_{rspq} = c_{pqrs}$. Equation (11.3.7) has now been cast in the form of (11.3.1) and so we may recognise the set of equivalent forces **e**. For the traction jump we have force elements distributed along Σ weighted by the size of the discontinuity. The displacement jump leads to force doublets along Σ which are best represented in terms of a moment tensor density m_{pq}.

We emphasise the difference in character between the two classes of discontinuity by writing

$$u_k(\mathbf{x},\omega) = \int_V \mathrm{d}^3\boldsymbol{\eta}\,\{G_{kq}(\mathbf{x},\boldsymbol{\eta},\omega)\epsilon_q(\boldsymbol{\eta},\omega) + \partial_p G_{kq}(\mathbf{x},\boldsymbol{\eta},\omega)m_{pq}(\boldsymbol{\eta},\omega)\}. \quad (11.3.8)$$

The forces $\boldsymbol{\epsilon}$ are then determined by the traction jump

$$\epsilon_q(\boldsymbol{\eta},\omega) = -n_r(\boldsymbol{\xi})[\tau_{qr}(\boldsymbol{\xi},\omega)]_-^+\delta_\Sigma(\boldsymbol{\xi},\boldsymbol{\eta}). \quad (11.3.9)$$

The moment tensor density is specified by the displacement jump $[u]$ across the surface Σ

$$m_{pq}(\boldsymbol{\eta},\omega) = n_r(\boldsymbol{\xi})c_{pqrs}(\boldsymbol{\xi},\omega)[u_s(\boldsymbol{\xi},\omega)]_-^+\delta_\Sigma(\boldsymbol{\xi},\boldsymbol{\eta}), \quad (11.3.10)$$

which has the required symmetry for an indigenous source contribution (11.1.7).

The seismic radiation predicted by (11.3.8) is determined purely by the displacement and traction jumps across the surface Σ and the properties of the material surrounding the fault appear only indirectly through the nature of the Green's tensor G_{kq}. Frequently, some assumed model of the slip behaviour on the fault is used to specify $[\mathbf{u}]_-^+$. However, a full solution for $[\mathbf{u}]_-^+$ requires a calculation in which the propagating fault interacts with its surroundings (see, e.g., Kostrov & Das, 1988).

Generally earthquake models prescribe only tangential displacement jumps and then $n_r[\mathbf{u}]_-^+ = 0$. However, an opening crack model is appropriate to other classes of observable events, e.g., rock bursts in mines.

11.4 The moment tensor and source radiation

In Section 11.3 we have adopted a representation of the seismic radiation in terms of a combination of distributed force elements $\boldsymbol{\epsilon}$ and a moment tensor density m_{pq}

$$u_k(\mathbf{x},\omega) = \int_V \mathrm{d}^3\boldsymbol{\eta}\,\{G_{kq}(\mathbf{x},\boldsymbol{\eta},\omega)\epsilon_q(\boldsymbol{\eta},\omega) \\ +\partial_p G_{kq}(\mathbf{x},\boldsymbol{\eta},\omega)m_{pq}(\boldsymbol{\eta},\omega)\}. \quad (11.4.1)$$

We will henceforth restrict our attention to a 'stress-glut' moment tensor density, so that the integration in (11.4.1) can be restricted to the source region Y in which our model stress $\boldsymbol{\tau}$ differs from the actual physical stress.

For many purposes we are interested in the dominant effect of a source rather than examining the fine details of the dislocation process. We would like therefore to use a simple form such as a point source to characterise the radiation pattern. As we shall see we can indeed create a point multipole representation and, provided that the frequency is not too high, will be able to work with a point "Moment Tensor" representing the weights for a set of dipoles and couples (cf. figure 4.1).

For moderate frequencies the source region will not be large compared with seismic wavelengths and then for a distant observation point $\mathbf{x}$ we can make a Taylor series expansion for the variation of the Green's tensor $G_{kq}(\mathbf{x}, \boldsymbol{\eta}, \omega)$ with $\boldsymbol{\eta}$. Expanding about a suitable point $\mathbf{x}_s$ we have

$$G_{kq}(\mathbf{x}, \boldsymbol{\eta}, \omega) = G_{kq}(\mathbf{x}, \mathbf{x}_s, \omega) + (\eta_i - x_{si})\partial_i G_{kq}(\mathbf{x}, \mathbf{x}_s, \omega)$$
$$+ \tfrac{1}{2}(\eta_i - x_{si})(\eta_j - x_{sj})\partial_{ij} G_{kq}(\mathbf{x}, \boldsymbol{\eta}, \omega) + \dots , \tag{11.4.2}$$

and we expect the number of significant terms will reduce as the frequency diminishes. For a small fault we can use the point of initiation of the disturbance, the 'hypocentre', as $\mathbf{x}_s$. For extended faults the accuracy of the expansion (11.4.2) may be improved by making the expansion about the centroid of the disturbance $\mathbf{x}_S(\omega)$ rather than the hypocentre, which is just the point of initiation (see Dziewonski & Woodhouse, 1983).

The expansion (11.4.2) enables us to approximate the seismic radiation $\mathbf{u}$ through a sequence of terms which represent increasingly detailed aspects of the source behaviour.

We consider first the force contribution to (11.4.1), with the Taylor's series expansion of the Green's tensor we obtain

$$u_k^\epsilon(\mathbf{x}, \omega) = \int_Y \mathrm{d}^3 \boldsymbol{\eta}\, G_{kq}(\mathbf{x}, \boldsymbol{\eta}, \omega)\epsilon_q(\boldsymbol{\eta}, \omega),$$
$$= G_{kq}(\mathbf{x}, \mathbf{x}_s, \omega) \int_Y \mathrm{d}^3 \boldsymbol{\eta}\, \epsilon_q(\boldsymbol{\eta}, \omega)$$
$$+ \partial_i G_{kq}(\mathbf{x}, \mathbf{x}_s, \omega) \int_Y \mathrm{d}^3 \boldsymbol{\eta}\, (\eta_i - x_{si})\epsilon_q(\boldsymbol{\eta}, \omega) + \dots . \tag{11.4.3}$$

The displacement $\mathbf{u}^\epsilon$ is thereby represented in terms of the polynomial moments of the distributed forces,

$$u_k^\epsilon(\mathbf{x}, \omega) = G_{kq}(\mathbf{x}, \mathbf{x}_s, \omega)\mathcal{E}_q(\omega) + \partial_i G_{kq}(\mathbf{x}, \mathbf{x}_s, \omega)\mathcal{E}_{q,i}^{(1)} + \dots . \tag{11.4.4}$$

$\mathcal{E}$ will be the total force exerted on the source region Y and $\mathcal{E}_{q,i}^{(1)}$ the tensor of force moments about the point of expansion $\mathbf{x}_s$. All of the source elements in the series (11.4.4) will appear to be situated at the point $\mathbf{x}_s$. Thus we have represented the distributed force system in Y by a compound point source composed of a delta function and its derivatives at $\mathbf{x}_s$. For an *indigenous* source the total force $\mathcal{E}$ and the moments $\mathcal{E}_{q,i}^{(1)}$ will vanish. It is however appropriate to retain these terms since a number of practical sources, e.g. those depending on surface impact, are not indigenous.

The radiation associated with the moment tensor density may also be expanded as in (11.4.3),

$$\begin{aligned} u_k^m(\mathbf{x},\omega) &= \int_Y \mathrm{d}^3\boldsymbol{\eta}\, \partial_p G_{kq}(\mathbf{x},\boldsymbol{\eta},\omega) m_{pq}(\boldsymbol{\eta},\omega), \\ &= \partial_p G_{kq}(\mathbf{x},\mathbf{x}_\mathrm{S},\omega) \int_Y \mathrm{d}^3\boldsymbol{\eta}\, m_{pq}(\boldsymbol{\eta},\omega) \\ &\quad + \partial_{ip} G_{kq}(\mathbf{x},\mathbf{x}_\mathrm{S},\omega) \int_Y \mathrm{d}^3\boldsymbol{\eta}\, (\eta_i - x_{\mathrm{S}i}) m_{pq}(\boldsymbol{\eta},\omega) + \dots . \end{aligned} \tag{11.4.5}$$

This representation of $\mathbf{u}^m$ in terms of the polynomial moments of the tensor m_{pq} means that we may write

$$\begin{aligned} u_k^m(\mathbf{x},\omega) &= \partial_p G_{kq}(\mathbf{x},\mathbf{x}_\mathrm{S},\omega) M_{pq}(\omega) \\ &\quad + \partial_{ip} G_{kq}(\mathbf{x},\mathbf{x}_\mathrm{S},\omega) M^{(1)}_{pq,i}(\omega) + \dots . \end{aligned} \tag{11.4.6}$$

The integral of the moment tensor density across the source region M_{pq} is the quantity which is frequently referred to as *the* moment tensor

$$M_{pq}(\omega) = \int_Y \mathrm{d}^3\boldsymbol{\eta}\, m_{pq}(\boldsymbol{\eta},\omega). \tag{11.4.7}$$

We see that we can interpret $m_{pq}(\boldsymbol{\eta},\omega)\mathrm{d}^3\boldsymbol{\eta}$ as the moment tensor for the element of volume $\mathrm{d}^3\boldsymbol{\eta}$. The third order tensor $M^{(1)}_{pq,i}$ preserves more information about the spatial distribution of the moment tensor density $m_{pq}(\boldsymbol{\eta},\omega)$, and the inclusion of further terms in the series (11.4.6) gives a higher resolution of the source behaviour.

The sequence of localised moment tensor terms in (11.4.6) represent an equivalent point source description of the radiation from the original source volume Y. The leading order source system is a superposition of first derivatives of a delta function, describing force doublets. The moment tensor M_{pq} may therefore be regarded as the weighting factor to be applied to the nine elements of the array of dipoles and couples illustrated earlier in figure 4.1. The diagonal elements of $\mathbf{M}$ correspond to dipoles and the off-diagonal elements to pure couples. The higher order tensors $\mathbf{M}^{(j)}$ are the weighting factors in a multipole expansion.

11.4.1 Point dislocation sources

We now specialise our results for a dislocation source to the case of a small fault embedded in a uniform medium.

Using (11.3.9) and (11.4.3), we see that for low frequencies any traction jump across the fault Σ can be regarded as equivalent to the point force components

$$\mathcal{E}_j(\omega) = \int_\Sigma \mathrm{d}^2\boldsymbol{\xi}\, n_i(\boldsymbol{\xi})[\tau_{ij}(\boldsymbol{\xi},\omega)]_-^+. \tag{11.4.8}$$

For a small fault we may represent $\mathcal{E}_j$ as an averaged traction jump over the surface area A of the fault

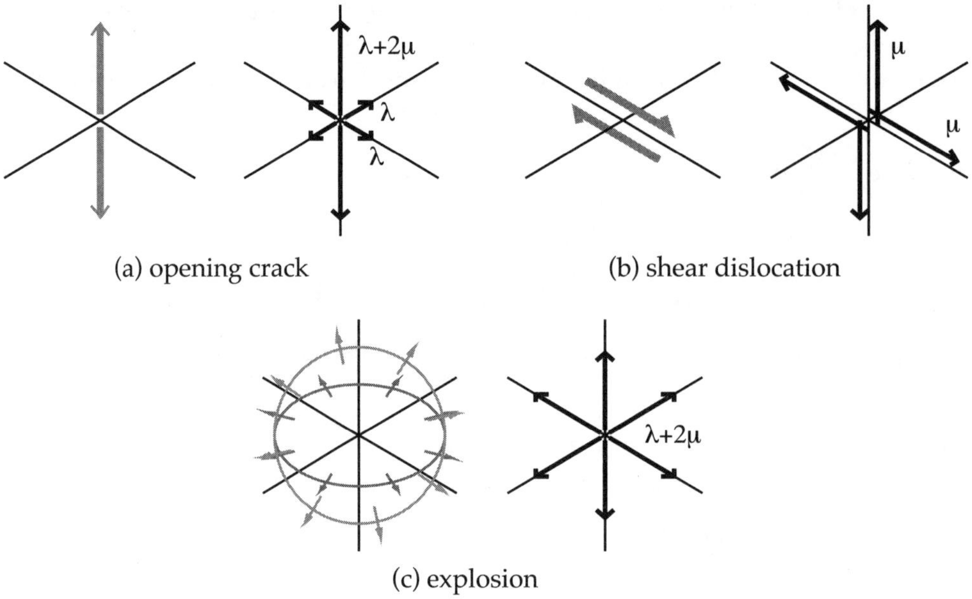

Figure 11.3. Force equivalents for simple source mechanisms: (a) an opening crack; (b) tangential slip on a fault; (c) an explosion.

$$\mathcal{I}_j(\omega) = An_i[\overline{\tau_{ij}(\omega)}]_-^+. \tag{11.4.9}$$

For a displacement discontinuity the equivalent leading order moment tensor is given by

$$M_{ij} = -\int_\Sigma \mathrm{d}^2\boldsymbol{\xi}\, n_k(\boldsymbol{\xi})c_{ijkl}(\boldsymbol{\xi})[u_l(\boldsymbol{\xi},\omega)]_-^+. \tag{11.4.10}$$

Consider now a displacement discontinuity in the direction of the unit vector $\boldsymbol{\nu}$ with jump $[u]$ in an isotropic medium. The moment tensor for a small fault is then

$$M_{ij} = A[\overline{u(\omega)}]\{\lambda n_k \nu_k + \mu(n_i\nu_j + n_j\nu_i)\}, \tag{11.4.11}$$

where $[\overline{u(\omega)}]$ is the averaged slip spectrum on the fault surface.

When the direction $\boldsymbol{\nu}$ lies along the normal $\mathbf{n}$, as in an opening crack - figure 11.3(a),

$$M_{ij} = A[\overline{u(\omega)}]\{\lambda\delta_{ij} + 2\mu n_i n_j\}, \tag{11.4.12}$$

and if we rotate the coordinate system so that the 3 axis lies along the normal, M_{ij} is diagonal with components

$$M_{ij} = A[\overline{u(\omega)}]\mathrm{diag}\{\lambda, \lambda, \lambda + 2\mu\}. \tag{11.4.13}$$

The equivalent force system is thus a centre of dilatation of strength λ with a dipole of strength 2μ along the normal to the fault.

For a purely tangential slip along the fault plane [figure 11.3(b)], as in the displacement to be expected in an earthquake, the slip vector is perpendicular to the normal to the fault plane

$$\boldsymbol{\nu}.\mathbf{n} = 0.$$

The moment tensor components now depend only on the shear modulus μ:

$$M_{ij} = A\mu[\overline{u(\omega)}]\{n_i\nu_j + n_j\nu_i\}. \qquad (11.4.14)$$

In (11.4.14) $\mathbf{n}$ and $\boldsymbol{\nu}$ appear in a completely symmetric role and so the zeroth order moment tensor M_{ij} does not allow one to distinguish between the fault plane with normal $\mathbf{n}$ and the perpendicular auxiliary plane with normal $\boldsymbol{\nu}$. Since $\mathbf{n}.\boldsymbol{\nu}$ vanishes, the trace of the moment tensor is zero

$$\mathrm{tr}\,\mathbf{M} = \sum_i M_{ii} = 0. \qquad (11.4.15)$$

In a coordinate frame with the 3 axis along $\mathbf{n}$ and the slip $\boldsymbol{\nu}$ in the direction of the 1-axis, the moment tensor has only two non-zero components:

$$M_{13} = M_{31} = M_0(\omega), \qquad \text{all other } M_{ij} = 0. \qquad (11.4.16)$$

Here we have introduced the moment spectrum

$$M_0(\omega) = A\mu[\overline{u(\omega)}], \qquad (11.4.17)$$

which defines the source characteristics as a function of time. The equivalent force distribution specified by (11.4.16) will be two couples of equal and opposite moment – the familiar 'double-couple' model of fault radiation.

These equivalent force systems apply to faults which are small compared with the wavelengths of the recorded seismic waves. For larger faults, the zeroth order point source contributions we have just discussed do not, by themselves, provide an adequate representation of the seismic radiation. In addition we need the first order contributions which give quadrupole sources and perhaps even higher order terms.

To a similar level of approximation, we can simulate the effect of a point explosion or implosion by using an isotropic moment tensor

$$M_{ij} = M_0(\omega)\delta_{ij}. \qquad (11.4.18)$$

Once again it is convenient to work in terms of the moment spectrum, which is now equal to $(\lambda + 2\mu)A_e\overline{u_r(\omega)}$ where A_e is the surface area of a sphere with the 'elastic radius' r_e and $\overline{u_r(\omega)}$ is the average radial displacement spectrum at this radius (Müller, 1973). The equivalent force system consists of three perpendicular dipoles of equal strength [figure 11.3(c)].

11.4.2 Radiation into an unbounded medium

With the approximation of a source of small spatial extent, the seismic displacement generated by the presence of a dislocation is given by

$$u_i(\mathbf{x},\omega) = G_{ij}(\mathbf{x},\mathbf{x}_S,\omega)\mathcal{F}_j(\omega) + \partial_k G_{ij}(\mathbf{x},\mathbf{x}_S,\omega)M_{jk}(\omega). \tag{11.4.19}$$

We specialise to a situation with $\mathbf{x}_S$ at the origin and consider an observation point at $\mathbf{R}$ in a direction specified by direction cosines γ_i. In the frequency domain we see that we have a product between the Green's tensor and the spectrum of the force or moment tensor; on transformation back into the time domain this produces a convolution of the temporal Green's tensor and the force or moment tensor time functions.

The displacement produced by a point force in an unbounded isotropic elastic medium was first given by Stokes (1849) and the results were extended to couple sources by Love (1903). These results enable us to write explicit expressions for the source radiation.

The force contribution to the displacement is

$$\begin{aligned} 4\pi\rho u_i^{\mathfrak{f}}(\mathbf{R},t) = (3\gamma_i\gamma_j - \delta_{ij})R^{-3}&\int_{R/\alpha}^{R/\beta} \mathrm{d}s\, s\mathcal{F}_j(t-s) \\ &+\gamma_i\gamma_j(\alpha^2 R)^{-1}\mathcal{F}_j(t-R/\alpha) \\ &-(\gamma_i\gamma_j - \delta_{ij})(\beta^2 R)^{-1}\mathcal{F}_j(t-R/\beta). \end{aligned} \tag{11.4.20}$$

The 'far-field' contribution decaying as R^{-1} follows the same time dependence as the force $\mathcal{F}$. Between the P and S wave arrivals is a disturbance with decays as R^{-3} and so becomes steadily less important as the observation point moves away from the source.

The displacement associated with the moment tensor M_{ij} depends on the spatial derivative of the Green's tensor and is given by

$$\begin{aligned} 4\pi\rho u_i^{m}(\mathbf{R},t) = 3(5\gamma_i\gamma_j\gamma_k - l_{ijk})R^{-4}&\int_{R/\alpha}^{R/\beta} \mathrm{d}s\, sM_{jk}(t-s) \\ &-(6\gamma_i\gamma_j\gamma_k - l_{ijk})(\alpha^2R^2)^{-1}M_{jk}(t-R/\alpha) \\ &+(6\gamma_i\gamma_j\gamma_k - l_{ijk} - \delta_{ij}\gamma_k)(\beta^2R^2)^{-1}M_{jk}(t-R/\beta) \\ &+\gamma_i\gamma_j\gamma_k(\alpha^3R)^{-1}\partial_t M_{jk}(t-R/\alpha) \\ &-(\gamma_i\gamma_j - \delta_{ij})\gamma_k(\beta^3R)^{-1}\partial_t M_{jk}(t-R/\beta), \end{aligned} \tag{11.4.21}$$

where

$$l_{ijk} = \gamma_i\delta_{jk} + \gamma_j\delta_{ik} + \gamma_k\delta_{ij}. \tag{11.4.22}$$

The 'far-field' terms which decay least rapidly with distance now behave like the time derivative of the moment tensor components.

The radiation predicted in the far field from (11.4.20) and (11.4.21) is rather different for P and S waves. In figure 11.4 we illustrate the radiation patterns in three dimensions for some simple sources. For a single force in the 3 direction the P wave radiation is two-lobed with a $\cos\theta$ dependence in local spherical polar coordinates. The S radiation resembles a doughnut with a $\sin\theta$ dependence. The patterns for a 33 dipole are modulated with a further $\cos\theta$ factor arising from differentiating in the 3 direction; and so we have a two-lobed $\cos^2\theta$ behaviour for P. The corresponding S wave radiation pattern, depending on $\sin\theta\cos\theta$, has an attractive waisted shape. In a

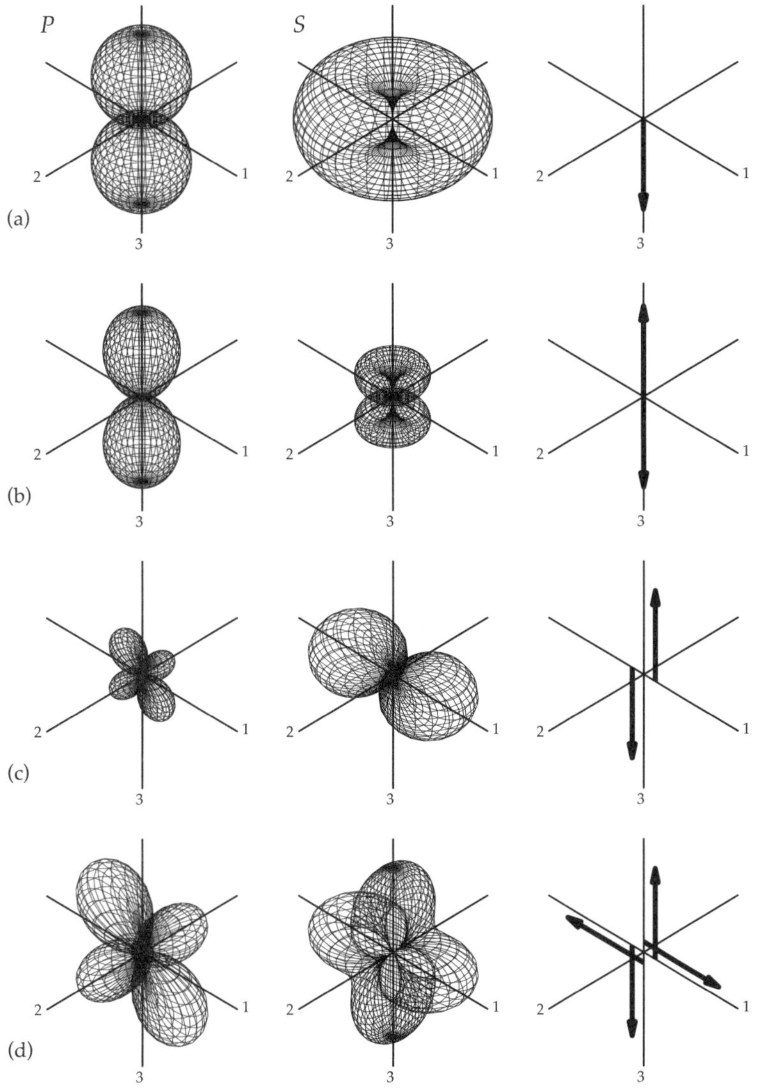

Figure 11.4. Radiation patterns for simple point sources in a uniform medium: a) single 3 force; b) 33 dipole; c) 31 couple; d) 31 double couple.

31 couple a horizontal derivative is applied to the force behaviour which leads to rather different radiation. The *P* wave pattern is here four-lobed with angular dependence $\sin\theta\cos\theta\cos\phi$; displacements of the same sign lie in opposing quadrants and the radiation maxima lie in the 13 plane. There is no *P* wave radiation in the 12 or 23 planes. The *S* wave radiation is proportional to $\sin^2\theta\cos\phi$ and this gives a two-lobed pattern with maxima along the 1 axis and a null on the 23 plane; the two lobes give an opposite sense of displacement to the medium and so impart a couple.

For an explosion, with an isotropic moment tensor $M_{jk} = M_0\delta_{jk}$, (11.4.21) reduces to

$$\begin{aligned} 4\pi\rho u_i(\mathbf{R},t) = \gamma_i\{&(\alpha^2R^2)^{-1}M_0(t-R/\alpha) \\ &+(\alpha^3R)^{-1}\partial_t M_0(t-R/\alpha)\}. \end{aligned} \tag{11.4.23}$$

This is a spherically symmetric P disturbance with purely radial displacement and no S wave part.

For a double couple without moment, simulating slip on a fault plane, $M_{jk} = M_0[n_i\nu_j + n_j\nu_i]$, the displacement at the receiver point $\mathbf{R}$ is given by

$$\begin{aligned} 4\pi\rho u_i(\mathbf{R},t) = &(9Q_i - 6T_i)R^{-4}\int_{R/\alpha}^{R/\beta} \mathrm{d}s\, sM_0(t-s) \\ &+(4Q_i - 2T_i)(\alpha^2R^2)^{-1}M_0(t-R/\alpha) \\ &-(3Q_i - 3T_i)(\beta^2R^2)^{-1}M_0(t-R/\beta) \\ &+Q_i(\alpha^3R)^{-1}\partial_t M_0(t-R/\alpha) \\ &+T_i(\beta^3R)^{-1}\partial_t M_0(t-R/\beta), \end{aligned} \tag{11.4.24}$$

where

$$\begin{aligned} Q_i &= 2\gamma_i(\gamma_j\nu_j)(\gamma_k n_k), \\ T_i &= n_i(\gamma_j\nu_j) + \nu_i(\gamma_k n_k) - Q_i. \end{aligned} \tag{11.4.25}$$

The vector $\mathbf{Q}$ which specifies the far-field P wave radiation is purely radial, along the direction of $\mathbf{R}$. Further $\mathbf{Q}.\mathbf{T} = Q_iT_i = 0$, so that the far-field S wave radiation which depends on $\mathbf{T}$ is purely transverse.

For both the explosion and the double couple, the radiation for the whole field may be expressed in terms of the far-field factors.

It is interesting to compare the radiation pattern for a double couple with that for a single couple with moment. If we take $\mathbf{n}$ in the 3 direction and $\boldsymbol{\nu}$ in the 1 direction then

$$Q_i = 2\gamma_i\gamma_1\gamma_3, \quad T_i = \gamma_1\delta_{i3} + \gamma_3\delta_{i1} - 2\gamma_i\gamma_1\gamma_3. \tag{11.4.26}$$

The far-field P wave radiation pattern is given by

$$|\mathbf{Q}|_{31} = 2\gamma_1\gamma_3 = \sin 2\theta\cos\phi, \tag{11.4.27}$$

and the far-field S radiation depends on

$$\begin{aligned} |\mathbf{T}|_{31} &= (\gamma_1^2 + \gamma_3^2 - 4\gamma_1^2\gamma_3^2)^{1/2} \\ &= (\cos^2 2\theta\cos^2\phi + \cos^2\theta\sin^2\phi)^{1/2}. \end{aligned} \tag{11.4.28}$$

The P wave radiation pattern for the double couple is just twice that for either of the constituent couples, so that there are still nodes in the pattern on the fault plane (12) and the auxiliary plane (23). The S wave radiation is now four lobed rather than two lobed and the net torque vanishes.

11.4.3 Influences on radiation patterns

The classical method of source characterisation, the fault-plane solution, relies on the assumption that the radiation leaving the source is indeed that for the same source in a uniform medium. Observations of P-wave polarity at distant stations are projected back onto the surface of a notional focal sphere surrounding the source, with allowance for the main effects of the Earth's structure using ray theory. The method requires that we can recognise the nodal planes separating dilatations and compressions, which should correspond to the fault plane and auxiliary plane. For a small source for which the localised moment tensor model is appropriate, the radiation pattern itself is not sufficient to distinguish the fault plane and other, usually geological, criteria have to be used.

In recent years centroid moment tensors have become the main form of representation of seismic sources. This is a convenient summary of point source properties but we need to recognise the limitations of this form. The point source will only be appropriate for frequencies low enough that the spatial extent is small compared with the source region. If there is a significant displacement between the location of the centroid and the hypocentre, the source model is unlikely to be a good representation at higher frequencies. Further, most centroid moment tensors do not correspond directly to a simple dislocation model and a variety of ways have been introduced to describe the nature of the source representation. Lay & Wallace (1995, p344-346) give a good summary of the different ways in which the deviatoric component of the moment tensor can be represented in terms of double couples and dipoles. Hudson, Pearce & Rogers (1989) have introduced an ingenious plot to represent the full range of different moment tensors within a single planar diagram.

Large earthquakes can have complex rupture on multiple fault systems, and then a single point moment tensor representation will then deviate significantly from a double-couple (see e.g. Bowman, 1992; Kuge & Kawakatsu, 1990). The details can sometimes be resolved by using higher frequency observations and distinct sub-event models.

Whatever the form of source representation, a number of factors can conspire to modify the apparent radiation pattern. We have assumed that the source lies in a uniform medium, and it is important that the correct seismic velocities are used at the source so that the angles of emergence for rays from the focal sphere are correctly positioned. If mantle velocities are used for a crustal double-couple event the apparent nodal planes will not be orthogonal; however this problem disappears if the correct velocities are employed.

When a source occurs in a region of velocity gradient, the radiation patterns will be modified by the presence of the structure. The effects will be frequency dependent. At low frequencies the radiation pattern will have the same form as in a locally uniform medium, but as the frequency increases the departures from the simple theory given above can become significant.

Many small earthquakes occur within regions which are under strain before a major earthquake. Such a prestrain has a number of effects (Walton, 1974). The presence of the prestrain will lead to anisotropy due to modification of existing crack systems in the rock, in addition to weak anisotropy associated with the strain itself. These combined effects will modify the radiation pattern, in particular in the neighbourhood of the crossing of the nodal planes. The anisotropy will lead to splitting of the S wave degeneracy, two quasi-S waves will exist with different velocity. Observations of such S wave splitting in an active earthquake zone have been reported by Crampin et al (1980).

The radiation patterns we have calculated depend on the assumption of an unbounded medium. As pointed out by Burridge, Lapwood & Knopoff (1964), the presence of a free surface close to the source significantly modifies the radiation pattern. If we insist on using an unbounded medium Green's tensor we cannot use the usual moment tensor representation (11.4.4). For sources which are shallow compared with the recorded wavelength, interference of P with the surface reflections pP, sP will modify the apparent first motion and give records which resemble those seen near nodal planes. The characteristic patterns of interference can themselves be exploited to provide information on source depth and the source mechanism with suitable modelling of higher frequency observations.

Appendix: Principal values of the moment tensor

We have seen above that we can construct the moment tensor corresponding to a point shear-dislocation source, but what are the required conditions for a given moment tensor to correspond to a shear dislocation?

Necessary and sufficient conditions for M_{pq} to have the required form of a multiple of $n_p\nu_q + \nu_p n_q$ are that
(i) one principal value of $\mathbf{M}$ vanishes, and
(ii) the trace of $\mathbf{M}$ is zero.
Thus for a double-couple model we require both $\det\mathbf{M}$ and $\operatorname{tr}\mathbf{M} = M_{xx} + M_{yy} + M_{zz}$ to be zero. In this case the principal directions are

$$\begin{aligned} \hat{\mathbf{p}} &= (\mathbf{n} - \boldsymbol{\nu})/\sqrt{2}, && \text{principal value } 1, \\ \hat{\mathbf{t}} &= (\mathbf{n} + \boldsymbol{\nu})/\sqrt{2}, && \text{principal value } -1, \\ \hat{\mathbf{b}} &= (\mathbf{n} \wedge \boldsymbol{\nu}), && \text{principal value } 0. \end{aligned} \tag{11.a.1}$$

The vector $\hat{\mathbf{p}}$ represents the compressional axis P, corresponding to the maximum compressional deformation, $\hat{\mathbf{t}}$ represents the tension axis T with minimum compressional deformation, and $\hat{\mathbf{b}}$ is the null axis B.

For a general moment tensor the principal values of $\mathbf{M}$ can also be found fairly easily. The first step is to remove any isotropic component by forcing the trace to vanish, thus we construct

$$M'_{ij} = M_{ij} - \tfrac{1}{3}\operatorname{tr}\mathbf{M}\,\delta_{ij}. \tag{11.a.2}$$

The principal values of $\mathbf{M}'$ then satisfy

$$\lambda^3 - K\lambda - \det\mathbf{M}' = 0, \tag{11.a.3}$$

where

$$K = M_{xy}^2 + M_{xz}^2 + M_{yz}^2 - M_{xx}M_{yy} - M_{yy}M_{zz} - M_{xx}M_{zz} \qquad (11.a.4)$$

with the following solutions:
(a) if $\det \mathbf{M}' = 0$,

$$\lambda = 0, \quad K^{1/2}, \quad -K^{-1/2} \qquad (11.a.5)$$

(b) if $\det \mathbf{M}' \neq 0$

$$\lambda = q\cos\theta, \quad q\cos(\theta + \tfrac{2}{3}\pi), \quad q\cos(\theta + \tfrac{4}{3}\pi), \qquad (11.a.6)$$

with

$$q = 2(\tfrac{1}{3}K)^{1/2}, \theta = \tfrac{1}{3}\cos^{-1}\left(\frac{3\det \mathbf{M}'}{Kq}\right). \qquad (11.a.7)$$

The principal directions may then be found by solving the equation system

$$(M'_{ij} - \lambda_{ij})u_j = 0, \quad i = 1, 2, 3, \qquad (11.a.8)$$

for the components of the vector $\boldsymbol{u}$.

When the trace of the original tensor $\mathbf{M}$ is non-zero, then $\frac{1}{3}\operatorname{tr}\mathbf{M}$ has to be added to each principal value.

12

Waves in Stratification

As we have previously noted the dominant dependence of the seismic wavespeeds is on depth (radius). As a result the special case of a stratified medium where the wavespeeds depend on just one coordinate provides a useful starting point from which to understand more complex situations.

For uniform media in both horizontal and spherical stratification we are able to work with simple dependencies on the depth x_3 or radial coordinate r. We are able to identify solutions corresponding to upgoing and downgoing waves by the nature of their phase dependence at fixed frequency. When the phase increases with increasing depth x_3 (decreasing radius r) we have a downgoing wave, which transports energy towards greater depth. If on the other hand the phase increases with decreasing depth x_3 (increasing radius r) we have an upgoing wave transporting energy to shallower depth.

When the slowness of the wave is such that the solution of the seismic wave equations does not correspond to a propagating wave we are able to make a link to suitable forms. For horizontal stratification, when $p > 1/v(x_3)$ we associate the "downgoing wave" solution with the solution that decays with depth. For spherical stratification, when $\wp > r/v(r)$ we link the solution which is regular at the origin $r = 0$ with the "downgoing wave". As we shall see below the nature of the analytical relation between propagating and evanescent solutions is somewhat different in the two cases.

12.1 Horizontal stratification

Consider an isotropic medium for which the Lamé moduli λ, μ and the density ρ are just functions of depth x_3, and attenuative effects are included by allowing an imaginary part to the moduli. The equations of motion (8.1.4) take the explicit form:

$$\begin{aligned}
\partial_1\tau_{11} + \partial_2\tau_{12} + \partial_3\tau_{13} &= -\rho\omega^2 u_1 - f_1,\\
\partial_1\tau_{21} + \partial_2\tau_{22} + \partial_3\tau_{23} &= -\rho\omega^2 u_2 - f_2,\\
\partial_1\tau_{31} + \partial_2\tau_{32} + \partial_3\tau_{33} &= -\rho\omega^2 u_3 - f_3.
\end{aligned} \tag{12.1.1}$$

The stress-strain relations are

$$\begin{aligned}
\tau_{11} &= (\lambda + 2\mu)\partial_1 u_1 + \lambda\partial_2 u_2 + \lambda\partial_3 u_3, \\
\tau_{22} &= \lambda\partial_1 u_1 + (\lambda + 2\mu)\partial_2 u_2 + \lambda\partial_3 u_3, \\
\tau_{33} &= \lambda\partial_1 u_1 + \lambda\partial_2 u_2 + (\lambda + 2\mu)\partial_3 u_3, \\
\tau_{13} &= \mu\left(\partial_3 u_1 + \partial_1 u_3\right), \\
\tau_{23} &= \mu\left(\partial_3 u_2 + \partial_2 u_3\right), \\
\tau_{12} &= \mu\left(\partial_2 u_1 + \partial_1 u_2\right).
\end{aligned} \tag{12.1.2}$$

12.1.1 Coupled equations for displacement and traction

We now rewrite the equation of motion and stress-strain relations in a form which emphasises the dependence on the x_3 coordinate by extracting all derivatives with respect to x_3 to the left hand side of the equations. We work in terms of the displacement $\mathbf{u}$ and the components of the traction $\mathbf{t}_3$ and recall the relation of the Lamé moduli to the isotropic seismic wavespeeds

$$\lambda + 2\mu = \rho\alpha^2, \qquad \mu = \rho\beta^2. \tag{12.1.3}$$

For simplicity in notation we also introduce a modulus ratio and three composite moduli

$$\begin{aligned}
\varsigma &= \lambda/(\lambda + 2\mu) = 1 - 2\beta^2/\alpha^2, \\
\rho\nu &= 4\mu(\lambda + \mu)/(\lambda + 2\mu) = 4\rho\beta^2(1 - 2\beta^2/\alpha^2), \\
\rho\upsilon &= 4\mu(3\lambda + 2\mu)/(\lambda + 2\mu) = 4\rho\beta^2(3 - 4\beta^2/\alpha^2), \\
\rho\varpi &= 2\mu\lambda/(\lambda + 2\mu) = 2\rho\beta^2(1 - 2\beta^2/\alpha^2).
\end{aligned} \tag{12.1.4}$$

The coupled equations for the components of displacement and traction are

$$\begin{aligned}
\partial_3 u_1 &= (\rho\beta^2)^{-1}\tau_{13} - \partial_1 u_3, \\
\partial_3 u_2 &= (\rho\beta^2)^{-1}\tau_{23} - \partial_2 u_3, \\
\partial_3 u_3 &= (\rho\alpha^2)^{-1}\tau_{33} - \varsigma\left(\partial_1 u_1 + \partial_2 u_2\right), \\
\partial_3 \tau_{13} &= -\rho\omega^2 u_1 - \partial_1\tau_{11} - \partial_2\tau_{12} - f_1, \\
\partial_3 \tau_{23} &= -\rho\omega^2 u_2 - \partial_1\tau_{21} - \partial_2\tau_{22} - f_2, \\
\partial_3 \tau_{33} &= -\rho\omega^2 u_3 - \partial_1\tau_{31} - \partial_2\tau_{32} - f_3.
\end{aligned} \tag{12.1.5}$$

We can use the remaining stress-strain relations and the symmetry properties of the stress tensor to convert the right hand side of the equation set (12.1.5) entirely in terms of the displacement and the 3-traction. The required terms are

$$\begin{aligned}
\partial_1\tau_{11} &= \rho\nu\partial_{11}u_1 + \rho\varpi\partial_{12}u_2 + \varsigma\partial_1\tau_{33}, \\
\partial_2\tau_{22} &= \rho\nu\partial_{22}u_2 + \rho\varpi\partial_{12}u_1 + \varsigma\partial_2\tau_{33}, \\
\partial_1\tau_{12} &= \rho\beta^2\partial_{12}u_1 + \rho\beta^2\partial_{11}u_2, \\
\partial_2\tau_{12} &= \rho\beta^2\partial_{11}u_1 + \rho\beta^2\partial_{12}u_2.
\end{aligned} \tag{12.1.6}$$

Now substituting (12.1.6) back into the equation set (12.1.5) we obtain

$$
\begin{aligned}
\partial_3 u_1 &= (\rho\beta^2)^{-1}\tau_{13} - \partial_1 u_3, \\
\partial_3 u_2 &= (\rho\beta^2)^{-1}\tau_{23} - \partial_2 u_3, \\
\partial_3 u_3 &= (\rho\alpha^2)^{-1}\tau_{33} - \varsigma\,(\partial_1 u_1 + \partial_2 u_2), \\
\partial_3 \tau_{13} &= -\rho\omega^2 u_1 - \rho\nu\partial_{11}u_1 - \rho\beta^2\partial_{22}u_2 - \rho\upsilon\partial_{12}u_2 - \varsigma\partial_1\tau_{33} - f_1, \\
\partial_3 \tau_{23} &= -\rho\omega^2 u_2 - \rho\nu\partial_{22}u_2 - \rho\beta^2\partial_{11}u_2 - \rho\upsilon\partial_{12}u_1 - \varsigma\partial_2\tau_{33} - f_2, \\
\partial_3 \tau_{33} &= -\rho\omega^2 u_3 - \partial_1\tau_{13} - \partial_2\tau_{23} - f_3,
\end{aligned}
\tag{12.1.7}
$$

in terms of the ratio ς and the composite moduli ν, υ introduced in (12.1.4).

We consider solutions for displacement and traction with the form of a plane wave in the horizontal coordinates

$$
\begin{pmatrix} \mathbf{u} \\ \mathbf{t} \end{pmatrix} = \begin{pmatrix} \mathbf{u}(x_3) \\ \mathbf{t}(x_3) \end{pmatrix} \exp(i\omega[p_1x_1 + p_2x_2 - t]). \tag{12.1.8}
$$

In a uniform region each horizontal slowness pair (p_1, p_2) can be associated with solutions with a depth dependence

$$
\exp(i\omega q_n x_3), \quad n = 1, 6. \tag{12.1.9}
$$

The vertical slownesses q_n are to be determined from a matrix eigenvalue problem. The three solutions for which q_n is positive will have a phase which increases with increasing depth x_3 and so can be identified with downgoing waves. The three solutions with q_n negative correspond to upgoing waves. The displacement and tractions corresponding to the different q_n are the eigenvectors of the same matrix system.

For material with a vertical symmetry axis there is no loss of generality by considering a plane wave solution of the form

$$
\begin{pmatrix} \mathbf{u} \\ \mathbf{t} \end{pmatrix} = \begin{pmatrix} \mathbf{u}(x_3) \\ \mathbf{t}(x_3) \end{pmatrix} \exp(i\omega[px_1 - t]), \tag{12.1.10}
$$

since we can rotate the coordinates to bring the x_1-axis into coincidence with the propagation direction. With only a single horizontal slowness component there is a significant simplification in the form of the coupled equations for the displacement and traction components which can now be separated into two distinct sets.

The first set links the 2-component of displacement and traction

$$
\begin{aligned}
\partial_3 u_2 &= (\rho\beta^2)^{-1}\tau_{23}, \\
\partial_3 \tau_{23} &= -\rho\omega^2 u_2 + \rho\beta^2 p^2 u_2 - f_2,
\end{aligned}
\tag{12.1.11}
$$

where we have used the fact that the derivative ∂_1 extracts the term ip from the plane wave solution, and also the absence of any dependence on x_2. This set of equations describes horizontally polarised shear waves (*SH*).

The second set couples the remaining displacement and traction terms, $(u_1, u_3, \tau_{13}, \tau_{33})$ corresponding to linked *P* and vertically polarised shear waves (*SV*):

$$\begin{aligned}
\partial_3 u_1 &= (\rho\beta^2)^{-1}\tau_{13} - \mathrm{i}pu_3, \\
\partial_3 u_3 &= (\rho\alpha^2)^{-1}\tau_{33} - \mathrm{i}p\varsigma u_1, \\
\partial_3 \tau_{13} &= -\rho\omega^2 u_1 + \rho\nu p^2 u_1 - \mathrm{i}p\varsigma\tau_{33} - f_1, \\
\partial_3 \tau_{33} &= -\rho\omega^2 u_3 - \mathrm{i}p\tau_{13} - f_3.
\end{aligned} \tag{12.1.12}$$

We have derived the equations above in terms of cartesian coordinates, but with a slight change of notation we can generate a form which is also suitable for a cylindrical coordinate system with a vertical symmetry axis (see, e.g., Kennett, 1983, Chapter 2). We set

$$\begin{aligned}
U &= \mathrm{i}u_3, \qquad & P &= \mathrm{i}\omega^{-1}\tau_{33}, \\
V &= u_1, & S &= \omega^{-1}\tau_{13}, \\
W &= u_2, & T &= \omega^{-1}\tau_{23}.
\end{aligned} \tag{12.1.13}$$

In terms of these new variables the equation systems (12.1.11), (12.1.12) can be written as:
(a) for *SH* waves,

$$\frac{\partial}{\partial z}\begin{pmatrix} W \\ T \end{pmatrix} = \omega \begin{pmatrix} 0 & (\rho\beta^2)^{-1} \\ \rho[\beta^2 p^2 - 1] & 0 \end{pmatrix} \begin{pmatrix} W \\ T \end{pmatrix} - \begin{pmatrix} 0 \\ F_H \end{pmatrix}, \tag{12.1.14}$$

and (b) for *P-SV* waves

$$\frac{\partial}{\partial z}\begin{pmatrix} U \\ V \\ P \\ S \end{pmatrix} = \omega \begin{pmatrix} 0 & p\varsigma & (\rho\alpha^2)^{-1} & 0 \\ -p & 0 & 0 & (\rho\beta^2)^{-1} \\ -\rho & 0 & 0 & p \\ 0 & \rho[\nu p^2 - 1] & -p\varsigma & 0 \end{pmatrix} \begin{pmatrix} U \\ V \\ P \\ S \end{pmatrix} - \begin{pmatrix} 0 \\ 0 \\ F_z \\ F_V \end{pmatrix}, \tag{12.1.15}$$

where we have also written z for the x_3 coordinate.

12.1.2 Upgoing and downgoing waves

We now specialise to the case of a uniform medium in the absence of any sources. We look for displacement and traction solutions, in terms of the depth coordinate x_3, of the form

$$\exp(\mathrm{i}\omega q x_3), \tag{12.1.16}$$

from which we identify upgoing and downgoing wave solutions via their depth dependence.

12.1.2.1 SH waves:

From the coupled equations (12.1.14) for horizontally polarised shear waves (SH) whose displacement lies in the x_2-direction we generate a 2×2 matrix eigenvalue problem for the vertical slowness q_i,

$$\begin{pmatrix} \mathrm{i}q_i & (\rho\beta^2)^{-1} \\ \rho[\beta^2 p^2 - 1] & \mathrm{i}q_i \end{pmatrix} \begin{pmatrix} W_i \\ T_i \end{pmatrix} = 0. \tag{12.1.17}$$

This eigenvalue equation requires the vertical slownesses q_i to satisfy

$$q_i^2 = \beta^{-2} - p^2, \tag{12.1.18}$$

so that we have two solutions for the vertical wavenumber with opposite signs, and thus opposite phase behaviour with respect to x_3. We set

$$q_\beta = [\beta^{-2} - p^2]^{1/2} \tag{12.1.19}$$

and this positive square root will corresponds to downgoing waves (when $p < 1/\beta$), since the phase will increase with increasing depth. We associate the negative square root with the upgoing waves.

We can then think of the *SH* wavefield as composed of a combination of both upgoing and downgoing *SH* waves

$$\begin{aligned} \chi_D &= H_D \exp(+i\omega q_\beta x_3) \exp(i\omega[px_1 - t]), && \text{downgoing } SH, \\ \chi_U &= H_U \exp(-i\omega q_\beta x_3) \exp(i\omega[px_1 - t]), && \text{upgoing } SH. \end{aligned} \tag{12.1.20}$$

Associated with these wave 'potentials' we will have displacement and traction fields which are the eigenvectors of (12.1.17). For an SH wave travelling at an angle j to the vertical $p = \sin j/\beta$, $q_\beta = \cos j/\beta$ and the associated displacements are

$$\mathbf{u}_{U,D} = C(0, 1, 0). \tag{12.1.21}$$

The displacement-traction vectors for the upgoing and downgoing SH waves are

$$\mathbf{b}_U^H = \begin{pmatrix} W \\ T \end{pmatrix}_U = \epsilon_H \begin{pmatrix} 1 \\ -i\mu q_\beta \end{pmatrix}, \quad \mathbf{b}_D^H = \begin{pmatrix} W \\ T \end{pmatrix}_D = \epsilon_H \begin{pmatrix} 1 \\ i\mu q_\beta \end{pmatrix}, \tag{12.1.22}$$

the normalisation can be conveniently chosen so that the solutions each carry unit energy flux across planes x_3 = const, for which

$$\epsilon_H = 1/(2\mu q_\beta)^{1/2}. \tag{12.1.23}$$

When the horizontal slowness $p > 1/\beta$, we have a complex root for q_β. We will make the choice of branch cut in the complex plane so that the imaginary part of ωq_β will be positive, i.e.,

$$\operatorname{Im} \omega q_\beta \geq 0, \tag{12.1.24}$$

the frequency ω enters from our choice (12.1.16) of trial solution in terms of depth. With the choice (12.1.24), downward propagating waves χ_D in an attenuative medium map into evanescent waves which decay with depth. We can illustrate the behaviour most easily for a perfectly elastic medium for which

$$\exp(i\omega q_\beta x_3) = \exp(-\omega|q_\beta|x_3), \quad p > 1/\beta. \tag{12.1.25}$$

In a similar way, upgoing waves χ_U map to evanescent waves which increase exponentially with increasing depth x_3.

The full wavefield can be constructed by linear superposition of the upgoing and downgoing waves. The motion-traction vector **b** for slowness p and frequency ω at a particular level ($x_3 = z$) can be expressed as

$$\begin{pmatrix} W(z) \\ T(z) \end{pmatrix} = \epsilon_H \begin{pmatrix} 1 \\ -i\mu q_\beta \end{pmatrix} H_U e^{-i\omega q_\beta z} + \epsilon_H \begin{pmatrix} 1 \\ i\mu q_\beta \end{pmatrix} H_D e^{i\omega q_\beta z}, \tag{12.1.26}$$

where H_U, H_D represent the weighting factors for upgoing and downgoing waves respectively. Equation (12.1.26) can be conveniently rewritten as a matrix relation

$$\begin{pmatrix} W(z) \\ T(z) \end{pmatrix} = \epsilon_H \begin{pmatrix} 1 & 1 \\ -i\mu q_\beta & i\mu q_\beta \end{pmatrix} \begin{pmatrix} H_U e^{-i\omega q_\beta z} \\ H_D e^{+i\omega q_\beta z} \end{pmatrix}, \tag{12.1.27}$$

which can be represented schematically as

$$\mathbf{b}(z) = \mathbf{D}\mathbf{v}(z), \tag{12.1.28}$$

where the wavevector

$$\mathbf{v}(z) = [H_U e^{-i\omega q_\beta z}, H_D e^{i\omega q_\beta z}]^T \tag{12.1.29}$$

contains the weighting factors for the upgoing and downgoing waves and their associated vertical phase terms.

12.1.2.2 P-SV waves

Once again using a trial solution (12.1.16) in terms of a propagating wave with respect to x_3, we obtain a 4×4 eigenvalue equation for *P* waves and vertically polarised shear waves (*SV*) waves.

For an isotropic medium, this eigenvalue equation factorises to yield distinct vertical slownesses for P and SV waves

$$q_\alpha^2 = \alpha^{-2} - p^2, \qquad q_\beta^2 = \beta^{-2} - p^2. \tag{12.1.30}$$

In a uniform layer we will have four waves with displacement in the vertical plane

$$\begin{aligned}
\phi_D &= P_D \exp(+i\omega q_\alpha x_3) \exp(i\omega[px_1 - t]) && \text{downgoing } P, \\
\phi_U &= P_U \exp(-i\omega q_\alpha x_3) \exp(i\omega[px_1 - t]) && \text{upgoing } P, && (12.1.31) \\
\psi_D &= S_D \exp(+i\omega q_\beta x_3) \exp(i\omega[px_1 - t]) && \text{downgoing } SV, \\
\psi_U &= S_U \exp(-i\omega q_\beta x_3) \exp(i\omega[px_1 - t]) && \text{upgoing } SV, && (12.1.32)
\end{aligned}$$

with associated displacement and traction fields.

The displacement-traction vectors corresponding to the up and downgoing waves are for *P*:

$$\mathbf{b}^P_{U,D} = \begin{pmatrix} U \\ V \\ P \\ S \end{pmatrix}^P_{U,D} = \epsilon_\alpha \begin{pmatrix} \mp i q_\alpha \\ p \\ \rho(2\beta^2 p^2 - 1) \\ \mp i 2\rho\beta^2 p q_\alpha \end{pmatrix}, \quad \epsilon_\alpha = 1/(2\rho q_\alpha)^{1/2} \tag{12.1.33}$$

where the upper sign corresponds to upgoing waves and the lower sign to downgoing waves. For a *P* wave travelling at an angle i to the vertical, the horizontal slowness $p = \sin i/\alpha$ and the vertical slowness $q_\alpha = \cos i/\alpha$, and the displacement for upgoing and downgoing *P* waves take the form

$$\mathbf{u}^P_{U,D} = A(\sin i, 0, \mp\cos i). \tag{12.1.34}$$

For *SV* waves

$$\mathbf{b}^S_{U,D} = \begin{pmatrix} U \\ V \\ P \\ S \end{pmatrix}^S_{U,D} = \epsilon_\beta \begin{pmatrix} p \\ \pm\mathrm{i}q_\beta \\ \pm\mathrm{i}2\rho\beta^2 pq_\beta \\ \rho(2\beta^2p^2-1) \end{pmatrix}, \quad \epsilon_\beta = 1/(2\rho q_\beta)^{1/2}, \tag{12.1.35}$$

with the sign convention that the upper sign represents upgoing waves. For an *SV* wave travelling at an angle j to the vertical $p = \sin j/\beta$, $q_\beta = \cos j/\beta$ and the associated displacements are

$$\mathbf{u}^S_{U,D} = B(\mp\cos j, 0, \sin j). \tag{12.1.36}$$

For a common horizontal slowness p we have a Snell's law relation between the inclination of P and S waves

$$p = \frac{\sin i}{\alpha} = \frac{\sin j}{\beta}. \tag{12.1.37}$$

The P wave always has the greater inclination to the vertical at fixed slowness.

For *P-SV* waves in an isotropic medium, we can represent the displacement and traction fields in a similar way to (12.1.26), (12.1.27) by making a linear combination of the solutions for upgoing and downgoing P and *SV* waves

$$\begin{pmatrix} U(z) \\ V(z) \\ P(z) \\ S(z) \end{pmatrix} = \mathbf{D} \begin{pmatrix} P_U e^{-\mathrm{i}\omega q_\alpha z} \\ S_U e^{-\mathrm{i}\omega q_\beta z} \\ P_D e^{+\mathrm{i}\omega q_\alpha z} \\ S_D e^{+\mathrm{i}\omega q_\beta z} \end{pmatrix}, \tag{12.1.38}$$

where the columns of $\mathbf{D}$ are the motion-stress vectors for the different styles of waves

$$\mathbf{D} = \left(\mathbf{b}^P_U, \mathbf{b}^S_U, \mathbf{b}^P_D, \mathbf{b}^S_D\right). \tag{12.1.39}$$

This has the same formal structure as (12.1.27), (12.1.28) and we can emphasise the relation by partitioning the wavevector $\mathbf{v}$ into upgoing and downgoing parts, and making a corresponding partition of the eigenvector matrix $\mathbf{D}$ so that

$$\mathbf{b}(z) = \begin{pmatrix} \boldsymbol{w}(z) \\ \boldsymbol{t}(z) \end{pmatrix} = \begin{pmatrix} \boldsymbol{m}_U & \boldsymbol{m}_D \\ \boldsymbol{n}_U & \boldsymbol{n}_D \end{pmatrix} \begin{pmatrix} \boldsymbol{v}_U(z) \\ \boldsymbol{v}_D(z) \end{pmatrix}, \tag{12.1.40}$$

where the action of $\boldsymbol{m}_U$ is to generate the displacement associated with upgoing waves and $\boldsymbol{n}_U$ the corresponding tractions.

The matrices $\boldsymbol{m}$ and $\boldsymbol{n}$ have the explicit forms

$$\begin{aligned} \boldsymbol{m}_{U,D} &= \begin{pmatrix} \mp\mathrm{i}q_\alpha\epsilon_\alpha & p\epsilon_\beta \\ p\epsilon_\alpha & \mp\mathrm{i}q_\beta\epsilon_\beta \end{pmatrix}, \\ \boldsymbol{n}_{U,D} &= \begin{pmatrix} \rho(2\beta^2p^2-1)\epsilon_\alpha & \mp2\mathrm{i}\rho\beta^2pq_\beta\epsilon_\beta \\ \mp2\mathrm{i}\rho\beta^2pq_\alpha\epsilon_\alpha & \rho(2\beta^2p^2-1)\epsilon_\beta \end{pmatrix}. \end{aligned} \tag{12.1.41}$$

We can recover the wave components $\boldsymbol{v}_U$, $\boldsymbol{v}_D$ from the displacement and traction field by inverting (12.1.28)

$$\mathbf{v} = \mathbf{D}^{-1}\mathbf{b}. \tag{12.1.42}$$

The inverse $\mathbf{D}^{-1}$ can be written explicitly in terms of the partitions of $\mathbf{D}$ so that

$$\begin{pmatrix} \boldsymbol{v}_U \\ \boldsymbol{v}_D \end{pmatrix} = \mathrm{i}\begin{pmatrix} -\boldsymbol{n}_D^T & \boldsymbol{m}_D^T \\ \boldsymbol{n}_U^T & -\boldsymbol{m}_U^T \end{pmatrix}\begin{pmatrix} \boldsymbol{w} \\ \boldsymbol{t} \end{pmatrix}. \tag{12.1.43}$$

This relation applies to *SH* waves and *P-SV* waves in both isotropic and transversely isotropic media.

12.2 Spherical stratification

Although the propagation of seismic waves in a uniform zone within spherical stratification can be described in terms of plane waves, this class of representation is not well adapted to dealing with boundary conditions on surfaces $r = const$. Instead of Fourier transforms over the coordinates x_1, x_2 we work directly with spherical polar coordinates (r, θ, ϕ).

When the seismic properties vary only with radius r, the seismic wave equations are separable and we can extract the angular dependence in terms of vector spherical harmonics. Following Aki & Richards (1980) we work in terms of the normalised surface harmonic

$$Y_l^m(\theta, \phi) = (-1)^m\left(\frac{2l+1}{4\pi}\right)^{1/2}\left[\frac{(l-m)!}{(l+m)!}\right]^{1/2} P_l^m(\cos\theta)\mathrm{e}^{\mathrm{i}m\phi}, \tag{12.2.1}$$

where l, m are integers, $-l \le m \le l$, and $P_l^m(\cos\theta)$ is the associated Legendre function. We also introduce the quantity

$$\text{Ł} = [l(l+1)]^{1/2}, \tag{12.2.2}$$

which, as we shall see later, corresponds to the asymptotic angular wavenumber of travelling surface or body waves in the limit $l \to \infty$. In terms of unit vectors $\hat{\mathbf{e}}_r$, $\hat{\mathbf{e}}_\theta$, $\hat{\mathbf{e}}_\phi$ associated with each of the spherical coordinates, we can represent the gradient across the surface of a spherical shell $r = const$ through

$$\nabla_1 = \hat{\mathbf{e}}_\theta\partial_\theta + \hat{\mathbf{e}}_\phi\frac{1}{\sin\theta}\partial_\phi, \tag{12.2.3}$$

and the surface curl operator as

$$\hat{\mathbf{e}}_r \wedge \nabla_1 = -\hat{\mathbf{e}}_\theta\frac{1}{\sin\theta}\partial_\phi + \hat{\mathbf{e}}_\phi\partial_\theta, \tag{12.2.4}$$

acting in a direction perpendicular to ∇_1. The set of vector spherical harmonics

$$\mathbf{P}_l^m = \hat{\mathbf{e}}_r Y_l^m(\theta, \phi), \quad \mathbf{B}_l^m = \frac{1}{\text{Ł}}\nabla_1 Y_l^m(\theta, \phi) \quad \mathbf{C}_l^m = \frac{1}{\text{Ł}}\hat{\mathbf{e}}_r \wedge \nabla_1 Y_l^m(\theta, \phi), \tag{12.2.5}$$

provide a complete basis for the representation of the wavefield. Note that $\mathbf{B}_0^0 = 0$, $\mathbf{C}_0^0 = 0$ since Y_0^0 is independent of θ and ϕ. A wide variety of different choices can be made for the surface harmonics, both in normalisation and form. For example, Dahlen & Tromp (1998, Chapter 8) favour the use of real harmonics $\mathcal{Y}_{lm}(\theta, \phi)$ as more suitable for the representation of the free oscillations of the Earth.

We can express the displacement vector $\mathbf{u}$ and the r-component of traction $\boldsymbol{\tau}_r = (\tau_{rr}, \tau_{r\theta}, \tau_{r\phi})$ in terms of the vector spherical harmonics as

$$\mathbf{u} = U(r)\mathbf{P}_l^m + V(r)\mathbf{B}_l^m + W(r)\mathbf{C}_l^m, \tag{12.2.6}$$

and

$$\omega^{-1}\boldsymbol{\tau}_r = P(r)\mathbf{P}_l^m + S(r)\mathbf{B}_l^m + T(r)\mathbf{C}_l^m, \tag{12.2.7}$$

where for simplicity we have suppressed the dependence of the variables U, V on l,m and emphasised the radial variation. In terms of the displacement and traction elements we have just introduced, we can recast the seismic equations with neglect of gravitation into a form where derivatives with respect to radius are collected on the left hand side of the equation (cf 12.1.14, 12.1.15 in terms of depth). As in the case of horizontal stratification, the coupled first-order differential equations separate into two groups. The first group is associated with *SH* waves and links only the W, T coefficients of the $\mathbf{C}_l^m$ harmonic. The second group links the coefficients of the $\mathbf{P}_l^m$ and $\mathbf{B}_l^m$ harmonics and represents the *P-SV* wave system where the polarisation lies in a plane through the centre of the Earth.

12.2.1 SH waves

For the *SH* system

$$\frac{\partial}{\partial r}\begin{pmatrix} W \\ T \end{pmatrix} = \omega \begin{pmatrix} \dfrac{1}{\omega r} & \dfrac{1}{\mu} \\ \dfrac{\mu}{\omega^2 r^2}(l+2)(l-1) - \rho & -\dfrac{3}{\omega r} \end{pmatrix} \begin{pmatrix} W \\ T \end{pmatrix}. \tag{12.2.8}$$

By analogy with the horizontally stratified case we introduce a displacement-traction vector

$$\mathfrak{b}^H = [W, T]^T, \tag{12.2.9}$$

and we note that $\mathfrak{b}^H$ will be continuous across a spherical surface $r = const.$

In a uniform region we can construct solutions in terms of spherical Bessel functions, e.g. a solution which asymptotically corresponds to an upgoing wave is given by

$$\mathfrak{b}_u^H = \epsilon_H \begin{pmatrix} h_l^{(1)}(k_\beta r) \\ \rho\beta \left\{ h_l^{(1)\prime}(k_\beta r) - \dfrac{1}{k_\beta r} h_l^{(1)}(k_\beta r) \right\} \end{pmatrix}, \tag{12.2.10}$$

where $k_\beta = \omega/\beta$ and $'$ denotes differentiation with respect to the argument. The corresponding solution for a downgoing wave $\mathfrak{b}_d^H$ has $h_l^{(1)}$ replaced by $h_l^{(2)}$,

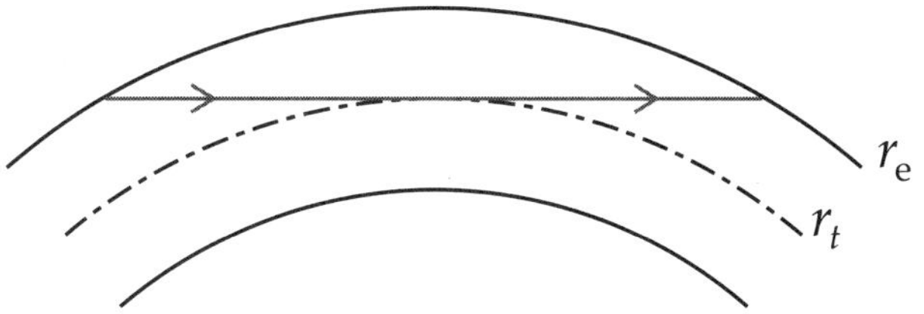

Figure 12.1. A ray in a uniform medium, with an effective turning point at r_t.

$$\mathfrak{b}_d^H = \epsilon_H \left(\begin{array}{c} h_l^{(2)}(k_\beta r) \\ \rho\beta \left\{ h_l^{(2)\prime}(k_\beta r) - \dfrac{1}{k_\beta r} h_l^{(2)}(k_\beta r) \right\} \end{array} \right). \tag{12.2.11}$$

The normalisation constant ϵ_H is chosen so that the time-averaged energy flux across a spherical surface is unity

$$\epsilon_H = \sqrt{2/\rho\beta^3}. \tag{12.2.12}$$

The asymptotic forms of the spherical Hankel functions $h_l^{(1)}$, $h_l^{(2)}$ for large real argument z are

$$h_l^{(1)}(z) = f_l(z)\exp(\mathrm{i}z), \quad h_l^{(2)}(z) = f_l^*(z)\exp(-\mathrm{i}z), \tag{12.2.13}$$

where $f_l(z)$ is a polynomial in $1/z$ and f^* denotes a complex conjugate. We recall that we have assumed a time dependence $\exp(-\mathrm{i}\omega t)$, and so $h_l^{(1)}(k_\beta r)$ represents an upgoing *SH* wave and $h_l^{(2)}(k_\beta r)$ a downgoing wave. Furthermore the $\mathfrak{b}_d^H$ solution has energy flux directed towards the origin and $\mathfrak{b}_u^H$ directed away from the origin, confirming our identification of the physical character of the solutions.

However, although $h_l^{(1)}(k_\beta r)$, $h_l^{(2)}(k_\beta r)$ have the character of propagating wave solutions at large argument, they have very different character at small argument when they are large and no longer numerically independent. The solution of the seismic wave equations when $k_\beta r$ is small is more naturally written in terms of the spherical Bessel functions $j_l(k_\beta r)$ and $y_l(k_\beta r)$,

$$j_l(z) = \tfrac{1}{2}\left(h_l^{(1)}(z) + h_l^{(2)}(z)\right), \quad y_l(z) = -\tfrac{1}{2}\mathrm{i}\left(h_l^{(1)}(z) - h_l^{(2)}(z)\right). \tag{12.2.14}$$

$j_l(z)$ is regular at the origin, whereas y_l is not. At some level then it will be appropriate to switch from a representation of the wavefield in terms of the spherical Hankel functions to a form in terms of the spherical Bessel functions.

We can view this change of representation in an alternative way. A ray in a uniform medium is a straight line so that within a uniform spherical zone there can be an effective turning point as at r_t in figure 12.1. The ray behaviour corresponds to the high frequency limit $\omega \to \infty$ of the spherical Hankel functions with $\wp = Ł/\omega$ fixed. Strictly $\wp = (l + \frac{1}{2})$, but for large l the difference between $Ł$ and $l + \frac{1}{2}$ is irrelevant.

The asymptotic form of the spherical Hankel functions can be expressed in terms of the radial waveslowness $\bar{q}$ (cf 10.2.16),

$$\bar{q}_\beta = \left(\frac{1}{\beta^2} - \frac{\wp^2}{r^2}\right)^{1/2}. \tag{12.2.15}$$

The turning point radius $r_t = \wp\beta$ where $\bar{q}_\beta^2(r_t) = 0$. For $\bar{q}_\beta^2 > 0$, i.e., above the turning point r_t

$$\begin{aligned} h_l^{(1)}(k_\beta r) &\sim \frac{1}{\omega r \beta^{1/2} \bar{q}_\beta^{1/2}} \exp\left\{\mathrm{i}\int_r^{r_t} \mathrm{d}r\, \omega\bar{q}_\beta - \mathrm{i}\frac{\pi}{4}\right\}, \\ h_l^{(2)}(k_\beta r) &\sim \frac{1}{\omega r \beta^{1/2} \bar{q}_\beta^{1/2}} \exp\left\{-\mathrm{i}\int_r^{r_t} \mathrm{d}r\, \omega\bar{q}_\beta + \mathrm{i}\frac{\pi}{4}\right\}, \end{aligned} \tag{12.2.16}$$

while for $\bar{q}_\beta^2 < 0$, i.e. below the turning point r_t

$$\left.\begin{matrix} h_l^{(1)}(k_\beta r) \\ h_l^{(2)}(k_\beta r) \end{matrix}\right. \sim \mp\mathrm{i}\frac{1}{\omega r \beta^{1/2} |\bar{q}_\beta|^{1/2}} \exp\left\{\int_r^{r_t} \mathrm{d}r\, \omega|\bar{q}_\beta|\right\}. \tag{12.2.17}$$

These WKBJ approximations are not uniform and so are not valid in the vicinity of the turning point. From (12.2.16), (12.2.17) we see that above the turning point $h_l^{(1)}$, $h_l^{(2)}$ describe propagating waves, but below the turning point they both have an asymptotic form that increases exponentially with the separation from the turning point.

The turning point level itself is a good point at which to switch to the spherical Bessel functions, as can be seen from the comparable asymptotic approximations to (12.2.16), (12.2.17). Far above the turning point ($\bar{q}_\beta^2 > 0$),

$$\begin{aligned} j_l(k_\beta r) &\sim \frac{1}{\omega r \beta^{1/2} \bar{q}_\beta^{1/2}} \cos\left\{\int_r^{r_t} \mathrm{d}r\, \omega\bar{q}_\beta - \mathrm{i}\frac{\pi}{4}\right\}, \\ y_l(k_\beta r) &\sim \frac{1}{\omega r \beta^{1/2} \bar{q}_\beta^{1/2}} \sin\left\{\int_r^{r_t} \mathrm{d}r\, \omega\bar{q}_\beta - \mathrm{i}\frac{\pi}{4}\right\}, \end{aligned} \tag{12.2.18}$$

whereas far below the turning point ($\bar{q}_\beta^2 < 0$)

$$\begin{aligned} j_l(k_\beta r) &\sim \frac{1}{2\omega r \beta^{1/2} |\bar{q}_\beta|^{1/2}} \exp\left\{-\int_r^{r_t} \mathrm{d}r\, \omega|\bar{q}_\beta|\right\}, \\ y_l(k_\beta r) &\sim \frac{1}{\omega r \beta^{1/2} \bar{q}_\beta^{1/2}} \exp\left\{\int_r^{r_t} \mathrm{d}r\, \omega|\bar{q}_\beta|\right\}. \end{aligned} \tag{12.2.19}$$

Thus, above the turning point, y_l and j_l asymptotically describe standing waves. Below the turning point j_l behaves like a downgoing evanescent wave (it decreases exponentially as r decreases below the turning point) and y_l describes an upgoing evanescent wave. We can construct displacement-traction vectors $\hat{\mathfrak{b}}_d^H$ by replacing $h_l^{(2)}$ in (12.2.11) by j_l and this will be a regular solution at the origin. In a similar way we can construct $\hat{\mathfrak{b}}_u^H$ by replacing $h_l^{(2)}$ in (12.2.11) by y_l.

The situation we have encountered here where we need two different styles of representation on the two sides of the turning point is in fact common to all situations where we have turning point phenomena. We will want to use forms for the wavefield which mirror the expected physical behaviour and will need to match these at the turning level r_t. At the turning point level we can work with either the $\mathfrak{b}^H_{u,d}$ or $\hat{\mathfrak{b}}^H_{u,d}$ forms. However, when later we wish to construct reflection and transmission properties for seismic waves we will need to recognise that in the evanescent regime the connection is to the solutions described by the spherical Bessel functions $\hat{\mathfrak{b}}^H_{u,d}$ rather than to the forms in terms of the Hankel functions which represent upgoing and downgoing waves above the turning point.

The analytic continuation for the solutions in horizontal stratification between propagating forms for $\bar{q}_\beta^2 > 0$ and evanescent forms for $\bar{q}_\beta^2 < 0$ arises from the special case that there is no turning point and corresponds to a velocity distribution $\beta = ur$ in a sphere.

As in the case of horizontal stratification we can construct the full wavefield by the superposition of asymptotically up- and down-going waves with the difference that the fundamental solutions depend on radius r. Above the turning point,

$$\mathfrak{b}(r) = \begin{pmatrix} W(r) \\ T(r) \end{pmatrix} = \mathfrak{b}^H_u(r)H_u(r) + \mathfrak{b}^H_d(r)H_d(r), \tag{12.2.20}$$

Below the turning point we take,

$$\mathfrak{b}(r) = \begin{pmatrix} W(r) \\ T(r) \end{pmatrix} = \hat{\mathfrak{b}}^H_u(r)H_u(r) + \hat{\mathfrak{b}}^H_d(r)H_d(r). \tag{12.2.21}$$

The radial dependence of the wavefield, e.g. $\mathfrak{b}^H_u$, is linked to the angular dependence $P_l^m(\cos\theta)$ through the vector harmonics. In terms of the Legendre polynomials,

$$P_l(x) = \frac{1}{2^l l!}\frac{\mathrm{d}^l}{\mathrm{d}x^l}(x^2-1)^l, \tag{12.2.22}$$

the associated Legendre functions

$$P_l^m(x) = (1-x^2)^{m/2}\frac{\mathrm{d}^m}{\mathrm{d}x^m}P_l(x). \tag{12.2.23}$$

The $P_l(\cos\theta)$ represent stationary solutions as a function of θ, with an asymptotic form for large positive l,

$$P_l(\cos\theta) \sim \frac{1}{(2\pi l\sin\theta)^{1/2}}\cos\left[(l+\tfrac{1}{2})\theta - \frac{\pi}{4}\right], \tag{12.2.24}$$

which is valid away from the poles $\theta = 0$ or π, where $\sin\theta = 0$. Appendix B of Dahlen & Tromp (1998) provides a succinct summary of the useful properties of the various classes of Legendre function.

For body waves and surface waves we want to follow travelling wave solutions with respect to θ. It is convenient to follow Nussenveig (1965) and employ the decomposition

$$P_l(\cos\theta) = Q_l^{(1)}(\theta) + Q_l^{(2)}(\theta), \tag{12.2.25}$$

in terms of two oppositely directed travelling disturbances. For large l we have the asymptotic representations

$$Q_l^{(1,2)}(\theta) \sim \frac{1}{(2\pi l \sin\theta)^{1/2}} \exp\left[\mp i\left((l+\tfrac{1}{2})\theta - \frac{\pi}{4}\right)\right], \tag{12.2.26}$$

which have the desired character away from the poles. A comparable decomposition can be extended to the associated Legendre function

$$P_l^m(\cos\theta) = Q_{lm}^{(1)}(\theta) + Q_{lm}^{(2)}(\theta), \tag{12.2.27}$$

with asymptotic forms

$$Q_{lm}^{(1,2)}(\theta) \sim \frac{2(-l)^m}{(\pi l \sin\theta)^{1/2}} \exp\left[\mp i\left((l+\tfrac{1}{2})\theta + m\frac{\pi}{2} - \frac{\pi}{4}\right)\right], \tag{12.2.28}$$

for large l and small positive $m = 1, 2, \ldots$. In chapter 16 we will see how these travelling wave forms can be used in the construction of seismograms.

12.2.2 P-SV waves

We can make a comparable development for *P-SV* waves, which correspond to high order spheroidal oscillations of a sphere. The coupled equations for displacement and traction can again be expressed as a set of coupled first-order differential equations in terms of the displacement-traction vector

$$\mathfrak{b}^s = [U, V, P, S]^T, \tag{12.2.29}$$

in the form

$$\frac{\mathrm{d}}{\mathrm{d}r}\mathfrak{b}^s = \omega \mathfrak{A}^s \mathfrak{b}^s. \tag{12.2.30}$$

The coefficient matrix $\mathfrak{A}^s$ has the explicit representation

$$\mathfrak{A}^s = \begin{pmatrix} -\dfrac{2\varsigma}{\omega r} & \dfrac{\text{Ł}\varsigma}{\omega r} & \dfrac{1}{\rho\alpha^2} & 0 \\ -\dfrac{\text{Ł}}{\omega r} & \dfrac{1}{\omega r} & 0 & \dfrac{1}{\rho\beta^2} \\ -\rho + \dfrac{\rho\upsilon}{\omega^2 r^2} & -\dfrac{\text{Ł}\rho\upsilon}{2\omega^2 r^2} & -\dfrac{2(1-\varsigma)}{\omega r} & \dfrac{\text{Ł}}{\omega r} \\ -\dfrac{\text{Ł}\rho\upsilon}{2\omega^2 r^2} & -\rho + \dfrac{\rho(\nu \text{Ł}^2 - 2\beta^2)}{\omega^2 r^2} & -\dfrac{\text{Ł}\varsigma}{\omega r} & -\dfrac{3}{\omega r} \end{pmatrix} \tag{12.2.31}$$

in terms of the combinations of moduli introduced in (12.1.4).

In a uniform medium we can extract solutions associated with the independent propagation of *P* and *SV* waves which once again depend on spherical Bessel functions. Thus, the displacement-traction vector for upward propagating *P* waves is

$$\mathfrak{b}_u^P = \epsilon_\alpha \begin{pmatrix} h_l^{(1)\prime}(k_\alpha r) \\ \dfrac{\text{Ł}}{k_\alpha r} h_l^{(1)\prime}(k_\alpha r) \\ \rho\left\{\alpha + \dfrac{2\beta^2\text{Ł}^2}{\alpha k_\alpha^2 r^2}\right\} h_l^{(1)}(k_\alpha r) - \dfrac{4\rho\beta^2}{\omega r} h_l^{(1)\prime}(k_\alpha r) \\ \dfrac{\rho\beta^2\text{Ł}}{\omega r}\left\{h_l^{(1)\prime}(k_\alpha r) - \dfrac{1}{k_\alpha r} h_l^{(1)}(k_\alpha r)\right\} \end{pmatrix}, \tag{12.2.32}$$

where $k_\alpha = \omega/\alpha$. The corresponding form for an upward propagating *SV* wave is

$$\mathfrak{b}_u^S = \epsilon_\beta \begin{pmatrix} \dfrac{\L}{k_\beta r} h_l^{(1)}(k_\beta r) \\ \dfrac{1}{k_\beta r} h_l^{(1)}(k_\beta r) + h_l^{(1)\prime}(k_\beta r) \\ \dfrac{2\rho\beta^2 \L}{\omega r} \left\{ h_l^{(1)}(k_\beta r) - \dfrac{1}{k_\beta r} h_l^{(1)}(k_\beta r) \right\} \\ \rho\beta \left\{ \left(\dfrac{2\L^2 - 1}{k_\beta^2 r^2} - 1 \right) h_l^{(1)}(k_\beta r) - \dfrac{2}{k_\beta r} h_l^{(1)\prime}(k_\beta r) \right\} \end{pmatrix}, \tag{12.2.33}$$

with $k_\beta = \omega/\beta$. The two sets of solutions associated with downward propagating waves $\mathfrak{b}_d^P$, $\mathfrak{b}_d^S$ are obtained, as in the *SH* case, by replacing $h_l^{(1)}$ by $h_l^{(2)}$. The factors ϵ_α, ϵ_β are once again chosen to normalise energy flux with respect to r,

$$\epsilon_\alpha = \sqrt{2/\rho\alpha^3}, \quad \epsilon_\beta = \epsilon_H = \sqrt{2/\rho\beta^3}. \tag{12.2.34}$$

For each of the wavetypes we can also introduce the forms $\hat{\mathfrak{b}}_{u,d}^P$, $\hat{\mathfrak{b}}_{u,d}^P$, depending on j_l and y_l, appropriate to the zone below the turning point for that wavetype. Because of the differences in the wavespeed distributions, the turning points for *P* and *S* for common slowness $\wp$ occur at very different levels, with the *S* turning point at much greater depth (smaller radius). Different forms of solution will therefore be needed for *P* and *S* in the interval between the two turning points.

12.2.3 A fluid zone

In a fluid layer, such as the outer core, the shear strength vanishes and we only have *P* wave solutions. We can use the same vector harmonic representation as before but now $S \equiv 0$, $V = -\L P/\rho\omega r$, and

$$\frac{\mathrm{d}}{\mathrm{d}r}\begin{pmatrix} U \\ P \end{pmatrix} = \omega \begin{pmatrix} -\dfrac{2}{\omega r} & \dfrac{1}{\lambda} - \dfrac{\L}{\rho\omega^2 r^2} \\ -\rho & 0 \end{pmatrix} \begin{pmatrix} U \\ P \end{pmatrix}. \tag{12.2.35}$$

In a uniform region, an upward propagating solution is

$$\mathfrak{b}_u^f = \begin{pmatrix} U \\ P \end{pmatrix}_u^f = \begin{pmatrix} h_l^{(1)}(k_\alpha r) \\ -\dfrac{1}{\rho\alpha\omega} h_l^{(1)\prime}(k_\alpha r) \end{pmatrix}, \tag{12.2.36}$$

where now $\alpha = (\lambda/\rho)^{1/2}$ is the *P* wavespeed in the fluid. Comparable downgoing and evanescent solutions can be found by substitution of the requisite spherical Bessel functions.

12.3 The influence of velocity gradients

For uniform media we are able to find representations of the wavefield in terms of the independent propagation of the different wavetypes and to classify the contributions through their direction of propagation relative to the stratification. In the spherical case we have seen the need to use different forms above and below the effective turning point associated with the straight rays.

For modest gradients in seismic wavespeeds we expect that the behaviour will retain much of the character of the uniform medium, but that there will be different effects arising from the presence of the gradients. As soon as we encounter a wavespeed gradient we introduce the possibility of turning points, even for horizontal stratification, and so we will establish a framework based on uniform asymptotic approximations which allows a high frequency treatment of the dominant part of the propagation process and then include additional gradient (and sphericity) effects through a sequence of interaction terms.

12.3.1 Uniform approximations for a smoothly varying medium

For a zone of smooth variation in seismic wavespeeds we need to consider the wave propagation effects which correspond to the presence of a turning point in a ray description. In a horizontally stratified medium at high frequencies, we would expect the behaviour of a propagating downgoing wave in a smoothly varying medium to be of the form

$$\exp\left(\int_{z_1}^{z_2} \mathrm{d}z\, \mathrm{i}\omega q_\beta(z)\right),$$

and in the evanescent regime

$$\exp\left(-\int_{z_1}^{z_2} \mathrm{d}z\, \omega |q_\beta(z)|\right).$$

These results can be justified by extending the wavefield decomposition for a uniform medium to the case of slow variation. But, these simple exponential terms are accompanied by additional terms which are sensitive to the gradients in seismic parameters and are singular at the turning level.

It is however possible to generate uniform asymptotic representations which can be used in the neighbourhood of the turning level and which have a simple and recognisable character well away from this level. Our aim here is to generate expressions for the seismic wavefield that can be related directly to the ray theory results of Chapter 10 at high frequency, but which also include an improved description of turning point phenomena and the subtler effects of gradients and sphericity.

For a scalar wave, a linear slowness profile is a simple case which exhibits turning point behaviour. Gans (1915) found a solution for this profile which represents both the behaviour in the propagating regime and the exponential decay below the turning level. This solution may be written in terms of an Airy function $\mathrm{Ai}(x)$. Langer (1937) recognised that by a mapping of the argument of the Airy function a uniform asymptotic solution across a turning point can be found for a general monotonic wavespeed distribution.

With the aid of the Langer approach, it is possible to cast the leading order asymptotic solutions for seismic waves in a smoothly varying medium in a form which allows the identification of wave components with upgoing and downgoing character, well away from turning points. This representation proves particularly convenient for reflection and transmission problems, as we shall see in Chapter 14.

The first step is to transform the evolution equation for the vector $\mathbf{b}$,

$$\partial_z \mathbf{b} = \omega \mathbf{A}(z)\mathbf{b} \tag{12.3.1}$$

by writing

$$\mathbf{b} = \mathbf{C}\mathbf{u}, \tag{12.3.2}$$

with a transformation matrix $\mathbf{C}$ chosen so that $\mathbf{u}$ satisfies an equation whose leading terms can be cast into the form needed for the application of Langer's approach. The vector $\mathbf{u}$ will satisfy

$$\partial_z \mathbf{u} = [\omega \mathbf{C}^{-1}\mathbf{A}\mathbf{C} - \mathbf{C}^{-1}\partial_z\mathbf{C}]\mathbf{u}. \tag{12.3.3}$$

and we seek to choose $\mathbf{C}$ so that $\mathbf{C}^{-1}\mathbf{A}\mathbf{C}$ has a suitable form.

For *SH* waves the Langer approach can be used with little modification. However, for *P-SV* waves the Langer approach can only be applied to one wavetype at a time and complications arise from the separation of the turning levels of *P* and *SV* waves. The analysis is somewhat involved but the end results are not difficult to implement.

12.3.1.1 SH waves

We choose a transformation matrix $\mathbf{C}_H$,

$$\mathbf{C}_H = \frac{1}{(\rho p)^{1/2}} \begin{pmatrix} \beta^{-1} & 0 \\ 0 & \rho p \beta \end{pmatrix}, \tag{12.3.4}$$

so that $\mathbf{C}_H^{-1}\mathbf{A}_H\mathbf{C}_H$ has only off-diagonal elements,

$$\boldsymbol{H}_\beta = \mathbf{C}_H^{-1}\mathbf{A}_H\mathbf{C}_H = \begin{pmatrix} 0 & p \\ -q_\beta^2/p & 0 \end{pmatrix}. \tag{12.3.5}$$

The coupling matrix $-\mathbf{C}_H^{-1}\partial_z\mathbf{C}_H$ depends only on the elastic parameter gradients

$$-\mathbf{C}_H^{-1}\partial_z\mathbf{C}_H = \begin{pmatrix} \frac{1}{2}\partial_z\mu/\mu & 0 \\ 0 & \frac{1}{2}\partial_z\mu/\mu \end{pmatrix}, \tag{12.3.6}$$

and is well behaved at the turning level for *SH* waves.

The vector $\mathbf{u}_H$ thus satisfies,

$$\begin{aligned} \partial_z \mathbf{u}_H &= \left\{\omega \boldsymbol{H}_\beta - \mathbf{C}_H^{-1}\partial_z\mathbf{C}_H\right\}\mathbf{u}_H, \\ &= \left\{\frac{\omega}{(\rho p)^{1/2}} \begin{pmatrix} \beta^{-1} & 0 \\ 0 & \rho p \beta \end{pmatrix} + \begin{pmatrix} \frac{1}{2}\partial_z\mu/\mu & 0 \\ 0 & \frac{1}{2}\partial_z\mu/\mu \end{pmatrix}\right\}\mathbf{u}_H. \end{aligned} \tag{12.3.7}$$

We now seek a matrix $\hat{\mathbf{E}}_\beta$ which can provide a good asymptotic representation of the solution of (12.3.7) at high frequencies. Following Woodhouse (1978) we construct $\hat{\mathbf{E}}_\beta$ from the solutions of Airy's equation

$$\frac{d^2y}{dx^2} - xy = 0. \tag{12.3.8}$$

The two linearly independent Airy functions $\mathrm{Ai}(x)$, $\mathrm{Bi}(x)$ have therefore the property that their second derivative is just a multiple of the original function. This property allows us to match the off-diagonal high frequency part of (12.3.7) by a suitable mapping of the dependent variable in Airy's equation. For a slightly dissipative medium we take

$$\hat{\mathbf{E}}_\beta(\omega, p, z) = \pi^{1/2}\begin{pmatrix} s_\beta\omega^{1/6}r_\beta^{1/2}\mathrm{Bi}(-\omega^{2/3}\phi_\beta) & s_\beta\omega^{1/6}r_\beta^{1/2}\mathrm{Ai}(-\omega^{2/3}\phi_\beta) \\ \omega^{-1/6}r_\beta^{-1/2}\mathrm{Bi}'(-\omega^{2/3}\phi_\beta) & \omega^{-1/6}r_\beta^{-1/2}\mathrm{Ai}'(-\omega^{2/3}\phi_\beta) \end{pmatrix}, \tag{12.3.9}$$

with

$$s_\beta = -\partial_z\phi_\beta/|\partial_z\phi_\beta|, \quad r_\beta = p/|\partial_z\phi_\beta|. \tag{12.3.10}$$

The argument of the Airy functions is chosen so that the derivative of $\hat{\mathbf{E}}_\beta$ can be brought into the same form as $\mathbf{H}_\beta$ (12.3.5), so we require

$$\phi_\beta(\partial_z\phi_\beta)^2 = q_\beta^2 = \beta^{-2} - p^2. \tag{12.3.11}$$

The solution for ϕ_β can then be written as

$$\omega^{2/3}\phi_\beta = \mathrm{sgn}\{\mathrm{Re}\, q_\beta^2\}[\tfrac{3}{2}\omega\tau_\beta]^{2/3}, \tag{12.3.12}$$

where

$$\begin{aligned} \omega\tau_\beta &= \int_z^{z_\beta} d\zeta\, \omega q_\beta(\zeta), \quad \mathrm{Re}\, q_\beta^2 > 0, \\ &= \int_z^{z_\beta} d\zeta\, i\omega q_\beta(\zeta), \quad \mathrm{Re}\, q_\beta^2 < 0, \end{aligned} \tag{12.3.13}$$

for positive frequency ω; we have here made use of our choice of branch cut (12.1.24) $\{\mathrm{Im}\,\omega q_\beta \geq 0\}$ for the radical q_β. We take the principal value of the 2/3 power in (12.3.13). Note that τ_β is simply an integral of the vertical slowness and so relates directly to the ray theory definition (10.1.10) above the turning level.

A convenient reference level needs to be chosen for z_β so that the character of the solution should have a direct correspondence with the physical situation. For a perfectly elastic medium and a p value such that a turning point exists, we would choose z_β to lie at the depth Z_β for which $q_\beta(Z_\beta) = 0$. For small dissipation Q_β^{-1} we would choose z_β to be the depth at which $\mathrm{Re}\, q_\beta = 0$. If the slowness p is such that no turning point occurs for S waves in the region of interest, ϕ_β is non-unique since any choice of z_β may be taken. However, a suitable choice is extrapolate $\beta(z)$ so that a turning point Z_β is created and this can then be used as the reference level.

The functional form of the elements of $\hat{\boldsymbol{E}}_\beta$ is quite complex and it is convenient to adopt a simplified notation, (Kennett & Woodhouse, 1978; Kennett & Illingworth, 1981),

$$\hat{\boldsymbol{E}}_\beta = \begin{pmatrix} s_\beta \mathrm{Bj}(\omega\tau_\beta) & s_\beta \mathrm{Aj}(\omega\tau_\beta) \\ \mathrm{Bk}(\omega\tau_\beta) & \mathrm{Ak}(\omega\tau_\beta) \end{pmatrix}. \tag{12.3.14}$$

The inverse of $\hat{\boldsymbol{E}}_\beta$ is readily constructed by using the result that the Wronskian of Ai and Bi is π^{-1}, and has the form

$$\hat{\boldsymbol{E}}_\beta^{-1} = \begin{pmatrix} -s_\beta \mathrm{Ak}(\omega\tau_\beta) & \mathrm{Aj}(\omega\tau_\beta) \\ s_\beta \mathrm{Bk}(\omega\tau_\beta) & -\mathrm{Bj}(\omega\tau_\beta) \end{pmatrix}. \tag{12.3.15}$$

The 'phase matrix' $\hat{\boldsymbol{E}}_\beta$ satisfies the evolution equation with depth,

$$\partial_z \hat{\boldsymbol{E}}_\beta = [\omega \boldsymbol{H}_\beta + \partial_z \boldsymbol{\Phi}_\beta \boldsymbol{\Phi}_\beta^{-1}]\hat{\boldsymbol{E}}_\beta; \tag{12.3.16}$$

so that the match of the leading term in (12.3.7) is achieved. However, we have an additional contribution through the diagonal matrix $\partial_z \boldsymbol{\Phi}_\beta \boldsymbol{\Phi}_\beta^{-1}$ which depends on the Airy function argument ϕ_β

$$\partial_z \boldsymbol{\Phi}_\beta \boldsymbol{\Phi}_\beta^{-1} = \tfrac{1}{2}(\partial_{zz}\phi_\beta / \partial_z \phi_\beta) \begin{pmatrix} -1 & 0 \\ 0 & 1 \end{pmatrix}, \tag{12.3.17}$$

and is well behaved even at turning points, where

$$\partial_{zz}\phi_\beta / \partial_z \phi_\beta = \{\partial_{zz}\beta - 3p(\partial_z\beta)^2\}/4\partial_z\beta.$$

From the columns of $\mathbf{C}_H(p,z)\hat{\mathbf{E}}(\omega,p,z)$ we are able to construct approximations to the motion-stress vector $\mathbf{b}$, which will be effective at high frequencies when $\omega\boldsymbol{H}_\beta$ is the dominant term in (12.3.7). But, we have not included the contribution from the gradient $\mathbf{C}_H^{-1}\partial_z\mathbf{C}_H$ or the additional part from the phase dependence $\partial_z \boldsymbol{\Phi}_\beta \boldsymbol{\Phi}_\beta^{-1}$. Correction terms need to be applied to $\mathbf{CE}$ to allow for these effects.

At high frequencies the argument of the Airy functions becomes large, and we may use the asymptotic representations of the functions. Below an S wave turning point in a perfectly elastic medium (i.e., $q_\beta^2 < 0$), the entries of $\hat{\boldsymbol{E}}_\beta$ are asymptotically

$$\begin{aligned}
\mathrm{Aj}(\omega\tau_\beta) &\sim \tfrac{1}{2}\left(\frac{p}{|q_\beta|}\right)^{1/2} \exp(-\omega|\tau_\beta|), \\
\mathrm{Ak}(\omega\tau_\beta) &\sim -\tfrac{1}{2}\left(\frac{|q_\beta|}{p}\right)^{1/2} \exp(-\omega|\tau_\beta|), \\
\mathrm{Bj}(\omega\tau_\beta) &\sim \left(\frac{p}{|q_\beta|}\right)^{1/2} \exp(\omega|\tau_\beta|), \\
\mathrm{Bk}(\omega\tau_\beta) &\sim \left(\frac{|q_\beta|}{p}\right)^{1/2} \exp(\omega|\tau_\beta|).
\end{aligned} \tag{12.3.18}$$

In a perfectly elastic medium the argument of the exponential term is

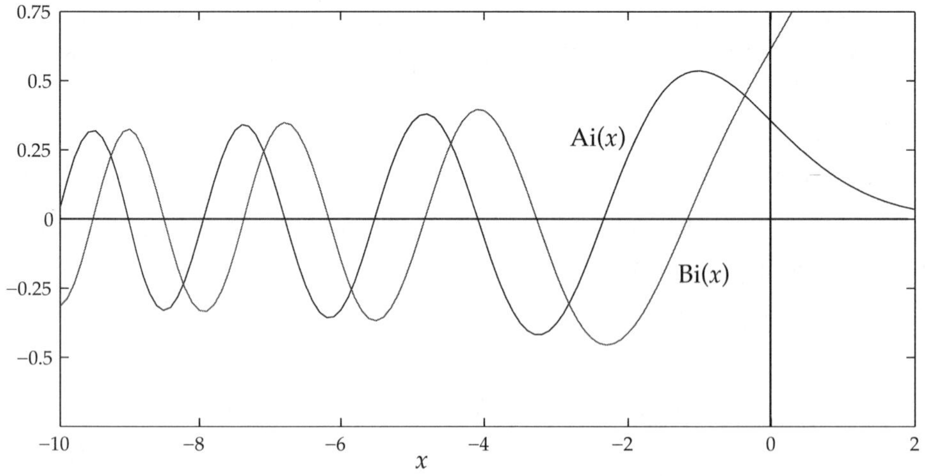

Figure 12.2. The Airy functions Ai(x), Bi(x) which provide a uniform approximation linking propagating and evanescent regimes

$$\omega|\tau_\beta| = \omega \left| \int_z^{Z_\beta} \mathrm{d}\zeta \left(p^2 - \frac{1}{\beta^2(\zeta)} \right)^{1/2} \right|, \tag{12.3.19}$$

so that we have a similar functional form to (12.1.25) for a uniform medium in the evanescent regime.

Well below the turning level $\hat{\boldsymbol{E}}_\beta$ gives a good description of the evanescent wavefields. Above the turning point, the asymptotic behaviour of Aj and Bk is as $\cos(\omega\tau_\beta - \pi/4)$, and for Ak and Bj as $\sin(\omega\tau_\beta - \pi/4)$. The Airy functions therefore provide the desired link across the turning level in a single representation, as illustrated in figure 12.2. The asymptotic forms fail just in the neighbourhood of the origin, which in our implementation falls at the turning level.

The elements Aj, Bj etc describe standing waves above the turning level and in this form are directly suitable for use for describing surface wave fields. However, when we want to track propagation processes for body waves we would prefer to consider travelling wave forms. We can find a suitable representation by constructing a new matrix $\boldsymbol{E}_\beta$, whose columns are a linear combination of those of $\hat{\boldsymbol{E}}_\beta$,

$$\boldsymbol{E}_\beta = \hat{\boldsymbol{E}}_\beta . \frac{1}{\sqrt{2}} \begin{pmatrix} \mathrm{e}^{-\mathrm{i}\pi/4} & \mathrm{e}^{\mathrm{i}\pi/4} \\ \mathrm{e}^{\mathrm{i}\pi/4} & \mathrm{e}^{-\mathrm{i}\pi/4} \end{pmatrix}. \tag{12.3.20}$$

The new matrix $\boldsymbol{E}_\beta$ will also satisfy a differential equation of the form (12.3.16), and so, at high frequencies, is an equally good candidate for approximating the motion-stress vector **b**.

In terms of the Airy function entries

$$\boldsymbol{E}_\beta = \frac{1}{\sqrt{2}} \begin{pmatrix} s_\beta \mathrm{e}^{\mathrm{i}\pi/4}(\mathrm{Aj} - \mathrm{iBj}) & s_\beta \mathrm{e}^{-\mathrm{i}\pi/4}(\mathrm{Aj} + \mathrm{iBj}) \\ \mathrm{e}^{\mathrm{i}\pi/4}(\mathrm{Ak} - \mathrm{iBk}) & \mathrm{e}^{-\mathrm{i}\pi/4}(\mathrm{Ak} + \mathrm{iBk}) \end{pmatrix},$$

$$\boldsymbol{E}_\beta = \frac{1}{\sqrt{2}} \begin{pmatrix} s_\beta \mathrm{Ej}(\omega\tau_\beta) & s_\beta \mathrm{Fj}(\omega\tau_\beta) \\ \mathrm{Ek}(\omega\tau_\beta) & \mathrm{Fk}(\omega\tau_\beta) \end{pmatrix}. \tag{12.3.21}$$

The high frequency asymptotic forms well above a turning point ($q_\beta^2 > 0$) are:

$$\begin{aligned} \mathrm{Ej}(\omega\tau_\beta) &\sim \left(\frac{p}{q_\beta}\right)^{1/2} \exp(\mathrm{i}\omega\tau_\beta), \\ \mathrm{Ek}(\omega\tau_\beta) &\sim -\mathrm{i}\left(\frac{q_\beta}{p}\right)^{1/2} \exp(\mathrm{i}\omega\tau_\beta), \\ \mathrm{Fj}(\omega\tau_\beta) &\sim \left(\frac{p}{q_\beta}\right)^{1/2} \exp(-\mathrm{i}\omega\tau_\beta), \\ \mathrm{Fk}(\omega\tau_\beta) &\sim \mathrm{i}\left(\frac{q_\beta}{p}\right)^{1/2} \exp(-\mathrm{i}\omega\tau_\beta). \end{aligned} \tag{12.3.22}$$

The argument of the complex exponentials is now

$$\mathrm{i}\omega\tau_\beta = \mathrm{i}\omega \int_z^{Z_\beta} \mathrm{d}\zeta\, q_\beta(\zeta) = \mathrm{i}\omega \int_z^{Z_\beta} \mathrm{d}\zeta \left(\frac{1}{\beta^2(\zeta)} - p^2\right)^{1/2}. \tag{12.3.23}$$

With our convention that z increases downwards, τ_β (12.3.23) is a decreasing function of z above the turning level for the S waves Z_β. Thus we can recognise that Ej, Ek asymptotically have the character of upgoing waves; similarly Fj, Fk have the character of downgoing waves. These interpretations for large arguments $\omega\tau_\beta$ are misleading if extrapolated too close to the turning point. Below the turning point all the entries $\boldsymbol{E}_\beta$ increase exponentially with depth because of the dominance of the Bi terms, and so $\boldsymbol{E}_\beta$ is not useful in the evanescent regime.

Once again $\boldsymbol{E}_\beta$ has a simple inverse which may be expressed in terms of the entries of $\boldsymbol{E}_\beta$:

$$\boldsymbol{E}_\beta^{-1} = \mathrm{i}2^{-1/2} \begin{pmatrix} -s_\beta \mathrm{Fk} & \mathrm{Fj} \\ s_\beta \mathrm{Ek} & -\mathrm{Ej} \end{pmatrix}. \tag{12.3.24}$$

In the propagating regime, above a turning point we will use $\boldsymbol{E}_\beta$ as the basis for our $\mathbf{b}$ vector representation, and in the evanescent region we will use $\hat{\boldsymbol{E}}_\beta$. Both of these matrices were constructed to take advantage of the uniform approximations afforded by the Airy function for an isolated turning point. A different choice of phase matrix with parabolic cylinder function entries is needed for a uniform approximation with two close turning points (Woodhouse, 1978).

12.3.1.2 *P-SV waves*

When we consider the *P-SV* wave system we are only able to apply the Langer approach to one wave type at a time, and so we must make a transformation as in (3.54) to bring $\mathbf{C}_s^{-1}\mathbf{A}_s\mathbf{C}_s$ into block diagonal form where the entry for each wave type has the structure (3.56). Thus we seek

$$\mathbf{H} = \mathbf{C}_s^{-1}\mathbf{A}_s\mathbf{C}_s = \begin{pmatrix} \boldsymbol{H}_\alpha & \mathbf{0} \\ & \\ \mathbf{0} & \boldsymbol{H}_\beta \end{pmatrix}, \tag{12.3.25}$$

and guided by the work of Chapman (1974b) and Woodhouse (1978) we take

$$\mathbf{C}_s = (\rho p)^{-1/2} \begin{pmatrix} 0 & p & p & 0 \\ p & 0 & 0 & p \\ \rho(2\beta^2p^2-1) & 0 & 0 & 2\rho\beta^2p^2 \\ 0 & 2\rho\beta^2p^2 & \rho(2\beta^2p^2-1) & 0 \end{pmatrix}. \tag{12.3.26}$$

As in the *SH* case we have avoided using the slownesses q_α, q_β but there is no longer such a simple relation between **b** and **u**. It is interesting to note that we can construct the columns of $\mathbf{C}_s$ by taking the sum and difference of the columns of **D** and rescaling; indicating a possible standing wave interpretation for **u**. The *P-SV* coupling terms are given by

$$-\mathbf{C}_s^{-1}\partial_z\mathbf{C}_s = \begin{pmatrix} \gamma_A & 0 & 0 & -\gamma_C \\ 0 & -\gamma_A & -\gamma_B & 0 \\ 0 & \gamma_C & \gamma_A & 0 \\ \gamma_C & 0 & 0 & -\gamma_A \end{pmatrix}, \tag{12.3.27}$$

where

$$\gamma_A = 2\beta^2p^2\partial_z\mu/\mu - \tfrac{1}{2}\partial_z\rho/\rho, \tag{12.3.28}$$

controls the rate of change of both the *P* and *S* wave coefficients. The off-diagonal terms in (12.3.27)

$$\gamma_B = 2\beta^2p^2\partial_z\mu/\mu - \partial_z\rho/\rho, \quad \gamma_C = 2\beta^2p^2\partial_z\mu/\mu, \tag{12.3.29}$$

lead to cross-coupling between *P* and *S* elements.

For the *P-SV* wave system we have constructed $\mathbf{C}_s$ to give a high frequency block diagonal structure (12.3.25) and so the corresponding phase matrix **E** has a block diagonal form. Above all turning points **E** has $\boldsymbol{E}_\alpha$ for the P wave contribution, and $\boldsymbol{E}_\beta$ for the *SV* contribution. Below both P and S turning levels $\hat{\mathbf{E}}$ is constructed from $\hat{\boldsymbol{E}}_\alpha$, $\hat{\boldsymbol{E}}_\beta$. In the intermediate zone below the *P* turning level, so that *P* waves are evanescent, but with *S* waves still propagating, we take the block diagonal form

$$\bar{\mathbf{E}} = \begin{pmatrix} \hat{\boldsymbol{E}}_\alpha & \mathbf{0} \\ & \\ \mathbf{0} & \boldsymbol{E}_\beta \end{pmatrix}. \tag{12.3.30}$$

In the high frequency limit we ignore any coupling between *P* and *SV* waves.

12.3.2 Relation to upgoing and downgoing wave decomposition

The asymptotic forms developed in the previous section provide high frequency approximations to the wavefield that do not allow for any coupling between *P* and *S* waves. This form of the uniform approximations provides a description of the depth dependence of the wavefield that is able to cope with the situation at a turning level, where the distinction between upgoing and downgoing waves is eliminated. However, when we study travelling waves our aim is to follow the character of the elements of the wavefield and so distinguish the direction of propagation. This can be achieved by rearranging the leading order approximation **CE** into a form with only diagonal phase terms.

(a) Above the turning level

Based on the approach of Richards (1976) we introduce generalised vertical slownesses $\eta_{\beta u,d}$ derived from the uniform asymptotic approximations, which asymptotically reduce to the conventional vertical slowness q_β far above the turning level. In terms of the Airy function entries of the matrix $\boldsymbol{E}_\beta$ the generalised vertical slownesses are

$$\mathrm{i}\eta_{\beta u}(p,\omega,z) = -p\frac{\mathrm{Ek}(\omega\tau_\beta(z))}{\mathrm{Ej}(\omega\tau_\beta(z))}, \qquad \mathrm{i}\eta_{\beta d}(p,\omega,z) = p\frac{\mathrm{Fk}(\omega\tau_\beta(z))}{\mathrm{Fj}(\omega\tau_\beta(z))}. \tag{12.3.31}$$

We recall that

$$\tau_\beta = \int_z^{Z_\beta} \mathrm{d}\zeta\, \mathrm{i}\omega q_\beta(\zeta), \tag{12.3.32}$$

and so depends on the slowness structure between z and the turning level Z_β. The generalised slownesses $\eta_{\beta u,d}$ are therefore not just defined by the local elastic properties. In (12.3.31) the use of the subscripts u,d again reminds us that the Airy elements have up and downgoing character only in the asymptotic regime, far from a turning level. Well above the turning point, $q_\beta^2 > 0$, and asymptotically

$$\eta_{\beta u}(p,\omega) \sim q_\beta(p), \qquad \eta_{\beta d}(p,\omega) \sim q_\beta(p), \tag{12.3.33}$$

and also

$$\begin{aligned} (2\rho p)^{-1/2}\mathrm{Ej}(\omega\tau_\beta) &\sim \epsilon_\beta \exp[-\mathrm{i}\omega\int_{Z_\beta}^{z} \mathrm{d}\zeta\, q_\beta(\zeta)], \\ (2\rho p)^{-1/2}\mathrm{Fj}(\omega\tau_\beta) &\sim \epsilon_\beta \exp[\mathrm{i}\omega\int_{Z_\beta}^{z} \mathrm{d}\zeta\, q_\beta(\zeta)], \end{aligned} \tag{12.3.34}$$

We now use just the Ej, Fj terms to describe the phase behaviour and include the Ek, Fk components by employing the generalised slownesses $\eta_{\beta u,d}$.

The leading order approximation for *SH* waves can then be written as

$$\mathbf{C}_H(z)\mathbf{E}_H(z) = \mathbf{D}_H(z)\mathbf{E}_H(z) = [\mathbf{b}_u^H, \mathbf{b}_d^H]\mathbf{E}_H(z), \tag{12.3.35}$$

where $\mathbf{E}_H(z)$ is the diagonal matrix

$$\mathbf{E}_H(z) = \begin{pmatrix} \mathrm{E}_{Hu} & 0 \\ 0 & \mathrm{E}_{Hd} \end{pmatrix} = \frac{1}{(2\rho p)^{1/2}} \begin{pmatrix} \mathrm{Ej}(\omega\tau_\beta(z)) & 0 \\ 0 & \mathrm{Fj}(\omega\tau_\beta(z)) \end{pmatrix}. \tag{12.3.36}$$

Ej is associated with upgoing waves and Fj with downgoing waves. The column vectors of $\mathbf{D}_H$ are given by

$$\mathbf{b}_{u,d}^H(z) = [s_\beta\beta^{-1}, \mp\mathrm{i}\rho\beta\eta_{\beta u,d}]^T, \tag{12.3.37}$$

The upper sign is associated with the suffix u, and for a model in which the shear velocity increases with depth $s_\beta = 1$. The form of representation (12.3.35) is designed to provide as close a link as possible to a viewpoint in terms of upgoing and downgoing waves. For high frequencies, the columns of $\mathbf{D}(z)$ are asymptotically equivalent to the upgoing and downgoing motion-stress vectors for a locally uniform region at z, and the phase terms $\mathsf{E}_{Hu,d}$ also have the requisite character for up- and down-going propagation.

For *P-SV* waves we reorganise the leading order approximation away from a block-diagonal representation by wavetype and instead emphasise the character of the wavefield. Thus we write,

$$\mathsf{C}_S\mathsf{E}_S = \mathbf{D}_S\mathbf{E}_S = [\mathbf{b}_u^P, \mathbf{b}_u^S, \mathbf{b}_d^P, \mathbf{b}_d^S]\mathbf{E}_S, \tag{12.3.38}$$

with

$$\mathbf{E}_S = \frac{1}{(2\rho p)^{1/2}}\mathrm{diag}[\mathrm{Ej}(\omega\tau_\alpha), \mathrm{Fj}(\omega\tau_\alpha), \mathrm{Ej}(\omega\tau_\beta), \mathrm{Fj}(\omega\tau_\beta)], \tag{12.3.39}$$

and column vectors

$$\begin{aligned}\mathbf{b}_{u,d}^P &= [\mp \mathrm{i}\eta_{\alpha u,d}, s_\alpha p, s_\alpha \rho(2\beta^2p^2 - 1), \mp 2\mathrm{i}\rho\beta^2\eta_{\alpha u,d}]^T,\\ \mathbf{b}_{u,d}^S &= [s_\beta p, \mp \mathrm{i}\eta_{\beta u,d}, \mp 2\mathrm{i}\rho\beta^2\eta_{\beta u,d}, s_\beta\rho(2\beta^2p^2 - 1)]^T.\end{aligned} \tag{12.3.40}$$

Once again we recover the desired physical character in the high-frequency asymptotic limit.

(b) Below the turning level

Below a turning point we have evanescent waves and then will employ the Ai, Bi Airy functions. We therefore introduce modified forms for the generalised vertical slownesses

$$\mathrm{i}\hat{\eta}_{\beta u} = -p\frac{\mathrm{Bk}(\omega\tau_\beta)}{\mathrm{Bj}(\omega\tau_\beta)}, \qquad \mathrm{i}\hat{\eta}_{\beta d} = p\frac{\mathrm{Ak}(\omega\tau_\beta)}{\mathrm{Aj}(\omega\tau_\beta)}. \tag{12.3.41}$$

In the far evanescent regime we have the high-frequency asymptotic properties

$$\hat{\eta}_{\beta u} \sim \mathrm{i}|q_\beta|, \qquad \hat{\eta}_{\beta d} \sim \mathrm{i}|q_\beta|, \tag{12.3.42}$$

and now

$$\begin{aligned}(\rho p)^{-1/2}\mathrm{Aj}(\omega\tau_\beta) &\sim (\mathrm{i}/2)^{1/2}\epsilon_\beta \exp[-\omega|\tau_\beta|],\\ (\rho p)^{-1/2}\mathrm{Bj}(\omega\tau_\beta) &\sim (2\mathrm{i})^{1/2}\epsilon_\beta \exp[-\omega|\tau_\beta|].\end{aligned} \tag{12.3.43}$$

In this region we express the leading order approximation for *SH* waves in a form which is comparable to the propagating case

$$\mathsf{C}_H\hat{\mathsf{E}}_H = \hat{\mathbf{D}}_H\hat{\mathbf{E}}_H = [\hat{\mathbf{b}}_u^H, \hat{\mathbf{b}}_d^H]\hat{\mathbf{E}}_H. \tag{12.3.44}$$

The column vectors $\hat{\mathbf{b}}^H$ differ from $\mathbf{b}^H$ by using the modified slownesses $\hat{\eta}$, and

$$\hat{\mathbf{E}}_H(z) = \begin{pmatrix}\hat{\mathsf{E}}_{Hu} & 0\\ 0 & \hat{\mathsf{E}}_{Hd}\end{pmatrix} = \frac{1}{(2\rho p)^{1/2}}\begin{pmatrix}\mathrm{Bj}(\omega\tau_\beta(z)) & 0\\ 0 & \mathrm{Aj}(\omega\tau_\beta(z))\end{pmatrix}. \tag{12.3.45}$$

For *P-SV* waves we make a similar development to the above when both *P* and *S* waves are evanescent. When only *P* waves are evanescent we use the $\hat{\mathbf{b}}^P$ forms and retain the propagating $\mathbf{b}^S$ vectors for *S* waves.

12.3.3 Gradient corrections

We now wish to improve on our high frequency approximations for the seismic wavefield by introducing suitable corrections for the influence of the gradients. We introduce a fundamental stress-displacement matrix $\mathbf{B}(p, z, \omega)$ whose columns are independent solutions of the evolution equation with depth (12.3.1). Any motion stress vector can then be constructed as a linear combination of the columns of $\mathbf{B}$: $\mathbf{b}(z) = \mathbf{B}(z)\mathbf{c}$, in terms of a constant vector $\mathbf{c}$.

We can regard the combination $\mathbf{CE}$ as a high frequency approximation to $\mathbf{B}(p, z, \omega)$ with suitable phase behaviour and now seek suitable correction terms to compensate for the limitations of the approximation. The most convenient development for describing travelling waves is to look for a solution for $\mathbf{B}(p, z, \omega)$ in the form,

$$\mathbf{B}(p, z, \omega) = \mathbf{C}(p, z)\mathbf{E}(p, z, \omega)\mathbf{L}(p, z, \omega). \tag{12.3.46}$$

The matrix $\mathbf{L}$ then satisfies

$$\partial_z \mathbf{L} = -\mathbf{E}^{-1}[\mathbf{C}^{-1}\partial_z \mathbf{C} + \partial_z \mathbf{\Phi}\mathbf{\Phi}^{-1}]\mathbf{E}\mathbf{L} = \{\mathbf{E}^{-1}\mathbf{j}\mathbf{E}\}\mathbf{L}. \tag{12.3.47}$$

The frequency dependence of $\mathbf{L}$ arises from the phase terms $\mathbf{E}$, $\mathbf{E}^{-1}$ and the form of $\mathbf{L}$ is controlled by the choice of $\mathbf{E}$. The equation for $\mathbf{L}$ is equivalent to the integral equation

$$\mathbf{L}(z; z_{ref}) = \mathbf{I} + \int_{z_{\text{ref}}}^{z} \mathrm{d}\zeta \,\{\mathbf{E}^{-1}(\zeta)\mathbf{j}(\zeta)\mathbf{E}(\zeta)\}\mathbf{L}(\zeta), \tag{12.3.48}$$

where we would choose the lower limit of integration z_{ref} to correspond to the physical situation. For a region containing a turning point we take z_{ref} at the turning level and use different forms of the phase matrix above and below this level; otherwise we may take any convenient reference level. We may now make an iterative solution of (12.3.48) (Chapman, 1981; Kennett & Illingworth, 1981) in terms of an 'interaction series'. We construct successive estimates as

$$\mathbf{L}_r(z; z_{\text{ref}}) = \mathbf{I} + \int_{z_{\text{ref}}}^{z} \mathrm{d}\zeta \,\{\mathbf{E}^{-1}(\zeta)\mathbf{j}(\zeta)\mathbf{E}(\zeta)\}\mathbf{L}_{r-1}(\zeta), \tag{12.3.49}$$

with $\mathbf{L}_0 = \mathbf{I}$. At each stage we bring in a further coupling into the parameter gradients present in $\mathbf{j}$ through $\mathbf{C}^{-1}\partial_z\mathbf{C}$ and $\partial_z\mathbf{\Phi}\mathbf{\Phi}^{-1}$. All the elements of $\mathbf{E}^{-1}\mathbf{j}\mathbf{E}$ are bounded and so the recursion (12.3.49) will converge. The interaction series is therefore of the form

$$\begin{aligned}\mathbf{L}(z; z_{\text{ref}}) = \mathbf{I} &+ \int_{z_{\text{ref}}}^{z} \mathrm{d}\zeta \,\{\mathbf{E}^{-1}\mathbf{j}\mathbf{E}\}(\zeta) \\ &+ \int_{z_{\text{ref}}}^{z} \mathrm{d}\zeta \,\{\mathbf{E}^{-1}\mathbf{j}\mathbf{E}\}(\zeta) \int_{z_{\text{ref}}}^{\zeta} \mathrm{d}\eta \,\{\mathbf{E}^{-1}\mathbf{j}\mathbf{E}\}(\eta) + \ldots.\end{aligned} \tag{12.3.50}$$

and the terms may be identified with successive interactions of the seismic waves with the parameter gradients. For slowly varying media the elements of **j** will be small, and the series (12.3.50) will converge rapidly. In this case it may be sufficient to retain only the first integral term.

For *SH* waves the total gradient effects are controlled by the diagonal matrix

$$\mathbf{j}_H = (\tfrac{1}{2}\partial_z\mu/\mu + \tfrac{1}{2}\partial_{zz}\phi_\beta/\partial_z\phi_\beta)\mathrm{diag}[1,-1], \tag{12.3.51}$$

and

$$\begin{aligned} \boldsymbol{E}_\beta^{-1}\mathbf{j}_H\boldsymbol{E}_\beta = & -\mathrm{i}\tfrac{1}{4}(\partial_z\mu/\mu + \partial_{zz}\phi_\beta/\partial_z\phi_\beta) \\ & \times \begin{pmatrix} \mathrm{Ej}_\beta\mathrm{Fk}_\beta + \mathrm{Ek}_\beta\mathrm{Fj}_\beta & 2\mathrm{Fj}_\beta\mathrm{Fk}_\beta \\ -2\mathrm{Ej}_\beta\mathrm{Ek}_\beta & -(\mathrm{Ej}_\beta\mathrm{Fk}_\beta + \mathrm{Ek}_\beta\mathrm{Fj}_\beta) \end{pmatrix}, \end{aligned} \tag{12.3.52}$$

where we have written Ej_β for $\mathrm{Ej}(\omega\tau_\beta)$. This representation is well behaved at a turning point. Asymptotically, well away from any turning level, in the propagating and evanescent regimes

$$\partial_{zz}\phi_\beta/\partial_z\phi_\beta \approx \partial_z q_\beta/q_\beta, \tag{12.3.53}$$

and then the gradient term behaves like

$$\gamma_H = \tfrac{1}{2}\partial_z q_\beta/q_\beta + \tfrac{1}{2}\partial_z\mu/\mu. \tag{12.3.54}$$

In the propagating regime, for example,

$$\boldsymbol{E}_\beta^{-1}\mathbf{j}_H\boldsymbol{E}_\beta \approx \gamma_H \begin{pmatrix} 0 & \exp[-2\mathrm{i}\omega\tau_\beta] \\ \exp[+2\mathrm{i}\omega\tau_\beta] & 0 \end{pmatrix}, \tag{12.3.55}$$

and the interaction series can be interpreted in terms of successive reflections from the gradients.

For the *P-SV* system, the structure of the block diagonal entries of $\mathbf{j}_P$ for *P* and *S* parallel our discussion for the *SH* wave case and asymptotically the diagonal matrices control the interaction with the gradients. The off-diagonal matrices lead to interconversion of *P* and *SV* waves. The structure of the interaction terms **L** for *P-SV* waves is

$$\mathbf{L}_S = \begin{pmatrix} \boldsymbol{I} + \boldsymbol{L}_{\alpha\alpha} & \boldsymbol{L}_{\alpha\beta} \\ \boldsymbol{L}_{\beta\alpha} & \boldsymbol{I} + \boldsymbol{L}_{\beta\beta} \end{pmatrix}. \tag{12.3.56}$$

$\boldsymbol{L}_{\alpha\alpha}$, $\boldsymbol{L}_{\beta\beta}$ represent multiple interactions with the parameter gradients without change of wave type. $\boldsymbol{L}_{\alpha\beta}$, $\boldsymbol{L}_{\beta\alpha}$ allow for the coupling between *P* and *SV* waves that is not present in our choice of phase matrix and which only appears with the first integral contribution in (12.3.50).

From the series (12.3.50) we may find $\mathbf{L}(p,z;z_{\mathrm{ref}},\omega)$ to any desired level of interaction with the medium and then construct an approximate **B** matrix as

$$\mathbf{B}_I(p,z,\omega) \approx \mathbf{C}(p,z)\mathbf{E}(p,z,\omega)\mathbf{L}(p,z;z_{\mathrm{ref}},\omega). \tag{12.3.57}$$

The interaction series for **L** is not restricted in its frequency coverage. Although our starting point is a high frequency approximation to the solution, this is compensated for by the presence of the same term in the kernel of the interaction series development. The number of terms required to get an adequate approximation of the wavefield depends on $\{\mathbf{E}^{-1}\mathbf{j}\mathbf{E}\}$ and thus on the size of the parameter gradients and the frequency. At low frequencies we need more terms in the interaction series to counteract the high frequency character of **E**.

With the use of the interaction series, all the wavefield representations depend on the *leading order* approximation $\mathbf{C}(p,z)\mathbf{E}(p,z,\omega)$ in which there is no coupling between P and S waves. The gradient induced coupling will be introduced by the interaction term $\mathbf{L}(p,z,\omega)$.

The approximate fundamental matrix $\mathbf{B}_I$ in (12.3.57) may alternatively be expressed in terms of the upgoiong and downgoing wave representation as

$$\mathbf{B}_I(p,z,\omega) \approx \mathsf{D}(p,z)\mathbf{E}(p,z,\omega)\mathbf{L}_{ud}(p,z;z_{\text{ref}},\omega). \tag{12.3.58}$$

A consequence of the arrangement by wave character is that for *P-SV* waves the interaction terms **L** must also be rearranged to suit the new formulation. Thus

$$\mathbf{L}_{ud} = \Xi\mathbf{L}_s \tag{12.3.59}$$

where the symmetric matrix

$$\Xi = \begin{pmatrix} 1 & 0 & 0 & 0 \\ 0 & 0 & 1 & 0 \\ 0 & 1 & 0 & 0 \\ 0 & 0 & 0 & 1 \end{pmatrix}, \quad \Xi = \Xi^{-1}. \tag{12.3.60}$$

For *SH* waves no adjustment need be applied and $\mathbf{L}_{ud} = \mathbf{L}$.

12.3.4 Spherical stratification

In spherical stratification we may again make a uniform asymptotic approximation using Airy functions for an isolated turning point. Much of the approach parallels the treatment for horizontal stratification but we will see that there are some extra features that arise from the spherical structure.

Within a region $r_2 < r < r_1$ where the elastic parameters are smoothly varying we can rewrite the equations (12.2.8) and (12.2.30) in terms of a sequence of terms in inverse powers of frequency,

$$\partial_r \mathfrak{b} = \omega[\mathfrak{A}_0 + \omega^{-1}\mathfrak{A}_1 + \omega^{-2}\mathfrak{A}_2]\mathfrak{b}. \tag{12.3.61}$$

In the process we recognise a spherical slowness parameter $\wp = \L/\omega$.

For *SH* waves

$$\mathfrak{A}_0^H = \begin{pmatrix} 0 & \dfrac{1}{\mu} \\ \dfrac{\mu\wp^2}{r^2} & 0 \end{pmatrix}, \quad \mathfrak{A}_1^H = \begin{pmatrix} \dfrac{1}{r} & 0 \\ 0 & -\dfrac{3}{r} \end{pmatrix}, \quad \mathfrak{A}_2^H = \begin{pmatrix} 0 & 0 \\ \dfrac{-2\mu}{r^2} & 0 \end{pmatrix}, \tag{12.3.62}$$

and for *P-SV* waves

$$\mathfrak{A}_0^s = \begin{pmatrix} 0 & \frac{\wp\varsigma}{r} & \frac{1}{\rho\alpha^2} & 0 \\ -\frac{\wp}{r} & \frac{1}{\omega r} & 0 & \frac{1}{\rho\beta^2} \\ -\rho & 0 & 0 & \frac{\wp}{r} \\ 0 & -\rho + \frac{\rho(\nu\wp^2)}{r^2} & -\frac{\wp\varsigma}{r} & 0 \end{pmatrix},$$

$$\mathfrak{A}_1^s = \begin{pmatrix} -\frac{2\varsigma}{r} & 0 & 0 & 0 \\ 0 & \frac{1}{\omega r} & 0 & 0 \\ 0 & -\frac{\wp\rho\upsilon}{2r^2} & -\frac{2(1-\varsigma)}{r} & 0 \\ -\frac{\wp\rho\upsilon}{2r^2} & 0 & 0 & -\frac{3}{\omega r} \end{pmatrix}, \tag{12.3.63}$$

$$\mathfrak{A}_2^s = \begin{pmatrix} 0 & 0 & 0 & 0 \\ 0 & 0 & 0 & 0 \\ \frac{\rho\upsilon}{r^2} & 0 & 0 & 0 \\ 0 & -\frac{2\rho\beta^2}{r^2} & 0 & 0 \end{pmatrix}.$$

The leading order matrices $\mathfrak{A}_0$ have the same functional form as in the case of horizontal stratification but with the slowness p replaced by $\wp/r$. We again use a representation of the displacement traction vector of the type

$$\mathfrak{b} = \mathfrak{C}\mathfrak{u}, \tag{12.3.64}$$

where the matrix $\mathfrak{C}$ is arranged to bring the evolution equation for the vector $\mathfrak{u}$ into a form where the leading order term can be approximated via a combination of Airy function terms. The transformation matrices $\mathfrak{C}_H$, $\mathfrak{C}_s$ can be derived from the forms for horizontal stratification (12.3.4), (12.3.26) with the same slowness substitution as for the $\mathfrak{A}_0$ matrices. The leading order analysis therefore mirrors the development in the previous section with the vertical waveslowness q_β replaced by the radial waveslowness

$$\bar{q}_\beta^2 = \frac{1}{\beta^2} - \frac{\wp^2}{r^2}. \tag{12.3.65}$$

The equation to be satisfied by the vector $\mathfrak{u}$ is now

$$\partial_r \mathfrak{u} = \left\{\omega\mathfrak{C}^{-1}\mathfrak{A}_0\mathfrak{C} + [\mathfrak{C}^{-1}\mathfrak{A}_1\mathfrak{C} - \mathfrak{C}^{-1}\partial_r\mathfrak{C}^{-1}] + \omega^{-1}\mathfrak{C}^{-1}\mathfrak{A}_2\mathfrak{C}\right\}\mathfrak{u}. \tag{12.3.66}$$

The leading order term can be matched using Airy function solutions chosen to match the radial position relative to the turning point for the particular wavetype. For example, for *SH* waves above the turning level r_β we would work in terms of the matrix

$$\boldsymbol{E}_\beta = \frac{1}{\sqrt{2}}\begin{pmatrix} s_\beta \mathrm{Ej}(\omega\tau_\beta) & s_\beta \mathrm{Fj}(\omega\tau_\beta) \\ \mathrm{Ek}(\omega\tau_\beta) & \mathrm{Fk}(\omega\tau_\beta) \end{pmatrix}. \tag{12.3.67}$$

where now

$$\tau_\beta = \int_{r_\beta}^{r} \mathrm{d}r\, \bar{q}_\beta, \quad \phi_\beta = \mathrm{sgn}(\bar{q}_\beta^2)\left|\tfrac{3}{2}\tau_\beta\right|^{2/3}, \quad s_\beta = \mathrm{sgn}(\partial_r\phi_\beta), \tag{12.3.68}$$

The pattern of application of the phase matrices matches the case of horizontal stratification. The forms should be chosen to match the physical wavefield for each wavetype. Thus, between the turning levels of P and S waves we would use the evanescent form $\hat{\boldsymbol{E}}_\alpha$ for P waves and the propagating form $\boldsymbol{E}_\beta$ for S waves.

The leading order approximation will be a good representation at some distance form the origin, when the frequency is high enough for the sphericity contributions $\omega^{-1}\mathfrak{A}_1$, $\omega^{-2}\mathfrak{A}_2$ to be negligible compared with $\mathfrak{A}_0$, and also the parameter gradients are small so that we can ignore partial reflections. Under these conditions the leading order term is essentially equivalent to considering horizontal stratification derived by the earth flattening transformation (10.2.56) from a spherical model. However, the direct use of the spherical model will provide an improved treatment of density effects because there is no exact flattening transformation for density ρ (Chapman, 1973; Müller, 1977; Illingworth, 1982).

The $\boldsymbol{E}_\beta$ matrix and the comparable form $\hat{\boldsymbol{E}}_\beta$ depending on Ai, Bi to be used below the turning level r_β both satisfy equations of the form

$$\partial_r \boldsymbol{E}_\beta = \left\{\omega \boldsymbol{H}_\beta + \partial_r \Phi_\beta \Phi_\beta^{-1}\right\} \boldsymbol{E}_\beta, \tag{12.3.69}$$

where

$$\partial_r \Phi_\beta \Phi_\beta^{-1} = \tfrac{1}{2}\left(\frac{1}{r} + \frac{\partial_{rr}\phi_\beta}{\partial_r\phi_\beta}\right)\begin{pmatrix} -1 & 0 \\ 0 & 1 \end{pmatrix}. \tag{12.3.70}$$

The leading term $\omega\boldsymbol{H}_\beta$ matches $\omega\mathfrak{C}_H^{-1}\mathfrak{A}_{H0}\mathfrak{C}_H$ but the secondary term $\partial_r\Phi_\beta\Phi_\beta^{-1}$ has to be added to the set of correction terms.

For spherical stratification we can seek corrections to the leading order approximation in terms of a matrix $\mathfrak{L}$ generated by an 'interaction series'. The differential equation for $\mathfrak{L}$ is then

$$\partial_r \mathfrak{L} = -\mathfrak{E}^{-1}\left\{\left[\mathfrak{C}^{-1}\partial_r\mathfrak{C} + \partial_r \boldsymbol{\Phi}\boldsymbol{\Phi}^{-1}\right] - \left[\mathfrak{C}^{-1}\mathfrak{A}_1\mathfrak{C} + \omega^{-1}\mathfrak{C}^{-1}\mathfrak{A}_2\mathfrak{C}\right]\right\}\mathfrak{E}\mathfrak{L}, \tag{12.3.71}$$

in terms of the appropriate spherical phase matrix $\mathfrak{E}$. The first group of terms on the right hand side of (12.3.71) relate directly to the gradients of the seismic parameters and the second group arise from the sphericity of the model. We can again convert (12.3.71) into the form of an integral equation and seek an iterative solution with each term corresponding to different classes of partial reflection from the combined effects of parameter gradients and sphericity.

When the leading order approximation is adequate, we can make a transcription of the results in terms of upgoing and downgoing waves as in the previous section. The generalised radial slownesses has equivalent forms to the horizontally stratified case. Thus in the propagating regime we use

$$i\eta_{\beta u} = -p\frac{\mathrm{Ek}(\omega\tau_\beta)}{\mathrm{Ej}(\omega\tau_\beta)}, \qquad i\eta_{\beta d} = p\frac{\mathrm{Fk}(\omega\tau_\beta)}{\mathrm{Fj}(\omega\tau_\beta)}, \tag{12.3.72}$$

where τ_β is given by (12.3.68). In the evanescent regime

$$i\hat{\eta}_{\beta u} = -p\frac{\mathrm{Bk}(\omega\tau_\beta)}{\mathrm{Bj}(\omega\tau_\beta)}, \qquad i\hat{\eta}_{\beta d} = p\frac{\mathrm{Ak}(\omega\tau_\beta)}{\mathrm{Aj}(\omega\tau_\beta)}. \tag{12.3.73}$$

These generalised slownesses are equivalent to the formulation of Richards (1976) but have been derived by a very different route. In the high frequency asymptotic regime far from the turning points, the generalised slowness reduce to the expected form for the local radial slowness. However, as noted for horizontal stratification, the full forms for the generalised slowness do not just depend on local properties, but also involve the medium between the current level and the turning point.

Unfortunately the interaction series for $\mathfrak{L}$ diverges near the origin, which is a singular point for (12.3.61) through the presence of terms such as r^{-1}, r^{-2} in the $\mathfrak{A}$ matrices. However a modification of the Langer approach can be made to secure a uniform asymptotic approximation based on Bessel functions rather than Airy functions (Olver, 1974; Illingworth, 1982).

Appendix: Upgoing and downgoing waves in anisotropic media

12.a Transversely Isotropic Media

In a transversely isotropic medium where the properties depend on depth, there is again separation of the stress-displacement vector equations into *P-SV* and *SH* wave sets. A similar derivation can be made to that at the beginning of the chapter, or alternatively can be extracted from the results for general anisotropy below.

12.a.1 SH waves

The solutions take a similar form to the isotropic case with slight modifications introduced by the differences in vertical and horizontal wavespeeds. The vertical slowness q_H now satisfies

$$q_H^2 = \frac{\rho}{L} - \frac{N}{L}p^2 = (1 - p^2\beta_h^2)/\beta_v^2 \tag{12.a.1}$$

which, except at vertical incidence ($p = 0$), differs from the *SV* case. The main modification to our discussions of an isotropic medium is that the vertical slownesses for *SV* and *SH* are no longer equal.

The displacement-traction vectors for downgoing and upgoing waves are

$$\mathbf{b}_u^H = \begin{pmatrix} W \\ T \end{pmatrix}_u = \epsilon_H \begin{pmatrix} 1 \\ -\mathrm{i}Lq_H \end{pmatrix}, \qquad \mathbf{b}_d^H = \begin{pmatrix} W \\ T \end{pmatrix}_d = \epsilon_H \begin{pmatrix} 1 \\ \mathrm{i}Lq_H \end{pmatrix} \tag{12.a.2}$$

with normalisation factor

$$\epsilon_H = 1/(2Lq_H)^{1/2}. \tag{12.a.3}$$

12.a.2 P and SV waves

For a transversely isotropic medium the vertical slownesses satisfy a quartic equation

$$q_i^4 + \left[\frac{p^2L-\rho}{C} - \frac{p^2A-\rho}{L} - \frac{p^2(F+L)^2}{CL}\right] q_i^2 + \frac{(p^2L-\rho)(p^2A-\rho)}{CL} = 0. \qquad (12.a.4)$$

The qP waves have vertical slowness $\pm q_P$ where

$$q_P = \left[\frac{\rho}{2}\left(\frac{1}{L}+\frac{1}{C}\right) - Sp^2 - R\right]^{1/2}, \qquad (12.a.5)$$

where

$$S = \frac{1}{2LC}(AC - F^2 - 2LF), \qquad (12.a.6)$$

$$R = \left[(S^2 - \frac{A}{C})p^4 + \left\{\frac{\rho}{CL}(A+L) - \rho\left(\frac{1}{L}+\frac{1}{C}\right)S\right\}p^2 + \frac{\rho^2}{4}\left(\frac{1}{L}-\frac{1}{C}\right)^2\right]^{1/2}, \qquad (12.a.7)$$

$\operatorname{Re} R \geq 0, \quad \operatorname{Im} R \geq 0$

The associated displacement traction eigenvectors are

$$\mathbf{b}^P_{u,d} = \epsilon_P \begin{pmatrix} \pm \mathrm{i} p \Gamma_P \\ p \\ -(pF - C\Gamma_P q_P)p \\ \mp \mathrm{i} L(q_P - p\Gamma_P)p \end{pmatrix}, \qquad (12.a.8)$$

with

$$\Gamma_P = \frac{(Lq_P^2 + Ap^2 - \rho)}{pq_P(F+L)}, \quad \epsilon_P = \left[\frac{-(F+L)}{4p\Gamma_P CLR}\right]^{1/2}. \qquad (12.a.9)$$

As in the isotropic case the displacement traction vector is normalised to unit energy flux in the x_3-direction. Note that Γ_P is negative when q_P is real.

For the qSV waves, the vertical slowness

$$q_S = \left[\frac{\rho}{2}\left(\frac{1}{L}+\frac{1}{C}\right) - Sp^2 + R\right]^{1/2}, \qquad (12.a.10)$$

and the displacement-traction eigenvectors are

$$\mathbf{b}^S_{u,d} = \epsilon_S \begin{pmatrix} -q_S\Gamma_S \\ \mp \mathrm{i} q_S \\ \mp \mathrm{i}(C\Gamma_S q_S - pF)q_S \\ -L(q_S - p\Gamma_S)q_S \end{pmatrix}, \qquad (12.a.11)$$

where

$$\Gamma_S = \frac{(Lq_S^2 + Ap^2 - \rho)}{pq_S(F+L)}, \quad \epsilon_S = \left[\frac{p(F+L)}{4q_S^2\Gamma_S CLR}\right]^{1/2}. \qquad (12.a.12)$$

12.b General anisotropy

For a medium in which the dominant variation in elastic parameters is in depth, it is again convenient to rearrange the elastodynamic equations into a form in which the depth dependence is highlighted.

For a general anisotropic medium, we introduce the set of elastic modulus matrices $\boldsymbol{C}_{ij}$ (Woodhouse, 1974) such that

$$(\mathbf{C}_{ij})_{kl} = c_{ikjl}, \tag{12.b.1}$$

and the traction vectors $\boldsymbol{\tau}_i$ corresponding to normals along the coordinate axes,

$$(\boldsymbol{\tau}_i)_j = \tau_{ji}. \tag{12.b.2}$$

In terms of the displacement vector **u**,

$$\boldsymbol{\tau}_i = \mathbf{C}_{i3}\partial_3\mathbf{u} + \mathbf{C}_{i\nu}\partial_\nu\mathbf{u}, \quad \nu = 1, 2. \tag{12.b.3}$$

Here we have written ∂_ν for $\partial/\partial x_\nu$, and have adopted the convention that Greek suffices will always refer to summation over horizontal coordinates.

The expression for the traction in the 3-direction can be rewritten to extract the x_3 derivative of the displacement

$$\mathbf{C}_{33}\partial_3 = \tau_3 - \mathbf{C}_{\nu 3}\partial_\nu\mathbf{u}, \tag{12.b.4}$$

and we can use this to recast the expressions for the horizontal tractions $\boldsymbol{\tau}_1$, $\boldsymbol{\tau}_2$ as

$$\begin{aligned}\boldsymbol{\tau}_\nu &= \mathbf{C}_{\nu\sigma}\partial_\sigma\mathbf{u} + \mathbf{C}_{\nu 3}(\mathbf{C}_{33}^{-1}\boldsymbol{\tau}_3 - \mathbf{C}_{33}^{-1}\mathbf{C}_{3\sigma}\partial_\sigma\mathbf{u}),\\ &= \mathbf{Q}_{\nu\sigma}\partial_\sigma\mathbf{u} + \mathbf{C}_{\nu 3}\mathbf{C}_{33}^{-1}\boldsymbol{\tau}_3,\end{aligned} \tag{12.b.5}$$

where the matrix $\mathbf{Q}_{\nu\sigma}$ is given by

$$\mathbf{Q}_{\nu\sigma} = \mathbf{C}_{\nu\sigma} - \mathbf{C}_{\nu 3}\mathbf{C}_{33}^{-1}\mathbf{C}_{3\sigma}. \tag{12.b.6}$$

The equation of motion (12.1.1) can be recast as

$$\partial_3\boldsymbol{\tau}_3 = \rho\partial_{tt}\mathbf{u} - \partial_\nu\boldsymbol{\tau}_\nu. \tag{12.b.7}$$

On rearranging (12.b.4), (12.b.7) so that derivatives with respect to the x_3 coordinate (z) appear only on the left hand side of the equations we obtain

$$\partial_3\mathbf{u} = -\mathbf{C}_{33}^{-1}\mathbf{C}_{3\sigma}\partial_\sigma\mathbf{u} + \mathbf{C}_{33}^{-1}\boldsymbol{\tau}_3, \tag{12.b.8}$$

$$\partial_3\boldsymbol{\tau}_3 = \rho\partial_{tt}\mathbf{u} - \partial_\nu(\mathbf{Q}_{\nu\sigma}\partial_\sigma\mathbf{u}) - \partial_\sigma(\mathbf{C}_{\sigma 3}\mathbf{C}_{33}^{-1}\boldsymbol{\tau}_3), \tag{12.b.9}$$

In order to simplify subsequent notation we write

$$\mathbf{J}_{\sigma 3} = \mathbf{C}_{33}^{-1}\mathbf{C}_{3\sigma}, \quad \mathbf{K}_{33} = \mathbf{C}_{33}^{-1}, \tag{12.b.10}$$

and as a result of the symmetry of the elastic modulus tensor

$$\mathbf{J}_{\sigma 3}^T = \mathbf{C}_{\sigma 3}\mathbf{C}_{33}^{-1}, \tag{12.b.11}$$

so that (12.b.8), (12.b.9) depend on the elastic moduli through the expressions $\mathbf{Q}_{\nu\sigma}$, $\mathbf{J}_{\sigma 3}$, $\mathbf{K}_{33}$. We will also write **t** for the traction vector $\boldsymbol{\tau}_3$.

The two equations (12.b.8), (12.b.9) for the depth evolution of the displacement and traction fields are valid for a general medium. If the material does not vary across horizontal planes, we can write the two coupled equations for displacement and traction in the form

$$\partial_3\begin{pmatrix}\mathbf{u}\\ \mathbf{t}\end{pmatrix} = \begin{pmatrix}-\mathbf{J}_{\sigma 3}\partial_\sigma & \mathbf{K}_{33}\\ \rho\mathbf{I}\partial_{tt} - \mathbf{Q}_{\nu\sigma}\partial_{\nu\sigma} & -\mathbf{J}_{\sigma 3}^T\partial_\sigma\end{pmatrix}\begin{pmatrix}\mathbf{u}\\ \mathbf{t}\end{pmatrix}, \tag{12.b.12}$$

where $\mathbf{I}$ is a 3×3 unit matrix.

We now consider a plane wave solution for the displacement and traction

$$\begin{pmatrix}\mathbf{u}\\ \mathbf{t}\end{pmatrix} = \begin{pmatrix}\mathbf{u}(x_3)\\ \mathbf{t}(x_3)\end{pmatrix}\exp(i\omega[p_1x_1 + p_2x_2 - t]), \tag{12.b.13}$$

where p_1, p_2 are the horizontal components of the slowness vector. The depth dependent part of the field must then satisfy

$$\partial_3\begin{pmatrix}\mathbf{u}(x_3)\\ \omega^{-1}\mathbf{t}(x_3)\end{pmatrix} = \omega\begin{pmatrix}-i\mathbf{J}_{\sigma 3}p_\sigma & \mathbf{K}_{33}\\ -\rho\mathbf{I} + \mathbf{Q}_{\nu\sigma}p_\nu p_\sigma & -i\mathbf{J}_{\sigma 3}^T p_\sigma\end{pmatrix}\begin{pmatrix}\mathbf{u}(x_3)\\ \omega^{-1}\mathbf{t}(x_3)\end{pmatrix}. \tag{12.b.14}$$

For a locally uniform region the coupled equations (12.b.14) have solutions with a depth dependence of the form

$$\exp(\mathrm{i}\omega q_n x_3), \quad n = 1,6 \qquad (12.b.15)$$

With the common seismological coordinate system (x_1 - North, x_2 - East, x_3 - Down) we can identify the three solutions for which the phase increases with increasing depth x_3 (i.e. q_n positive) as downgoing waves. Similarly the three solutions with q_n negative correspond to upgoing waves. The vertical slownesses q_n can be found from the eigenvalues $\mathrm{i}q_n$ of the matrix on the right hand side of (12.b.14) and the associated displacement and traction vectors from the eigenvectors, since we require

$$\begin{pmatrix} -\mathrm{i}\boldsymbol{J}_{\sigma 3}p_\sigma - \mathrm{i}q_n\boldsymbol{I} & \boldsymbol{K}_{33} \\ -\rho\boldsymbol{I} + \boldsymbol{Q}_{\nu\sigma}p_\nu p_\sigma & -\mathrm{i}\boldsymbol{J}^T_{\sigma 3}p_\sigma - \mathrm{i}q_n\boldsymbol{I} \end{pmatrix} \begin{pmatrix} \mathbf{u}_n(x_3) \\ \omega^{-1}\mathbf{t}_n(x_3) \end{pmatrix} = \mathbf{0}. \qquad (12.b.16)$$

In a generally anisotropic medium (12.b.16) has to be solved for each horizontal slowness pair (p_1, p_2) to generate the vertical slownesses appropriate to that particular plane wave configuration.

When the material possesses some form of symmetry, the complexity of the eigenvalue problem is reduced. As we have seen, analytic solutions can readily found for isotropic and transversely isotropic media with a vertical symmetry axis. The properties of the eigenvector solutions for a fully anisotropic medium have been investigated in detail by Fryer & Frazer (1987).

13

Reflection and Transmission

In the previous chapter we described the properties of plane seismic waves travelling in an unbounded medium but the interaction of seismic waves with contrasts in elastic properties controls the energy returned to the surface.

We have given a brief discussion of reflection and transmission problems in Section 3.1 for both the free surface and an internal interface, making use of wavefront diagrams to represent the behaviour. In this chapter we show how to develop a more quantitative treatment based on descriptions of upgoing and downgoing waves. We will first consider a plane stratified medium and then indicate how the concepts can be extended to the case of spherical stratification.

In a uniform region we can represent the seismic wavefield as a superposition of upgoing and downgoing plane waves which propagate independently in the absence of material interfaces. Once boundaries are present we have to take account of interaction between different wavetypes and different directions of propagation.

We start by setting up a representation of the local wavefield in terms of upgoing and downgoing propagating components for the different wavetypes. This representation enables us to examine the process of reflection at a free surface where the traction vanishes. Such reflection can lead to very efficient conversion between wavetypes. Reflection also occurs at a material interface but is accompanied by transmission into the adjoining medium.

13.1 A free surface

At a free surface the traction must vanish, and in order to satisfy this boundary condition we have to look for a suitable combination of upward and downward travelling waves such that the total tractions sum to zero.

We will suppose that the free surface lies on the plane $x_3 = 0$, and in that case we can build a solution from components with a common slowness p and hence a common dependence on x_1 and t of the form

$$\exp(i\omega[px_1 - t]). \qquad (13.1.1)$$

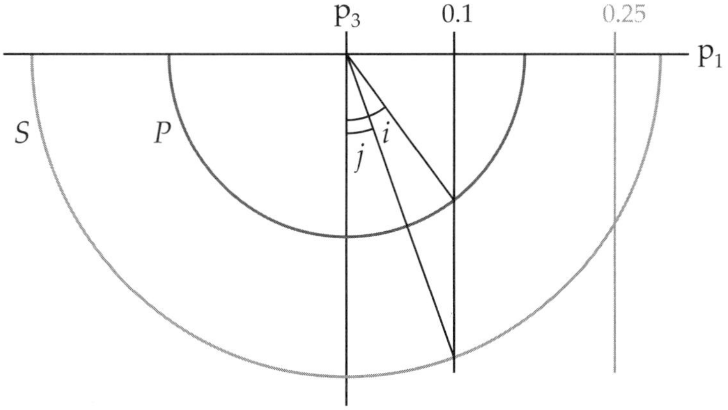

Figure 13.1. Slowness surface construction for reflection at the free surface of an isotropic medium, $\alpha_0 = 6.0$ km/s, $\beta_0 = 3.4$ km/s.

Consider the case of SH wave reflection with an incident upgoing wave H_U impinging on the surface. This will generate a downgoing SH wave H_D, but since the boundary is horizontal there will be no conversion to other wavetypes. Also for common slowness the downgoing plane wave must have the same inclination to the vertical as the upgoing wave.

We make use of the displacement-traction relation (12.1.6) in association with a superposition of upgoing and downgoing waves of the form (12.1.4) to write

$$\begin{pmatrix} W \\ T \end{pmatrix} = \epsilon_H \begin{pmatrix} 1 & 1 \\ -\mathrm{i}\mu_o q_{\beta 0} & \mathrm{i}\mu_o q_{\beta 0} \end{pmatrix} \begin{pmatrix} H_U \exp[-\mathrm{i}\omega q_\beta x_3] \\ H_D \exp[+\mathrm{i}\omega q_\beta x_3] \end{pmatrix}, \tag{13.1.2}$$

where we have suppressed the common exponential factor (13.1.1). When we impose the condition of vanishing traction T as $x_3 = 0$ the relation between the upgoing and downgoing components is determined by

$$\begin{pmatrix} W(0) \\ 0 \end{pmatrix} = \epsilon_H \begin{pmatrix} 1 & 1 \\ -\mathrm{i}\mu_o q_{\beta 0} & \mathrm{i}\mu_o q_{\beta 0} \end{pmatrix} \begin{pmatrix} H_U \\ H_D \end{pmatrix}, \tag{13.1.3}$$

and so from the traction equation

$$\mathrm{i}\mu_o q_{\beta 0}(H_U - H_D) = 0, \quad \text{i.e.} \quad H_U = H_D. \tag{13.1.4}$$

The downgoing component H_D is equal to the upgoing component H_U and so we have total reflection without change of phase, i.e., a reflection coefficient $\mathrm{R}_F^{HH} = 1$.

For *P-SV* waves, the situation is more complex. The requirement of common slowness means that where reflection occurs without change of wavetype, the angles of incidence and reflection are the same. Where conversion occurs a form of Snell's law operates, and the inclination to the vertical of the *P* waves (i) and *S* waves (j) in an isotropic medium are related by

$$\frac{\sin i}{\alpha_0} = \frac{\sin j}{\beta_0}. \tag{13.1.5}$$

This relation can also be derived by considering the intersection of a line of constant slowness with the slowness surfaces for P and SV waves, which for isotropic media are circles of radii $1/\alpha_0$ and $1/\beta_0$ as illustrated in figure 13.1 for two slownesses. For $p = 0.25$ s/km, shown in grey, the P wave is evanescent.

For a transversely isotropic medium, the qP and qSV velocities will depend on the horizontal slowness, but the appropriate angles can again be determined from the slowness surfaces by constructing points with a common projection onto the 1-axis (cf. figure 13.1).

We again adopt a representation of the wavefield in terms of upgoing and downgoing waves using (12.1.38)

$$\mathbf{b}(x_3) = \begin{pmatrix} \boldsymbol{w}(x_3) \\ \boldsymbol{t}(x_3) \end{pmatrix} = \begin{pmatrix} \boldsymbol{m}_U & \boldsymbol{m}_D \\ \boldsymbol{n}_U & \boldsymbol{n}_D \end{pmatrix} \begin{pmatrix} \boldsymbol{v}_U(x_3) \\ \boldsymbol{v}_D(x_3) \end{pmatrix}, \tag{13.1.6}$$

where we recall the partitions $\boldsymbol{m}_U$, $\boldsymbol{n}_U$ correspond to the displacements and tractions connected with upgoing P and S waves. The displacement and traction matrices have the forms

$$\boldsymbol{m}_{U,D} = \begin{pmatrix} \mp \mathrm{i} q_\alpha \epsilon_\alpha & p\epsilon_\beta \\ p\epsilon_\alpha & \mp \mathrm{i} q_\beta \epsilon_\beta \end{pmatrix},$$
$$\boldsymbol{n}_{U,D} = \begin{pmatrix} \rho(2\beta^2 p^2 - 1)\epsilon_\alpha & \mp 2\mathrm{i}\rho\beta^2 p q_\beta \epsilon_\beta \\ \mp 2\mathrm{i}\rho\beta^2 p q_\alpha \epsilon_\alpha & \rho(2\beta^2 p^2 - 1)\epsilon_\beta \end{pmatrix}, \tag{13.1.7}$$

for an isotropic medium.

At the free surface we will have an incident upgoing field $\boldsymbol{v}_U$ which will give rise to a reflected downgoing field $\boldsymbol{v}_D$ to satisfy the the condition of vanishing traction $\mathbf{t}$ at the free surface. The condition $\mathbf{t}(0) = \mathbf{0}$ (i.e., $\tau_{13} = \tau_{33} = 0$) requires the downgoing wave components $\boldsymbol{v}_D$ to be determined from

$$\begin{pmatrix} \boldsymbol{w}(0) \\ \mathbf{0} \end{pmatrix} = \begin{pmatrix} \boldsymbol{m}_U & \boldsymbol{m}_D \\ \boldsymbol{n}_U & \boldsymbol{n}_D \end{pmatrix} \begin{pmatrix} \boldsymbol{v}_U \\ \boldsymbol{v}_D \end{pmatrix}, \tag{13.1.8}$$

so that

$$\boldsymbol{n}_U \boldsymbol{v}_U + \boldsymbol{n}_D \boldsymbol{v}_D = 0. \tag{13.1.9}$$

We now introduce the concept of a matrix of reflection coefficients $\mathbf{R}_F$ which summarises the interrelations between the different wave components, so that

$$\boldsymbol{v}_D = \mathbf{R}_F \boldsymbol{v}_U, \tag{13.1.10}$$

which can be expanded into the explicit form

$$\begin{pmatrix} P_D \\ S_D \end{pmatrix} = \begin{pmatrix} \mathrm{R}_F^{PP} & \mathrm{R}_F^{PS} \\ \mathrm{R}_F^{SP} & \mathrm{R}_F^{SS} \end{pmatrix} \begin{pmatrix} P_U \\ S_U \end{pmatrix}, \tag{13.1.11}$$

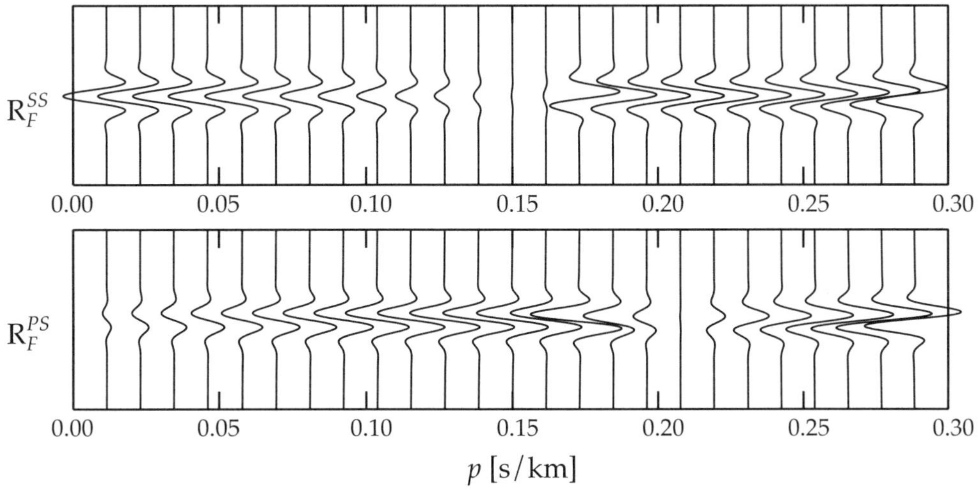

Figure 13.2. The behaviour of the reflected pulses from the free-surface for an incident S wave as a function of slowness: $\alpha_0 = 6.0$ km/s, $\beta_0 = 3.4$ km/s. The critical slowness for P is 0.167 s/km and the converted pulse for larger slowness has no direct physical meaning.

R_F^{PP} is the reflected amplitude for P waves and R_F^{SP} the amplitude of the reflected S wave from a unit incident P wave.

From (13.1.9), (13.1.10) we can identify the reflection matrix as

$$\mathbf{R}_F = -\boldsymbol{n}_D^{-1}\boldsymbol{n}_U, \tag{13.1.12}$$

in terms of the traction partitions of the matrix **D** comprised of the motion-stress vectors for the different styles of waves. Using the expressions (12.1.13) and (12.1.15) for the vector components of **D** in an isotropic medium we can express the reflection coefficients at the free-surface in the explicit form:

$$\begin{pmatrix} R_F^{PP} & R_F^{PS} \\ R_F^{SP} & R_F^{SS} \end{pmatrix} = \frac{1}{4p^2 q_{\alpha 0} q_{\beta 0} + \upsilon^2} \begin{pmatrix} 4p^2 q_{\alpha 0} q_{\beta 0} - \upsilon^2 & 4p\upsilon(q_{\alpha 0} q_{\beta 0})^{1/2} \\ 4p\upsilon(q_{\alpha 0} q_{\beta 0})^{1/2} & 4p^2 q_{\alpha 0} q_{\beta 0} - \upsilon^2 \end{pmatrix}, \tag{13.1.13}$$

where $\upsilon = (2p^2 - \beta_0^{-2})$. With the normalisation of the upgoing and downgoing waves by their energy transport in the x_3-direction, we have the same value for the R_F^{PP} and R_F^{SS} coefficients, and the same amplitude for the R_F^{PS} and R_F^{SP} coefficients. A similar structure occurs for transverse isotropy but the expressions are algebraically more complicated.

We have displayed the free surface reflection coefficients for *P-SV* waves as a function of slowness in figure 3.4, and will here look at a different aspect of the problem by examining the nature of reflected pulses from the surface. For very small slownesses, i.e. steep incidence angles, the dominant reflection process is in the same wavetype but the significance of the conversion process increases rapidly with increasing slowness and for shallow incidence conversion is the dominant process.

For slownesses p less than the critical slowness for P, α_0^{-1}, the unconverted pulses have the same form as the incident wave but with scaled amplitude, and the converted phases have a $-\pi/2$ phase shift and thus are related to the Hilbert transform of the incident pulse.

Once the horizontal slowness p exceeds α_0^{-1} (0.167 s/km for the example in figure 13.2), P waves are evanescent in the x_3-direction and $q_{\alpha 0}^2 < 0$. There is total reflection of S waves, but with a phase shift arising from the complex reflection coefficient. The phase shift leads to a distortion of the waveform of a plane SV pulse reflected at the surface for $p > \alpha_0^{-1}$ as shown in figure 13.2. The converted pulse only has physical reality for slownesses less than the critical slowness for P, α_0^{-1}. For slownesses larger than 0.167 s/km, the S waves are totally reflected and the converted pulses link only to evanescent waves.

The real part of the reflection coefficients gives rise to a scaled version of the original pulse and the imaginary part introduces a scaling of the Hilbert transform of the pulse (see e.g. Hudson, 1962). Consider a pulse $f(t)$ with transform $\bar{f}(\omega)$. The effect of the reflection process is introduced by multiplication of each frequency component by a reflection coefficient $R = R' + \mathrm{i}R''$, The reflected pulse spectrum is

$$\bar{g}(\omega) = \bar{f}(\omega)R' + \mathrm{i}\bar{f}(\omega)R'', \quad \omega > 0. \tag{13.1.14}$$

The condition that the resulting pulse is real requires $\bar{g}(-\omega) = \bar{g}^*(\omega)$ in terms of the complex conjugate of $\bar{g}$. Thus we can write

$$\bar{g}(\omega) = \bar{f}(\omega)R' + \mathrm{i}\,\mathrm{sgn}(\omega)\bar{f}(\omega)R'', \tag{13.1.15}$$

Then on inverting the Fourier transform we obtain

$$g(t) = R'f(t) - R''\mathcal{H}f(t), \tag{13.1.16}$$

where $\mathcal{H}f(t)$ is the Hilbert transform (or allied function) for $f(t)$, with spectrum $-\mathrm{i}\,\mathrm{sgn}(\omega)\bar{f}$,

$$\mathcal{H}f(t) = \frac{1}{\pi}\mathrm{P}\int_{-\infty}^{\infty} \mathrm{d}s \frac{f(s)}{s-t}, \tag{13.1.17}$$

where P denotes the Cauchy principal value of the integral. For an impulse the Hilbert transform has precursory effects.

The free-surface reflection coefficients become singular at a slowness p_R such that the denominator in (13.1.12) vanishes, i.e.,

$$4p_R^2 q_{\alpha 0} q_{\beta 0} + (2p_R^2 - \beta_0^{-2})^2 = 0, \tag{13.1.18}$$

which is the condition for the existence of free Rayleigh surface waves on a uniform half space with wavespeeds α_0, β_0. In such a Rayleigh wave both P and S waves decay with depth away from the surface since $p_R > \beta_0^{-1}$.

The reflection coefficients at the free surface are larger than from any internal interfaces except for grazing incidence and so play a very important role in the propagation processes for seismic waves.

From (13.1.8) the surface displacement is expressed as

$$\boldsymbol{w}_0 = \boldsymbol{m}_U \boldsymbol{v}_U + \boldsymbol{m}_D \boldsymbol{v}_D = [\boldsymbol{m}_U + \boldsymbol{m}_D \mathbf{R}_F]\, \boldsymbol{v}_U, \tag{13.1.19}$$

where we have employed the definition of the free-surface reflection matrix (13.1.10) to express the displacement in terms of the incident wavefield $\boldsymbol{v}_U$. We can rewrite (13.1.19) in terms of a matrix of amplification factors $\mathbf{W}_F$ so that

$$\boldsymbol{w}_0 = \mathbf{W}_F \boldsymbol{v}_U, \quad \text{with} \quad \mathbf{W}_F = [\boldsymbol{m}_U + \boldsymbol{m}_D \mathbf{R}_F]\,. \tag{13.1.20}$$

The amplification matrix $\mathbf{W}_F$ represents the effects of the interaction of the incident and reflected waves. Writing w_Z and w_R for the vertical and radial components on which the *P-SV* waves appear we can write (13.1.19) as

$$\boldsymbol{w}_0 = \begin{pmatrix} w_Z \\ w_R \end{pmatrix} = \begin{pmatrix} \mathrm{W}_F^{ZP} & \mathrm{W}_F^{ZS} \\ \mathrm{W}_F^{RP} & \mathrm{W}_F^{RS} \end{pmatrix} \begin{pmatrix} P_U \\ S_U \end{pmatrix} = \mathbf{W}_F \begin{pmatrix} P_U \\ S_U \end{pmatrix} = \mathbf{W}_F \boldsymbol{v}_U. \tag{13.1.21}$$

For an isotropic medium $\mathbf{W}_F$ has the explicit form:

$$\mathbf{W}_F(p) = \begin{pmatrix} -\mathrm{i}q_{\alpha 0}\epsilon_{\alpha 0}C_1 & p\epsilon_{\beta 0}C_2 \\ p\epsilon_{\alpha 0}C_2 & -\mathrm{i}q_{\beta 0}\epsilon_{\beta 0}C_1 \end{pmatrix}, \tag{13.1.22}$$

in which the elements of $\boldsymbol{m}_U$ are modified by the presence of the conversion factors from infinite medium to free-surface displacements

$$\begin{aligned} C_1 &= 2\beta_0^{-2}(2p^2 - \beta_0^{-2})/\{4p^2 q_{\alpha 0} q_{\beta 0} + \upsilon^2\}, \\ C_2 &= 4\beta_0^{-2} q_{\alpha 0} q_{\beta 0}/\{4p^2 q_{\alpha 0} q_{\beta 0} + \upsilon^2\}. \end{aligned} \tag{13.1.23}$$

These amplification factors are appropriate to both propagating and evanescent waves and were plotted in figure 3.5 for the same medium properties as employed in figures 3.4 and 13.2.

For *SH* waves, the reflection coefficient is unity and we have a simple amplification by a factor of 2 due to the interference of the incident and reflected waves,

$$\mathrm{W}_F^{HH} = 2. \tag{13.1.24}$$

When we have the ground-motion recorded at the free-surface it is useful to be able to disentangle the influence of the surface and return to the incident waves. For plane incident waves of slowness p we can recover the incident upgoing waves $\boldsymbol{v}_U$ by constructing the inverse of $\mathbf{W}_F$, thus

$$\boldsymbol{v}_U = \begin{pmatrix} P_U \\ S_U \end{pmatrix} = \begin{pmatrix} \mathrm{V}_F^{PZ} & \mathrm{V}_F^{PR} \\ \mathrm{V}_F^{RP} & \mathrm{V}_F^{RS} \end{pmatrix} \begin{pmatrix} w_Z \\ w_R \end{pmatrix} = \mathbf{W}_F^{-1} \begin{pmatrix} w_Z \\ w_R \end{pmatrix} = \mathbf{W}_F^{-1} \boldsymbol{w}_0. \tag{13.1.25}$$

Despite the complex form of $\mathbf{W}_F$, the inverse is rather simple:

$$\mathbf{W}_F^{-1}(p) = \begin{pmatrix} \mathrm{V}_F^{PZ} & \mathrm{V}_F^{PR} \\ \mathrm{V}_F^{RP} & \mathrm{V}_F^{RS} \end{pmatrix} = \begin{pmatrix} \mathrm{i}(2\beta_0^2 p^2 - 1)/q_{\alpha 0}\epsilon_{\alpha 0} & 2\beta_0^2 p/\epsilon_{\alpha 0} \\ 2\beta_0^2 p/\epsilon_{\beta 0} & \mathrm{i}(2\beta_0^2 p^2 - 1)/q_{\beta 0}\epsilon_{\beta 0} \end{pmatrix}. \tag{13.1.26}$$

The inverse relation (13.1.25) is thus singular at the critical slownesses for P and SV waves, but away from these regions the coefficients are relatively slowly varying functions of slowness as illustrated in figure 13.3.

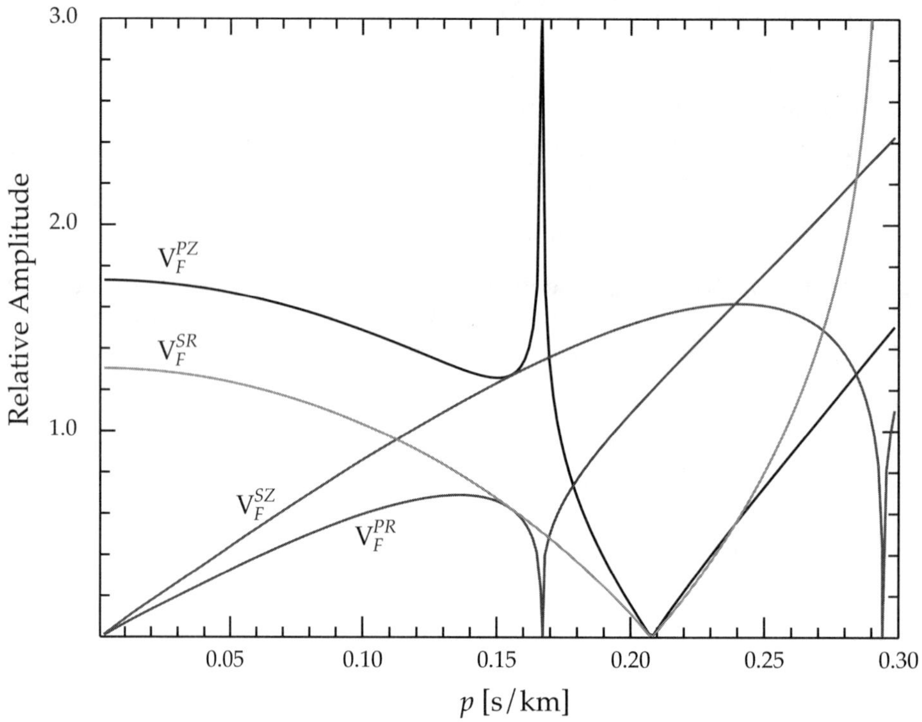

Figure 13.3. Dependence of the conversion factors between surface motion and incident *P* and *SV* wave amplitude on slowness.

The equivalent relation between the surface motion and the amplitude of an incident *SH* waves is a simple scaling by a factor of one half

$$V_F^{HT} = \tfrac{1}{2}. \tag{13.1.27}$$

13.2 A solid-solid interface

At an interface between two different media we require the two sides of the interface to move together so that the displacement must be continuous; further, in order to avoid stress concentrations at the interface, we require continuity of traction.

For a horizontal interface, we also need continuity of horizontal phase across the boundary so all the interacting plane waves must have the same horizontal slowness p. We can therefore relate the upgoing and downgoing waves on the two sides of the interface by making a decomposition of the form (13.1.2) or (13.1.8) on each side of the interface ($x_3 = z_I$).

13.2.1 SH waves

For SH waves, the requirement that the horizontal slowness be constant across the interface means that the angles of incidence and reflection are the same (j_1) and the angle of transmission (j_2) is related by Snell's law

$$\frac{\sin j_1}{\beta_1} = \frac{\sin j_2}{\beta_2}. \tag{13.2.1}$$

In medium 1, just above the interface, we can represent the displacement and traction in terms of upgoing and downgoing waves in the form

$$\begin{pmatrix} u_2 \\ \omega^{-1}\tau_{23} \end{pmatrix} = \epsilon_{H1} \begin{pmatrix} 1 & 1 \\ -\mathrm{i}\mu_1 q_{\beta 1} & \mathrm{i}\mu_1 q_{\beta 1} \end{pmatrix} \begin{pmatrix} H_{U1} \\ H_{D1} \end{pmatrix}, \tag{13.2.2}$$

whereas in medium 2 just below the interface, we can make a comparable decomposition

$$\begin{pmatrix} u_2 \\ \omega^{-1}\tau_{23} \end{pmatrix} = \epsilon_{H2} \begin{pmatrix} 1 & 1 \\ -\mathrm{i}\mu_2 q_{\beta 2} & \mathrm{i}\mu_2 q_{\beta 2} \end{pmatrix} \begin{pmatrix} H_{U2} \\ H_{D2} \end{pmatrix}. \tag{13.2.3}$$

The continuity of displacement and traction across the interface means that we must equate the two expressions (13.2.2), (13.2.3) so that

$$\epsilon_{H1} \begin{pmatrix} 1 & 1 \\ -\mathrm{i}\mu_1 q_{\beta 1} & \mathrm{i}\mu_1 q_{\beta 1} \end{pmatrix} \begin{pmatrix} H_{U1} \\ H_{D1} \end{pmatrix} = \epsilon_{H2} \begin{pmatrix} 1 & 1 \\ -\mathrm{i}\mu_2 q_{\beta 2} & \mathrm{i}\mu_2 q_{\beta 2} \end{pmatrix} \begin{pmatrix} H_{U2} \\ H_{D2} \end{pmatrix}, \tag{13.2.4}$$

and so the upgoing and downgoing wave elements on the two sides of the interface are related by

$$\begin{pmatrix} H_{U1} \\ H_{D1} \end{pmatrix} = \frac{1}{2(\mu_1\mu_2 q_{\beta 1} q_{\beta 2})^{1/2}} \begin{pmatrix} \mu_1 q_{\beta 1} + \mu_2 q_{\beta 2} & \mu_1 q_{\beta 1} - \mu_2 q_{\beta 2} \\ \mu_1 q_{\beta 1} - \mu_2 q_{\beta 2} & \mu_1 q_{\beta 1} + \mu_2 q_{\beta 2} \end{pmatrix} \begin{pmatrix} H_{U2} \\ H_{D2} \end{pmatrix}. \tag{13.2.5}$$

The combination μq_β plays the role of an impedance for obliquely travelling SH waves.

Consider an *downgoing wave* incident from medium 1. This will give rise to a reflected upgoing wave in the region $z < z_I$ and a transmitted downgoing wave in $z > z_I$. On the underside of the interface in medium 2 there will be no upgoing wave and so H_{U2} will be zero.

We define a reflection coefficient for downward propagation R_D^I, to connect the wave elements in medium 1,

$$H_{U1} = \mathrm{R}_D^I H_{D1}, \tag{13.2.6}$$

and a transmission coefficient T_D^I, to relate the downgoing wave components on the two sides of the interface,

$$H_{D2} = \mathrm{T}_D^I H_{D1}. \tag{13.2.7}$$

For the incident downgoing wave (13.2.5) can be written in the form

$$\begin{pmatrix} H_{U1} \\ H_{D1} \end{pmatrix} = \begin{pmatrix} Q_{UU} & Q_{UD} \\ Q_{DU} & Q_{DD} \end{pmatrix} \begin{pmatrix} 0 \\ H_{D2} \end{pmatrix}, \tag{13.2.8}$$

and so

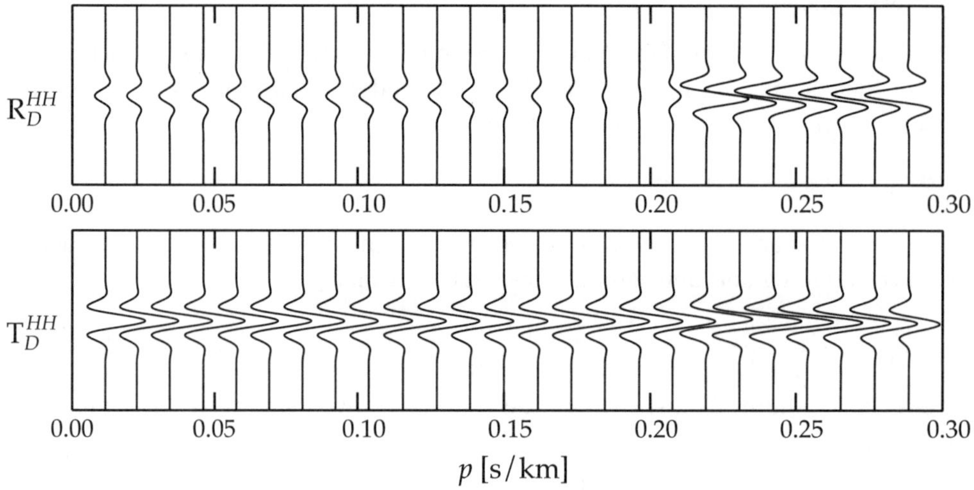

Figure 13.4. The behaviour of reflected and transmitted pulses for downward incidence of an *SH* wave on an interface with $\beta_1 = 3.4$ km/s, $\rho_1 = 2.7$ Mg/m^3, $\beta_2 = 4.6$ km/s, $\rho_2 = 3.3$ Mg/m^3.

$$R_D^I = Q_{UD}(Q_{DD})^{-1}, \quad T_D^I = (Q_{DD})^{-1}. \tag{13.2.9}$$

In terms of the SH wave impedances the reflection and transmission coefficients are

$$R_D^I = \frac{(\mu_1 q_{\beta 1} - \mu_2 q_{\beta 2})}{(\mu_1 q_{\beta 1} + \mu_2 q_{\beta 2})}, \qquad T_D^I = \frac{2(\mu_1 \mu_2 q_{\beta 1} q_{\beta 2})^{1/2}}{(\mu_1 q_{\beta 1} + \mu_2 q_{\beta 2})}. \tag{13.2.10}$$

For propagating waves in a perfectly elastic medium we have chosen the upgoing and downgoing waves to be normalised to unit energy flux in the z direction. The coefficients R_D^I, T_D^I are therefore measures of the reflected and transmitted energy in such propagating waves.

The structure of (13.2.10) is not affected if the half spaces are weakly dissipative so that we use complex S wavespeeds, or if the slowness p is such that we are considering evanescent waves, provided we take a consistent choice of branch cut for the radicals $q_{\beta 1}$, $q_{\beta 2}$. We will therefore refer to the expressions (13.2.10) as the reflection and transmission coefficients for any slowness p.

We will illustrate the behaviour of the *SH* interface coefficients by examining the character of reflected and transmitted waves for an incident downward *SH* waves on the interface. The reflection and transmission coefficients are real for slownesses p less than the critical slowness for S, β_2^{-1}, (0.217 s/km) so over most of the span displayed the effect of the interaction is just to produce a scaling of the incident pulse. Pulse distortion due to the presence of an imaginary part to the coefficients has a very slight effect on the post-critical reflections for which transmission links to evanescent waves. The phase shifts can be seen through the alignments of the pulses in reflection for $p > 0.21$ s/km.

As the contrast in properties across the interface becomes very small

$$\mathrm{R}_D^I \to \tfrac{1}{2}\frac{\Delta(\mu q_\beta)}{\mu q_\beta}, \quad \mathrm{T}_D^I \to 1, \tag{13.2.11}$$

where $\Delta(\mu q_\beta)$ is the contrast in the impedance across the interface.

For this scalar case the reflection and transmission coefficients for upward incidence from medium 2 are most easily obtained by exchanging the suffices 1 and 2, so that

$$\mathrm{R}_U^I = -\mathrm{R}_D^I, \quad \mathrm{T}_U^I = \mathrm{T}_D^I, \quad 1 - (\mathrm{R}_D^I)^2 = (\mathrm{T}_D^I)^2. \tag{13.2.12}$$

The structure of the SH wave results is preserved for a transversely isotropic medium with the replacement of the isotropic impedance μq_β by $L q_H$.

13.2.2 Coupled P and SV waves

Although P and SV waves propagate independently in a uniform medium, they are coupled by the boundary conditions at a horizontal interface. In addition to reflected and transmitted waves of the incident wave type, there will be conversion to the other wave type in both reflection and transmission. With the same slowness p for all the waves we ensure a common horizontal phase behaviour and so satisfy Snell's law both above and below the interface (see figure 3.7). The angles of propagation of the P and S waves above and below the interface are therefore related by

$$\frac{\sin i_1}{\alpha_1} = \frac{\sin i_2}{\alpha_2} = \frac{\sin j_1}{\beta_1} = \frac{\sin j_2}{\beta_2}. \tag{13.2.13}$$

The relations can also be determined by using the slowness surfaces for the media on the two sides of the interface as illustrated for two isotropic media in figure 13.3. The slowness surfaces are hemispheres on each side of the boundary and the propagation angles are linked by the intersection of a line of constant slowness p with the slowness surfaces for P and S waves.

The slowness surface diagram provides a useful way of categorising the behaviour at the interface in terms of the onset of evanescence for each of the different classes of waves. Within the appropriate slowness surface a wave will propagate, and outside it will be evanescent. The intersection of the wave surface with the horizontal axis thus indicates the critical slowness.

Consider then a P wave source in the upper medium: this will generate propagating P waves up to a slowness of 0.167 s/km. An incident P wave in the upper medium with a slowness of 0.15 s/km (indicated in grey in figure 13.5), will lie beyond the critical slowness for P in the lower medium but within the propagating regime for S waves in both media. There will be a postcritical reflection as well as conversion to S in reflection and transmission. The link to P in the lower medium will occur through three conical ('head') waves with the critical slowness (0.125 s/km) as is illustrated in the wavefront diagram for a source in the upper medium in figure 3.11.

In contrast for a P source in the lower medium, propagating P waves will be confined to $p < 0.125$ s/km and there is no possibility of head waves. The wavefront diagram for the reflected and transmitted waves for this case is much simpler (figure 3.12).

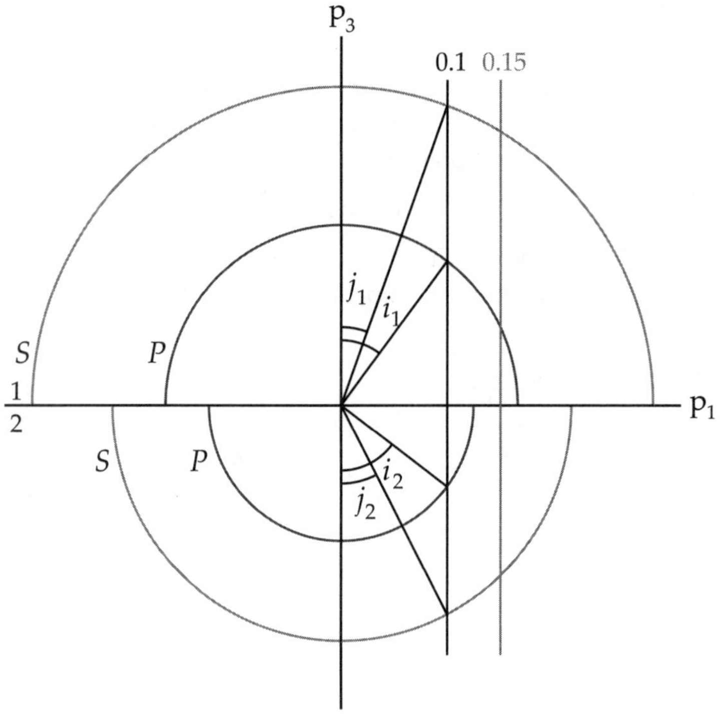

Figure 13.5. Slowness surfaces for the wavespeeds in the upper and lower media [α_1 = 6.0 km/s, β_1 = 3.4 km/s, α_2 = 8.0 km/s, β_2 = 4.6 km/s], and the connection between the propagation characteristics for a fixed horizontal slowness p.

The interaction between P and SV waves at the interface may be derived by using an extension of the matrix methods we have introduced in the case of free surface reflections. The displacement-traction vector just above the interface ($x_3 = z_I-$) can be represented in terms of upgoing and downgoing waves in medium 1 using (12.1.5)

$$\mathbf{b}(z_I-) = \begin{pmatrix} \boldsymbol{w}(z_I-) \\ \boldsymbol{t}(z_I-) \end{pmatrix} = \begin{pmatrix} \boldsymbol{m}_{U1} & \boldsymbol{m}_{D1} \\ \boldsymbol{n}_{U1} & \boldsymbol{n}_{D1} \end{pmatrix} \begin{pmatrix} \boldsymbol{v}_{U1}(z_I-) \\ \boldsymbol{v}_{D1}(z_I-) \end{pmatrix} = \mathbf{D}_1\mathbf{v}_1. \tag{13.2.14}$$

Similarly just below the interface in medium 2,

$$\mathbf{b}(z_I+) = \begin{pmatrix} \boldsymbol{w}(z_I+) \\ \boldsymbol{t}(z_I+) \end{pmatrix} = \begin{pmatrix} \boldsymbol{m}_{U2} & \boldsymbol{m}_{D2} \\ \boldsymbol{n}_{U2} & \boldsymbol{n}_{D2} \end{pmatrix} \begin{pmatrix} \boldsymbol{v}_{U2}(z_I+) \\ \boldsymbol{v}_{D2}(z_I+) \end{pmatrix} = \mathbf{D}_2\mathbf{v}_2. \tag{13.2.15}$$

The relation between the wave components above and below the interface can be obtained by equating the expressions (13.2.14) and (13.2.15) to force continuity of displacement and traction at z_I. Thus

$$\mathbf{v}_1 = \mathbf{D}_1^{-1}\mathbf{D}_2\mathbf{v}_2 = \mathbf{Q}\mathbf{v}_2. \tag{13.2.16}$$

We can split the wavevectors $\mathbf{v}_1$, $\mathbf{v}_2$ into their up and downgoing parts and partition the coupling matrix $\mathbf{Q}$ into 2×2 submatrices $\boldsymbol{Q}_{ij}$

$$\begin{pmatrix} \boldsymbol{v}_{U1} \\ \boldsymbol{v}_{D1} \end{pmatrix} = \begin{pmatrix} \boldsymbol{Q}_{UU} & \boldsymbol{Q}_{UD} \\ \boldsymbol{Q}_{DU} & \boldsymbol{Q}_{DD} \end{pmatrix} \begin{pmatrix} \boldsymbol{v}_{U2} \\ \boldsymbol{v}_{D2} \end{pmatrix},$$
$$= \mathrm{i} \begin{pmatrix} -\boldsymbol{n}_{D1}^T & \boldsymbol{m}_{D1}^T \\ \boldsymbol{n}_{U1}^T & -\boldsymbol{m}_{U1}^T \end{pmatrix} \begin{pmatrix} \boldsymbol{m}_{U2} & \boldsymbol{m}_{D2} \\ \boldsymbol{n}_{U2} & \boldsymbol{n}_{D2} \end{pmatrix} \begin{pmatrix} \boldsymbol{v}_{U2} \\ \boldsymbol{v}_{D2} \end{pmatrix}, \tag{13.2.17}$$

where we have employed (12.1.7) to represent the inverse $\mathbf{D}_1^{-1}$. The representation (13.2.17) shows a strong resemblance to the structure of the SH wave case. The partitions of the coupling matrix Q are given by

$$\begin{pmatrix} \boldsymbol{Q}_{UU} & \boldsymbol{Q}_{UD} \\ \boldsymbol{Q}_{DU} & \boldsymbol{Q}_{DD} \end{pmatrix} = \mathrm{i} \begin{pmatrix} -\boldsymbol{n}_{D1}^T\boldsymbol{m}_{U2} + \boldsymbol{m}_{D1}^T\boldsymbol{n}_{U2} & -\boldsymbol{n}_{D1}^T\boldsymbol{m}_{D2} + \boldsymbol{m}_{D1}^T\boldsymbol{n}_{D2} \\ \boldsymbol{n}_{U1}^T\boldsymbol{m}_{U2} - \boldsymbol{m}_{U1}^T\boldsymbol{n}_{U2} & \boldsymbol{n}_{U1}^T\boldsymbol{m}_{D2} - \boldsymbol{m}_{U1}^T\boldsymbol{n}_{D2} \end{pmatrix}. \tag{13.2.18}$$

Consider a downgoing wave system, comprising both P and SV waves, incident on the interface from medium 1. When this interacts with the interface we get reflected P and SV waves in medium 1 and transmitted downgoing waves in medium 2. No upward travelling waves will be generated in medium 2 and so $\boldsymbol{v}_{U2} = 0$. We can relate the upgoing and downgoing wave components in medium 1 by introducing a reflection matrix $\mathbf{R}_D^I$ for downward incidence, whose entries are reflection coefficients,

$$\boldsymbol{v}_{U1} = \mathbf{R}_D^I \boldsymbol{v}_{D1}, \quad \text{i.e.,} \quad \begin{pmatrix} P_{U1} \\ S_{U1} \end{pmatrix} = \begin{pmatrix} \mathrm{R}_D^{PP} & \mathrm{R}_D^{PS} \\ \mathrm{R}_D^{SP} & \mathrm{R}_D^{SS} \end{pmatrix} \begin{pmatrix} P_{D1} \\ S_{D1} \end{pmatrix}. \tag{13.2.19}$$

Similarly we can connect the downgoing wave components on the two sides of the interface by introducing a transmission matrix $\mathbf{T}_D^I$

$$\boldsymbol{v}_{D2} = \mathbf{T}_D^I \boldsymbol{v}_{D1}, \quad \text{i.e.,} \quad \begin{pmatrix} P_{D2} \\ S_{D2} \end{pmatrix} = \begin{pmatrix} \mathrm{T}_D^{PP} & \mathrm{T}_D^{PS} \\ \mathrm{T}_D^{SP} & \mathrm{T}_D^{SS} \end{pmatrix} \begin{pmatrix} P_{D1} \\ S_{D1} \end{pmatrix}. \tag{13.2.20}$$

Following Kennett (1983) we have chosen the convention for the conversion coefficients R_D^{PS} etc. so that the indexing of the reflection and transmission matrices follows the standard matrix pattern. This indexing simplifies manipulation of the results and checking of computer code but differs from the convention adopted by many authors. Aki & Richards (1980) use a different style of notation in which accents indicate the direction of propagation, thus (13.2.19) would be written as

$$\grave{\boldsymbol{v}}_1 = \check{\mathbf{R}}^I \acute{\boldsymbol{v}}_1. \tag{13.2.21}$$

The mnemonic convenience of this form is not as readily retained when we move to more complex situation or wish to consider the individual coefficients.

For an incident downgoing wave on the interface the set of wave elements in the upper and lower media are related by

$$\begin{pmatrix} \boldsymbol{v}_{U1} \\ \boldsymbol{v}_{D1} \end{pmatrix} = \begin{pmatrix} \boldsymbol{Q}_{UU} & \boldsymbol{Q}_{UD} \\ \boldsymbol{Q}_{DU} & \boldsymbol{Q}_{DD} \end{pmatrix} \begin{pmatrix} \mathbf{0} \\ \boldsymbol{v}_{D2} \end{pmatrix}, \tag{13.2.22}$$

and the reflection and transmission matrices can be found in terms of the partitions of $\mathbf{Q}$ as

$$\mathbf{T}_D^I = (\boldsymbol{Q}_{DD})^{-1}, \quad \mathbf{R}_D^I = \boldsymbol{Q}_{UD}(\boldsymbol{Q}_{DD})^{-1}, \tag{13.2.23}$$

cf (13.2.9) for *SH* waves.

An incident upgoing wave system from medium 2 will give reflected waves in medium 2 and transmitted waves in medium 1. No downgoing waves in medium 1 will be generated, so that $\boldsymbol{v}_{D1} = 0$, and now we have

$$\begin{pmatrix} \boldsymbol{v}_{U1} \\ \mathbf{0} \end{pmatrix} = \begin{pmatrix} \boldsymbol{Q}_{UU} & \boldsymbol{Q}_{UD} \\ \boldsymbol{Q}_{DU} & \boldsymbol{Q}_{DD} \end{pmatrix} \begin{pmatrix} \boldsymbol{v}_{U2} \\ \boldsymbol{v}_{D2} \end{pmatrix}. \tag{13.2.24}$$

We define the reflection and transmission matrices $\mathbf{R}_U^I$, $\mathbf{T}_U^I$ for these upward incident waves through the relations

$$\boldsymbol{v}_{D2} = \mathbf{R}_U^I \boldsymbol{v}_{U2}, \quad \boldsymbol{v}_{U1} = \mathbf{T}_U^I \boldsymbol{v}_{U2}. \tag{13.2.25}$$

Then from (13.2.24) we may construct $\mathbf{R}_U^I$, $\mathbf{T}_U^I$ from the partitions of $\mathbf{Q}$ as

$$\begin{aligned} \mathbf{R}_U^I &= -(\boldsymbol{Q}_{DD})^{-1}\boldsymbol{Q}_{DU}, \\ \mathbf{T}_U^I &= \boldsymbol{Q}_{UU} - \boldsymbol{Q}_{UD}(\boldsymbol{Q}_{DD})^{-1}\boldsymbol{Q}_{DU}. \end{aligned} \tag{13.2.26}$$

For this single interface we would, of course, obtain the same results by interchanging the suffices 1 and 2 in the expressions for $\mathbf{R}_D^I$, $\mathbf{T}_D^I$; but as we shall see the present method may be easily extended to more complex cases.

From the expressions for the reflection and transmission matrices in terms of the partitions of $\mathbf{Q}$ (13.2.23), (13.2.26) we can reconstruct the interface matrix itself as

$$\mathbf{Q} = \mathbf{D}_1^{-1}\mathbf{D}_2 = \begin{pmatrix} \mathbf{T}_U^I - \mathbf{R}_D^I(\mathbf{T}_D^I)^{-1}\mathbf{R}_U^I & \mathbf{R}_D^I(\mathbf{T}_D^I)^{-1} \\ -(\mathbf{T}_D^I)^{-1}\mathbf{R}_U^I & (\mathbf{T}_D^I)^{-1} \end{pmatrix}. \tag{13.2.27}$$

The matrices $\mathbf{D}_1$, $\mathbf{D}_2$ depend only on the slowness p and so $\mathbf{Q}$ is frequency independent, as are all the interface coefficients R_D^{PS}, T_U^{SP} etc.

The reflection and transmission matrices for the *P-SV* wave case can now be identified by using the expressions (13.2.18) for the partitions of the coupling matrix $\mathbf{Q}$ in terms of the elements of the displacement and stress transformation matrices $\boldsymbol{m}_{U1}$, $\boldsymbol{n}_{U2}$ etc.

All the interface coefficient matrices depend on $\boldsymbol{Q}_{DD}^{-1}$ and so the factor $\det \boldsymbol{Q}_{DD}$ will appear in the denominator of every reflection and transmission coefficient. The transmission coefficients are individual elements of $\mathbf{Q}$ divided by this determinant, but the reflection coefficients take the form of ratios of second order minors of $\mathbf{Q}$. The formal structure of the reflection and transmission results is the same for isotropic and transversely isotropic media; the differences come in the forms employed for the eigenvector matrices $\mathbf{D}$.

For isotropic media, the denominator in the expressions for the coefficients is

$$\begin{aligned} \Delta_{\text{st}} &= \det \boldsymbol{Q}_{DD} = \epsilon_{\alpha1}\epsilon_{\alpha2}\epsilon_{\beta1}\epsilon_{\beta2} \\ &\times \big\{[2p^2\Delta\mu(q_{\alpha1} - q_{\alpha2}) + (\rho_1 q_{\alpha2} + \rho_2 q_{\alpha1})][2p^2\Delta\mu(q_{\beta1} - q_{\beta2}) + (\rho_1 q_{\beta2} + \rho_2 q_{\beta1})] \\ &\quad + p^2\,[2\Delta\mu(q_{\alpha1}q_{\beta2} + p^2) - \Delta\rho][2\Delta\mu(q_{\beta1}q_{\alpha2} + p^2) - \Delta\rho]\big\}, \end{aligned} \tag{13.2.28}$$

where we have introduced the contrasts in shear modulus and density across the interface, $\Delta\mu = \mu_1 - \mu_2$, $\Delta\rho = \rho_1 - \rho_2$. It was pointed out by Stoneley (1924) that if the determinant (13.2.28) vanishes we have the possibility of free interface waves with evanescent decay away from the interface into the media on either side. These Stoneley waves have a rather restricted range of existence, for most reasonable density contrasts the shear velocities β_1 and β_2 must be nearly equal for (13.2.28) to be zero. The slowness of the Stoneley wave is always greater than $[\min(\beta_1, \beta_2)]^{-1}$.

For an isotropic medium the explicit form of the transmission and reflection coefficients for downward incidence are given by

$$\begin{pmatrix} \mathrm{T}_D^{PP} & \mathrm{T}_D^{PS} \\ \mathrm{T}_D^{SP} & \mathrm{T}_D^{SS} \end{pmatrix} = \frac{1}{\Delta_{\mathrm{st}}} \begin{pmatrix} q_{44} & -q_{34} \\ -q_{43} & q_{33} \end{pmatrix}, \tag{13.2.29}$$

$$\begin{pmatrix} \mathrm{R}_D^{PP} & \mathrm{R}_D^{PS} \\ \mathrm{R}_D^{SP} & \mathrm{R}_D^{SS} \end{pmatrix} = \frac{1}{\Delta_{\mathrm{st}}} \begin{pmatrix} q_{13}q_{44} - q_{14}q_{34} & q_{14}q_{33} - q_{13}q_{34} \\ q_{23}q_{44} - q_{24}q_{43} & q_{24}q_{33} - q_{23}q_{34} \end{pmatrix}, \tag{13.2.30}$$

where

$$\begin{aligned}
q_{13} &= \epsilon_{\alpha1}\epsilon_{\alpha2}[2p^2\Delta\mu(q_{\alpha1} + q_{\alpha2}) - (\rho_1 q_{\alpha2} - \rho_2 q_{\alpha1})], \\
q_{14} &= \mathrm{i}\epsilon_{\alpha1}\epsilon_{\beta2}p[2\Delta\mu(q_{\alpha1}q_{\beta2} - p^2) + \Delta\rho], \\
q_{23} &= \mathrm{i}\epsilon_{\beta1}\epsilon_{\alpha2}p[2\Delta\mu(q_{\beta1}q_{\alpha2} - p^2) + \Delta\rho], \\
q_{24} &= \epsilon_{\beta1}\epsilon_{\beta2}[2p^2\Delta\mu(q_{\beta1} + q_{\beta2}) - (\rho_1 q_{\beta2} - \rho_2 q_{\beta1})], \\
q_{33} &= \epsilon_{\alpha1}\epsilon_{\alpha2}[2p^2\Delta\mu(q_{\alpha1} - q_{\alpha2}) + (\rho_1 q_{\alpha2} + \rho_2 q_{\alpha1})], \\
q_{34} &= \mathrm{i}\epsilon_{\alpha1}\epsilon_{\beta2}p[2\Delta\mu(q_{\alpha1}q_{\beta2} + p^2) - \Delta\rho], \\
q_{43} &= \mathrm{i}\epsilon_{\beta1}\epsilon_{\alpha2}p[2\Delta\mu(q_{\beta1}q_{\alpha2} + p^2) - \Delta\rho], \\
q_{44} &= \epsilon_{\beta1}\epsilon_{\beta2}[2p^2\Delta\mu(q_{\beta1} - q_{\beta2}) + (\rho_1 q_{\beta2} + \rho_2 q_{\beta1})],
\end{aligned} \tag{13.2.31}$$

These expressions are rather complex and it is only by computing the dependence of the reflection and transmission coefficients on slowness that we are able to appreciate the full range of behaviour.

13.2.3 The variation of the interface coefficients with slowness

We have illustrated the amplitude and phase behaviour of the interface coefficients for a solid-solid boundary in figures 3.9, 3.10 as a function of slowness p. The use of slowness means that we can employ a common reference for P and S waves, which is not possible in terms of incidence angles. The details of the behaviour of the reflection coefficients are quite sensitive to the properties on the two sides of the interface and this forms the basis of recent interest in amplitude/slowness relations in reflection seismology.

The phase behaviour of the interface coefficients depends on the value of the slowness p relative to the critical slownesses for the different wave types. When $p < \alpha_2^{-1}$ the reflection and transmission coefficients for unconverted waves are real and for unconverted waves are pure imaginary. Beyond this critical slowness all the coefficients in the P-SV system are complex and so phase shifts are introduced in both

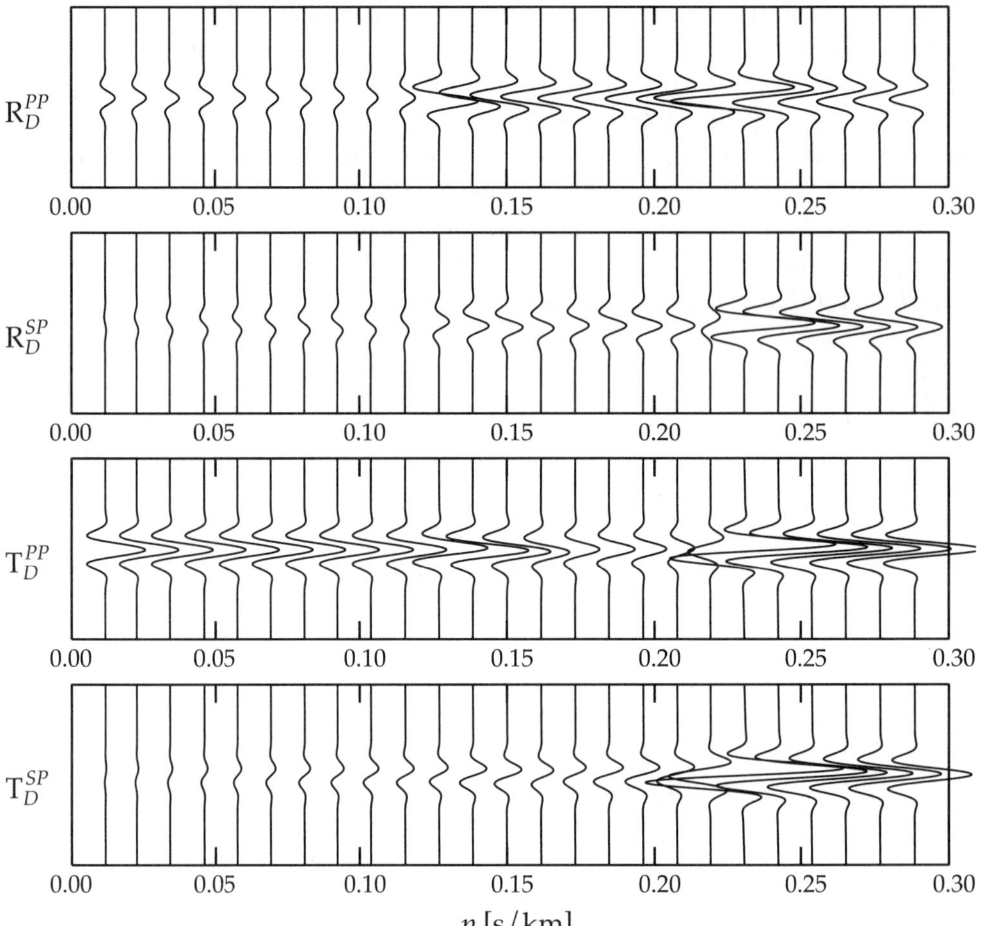

Figure 13.6. The behaviour of reflected and transmitted pulses for downward incidence of a P wave on an interface with $\alpha_1 = 6.0$ km/s, $\beta_1 = 3.4$ km/s, $\rho_1 = 2.7$ Mg/m^3, $\alpha_2 = 8.0$ km/s, $\beta_2 = 4.6$ km/s, $\rho_2 = 3.3$ Mg/m^3.

reflection and transmission. The critical slowness for S waves is β_2^{-1} and for p greater than this value both SV and SH waves are totally reflected. SH coefficients are real up to this critical slowness but the SV coefficients are complex for $p > \alpha_2^{-1}$.

When the interface coefficients are real, an incident plane wave pulse with slowness p is merely scaled in amplitude on reflection or transmission. For purely imaginary coefficients we get a scaling of the Hilbert transform of the incident pulse. Once the coefficients become complex, the shape of the reflected and transmitted pulses is modified with a phase shift, and as we have seen in (13.1.16) above, comprises a linear combination of the original pulse and its Hilbert transform.

In figures 13.6 and 13.7 we present the behaviour of the reflection and transmission

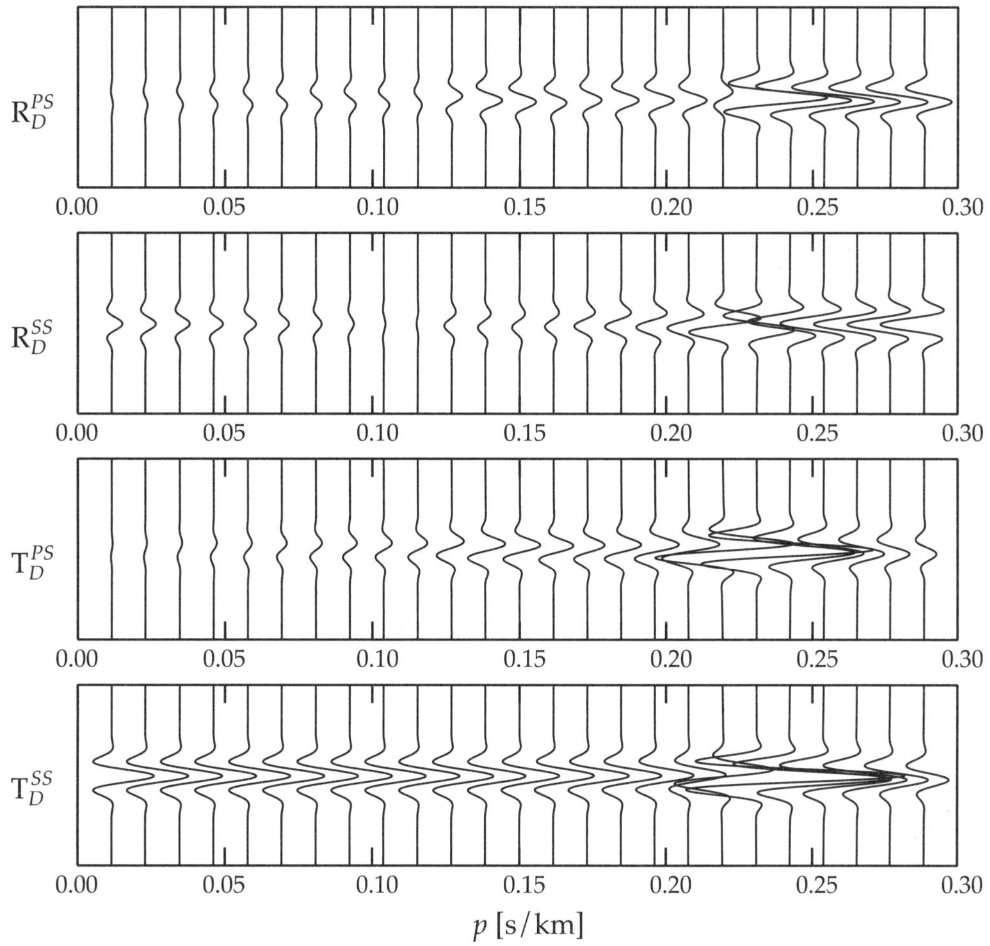

Figure 13.7. The behaviour of reflected and transmitted pulses for downward incidence of a S wave on an interface with $\alpha_1 = 6.0$ km/s, $\beta_1 = 3.4$ km/s, $\rho_1 = 2.7$ Mg/m^3, $\alpha_2 = 8.0$ km/s, $\beta_2 = 4.6$ km/s, $\rho_2 = 3.3$ Mg/m^3.

coefficients for downward incident P and S waves through their effect on the simple pulse as employed in figure 13.4. In this way we are able to present both the amplitude behaviour and the phase shifts associated with the complex coefficients for $p > \alpha_2^{-1}$. We use the same material properties as employed for the amplitude and phase diagrams in figures 3.9 and 3.10, so that the critical slowness for P waves in the lower medium is 0.125 km/s, and for S is 0.217 s/km. We present the pulses for the full range of slownesses even though for the larger values the propagation away from the interface would be in the form of evanescent P and SV waves.

Our choice of normalisation with respect to energy transport in the x_3 direction means that the converted reflection coefficients are symmetric,

$$\mathrm{R}_D^{PS}(p) = \mathrm{R}_D^{SP}(p), \tag{13.2.32}$$

and this is clearly reflected in figures 13.6 and 13.7. The process of conversion in both reflection and transmission is more efficient as the angles of incidence and slowness increase and can carry a significant fraction of the energy away from the interface. The amplitudes of the unconverted waves are enhanced in reflection near the critical points and at the onset of evanescence for the particular wavetype.

13.2.4 Small contrasts at an interface

When the contrast in elastic properties at an interface is small, there are useful approximations for the reflection and transmission coefficients. In terms of the contrasts in the seismic properties, $\Delta\rho = \rho_2 - \rho_1$, $\Delta\beta = \beta_2 - \beta_1$, $\Delta\alpha = \alpha_2 - \alpha_1$ the approximate reflection coefficients for downgoing waves take the form

$$\begin{aligned}
\mathrm{R}_D^{PP} &= \tfrac{1}{2}(1 - 4\beta^2 p^2)\frac{\Delta\rho}{\rho} + \tfrac{1}{2}\frac{1}{\alpha^2 q_\alpha^2}\frac{\Delta\alpha}{\alpha} - 4\beta^2 p^2 \frac{\Delta\beta}{\beta}, \\
\mathrm{R}_D^{PS} = \mathrm{R}_D^{SP} &= \frac{p}{(q_\alpha q_\beta)^{1/2}}\left[\tfrac{1}{2}\frac{\Delta\rho}{\rho} + \beta^2(q_\alpha q_\beta - p^2)\left(\frac{\Delta\rho}{\rho} + 2\frac{\Delta\beta}{\beta}\right)\right], \\
\mathrm{R}_D^{SS} &= \tfrac{1}{2}(1 - 4\beta^2 p^2)\frac{\Delta\rho}{\rho} + \left(\tfrac{1}{2}\frac{1}{\beta^2 q_\beta^2} - 4\beta^2 p^2\right)\frac{\Delta\beta}{\beta}.
\end{aligned} \tag{13.2.33}$$

These approximations are valid if the angles of incidence and reflection i, j are not close to 90°, and have often been used in studies of amplitude variation with angle of incidence.

The equivalent transmission coefficients for unconverted waves are close to unity and depend linearly on slowness p for conversion,

$$\begin{aligned}
\mathrm{T}_D^{PP} &= 1 - \tfrac{1}{2}\frac{\Delta\rho}{\rho} + \left(\tfrac{1}{2}\frac{1}{\alpha^2 q_\alpha^2} - 1)\right)\frac{\Delta\alpha}{\alpha}, \\
\mathrm{T}_D^{PS} = \mathrm{T}_D^{SP} &= \frac{p}{(q_\alpha q_\beta)^{1/2}}\left[\tfrac{1}{2}\frac{\Delta\rho}{\rho} + \beta^2(q_\alpha q_\beta + p^2)\left(\frac{\Delta\rho}{\rho} + 2\frac{\Delta\beta}{\beta}\right)\right], \\
\mathrm{T}_D^{SS} &= 1 - \tfrac{1}{2}\frac{\Delta\rho}{\rho} + \left(\tfrac{1}{2}\frac{1}{\beta^2 q_\beta^2} - 1\right)\frac{\Delta\beta}{\beta}.
\end{aligned} \tag{13.2.34}$$

We note that both in reflection and transmission the conversion coefficients depend on the contrast in ρ and shear modulus μ ($\rho\beta^2$), and the only dependence on P wave velocity enters through the vertical slowness q_α.

13.3 A fluid-solid interface

The reflection and transmission coefficients at a horizontal fluid-solid boundary can be obtained by a careful limiting process in which the shear wavespeed in medium 1 is forced to zero. The equivalent boundary conditions are the continuity of vertical

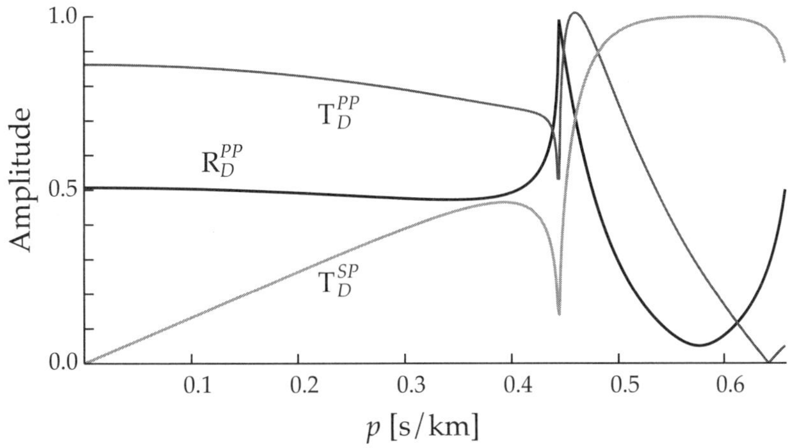

Figure 13.8. Reflection and transmission coefficients for a fluid-solid boundary with the S wavespeed in the solid less than the P wavespeed in the liquid. [α_1 = 1.50 km/s, ρ_1 = 1.03 Mg/m^3, α_2 = 2.25 km/s, β_2 = 1.10 km/s, ρ_2 = 2.10 Mg/m^3].

displacement and the τ_{33} component of traction. The τ_{13} component of traction must vanish at the boundary, since no shear traction can be sustained in the fluid. The horizontal component of displacement is discontinuous at the boundary.

For waves incident from the fluid side, wave conversion can only occur in transmission, but for P waves incident from the solid side conversion can occur in reflection. S waves within the solid can only generate a disturbance in the fluid by conversion in transmission. For small slownesses (near vertical wave propagation) the process of conversion is relatively inefficient but the amplitude of converted waves grows steadily with increasing slowness (figures 13.8, 13.9) and even for slownesses less than the critical point for P waves $p < \alpha_2^{-1}$ conversion can reach 20 per cent.

The reflection coefficient for P waves does not vary much with slowness until the neighbourhood of the critical slowness. Total reflection occurs just at the critical slowness (α_2^{-1}) but for larger slowness the reflection coefficient drops. This behaviour is in sharp contrast to the predictions for acoustic waves between two fluids for which total reflection occurs for all slowness greater than critical. The difference arises because in the elastic case energy can still be transmitted through the fluid-solid boundary to be converted into S waves within the solid. Such transmission in conversion can be very efficient and 90 percent or more of the energy in the slowness range ($p > \alpha_2^{-1}$) can be transmitted.

The dominant fluid-solid boundary of significance in seismology is the contact between seawater and the sea floor. In areas with soft sediments, the P wavespeeds at the sea floor are usually higher than the P velocity in the sea water (around 1500 m/s) but the shear wavespeeds are relatively low. However when the sea floor is composed of more consolidated material, the S wavespeed may be of the order of 1000

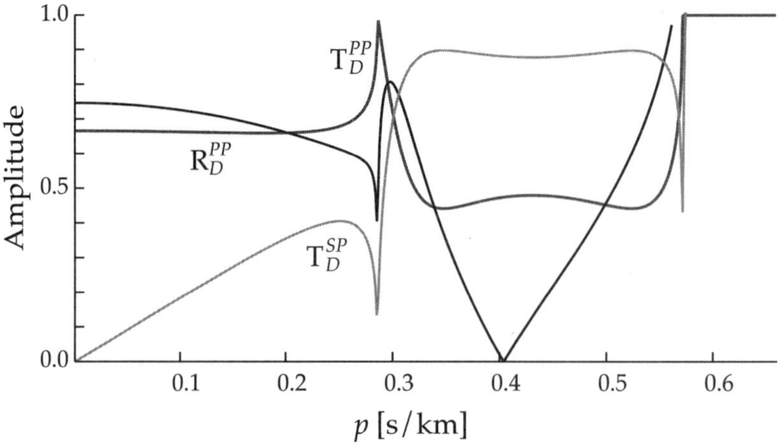

Figure 13.9. Reflection and transmission coefficients for a fluid-solid boundary with the S wavespeed in the solid greater than the P wavespeed in the liquid. [α_1 = 1.50 km/s, ρ_1 = 1.03 Mg/m^3, α_2 = 3.50 km/s, β_2 = 1.75 km/s, ρ_2 = 2.20 Mg/m^3].

m/s and in this case conversion in transmission can be particularly efficient (figure 13.8). In certain areas where the seafloor is composed of carbonates e.g. limestone platforms or cemented reef material, or where basaltic outcrop occurs at the surface the S wavespeed just below the sea bed may rise above 1500 m/s and then there are significant changes to the patterns of reflection coefficients (figure 13.9). With a higher shear wave velocity the postcritical reflection of P waves is enhanced at the expense of transmission into S, and at the critical slowness for S (β_2^{-1}) total reflection of P waves occurs since there is then no mode of propagation available within the seafloor. With such hard sea floors there can be high amplitude guided waves in shallow water built up from P waves which are successively totally reflected at the seafloor and at the sea surface.

13.4 Reflection and transmission at a spherical surface

In section 12.2 we have introduced representations of the seismic wavefield for uniform regions in spherical stratification in terms of components which depend on spherical Bessel functions. These components asymptotically have the character of upgoing and downgoing waves and can be used to set up a set of reflection and transmission coefficients at spherical surfaces.

Compared with the horizontally stratified case we have the added complication that the character of the solution for each wave type switches from oscillatory to exponential across a 'turning radius' which for P waves is

$$r_\alpha = \wp\alpha. \tag{13.4.1}$$

Above this level r_α we take solutions depending on $h_l^{(1)}(\omega r/\alpha)$ for upgoing P waves and $h_l^{(2)}(\omega r/\alpha)$ for downgoing P waves. Below r_α we switch to solutions which give a better representation of the evanescent character: $j_l(\omega r/\alpha)$ which decays away from r_α and $y_l(\omega r/\alpha)$ which grows exponentially. When we work with a span of slownesses at a fixed surface we will find that the location of the turning level varies with slowness. It will lie below the surface for propagating waves, touch the surface at the onset of evanescence, and lie above the surface for the evanescent field.

For a particular wave type we will designate as 'downgoing', the actual downgoing waves above the turning level and the evanescently decaying solution below this level. Similarly, we will use 'upgoing' to mean upward travelling waves above the turning level and the exponentially growing solution below. In this way we achieve the same specification as was possible in horizontal stratification by our choice of physical Riemann sheet (12.1.24).

We may now set up a representation of the displacement and traction fields at a radius r in a uniform medium with a character determined by the relative location of r and the turning levels r_α, r_β. The reflection and transmission matrices for a spherical interface (or the free surface) may then be found by following the development of the previous section with appropriate substitutions for the forms of the vectors corresponding to particular wave components.

14

Building the Response of a Model

We have just discussed the way in which the reflection and transmission properties of a single interface can be described in terms of reflection and transmission matrices at fixed slowness p. We now combine these results with the representations for the wavefield from Chapter 12 to represent reflection and transmission from a layer or a zone of gradients. We then indicate how to build up the reflection and transmission properties of a composite region from the properties of its constituents, and so are able to describe the response of major regions in the Earth. With this information we are able to assemble the response of an Earth model in the frequency-slowness domain, using a representation of the seismic source in terms of radiated waves. The approach we develop can be used for both horizontal and spherical stratification.

14.1 Reflection and transmission for a region

The reflection and transmission coefficients for a single interface are frequency independent, but as soon as we introduce further structure the properties are frequency dependent. We can therefore characterise the behaviour by the frequency-slowness dependence of the reflection or transmission of plane waves.

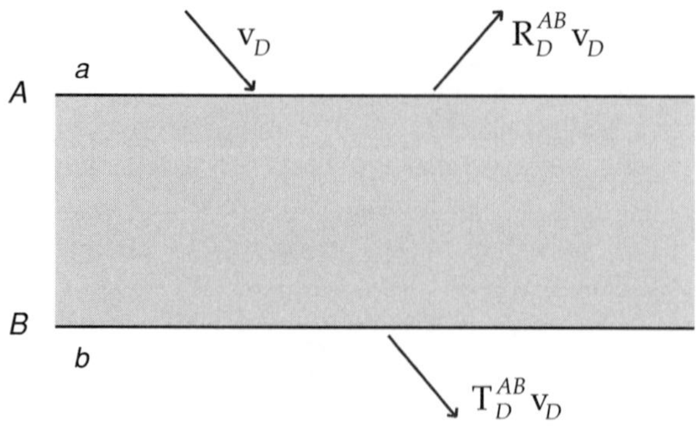

Figure 14.1. Reflection and transmission for the region (z_A, z_B).

We consider a portion of a stratified velocity model lying between depths z_A and z_B, and now abut uniform half spaces to achieve continuity of properties at z_A and z_B. We can then consider an incident downgoing wave from region a incident upon the zone AB at z_A. We define a reflection matrix $\mathbf{R}_D^{AB}$ for this region by relating the incident downgoing waves to the reflected upgoing waves passing into region a,

$$\boldsymbol{v}_U(z_A) = \mathbf{R}_D^{AB}\boldsymbol{v}_D(z_A), \tag{14.1.1}$$

and correspondingly a transmission matrix $\mathbf{T}_D^{AB}$ relating the downgoing waves passing into region b to the incident downgoing wave from region a,

$$\boldsymbol{v}_D(z_B) = \mathbf{T}_D^{AB}\boldsymbol{v}_D(z_A), \tag{14.1.2}$$

In a similar way we can introduce reflection and transmission matrices $\mathbf{R}_U^{AB}$, $\mathbf{T}_U^{AB}$ for upgoing waves incident from region b at z_B.

14.1.1 A uniform region

If the region between A and B is uniform there will be no reflection and so

$$\mathbf{R}_D^{AB} = \mathbf{R}_U^{AB} \equiv \mathbf{0}, \tag{14.1.3}$$

and the transmission terms reflect the change in phase introduced in the passage between z_A and z_B

$$\mathbf{T}_D^{AB} = \mathbf{E}_D^{AB}, \quad \mathbf{T}_U^{AB} = \mathbf{E}_U^{AB}, \tag{14.1.4}$$

where the phase terms $\mathbf{E}_D^{AB}$, $\mathbf{E}_U^{AB}$ represent 2×2 matrices for *P-SV* waves dependent on frequency ω and slowness p through the vertical slownesses q_α, q_β:

$$\mathbf{E}_D^{AB} = \mathbf{E}_U^{AB} = \begin{pmatrix} \exp[\mathrm{i}\omega q_\alpha(z_B - z_A)] & \mathbf{0} \\ \mathbf{0} & \exp[\mathrm{i}\omega q_\beta(z_B - z_A)] \end{pmatrix}, \tag{14.1.5}$$

For *SH* waves we have a simple frequency-dependent phase increment,

$$[\mathbf{E}_{DH}^{AB}]_H = [\mathbf{E}_U^{AB}]_H = \exp[\mathrm{i}\omega q_\beta(z_B - z_A)]. \tag{14.1.6}$$

Note that in the evanescent regime both the transmission terms $\mathbf{T}_D^{AB}$, $\mathbf{T}_U^{AB}$ are defined in terms of decaying exponentials, with the form $\exp[-\omega|q_\beta|(z_B - z_A)]$ for a perfectly elastic medium. The growing exponentials do not enter into this representation.

It is useful to make the distinction between upgoing and downgoing waves at this stage because although the phase income terms are equivalent in isotropic media in horizontal stratification, they are distinct for most anisotropic media and spherical stratification. We therefore employ forms which enable us to consider the spherical and anisotropic cases without further adaptation.

For models composed of uniform spherical shells it is worthwhile to follow Chapman & Phinney (1972) and extract the main spherical Bessel function term from the motion-stress vectors for the different wave components. For example, for propagating *SH* waves we would set

$$\mathfrak{b}_u^H = h_l^{(1)}(k_\beta r)\mathfrak{b}_u^{H'}, \quad \mathfrak{b}_d^H = h_l^{(2)}(k_\beta r)\mathfrak{b}_d^{H'}, \tag{14.1.7}$$

with a similar treatment for the P and SV wave components $\mathfrak{b}_u^P$, $\mathfrak{b}_d^P$, $\mathfrak{b}_d^S$, $\mathfrak{b}_d^S$.

Then across a uniform layer we can extract a diagonal term which represents the main dependence on r through a ratio of spherical Bessel functions evaluated at r_A and r_B. For propagating SH waves

$$[\mathbf{T}_U^{AB}]_H = \frac{h_l^{(1)}(k_\beta r_A)}{h_l^{(1)}(k_\beta r_B)} = [\mathbf{E}_{UH}^{AB}], \qquad [\mathbf{T}_D^{AB}]_H^{-1} = \frac{h_l^{(2)}(k_\beta r_A)}{h_l^{(2)}(k_\beta r_B)} = [\mathbf{E}_{DH}^{AB}]^{-1}, \tag{14.1.8}$$

asymptotically these ratios reduce to simple exponentials related as in (14.1.6). For the P-SV wave system we have 2×2 matrices of ratios of spherical Bessel functions:

$$\mathbf{T}_U^{AB} = \begin{pmatrix} \dfrac{h_l^{(1)}(k_\alpha r_A)}{h_l^{(1)}(k_\alpha r_B)} & 0 \\ 0 & \dfrac{h_l^{(1)}(k_\beta r_A)}{h_l^{(1)}(k_\beta r_B)} \end{pmatrix} = \mathbf{E}_U^{AB}, \tag{14.1.9}$$

$$[\mathbf{T}_D^{AB}]^{-1} = \begin{pmatrix} \dfrac{h_l^{(2)}(k_\alpha r_A)}{h_l^{(2)}(k_\alpha r_B)} & 0 \\ 0 & \dfrac{h_l^{(2)}(k_\beta r_A)}{h_l^{(2)}(k_\beta r_B)} \end{pmatrix} = [\mathbf{E}_D^{AB}]^{-1}. \tag{14.1.10}$$

Once again the high frequency asymptotes are diagonal matrices of exponentials related as in (14.1.5).

14.1.2 A uniform layer

As a simple example of reflection and transmission we consider a uniform layer bounded by interfaces at z_A and z_B with properties different from the surrounding half space a, b. Reflection can occur at the interfaces but within the layer there will only be transmission with a phase shift.

We illustrate the process for SH waves for which we have the simplicity of scalar coefficients. We designate the reflection and transmission coefficients at z_A with the superscript A and at z_B with the superscript B. The phase increments in transmission $E_U = E_D = \exp[i\omega q_\beta(z_B - z_A)]$. We will retain the distinction between E_U and E_D because it will make it easier to follow the sequence of interactions.

Consider then an incident downgoing wave at slowness p with unit amplitude impinging on the interface at z_A from the region a. There will be a reflection from the interface with amplitude R_D^A, accompanied by transmission into the uniform layer with amplitude T_D^A. The transmitted wave then reaches the interface B with a phase shift E_D and can be reflected back into the layer with amplitude $\mathrm{R}_D^B E_D \mathrm{T}_D^A$ or transmitted into region b with amplitude $\mathrm{T}_D^B E_D \mathrm{T}_D^A$ (see figure 14.2).

Following the internal reflection at B, we have a further phase increment E_U in the wave impinging on the upper boundary A. There is again partition with transmission through z_A into region a with amplitude $\mathrm{T}_U^A E_U \mathrm{R}_D^B E_D \mathrm{T}_D^A$ and internal reflection with

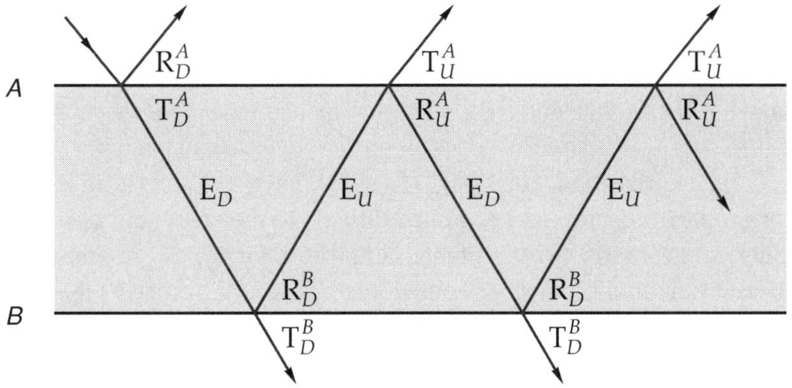

Figure 14.2. The sequence of reflection and transmission processes for a uniform layer bounded by interface at z_A and z_B.

amplitude $R_U^A E_U R_D^B E_D T_D^A$. The reflected wave continues to z_B where again both transmission into region b and internal reflection occur. The propagation process continues with a sequence of progressively higher numbers of internal reflections to produce a cumulative effect in reflection into region a of

$$\begin{aligned} R_D^{AB} &= R_D^A + T_U^A E_U R_D^B E_D T_D^A + T_U^A E_U R_D^B E_D R_U^A E_U R_D^B E_D T_D^A \\ &\quad + T_U^A E_U R_D^B E_D R_U^A E_U R_D^B E_D R_U^A E_U R_D^B E_D T_D^A + \ldots \\ &= R_D^A + T_U^A E_U R_D^B E_D \left[1 + R_U^A E_U R_D^B E_D + R_U^A E_U R_D^B E_D R_U^A E_U R_D^B E_D + \ldots\right] T_D^A, \end{aligned} \tag{14.1.11}$$

and transmission into region b of

$$\begin{aligned} T_D^{AB} &= T_D^A E_D T_D^A + T_D^A E_D R_U^A E_U R_D^B E_D T_D^A + T_D^A E_D R_U^A E_U R_D^B E_D R_U^A E_U R_D^B E_D T_D^A + \ldots \\ &= T_D^A E_D \left[1 + R_U^A E_U R_D^B E_D + R_U^A E_U R_D^B E_D R_U^A E_U R_D^B E_D + \ldots\right] T_D^A. \end{aligned} \tag{14.1.12}$$

The common sequence in the reflection and transmission terms arises from the sequence of internal reverberations with the layer and can be recognised as a geometric series

$$\left[1 + R_U^A E_U R_D^B E_D + R_U^A E_U R_D^B E_D R_U^A E_U R_D^B E_D + \ldots\right] = \left[1 - R_U^A E_U R_D^B E_D\right]^{-1}. \tag{14.1.13}$$

The inverse can thus be regarded as a reverberation operator reflecting the full set of internal reflections (figure 14.2).

With the aid of (14.1.13) we can recast the expression (14.1.11) for the reflection coefficient R_D^{AB} in the compact form

$$R_D^{AB} = R_D^A + T_U^A E_U R_D^B E_D \left[1 - R_U^A E_U R_D^B E_D\right] T_D^A; \tag{14.1.14}$$

with a comparable development from (14.1.12) for the transmission coefficient T_D^{AB}

$$T_D^{AB} = T_D^A E_D \left[1 - R_U^A E_U R_D^B E_D\right] T_D^A. \tag{14.1.15}$$

Since the individual coefficients are scalars for *SH* waves we can achieve some apparent simplification in (14.1.14), (14.1.15) by rearrangement of terms, e.g., combining the upward and downward phase shifts into a single term. Such forms of representation are common for the equivalent case in optics or acoustics (see, e.g., Brekhovskikh, 1960).

However, the expressions (14.1.14), (14.1.15) have the merit of following the sequence of propagation and can be applied directly to the analogous case of a uniform layer in spherical geometry (with suitable definitions for E_D, E_U in terms of ratios of spherical Bessel functions). When we consider coupled waves, such as the *P-SV* system, we will find that the order in which operations are applied are important because they are represented by matrices for which multiplication is not commutative. In general for two matrices **A**, **B** the product $\mathbf{AB} \neq \mathbf{AB}$.

Since we have carefully followed the sequence of physical processes, the forms (14.1.14), (14.1.15) we have just produced for *SH* waves can be immediately generalised to *P-SV* waves and general anisotropy.

14.2 Reflection and transmission for a composite region

We can build up the reflection and transmission response for a composite region when we know the reflection and transmission operators for its constituent parts by following the cumulative propagation processes. We consider now a region (z_A, z_C) which is split into two parts by the introduction of a level z_B. We can define reflection and transmission matrices for the two sub-regions *AB* and *BC* by introducing a notional homogeneous region at *B* with continuity of seismic properties at *B*.

We can then build up the reflection and transmission matrices at fixed frequency and slowness for the region *AC*, such as $\mathbf{R}_D^{AC}$, $\mathbf{T}_D^{AC}$ by looking at the sequence of propagation processes which follow the arrival of a downgoing wave $[\boldsymbol{D}^a]$ at z_A which we can illustrate by reference to figure 14.3.

The reflected wavefield returned from *AB* considered in isolation is given by

$$\mathbf{R}_D^{AB}[\boldsymbol{D}^a], \tag{14.2.1}$$

but this is only part of the total reflected field returned from *AC*, since we have the possibility of transmission through the region *AB* with subsequent reflection from the zone *BC*. The direct transmission through *AB*

$$\mathbf{T}_D^{AB}[\boldsymbol{D}^a], \tag{14.2.2}$$

will constitute the incident field for the region *BC*. The field (14.2.2) represents the wavefield transmitted into a notional uniform medium *b* below *B*, and will allow for all propagation paths within *AB* leading to energy ultimately leaving *AB* through the interface *B*. Part of this transmitted field will be reflected back from *BC*, so that *AB* will be 'insonified' from below by an upgoing field

$$\mathbf{R}_D^{BC}[\mathbf{T}_D^{AB}[\boldsymbol{D}^a]], \tag{14.2.3}$$

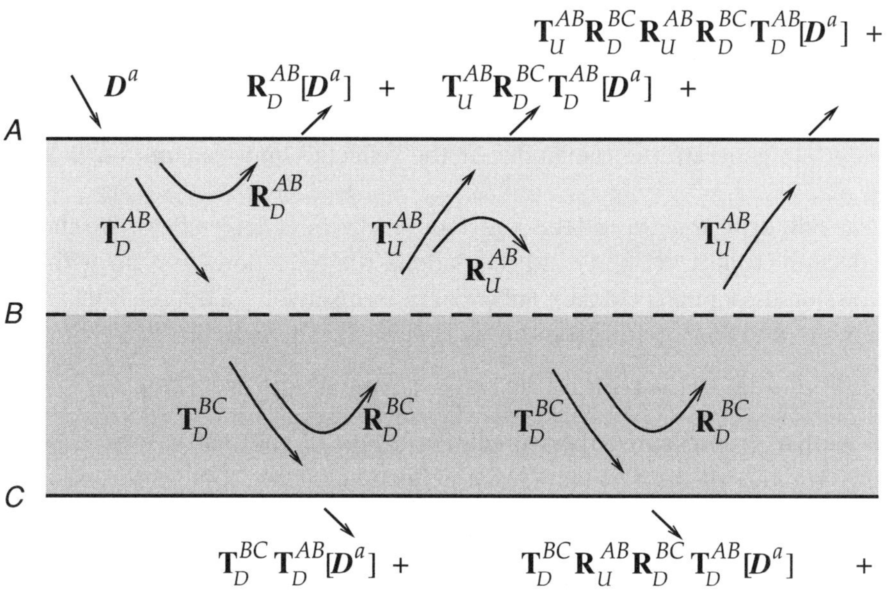

Figure 14.3. Configuration of a region *AC* divided into two parts *AB* and *BC* with a schematic representation of the construction of the reflection and transmission matrices for the region *AC* from the matrices for *AB* and *BC*.

and part will be transmitted through *BC* to give the contribution

$$\mathbf{T}_D^{BC}[\mathbf{T}_D^{AB}[\boldsymbol{D}^a]], \tag{14.2.4}$$

to the transmitted field in the medium *c*.

The effect of the field (14.2.3) incident from below on *B* is twofold: firstly there will be transmission through *AB* to give an additional contribution to the reflected field in *a* of

$$\mathbf{T}_U^{AB}[\mathbf{R}_D^{BC}[\mathbf{T}_D^{AB}[\boldsymbol{D}^a]]], \tag{14.2.5}$$

and, secondly some energy may be reflected back from *AB* to add the extra term

$$\mathbf{R}_U^{AB}[\mathbf{R}_D^{BC}[\mathbf{T}_D^{AB}[\boldsymbol{D}^a]]], \tag{14.2.6}$$

to the downgoing field emerging from the region *AB*.

Since both directions of wave propagation are now involved, we have to make a distinction between the reflection and transmission matrices for waves incident at *A* and *B*, e.g., $\mathbf{R}_D^{AB}$, $\mathbf{R}_U^{AB}$. Although a positional notation that distinguishes between $\mathbf{R}^{AB}$ and $\mathbf{R}^{BA}$ may be more elegant, the explicit use of suffices U, D indicating the sense of the incident wave minimises confusion.

The field (14.2.6) at the surface *B* will enhance the transmitted field at *C* by

$$\mathbf{T}_D^{BC}[\mathbf{R}_U^{AB}[\mathbf{R}_D^{BC}[\mathbf{T}_D^{AB}[\boldsymbol{D}^a]]]], \tag{14.2.7}$$

and return to B the contribution

$$\mathbf{R}_D^{BC}[\mathbf{R}_U^{AB}[\mathbf{R}_D^{BC}[\mathbf{T}_D^{AB}[\boldsymbol{D}^a]]]]. \tag{14.2.8}$$

The process of reflection and transmission from the regions AB and BC will continue alternately to generate the remainder of the reflected and transmitted field for the region AC.

We can now bring together all the contributions to the reflected field. We started with the reflection from AB (14.2.1) and then added the reflection from BC with two-way transmission through BC (14.2.6). Subsequent terms involve a double reflection: from above by AB and from below from BC as in (14.2.8). The total effect is

$$\mathbf{R}_D^{AC}[\boldsymbol{D}^a] = \mathbf{R}_D^{AB}[\boldsymbol{D}^a] + \mathbf{T}_U^{AB}\mathbf{R}_D^{BC}\mathbf{T}_D^{AB}[\boldsymbol{D}^a] + \mathbf{T}_U^{AB}\mathbf{R}_D^{BC}\mathbf{R}_U^{AB}\mathbf{R}_D^{BC}\mathbf{T}_D^{AB}[\boldsymbol{D}^a] + ..., \tag{14.2.9}$$

where we have represented the cumulative action of the matrices by a composite matrix. We can therefore identify the reflection matrix $\mathbf{R}_D^{AC}$ with a sequence of reflection terms

$$\mathbf{R}_D^{AC} = \mathbf{R}_D^{AB} + \mathbf{T}_U^{AB}\left(\mathbf{R}_D^{BC} + \mathbf{R}_D^{BC}\mathbf{R}_U^{AB}\mathbf{R}_D^{BC} + \mathbf{R}_D^{BC}\mathbf{R}_U^{AB}\mathbf{R}_D^{BC}\mathbf{R}_U^{AB}\mathbf{R}_D^{BC} + ...\right)\mathbf{T}_D^{AB}. \tag{14.2.10}$$

The entire set of possible propagation effects can be represented via a single formal expression:

$$\mathbf{R}_D^{AC} = \mathbf{R}_D^{AB} + \mathbf{T}_U^{AB}\mathbf{R}_D^{BC}\left(\mathbf{I} - \mathbf{R}_U^{AB}\mathbf{R}_D^{BC}\right)^{-1}\mathbf{T}_D^{AB}, \tag{14.2.11}$$

where $\mathbf{I}$ is the identity matrix. The inverse operator $(\mathbf{I} - \mathbf{R}_U^{AB}\mathbf{R}_D^{BC})^{-1}$ represents the cumulative effect of the entire sequence of internal reverberations in AC, since we have the identity

$$\left(\mathbf{I} - \mathbf{R}_U^{AB}\mathbf{R}_D^{BC}\right)^{-1} = \mathbf{I} + \mathbf{R}_U^{AB}\mathbf{R}_D^{BC} + \mathbf{R}_U^{AB}\mathbf{R}_D^{BC}\mathbf{R}_U^{AB}\mathbf{R}_D^{BC}\left(\mathbf{I} - \mathbf{R}_U^{AB}\mathbf{R}_D^{BC}\right)^{-1}, \tag{14.2.12}$$

in which we can recognize those parts of the field we have already encountered.

In transmission we obtain a similar result

$$\mathbf{T}_D^{AC} = \mathbf{T}_D^{BC}\left(\mathbf{I} + \mathbf{R}_U^{AB}\mathbf{R}_D^{BC} + \mathbf{R}_U^{AB}\mathbf{R}_D^{BC}\mathbf{R}_U^{AB}\mathbf{R}_D^{BC} + ...\right)\mathbf{T}_D^{AB}. \tag{14.2.13}$$

The overall transmission matrix $\mathbf{T}_D^{AC}$ can also be expressed in terms of the same reverberation operator

$$\mathbf{T}_D^{AC} = \mathbf{T}_D^{BC}\left(\mathbf{I} - \mathbf{R}_U^{AB}\mathbf{R}_D^{BC}\right)^{-1}\mathbf{T}_D^{AB}. \tag{14.2.14}$$

The matrix algebra implied by (14.2.10-14.2.14) is that all matrices act to their right and reflection and transmission matrices are non-commutative. As a result, the physical content of (14.2.10), (14.2.13) is obtained by reading from right to left, which may be readily seen by comparison with figure 14.3.

Now that we are able to add together the reflection and transmission effects of two regions we can extend the results to multiple regions by cascading the procedure. The addition of further regions to AC requires that the reflection and transmission matrices for AC be combined for those for the new region using relations comparable to (14.2.11) and (14.2.14).

In order to follow the reflection and transmission processes associated with individual wavetype we need to partition the reflection and transmission matrices. For isotropic elastic materials, we can make a split into the parts affecting P and S waves so that, e.g.

$$\mathbf{R}_D^{BC} = \begin{pmatrix} [\mathbf{R}_D^{BC}]_{PP} & [\mathbf{R}_D^{BC}]_{PS} \\ [\mathbf{R}_D^{BC}]_{SP} & [\mathbf{R}_D^{BC}]_{SS} \end{pmatrix}, \tag{14.2.15}$$

where $\mathbf{R}_{PS}$ is the operator which generates an upgoing P wave from a downgoing S wave. The full suite of P and S wave interactions can be followed by compounding the partitioned forms for the matrices. These combine by matrix multiplication, so that

$$[\mathbf{R}_D^{BC}\mathbf{T}_D^{AB}]_{PP} = [\mathbf{R}_D^{BC}]_{PP}[\mathbf{T}_D^{AB}]_{PP} + [\mathbf{R}_D^{BC}]_{PS}[\mathbf{T}_D^{AB}]_{SP}. \tag{14.2.16}$$

The interaction of multiple reflections and conversions can be determined by making an expansion of the reverberation operator (14.2.12). Kennett (1986) has shown that when coupling between P and S is restricted to a single conversion it is possible to obtain a very useful approximation for the reverberation operator (14.2.11) in which the P and S wave reverberation sequences can be identified separately

$$[\mathbf{I} - \mathbf{R}_U^{AB}\mathbf{R}_D^{BC}]^{-1} = \begin{pmatrix} [\mathrm{X}_{PP}]^{-1} & [\mathrm{X}_{PP}]^{-1}\mathrm{Y}_{PS}[\mathrm{X}_{SS}]^{-1} \\ [\mathrm{X}_{SS}]^{-1}\mathrm{Y}_{SP}[\mathrm{X}_{PP}]^{-1} & [\mathrm{X}_{SS}]^{-1} \end{pmatrix}, \tag{14.2.17}$$

where $[\mathrm{X}_{PP}]^{-1}$ and $[\mathrm{X}_{SS}]^{-1}$ are the reverberation operators for P and S waves respectively;

$$[\mathrm{X}_{PP}]^{-1} = \left(\mathrm{I} - [\mathbf{R}_U^{AB}]_{PP}[\mathbf{R}_D^{BC}]_{PP}\right)^{-1}. \tag{14.2.18}$$

The conversion terms Y_{SP}, Y_{PS} allow for one conversion;

$$\mathrm{Y}_{SP} = [\mathbf{R}_U^{AB}]_{SP}[\mathbf{R}_D^{BC}]_{PP} + [\mathbf{R}_U^{AB}]_{SS}[\mathbf{R}_D^{BC}]_{SP}. \tag{14.2.19}$$

With this expansion of the full reverberation operator it is relatively easily to follow the main conversion processes which are likely to contribute to seismic records.

The approximation (14.2.17) for the reverberation operator enables us to establish simple approximations for reflection and transmission coefficients with a limited allowance for conversions. Consider a composite transmission matrix in the form

$$\mathbf{T} = \mathbf{T}^B[\mathbf{I} - \mathbf{R}^A\mathbf{R}^B]^{-1}\mathbf{T}^C, \tag{14.2.20}$$

where, for example, $\mathbf{T}^C$ stands for a matrix such as $\mathbf{T}_D^{AB}$. The transmission coefficients including a single conversion are

$$\begin{aligned} \mathrm{T}_{PP} &= \mathrm{T}_{PP}^B[\mathrm{I} - \mathrm{R}_{PP}^A\mathrm{R}_{PP}^B]^{-1}\mathrm{T}_{PP}^C, \\ \mathrm{T}_{PS} &= \mathrm{T}_{PS}^B[\mathrm{I} - \mathrm{R}_{SS}^A\mathrm{R}_{SS}^B]^{-1}\mathrm{T}_{SS}^C + \mathrm{T}_{PP}^B[\mathrm{I} - \mathrm{R}_{PP}^A\mathrm{R}_{PP}^B]^{-1}\mathrm{T}_{PS}^C \\ &\quad + \mathrm{T}_{PP}^B[\mathrm{I} - \mathrm{R}_{PP}^A\mathrm{R}_{PP}^B]^{-1}(\mathrm{R}_{PS}^A\mathrm{R}_{SS}^B + \mathrm{R}_{PP}^A\mathrm{R}_{PS}^B)[\mathrm{I} - \mathrm{R}_{SS}^A\mathrm{R}_{SS}^B]^{-1}\mathrm{T}_{SS}^C. \end{aligned} \tag{14.2.21}$$

For a reflection matrix

$$\mathbf{R} = \mathbf{R}^C + \mathbf{T}^B\mathbf{R}^B[\mathbf{I} - \mathbf{R}^A\mathbf{R}^B]^{-1}\mathbf{T}^C, \tag{14.2.22}$$

the composite coefficients including just one conversion are

$$\begin{aligned} R_{PP} &= R^C_{PP} + T^B_{PP}R^B_{PP}[I - R^A_{PP}R^B_{PP}]^{-1}T^C_{PP}, \\ R_{PS} &= R^C_{PS} + T^B_{PP}R^B_{PP}[I - R^A_{PP}R^B_{PP}]^{-1}T^C_{PS} \\ &\quad +(T^B_{PP}R^B_{PS} + T^B_{PS}R^B_{SS})[I - R^A_{SS}R^B_{SS}]^{-1}T^C_{SS}. \end{aligned} \tag{14.2.23}$$

To this level of approximation the structure of the *PP*, *SS* coefficients is the same as if there was no coupling between the wavetypes, even though the individual coefficients such as R^A_{PP} will include an allowance for interconversion.

14.2.1 Reflection from a stack of uniform layers

The absence of reflections from within any uniform layer means that the reflection matrix for a stack of uniform layers depends heavily on the interface coefficients. Transmission through the layers gives phase terms which modulate the interface effects. For a stack of uniform layers the addition rules introduced in the previous section may be used to construct the reflection and transmission matrices in a two-stage recursive process. The phase delays through a layer and the interface terms are introduced alternately.

Consider a uniform layer $z_1 < z < z_2$ overlying a stack of such layers in the zone $z_2 < z < z_3$. We suppose the reflection and transmission matrices at z_2- just into this layer are known and write, e.g.,

$$\mathbf{R}_D(z_2-) = \mathbf{R}_D(z_2-, z_3+). \tag{14.2.24}$$

We may then add in the phase terms corresponding to transmission through the uniform layer using the addition rules (14.2.11), (14.2.14) to calculate the reflection and transmission matrices just below the interface at z_1+. Thus

$$\begin{aligned} \mathbf{R}_D(z_1+) &= \mathbf{E}^{12}_U\mathbf{R}_D(z_2-)\mathbf{E}^{12}_D, \\ \mathbf{T}_D(z_1+) &= \mathbf{T}_D(z_2-)\mathbf{E}^{12}_D, \\ \mathbf{T}_U(z_1+) &= \mathbf{E}^{12}_U\mathbf{T}_U(z_2-), \end{aligned} \tag{14.2.25}$$

where $\mathbf{E}^{12}_D$ is the phase increment for downward propagation through the uniform layer, and $\mathbf{E}^{12}_U$ the phase increment for upward propagation. For isotropic media $\mathbf{E}^{12}_D = \mathbf{E}^{12}_U$, but the distinction is necessary for anisotropic models and for the equivalent results in spherical stratification.

A further application of the addition rules allows us to include the reflection and transmission matrices for the interface z_1, e.g.

$$\mathbf{R}^1_D = \mathbf{R}_D(z_1-, z_1+). \tag{14.2.26}$$

The downward reflection and transmission matrices just above the interface at z_1- depend on the interface terms and the previously calculated quantities for downward propagation at z_1+

$$\begin{aligned} \mathbf{R}_D(z_1-) &= \mathbf{R}_D^1 + \mathbf{T}_U^1\mathbf{R}_D(z_1+)[1 - \mathbf{R}_U^1\mathbf{R}_D(z_1+)]^{-1}\mathbf{T}_D^1, \\ \mathbf{T}_D(z_1-) &= \mathbf{T}_D(z_1+)[1 - \mathbf{R}_U^1\mathbf{R}_D(z_1+)]^{-1}\mathbf{T}_D^1. \end{aligned} \qquad (14.2.27)$$

We note that, as we would expect, the combination of the elements (14.2.25) with (14.2.27) produces explicit results which correspond directly with the forms (14.2.14), (14.2.15) for *SH* waves in a single uniform layer.

The corresponding relations for reflection and transmission matrices for upward incidence at the base of the layering have a less simple form, since we are adding a layer at the most complex level of the wave propagation system

$$\begin{aligned} \mathbf{R}_U(z_1-) &= \mathbf{R}_U(z_2-) + \mathbf{T}_D(z_1+)\mathbf{R}_U^1[1 - \mathbf{R}_D(z_1+)\mathbf{R}_U^1]^{-1}\mathbf{T}_U(z_1+), \\ \mathbf{T}_U(z_1-) &= \mathbf{T}_U^1[1 - \mathbf{R}_D(z_1+)\mathbf{R}_U^1]^{-1}\mathbf{T}_U(z_1+). \end{aligned} \qquad (14.2.28)$$

These two applications of the addition rule may be used recursively to calculate the overall reflection and transmission matrices. We start at the base of the layering at z_3 and calculate the interfacial matrices e.g. $\mathbf{R}_D^3$ which will also be $\mathbf{R}_D(z_3-)$. We employ (14.2.25) to step the stack reflection and transmission matrices to the top of the lowest layer in the stack. Then we use the interfacial addition relations (14.2.27)-(14.2.28) to bring the stack matrices to the upper side of this interface. The cycle (14.2.25) followed by (14.2.27)-(14.2.28) allows us to work up the stack, a layer at a time, for an arbitrary number of layers.

For the downward matrices $\mathbf{R}_D$, $\mathbf{T}_D$ (14.2.25), (14.2.27) require only downward stack matrices to be held during the calculation. When upward matrices are needed it is often more convenient to calculate them separately starting at the top of the layering and working down a layer at a time. The resulting construction scheme has a similar structure to (14.2.25), (14.2.27) and is easily adapted to free-surface reflection matrices by starting with free-surface coefficients rather than those for an interface.

At fixed slowness p all the interfacial matrices $\mathbf{R}_D^I$, $\mathbf{T}_D^I$, etc. are frequency independent so that at each layer step frequency dependence enters via the phase terms $\mathbf{E}_{U,D}^{12}$. If the interfacial coefficients are stored, calculations may be performed rapidly at many frequencies for one slowness p.

When waves go evanescent in any layer our choice of branch cut for the vertical slowness q_α, q_β mean that the terms in $\mathbf{E}_D^{12}$ are such that

$$\exp\{i\omega q_\alpha(z_2 - z_1)\} = \exp\{-\omega|q_\alpha|(z_2 - z_1)\}, \qquad (14.2.29)$$

when $q_\alpha^2 < 0$. We always have $z_2 > z_1$ and so no exponential terms which grow with frequency will appear. This means that the recursive scheme is numerically stable even at high frequencies.

The scalar versions of the recursive forms for downward reflection and transmission coefficients have been known for a long time and are widely used in acoustics and physical optics. The extension to coupled waves seems first to have been used for transmission lines (Redheffer, 1961) and then in plasma studies (e.g., Altman & Cory, 1969; Denman, 1970). The recursive scheme was independently derived for the seismic case by Kennett (1974); alternative derivations have been given by Saastamoinen (1980) and Ursin (1983).

The recursive procedure provides an efficient means of constructing reflection and transmission responses for stratified models by dividing up structures into thin uniform layers. If desired, truncated reverberation sequences can be substituted for the matrix inverses to give restricted, approximate results.

A number of other recursive developments can be made for the calculation of the reflection and transmission properties of layered media (see, e.g., Müller, 1985), but the present forms have the advantage of a direct relation to physical processes.

14.2.2 The effects of velocity gradients

We have just seen how we can make a recursive development for the reflection and transmission properties of a stack of uniform layers. For much of the earth a more appropriate representation of the wavespeed distribution is to take regions of smoothly varying properties interrupted by only a few major discontinuities. Such a model can be approximated by a fine cascade of uniform layers but then the process of continuous refraction by parameter gradients is represented by high order multiple reflections within the uniform layers. With a large number of such layers the computational cost can be very high.

A more direct approach is to work with a model composed of zones of wavespeed gradients (figure 14.4). The stratification is then split with interfaces at the levels where is a discontinuity in the elastic parameters or their gradients (Kennett & Illingworth, 1981). This gives zones of wavespeed gradients sandwiched between uniform media and discrete interfaces. Once again a two-stage recursion can be made, with alternate interface and propagation cycles, but now the transmission delays for a uniform layer are replaced by the reflection and transmission effects of a gradient zone.

14.2.2.1 Reflection from a gradient zone

We now consider a region of monotonic wavespeed gradient in the depth interval (z_A, z_B), and arrange the properties of the bounding uniform half spaces to achieve continuity of the seismic wavespeeds and density at z_A and z_B. There will normally be a discontinuity in parameter gradient at these boundaries.

In the region a above z_A, and in the region b below z_B, we can build a representation of the wavefield in terms of upgoing and downgoing waves

$$\mathbf{b}(z_A) = \mathbf{D}_a\mathbf{v}(z_A), \qquad \mathbf{b}(z_B) = \mathbf{D}_b\mathbf{v}(z_B), \tag{14.2.30}$$

Within the gradient zone we employ the representations of Section 12.3.3 based on uniform asymptotic forms. Thus we represent the wavefield in (z_A, z_B) as

$$\mathbf{b}(z) = \mathbf{D}(z)\mathbf{E}(z)\mathbf{L}(z; z_r)\mathbf{c}, \tag{14.2.31}$$

where $\mathbf{c}$ represents the weighting for the asymptotically upgoing and downgoing waves components. The level z_r is the reference level for the calculation of the gradient correction terms $\mathbf{L}$.

The continuity condition for the motion-stress vector $\mathbf{b}$ at z_B enables us to determine the weighting vector c from the relation

$$\mathbf{b}(z_B) = \mathbf{D}(z_B)\mathbf{E}(z_B)\mathbf{L}_{ud}(z_B; z_r)\mathbf{c} = \mathbf{D}_b\mathbf{v}(z_B). \tag{14.2.32}$$

Similarly the continuity condition at z_A allows us to extract the wavevector $\boldsymbol{v}(z_A)$ from

$$\mathbf{b}(z_A) = \mathbf{D}_a\boldsymbol{v}(z_A) = \mathbf{D}(z_A)\mathbf{E}(z_A)\mathbf{L}_{ud}(z_A; z_r)\mathbf{c}. \tag{14.2.33}$$

When we combine (14.2.32), (14.2.33) we can obtain a relation between the wavectors at z_A and z_B

$$\begin{aligned}\mathbf{v}(z_A) &= \mathbf{Q}(z_A, z_B)\mathbf{v}(z_B)\\ &= \mathbf{D}_a^{-1}\mathbf{D}(z_A)\mathbf{E}(z_A)\mathbf{L}_{ud}(z_A; z_r)\mathbf{L}^{-1}(z_B; z_r)\mathbf{E}^{-1}(z_B)\mathbf{D}^{-1}(z_B)\mathbf{D}_b\mathbf{v}(z_B).\end{aligned} \tag{14.2.34}$$

Equation (14.2.34) is now in the same form as (13.2.16) for a single interface. Thus we can extract the reflection and transmission coefficients for the region AB from the partitions of $\mathbf{Q}(z_A, z_B)$ e.g.

$$\mathbf{T}_D^{AB} = (\mathbf{Q}_{DD})^{-1}, \quad \mathbf{R}_D^{AB} = \mathbf{Q}_{UD}(\mathbf{Q}_{DD})^{-1}. \tag{14.2.35}$$

It is however preferable to factor $\mathbf{Q}(z_A, z_B)$ into the terms governing the entry and exit of plane waves from the gradient zone $\mathbf{D}^{-1}(z_A)\mathbf{D}(z_A)$, $\mathbf{D}^{-1}(z_B)\mathbf{D}(z_B)$ and the matrix

$$\mathbf{F}(z_A, z_B) = \mathbf{E}(z_A)\mathbf{L}_{ud}(z_A, z_B; z_r)\mathbf{E}^{-1}(z_B), \tag{14.2.36}$$

which represents all the propagation characteristics within the gradient zone. The phase terms arise from the Airy function terms in $\mathbf{E}(z_A)$ and $\mathbf{E}(z_B)$ and the interaction sequence for the entire gradient zone

$$\mathbf{L}_{ud}(z_A, z_B; z_r) = \mathbf{L}_{ud}(z_A; z_r)\mathbf{L}_{ud}^{-1}(z_B; z_r). \tag{14.2.37}$$

Here z_r is the reference level from which the arguments of the Airy functions are calculated. $\mathbf{L}$ can, in principle, be found from the interaction series (12.3.50) to any required order of interaction within (z_A, z_B). $\mathbf{E}(z_A)$ and $\mathbf{E}(z_B)$ are diagonal matrices organised into blocks by asymptotic wave character. The interaction series for $\mathbf{L}_{ud}$ begins with the unit matrix and if all subsequent contributions are neglected, P and S waves appear to propagate independently within the gradient zone (cf. 12.3.56). Once the higher terms in the interaction series are included, the P and SV wave components with slowness p are coupled together. When the wavespeeds vary slowly with depth, this coupling is weak at moderate frequencies (Richards, 1974).

At the limits of the gradient zone we have introduced discontinuities in wavespeed gradient at z_A and z_B. The entry and exit terms in $\mathbf{Q}$ such as $\mathbf{D}^{-1}(z_A)\mathbf{D}(z_A)$ depend on the difference between the generalized slownesses in the gradient zone $\eta_{u,d}(\omega, p, z_A)$ and the corresponding radicals in the uniform region. For *SH* waves,

$$\mathbf{D}^{-1}(z_A)\mathsf{D}(z_A) = G_\beta(z_A) = \rho\epsilon_\beta \begin{pmatrix} q_\beta + \eta_{\beta u} & q_\beta - \eta_{\beta d} \\ q_\beta - \eta_{\beta u} & q_\beta + \eta_{\beta d} \end{pmatrix}. \tag{14.2.38}$$

For *P-SV* waves we will have a similar behaviour for each wavetype, and there will be no coupling between *P* and *SV* waves introduced by these interface terms. The interface matrix $\mathbf{G}(z_A)$ will be block-diagonal, with partitions organised by asymptotic wave character.

The coupling matrices $\mathbf{Q}$ may be expressed in terms of the interface matrices $\mathbf{G}$ as

$$\mathbf{Q}(z_A, z_B) = \mathbf{G}(z_A)\mathbf{F}(z_A, z_B)\mathbf{G}^{-1}(z_B). \tag{14.2.39}$$

For each of these matrices we introduce a set of *generalized* reflection and transmission matrices $\mathbf{r}_d$, $\mathbf{t}_d$ etc. such that, e.g.,

$$\bar{\mathbf{G}}(z_A) = \begin{pmatrix} \mathbf{t}_u^G - \mathbf{r}_d^G(\mathbf{t}_d^G)^{-1}\mathbf{r}_u^G & \mathbf{r}_d^G(\mathbf{t}_d^G)^{-1} \\ -(\mathbf{t}_d^G)^{-1}\mathbf{r}_u^G & (\mathbf{t}_d^G)^{-1} \end{pmatrix}. \tag{14.2.40}$$

The generalized elements for a matrix product $\bar{\mathbf{G}}\bar{\mathbf{F}}$ can be determined by an extension of the addition rules (14.2.11), (14.2.14), for example

$$\mathbf{r}_d^{GF} = \mathbf{r}_d^G + \mathbf{t}_u^G\mathbf{r}_d^F[\mathbf{I} - \mathbf{r}_u^G\mathbf{r}_d^F]^{-1}\mathbf{t}_d^G. \tag{14.2.41}$$

The reflection and transmission matrices $\mathbf{R}_D^{AB}$, $\mathbf{T}_D^{AB}$ etc. for the whole gradient zone (z_A, z_B) can therefore be built up from the $\mathbf{r}_d$, $\mathbf{t}_d$ matrices for the factors of the coupling matrix $\mathbf{Q}$.

For the interface matrix $\mathbf{G}$, the generalized elements $\mathbf{r}_d^G$, $\mathbf{r}_u^G$ have a simple form: for *SH* waves

$$\begin{aligned} \mathbf{r}_d^G|_{HH} &= (q_\beta - \eta_{\beta d})/(q_\beta + \eta_{\beta d}), \\ \mathbf{r}_u^G|_{HH} &= (\eta_{\beta u} - q_\beta)/(q_\beta + \eta_{\beta d}), \\ \mathbf{t}_u^G\mathbf{t}_d^G|_{HH} &= 2q_\beta(\eta_{\beta u} + \eta_{\beta d})/(q_\beta + \eta_{\beta d})^2. \end{aligned} \tag{14.2.42}$$

The individual transmission terms are not symmetric because D is not normalised in quite the same way as $\mathbf{D}$. Since the interface terms $\mathbf{G}$ does not couple *P* and *SV* waves, the *SS* elements of $\mathbf{r}_d$ etc. are equal to the *HH* elements and the *PP* elements are obtained by exchanging α for β. The reflection terms depend on the off-diagonal parts of G_α, G_β. We recall that $\eta_{\beta u,d}$ depend on the wavespeed distribution within the gradient zone and so the reflections should not be envisaged as just occurring at z_A. Chapman & Orcutt (1985, §3.6) point out that the terms (14.2.42) do not provide a full representation of possible reflection effects; for example, no reflection is predicted from a discontinuity in density gradient (see also Thomson & Chapman, 1984). The effects are subtle and can be compensated for by the inclusions of corrections from the interaction series (12.3.49).

In order to give a good approximation for the elements of the phase matrix E which depend on Airy functions, it is convenient to extrapolate the wavespeed distribution outside the gradient zone until turning points are reached and then to calculate the phase delays τ_α and τ_β from the P and S wave turning levels. When the turning points lie well outside the gradient zone (z_A, z_B), the generalized vertical slownesses $\eta_{\beta u,d}$ tend asymptotically to q_β and so $\boldsymbol{G}_\beta$ tends to a diagonal matrix. In this asymptotic regime there will therefore be no reflection associated with the discontinuities in wavespeed gradient. However, when turning levels lie close to z_A or z_B, we are not in the asymptotic regime for the Airy functions and so the differences $(\eta_{\beta u,d} - q_\beta)$ become significant. Thus, as pointed out by Doornbos (1981), when a discontinuity in velocity gradient occurs near the turning point of a wave there can be significant reflection.

The nature of the generalized reflection and transmission terms $\mathbf{r}_d^F$, $\mathbf{t}_d^F$ associated with the propagation matrix within the gradient zone $\mathbf{F}(z_A, z_B)$ depends strongly on the locations of P and S wave turning points relative to the gradient zone. A detailed discussion of the properties of $\mathbf{F}$ is presented in Kennett & Illingworth (1981) and Kennett (1983).

When the gradient zone lies above all turning points, the dominant transmission is supplemented by reflection terms from the gradients, which to first order are equivalent to the results of Chapman (1974) and Richards & Frasier (1976) based on the use of the WKBJ approximation. For moderate gradients the latter authors have shown that significant conversion of wave types can be generated by the coupling terms (12.3.56). We can visualise the effect by dividing a gradient zone up into thin layers; once the gradient is steep enough that significant contrast is introduced across a thin layer, there will be both noticeable reflection in the original wavetype and the possibility of conversion.

The most significant effects occur when the slowness p is such that turning points for either P or S waves occur within the zone (z_A, z_B). In this case we have to take account of the differences in the nature of the wavefield at the top and bottom of the gradient zone. At z_A we would wish to use the propagating elements for the wave type which has the turning point. However at z_B a better description is provided by using the evanescent forms.

We therefore split the gradient zone at the turning level z_r. We use the terms for propagating waves in the representation of the motion-stress vector $\mathbf{b}(z)$ above z_r, whereas in the region below z_r we use the appropriate terms for evanescence $\hat{\mathbf{E}}$ etc. The coupling matrix $\mathbf{Q}$ can then be built up using the chain rule by splitting the layering at the turning level z_r

$$\mathbf{Q}(z_A, z_B) = \mathbf{D}_a^{-1}\mathbf{D}(z_A)\mathbf{E}(z_A)\mathbf{L}_{ud}(z_A; z_r)\,\mathbf{H}\,\mathbf{L}^{-1}(z_B; z_r)\mathbf{E}^{-1}(z_B)\mathbf{D}^{-1}(z_B)\mathbf{D}_b. \quad (14.2.43)$$

The matrix $\mathbf{H}$ connects the differing functional forms for the motion-stress vector across z_r; the particular form of $\mathbf{H}$ depends on the character of the turning point.

With the split representation within the gradient zone we will use different forms for the interface matrices $\mathbf{G}(z_A)$, $\hat{\mathbf{G}}(z_B)$ at the top and bottom of the zone. Taking z_r

as the reference level for Airy function arguments, the propagation effects within the gradient zone are represented by

$$\mathbf{F}(z_A, z_B) = \mathbf{E}(z_A)\mathbf{L}(z_A; z_r)\,\mathbf{H}\hat{\mathbf{L}}^{-1}(z_B; z_r)\hat{\mathbf{E}}^{-1}(z_B), \tag{14.2.44}$$

and now the principal contribution to reflection from the zone will come from the presence of the linking matrix **H**.

The simplest situation is provided by a turning point for *SH* waves. The linking matrix has the explicit form

$$\boldsymbol{h} = \boldsymbol{E}_\beta^{-1}(z_r)\mathbf{C}_H^{-1}(z_r)\mathbf{C}_H(z_r)\hat{\boldsymbol{E}}_\beta(z_r), \tag{14.2.45}$$

in terms of the full phase matrices (12.3.14),(12.3.20). There is continuity of material properties at z_r and so

$$\boldsymbol{h} = \boldsymbol{E}_\beta^{-1}(z_r)\hat{\boldsymbol{E}}_\beta(z_r) = 2^{-1/2}\begin{pmatrix} e^{i\pi/4} & e^{-i\pi/4} \\ e^{-i\pi/4} & e^{i\pi/4} \end{pmatrix}. \tag{14.2.46}$$

If we neglect all gradient contributions by comparison with the coupling term $\boldsymbol{h}$ (i.e. take **L** and $\hat{\mathbf{L}}$ to be unit matrices) we have the approximate propagation matrix

$$\mathbf{F}_0(z_A, z_B) = \boldsymbol{E}_\beta(z_A)\boldsymbol{h}\hat{\boldsymbol{E}}_\beta^{-1}(z_B). \tag{14.2.47}$$

The matrix $\boldsymbol{h}$ accounts for total internal reflection at the turning point. The generalized reflection and transmission elements for $\mathbf{F}_0$ are then

$$\begin{aligned} \mathbf{r}_d^{F_0}|_{HH} &= \mathrm{Ej}_\beta(z_A)e^{-i\pi/2}[\mathrm{Fj}_\beta(z_A)]^{-1}, \quad & \mathbf{t}_d^{F_0}|_{HH} &= \sqrt{2}\mathrm{Aj}_\beta(z_B)e^{-i\pi/4}[\mathrm{Fj}_\beta(z_A)]^{-1}, \\ \mathbf{r}_u^{F_0}|_{HH} &= \mathrm{Aj}_\beta(z_B)e^{i\pi/2}[\mathrm{Bj}_\beta(z_B)]^{-1}, \quad & \mathbf{t}_u^{F_0}|_{HH} &= \sqrt{2}\mathrm{Ej}_\beta(z_A)e^{i\pi/4}[\mathrm{Bj}_\beta(z_B)]^{-1}. \end{aligned} \tag{14.2.48}$$

All these elements are well behaved numerically since the only exponentially increasing term, Bj_β, appears as an inverse.

When the limits of the gradient zone lie well away from the turning point we may make an asymptotic approximation to the Airy functions, and obtain the 'full-wave' approximation to the reflection (see, e.g., Budden, 1961)

$$\mathbf{r}_d^{F_0}|_{HH} \sim \exp\{2i\omega\int_{z_A}^{z_r} d\zeta\, q_\beta(\zeta) - i\pi/2\}. \tag{14.2.49}$$

This approximation corresponds to complete reflection with a phase shift of $\pi/2$ compared with the phase delay for propagation down to the turning level and back. This simple result forms the basis of much further work which seeks to extend ray theory (e.g., Richards, 1973; Chapman, 1978). The approximation will be most effective at high frequencies; and, for neglect of the interaction terms, requires only slight wavespeed gradients throughout (z_A, z_B). The corresponding approximation in transmission is

$$\mathbf{t}_d^{F_0}|_{HH} \sim \frac{1}{\sqrt{2}}\exp\{i\omega\int_{z_A}^{z_r} d\zeta\, q_\beta(\zeta) - i\pi/4\}\exp\{-\omega\int_{z_r}^{z_B} d\zeta\, |q_\beta(\zeta)|\}, \tag{14.2.50}$$

illustrating the damping of *SH* waves below the turning level.

For the *P-SV* wave system the situation is more complicated since we now have the possibility of both *P* and *S* wave turning levels. However, we note that the rather larger values of the *P* wavespeeds compared with *S* wavespeeds means that at the same slowness p the turning level for *P* will lie at shallower depths than for *S*. When there is just a *P* wave turning point at z_r, the main contribution will come from the link matrix for the *P* elements and we will have almost independent propagation of *P* and *SV* waves. The *PP* reflection and transmission elements will therefore have the form (14.2.48) with β replaced by α. The SS coefficient will just represent propagating waves.

When the turning point for *P* waves lies above z_A we have evanescent character for *P* throughout (z_A, z_B). The character of the *S* wave elements will now need to be modified across the *S* wave turning level at z_r. There will normally only be a very small reflection or transmission contribution from the evanescent *P* waves and coupling between *P* and *SV* waves is negligible. The SS coefficients will therefore match the HH coefficients in (14.2.48).

In the unusual circumstance when both *P* and *S* wave turning points occur within the same gradient zone, the calculation needs to be split at both turning levels, which requires some care with the coupling terms (Kennett & Illingworth, 1981).

We have assumed that both *P* and *S* wavespeeds increase with depth, so that s_α and s_β are positive. If either wavespeed is actually smoothly decreasing with depth we have a similar development with the roles of up and downgoing waves interchanged for that wave type.

We have discussed the properties of wave interactions with gradient for horizontal stratification in terms of a high frequency approximation to the seismic wavefield in terms of Airy functions with arguments determined by the delay times τ in the model. As we have shown in Section 12.3.4 we can make a comparable development for spherical stratification. The leading order approximation will be essentially the same since the earth flattening transformation (10.2.55) preserves delay times. To this level of approximation, it therefore makes no real difference whether we work with spherical stratification or a flattened wavespeed distribution. The structure of the correction terms in $\mathbf{L}$ will differ between the two cases, since there is additional frequency dependence in the spherical case, and there appears to be no density transformation which is optimum for both *P* and *SV* waves (Chapman, 1973).

14.2.2.2 Recursive construction of reflection and transmission matrices

For a model composed of smooth gradient zones interrupted by discontinuities in the elastic parameters or their gradients, such as illustrated in figure 14.4 we can build up the overall reflection and transmission matrices by a recursive application of the addition rules.

We suppose that the model is ultimately underlain by a uniform half space in $z > z_C$ with continuity of elastic properties at z_C (figure 14.4). We start by considering

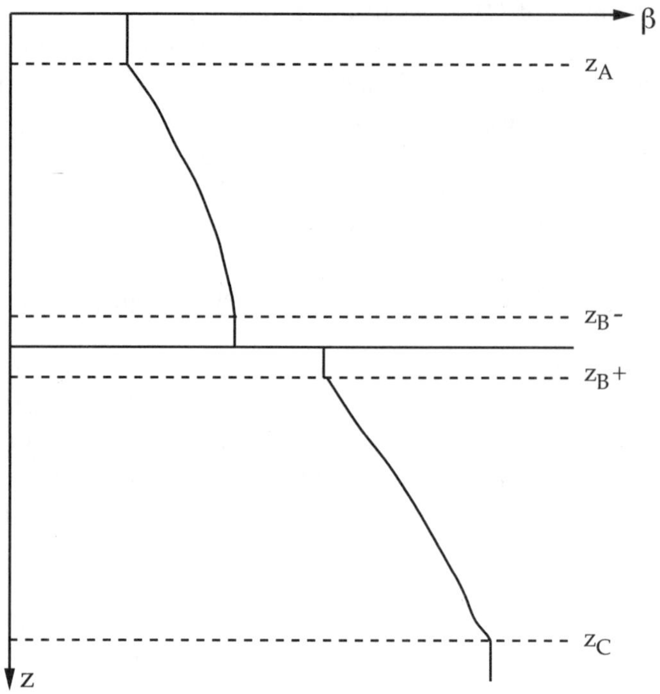

Figure 14.4. Division of a piecewise smooth medium in (z_A, z_C) into an interface zone at z_B between uniform media and gradient zones $z_A < z < z_B-$, $z_B+ < z < z_C$ bordered by uniform media.

the gradient zone in (z_B+, z_C), extended above by a uniform half space with the properties at z_B+, just below the next higher interface. For this gradient zone, we can construct the downward reflection and transmission matrices $\mathbf{R}_D^{B_+C}$, $\mathbf{T}_D^{B_+C}$ by successive applications of the addition rules for the $\mathbf{r}_d$, $\mathbf{t}_d$ matrices introduced in the previous section. $\mathbf{R}_D^{B_+C}$, $\mathbf{T}_D^{B_+C}$ can then be recognised as the reflection and transmission matrices as seen from a uniform half space at z_B+: $\mathbf{R}_D(z_B+)$, $\mathbf{T}_D(z_B+)$.

The effect of the interface can then be introduced, as in the uniform layer scheme, by adding in the interface matrices $\mathbf{R}_D^B$, $\mathbf{T}_D^B$ etc. for an interface between two uniform media. This yields the reflection and transmission matrices as seen at z_B- so that now, for example,

$$\mathbf{R}_D(z_B-) = \mathbf{R}_D^B + \mathbf{T}_U^B\mathbf{R}_D(z_B+)[\mathbf{I} - \mathbf{R}_U^B\mathbf{R}_D(z_B+)]^{-1}\mathbf{T}_D^B. \qquad (14.2.51)$$

The calculation can then be incremented to the top of the next gradient zone z_A by introducing the reflection and transmission effects of the region (z_A, z_B-). In the uniform layer case only phase delays in transmission were required, but for a gradient zone all the matrices $\mathbf{R}_D^{AB_-}$, $\mathbf{R}_U^{AB_-}$, $\mathbf{T}_D^{AB_-}$, $\mathbf{T}_U^{AB_-}$ are needed. These coefficients may be calculated by an inner recursion over the $\mathbf{r}_d$, $\mathbf{t}_d$ matrices for the gradient factors.

The reflection and transmission matrices for the region below z_A are then to be derived by using the addition rule between the matrices $\mathbf{R}_D(z_B-)$, $\mathbf{T}_D(z_B-)$ and those for the gradient zone (z_A, z_B-). We find

$$\begin{aligned}\mathbf{R}_D(z_A) &= \mathbf{R}_D^{AB-} + \mathbf{T}_U^{AB-}\mathbf{R}_D(z_B-)[\mathbf{I} - \mathbf{R}_U^{AB-}\mathbf{R}_D(z_B-)]^{-1}\mathbf{T}_D^{AB-},\\ \mathbf{T}_D(z_A) &= \mathbf{T}_D(z_B-)[\mathbf{I} - \mathbf{R}_U^{AB-}\mathbf{R}_D(z_B-)]^{-1}\mathbf{T}_D^{AB-}.\end{aligned} \tag{14.2.52}$$

If there is a further interface at z_A we once again add in the interface matrices via the addition rule, and then continue the calculation through the next gradient zone.

Over the main structural elements, we have a two-stage recursion over interfaces and gradient zones to build the overall reflection and transmission matrices. The reflection elements for the gradients are themselves found by recursive application of the same addition rules on the *generalized* coefficients.

A useful property of this approach to calculating the reflection properties is that we can separate the effect of an interface from the structure surrounding it.

At a discontinuity in parameter gradients at z_B, the elastic parameters are continuous and so

$$\mathbf{R}_U^B = \mathbf{R}_D^B = \mathbf{0}, \quad \mathbf{T}_U^B = \mathbf{T}_D^B = \mathbf{I}. \tag{14.2.53}$$

In the overall reflection and transmission coefficients the effect of the discontinuity in the gradient will arise from the contributions of the terms $\mathsf{D}^{-1}(z_B-)\mathbf{D}(z_B)$ and $\mathsf{D}^{-1}(z_B)\mathsf{D}(z_B+)$ to the reflection properties of the regions above and below the interface. We construct the discontinuity in gradient by superimposing the effect of the transition from the gradients on either side of the interface into a uniform medium. The procedure may be visualised by shrinking the jump in properties across z_B in figure 14.4 to zero.

An alternative development for the reflection and transmission properties of a piecewise continuous medium may be obtained by forming the coupling matrix for the entire region. For example for the case illustrated in figure 14.4.

$$\mathbf{Q}(z_A, z_C) = \mathsf{D}^{-1}(z_A)\mathbf{D}(z_A)\mathbf{F}(z_A, z_B-)\mathsf{D}^{-1}(z_B-)\mathbf{D}(z_B+)\mathbf{F}(z_B+, z_C)\mathsf{D}^{-1}(z_C)\mathbf{D}(z_C). \tag{14.2.54}$$

The overall reflection and transmission matrices can be constructed by using the addition rule to bring in alternately the effects of interfaces and propagation using the *generalized* matrices. At z_B the generalized interface coefficients are found from

$$\mathsf{D}^{-1}(z_B-)\mathsf{D}(z_+), \tag{14.2.55}$$

and depend on frequency ω as well as slowness p. Such generalized coefficients have been used by Richards (1976), Cormier & Richards (1977) and Choy (1977), when allowing for gradients near the core-mantle boundary.

Computationally there is little to choose between recursive schemes based on (14.2.51) and (14.2.52), and a comparable development from (14.2.54). The first scheme has the merit that the intermediate results at each stage are themselves reflection and transmission matrices.

14.3 Reflection from a zone including the free surface

The idea of a reflection matrix may also be used for waves incident from below on a region bounded by a free surface. To distinguish this situation from the usual case, we will use the notation $\mathbf{R}_U^{fB}$ for the reflection matrix from the region between the interface B and the free surface. The downgoing waves $\boldsymbol{D}^B$ returned from the region bounded below by B are then related to the upgoing waves $\boldsymbol{U}^B$ at B by

$$\boldsymbol{D}^B = \mathbf{R}_U^{fS}[\boldsymbol{U}^B]. \qquad (14.3.1)$$

The closest analogue to transmission in this case is the relation of the surface disturbance to the incident field at B. We can describe this by the action of a transfer operator $\mathbf{W}_U^{fB}$ so that the surface displacement can be found from

$$\mathbf{u}(0) = \mathbf{W}_U^{fB}[\boldsymbol{U}^B]. \qquad (14.3.2)$$

As we shall see later, the two free surface matrices $\mathbf{R}_U^{fB}$ and $\mathbf{W}_U^{fB}$ play an important role in the description of the seismic wavefield generated by a source.

The approach we have used to combine the reflection and transmission properties of two regions can readily be extended to include a zone bounded above by the free surface. We add the region (z_B, z_C) beneath B and then construct the upward reflection matrices for the region between z_C and the free surface in terms of the properties of the regions fB and BC. For the upward reflection matrix $\mathbf{R}_U^{fC}$,

$$\mathbf{R}_U^{fC} = \mathbf{R}_U^{BC} + \mathbf{T}_D^{BC}\mathbf{R}_U^{fB}\left(\mathbf{I} - \mathbf{R}_D^{BC}\mathbf{R}_U^{fB}\right)^{-1}\mathbf{T}_U^{BC}, \qquad (14.3.3)$$

and for the transfer operator $\mathbf{W}_U^{fC}$,

$$\mathbf{W}_U^{fC} = \mathbf{W}_U^{fB}\left(\mathbf{I} - \mathbf{R}_D^{BC}\mathbf{R}_U^{fB}\right)^{-1}\mathbf{T}_U^{BC}. \qquad (14.3.4)$$

An important special case occurs when we take the level B up to just below the free surface $z = 0+$ so that $\mathbf{R}_U^{fB} = \mathbf{R}_F$, $\mathbf{W}_U^{fB} = \mathbf{W}_F$, then

$$\mathbf{R}_U^{fC} = \mathbf{R}_U^{0C} + \mathbf{T}_D^{0C}\mathbf{R}_F\left(\mathbf{I} - \mathbf{R}_D^{0C}\mathbf{R}_F\right)^{-1}\mathbf{T}_U^{0C}, \qquad (14.3.5)$$

and for the transfer operator $\mathbf{W}_U^{fC}$

$$\mathbf{W}_U^{fC} = \mathbf{W}_F\left(\mathbf{I} - \mathbf{R}_D^{0C}\mathbf{R}_F\right)^{-1}\mathbf{T}_U^{0C}; \qquad (14.3.6)$$

the reverberation operator now operates explicitly on the whole structure between z_C and the free surface.

14.4 Source representation

We may include a point source in the equations for displacement and traction in stratified media discussed in Chapter 12 by including a discontinuity in the stress-displacement vector $\mathbf{b}$ at the level of the source (Hudson, 1969a). For force components $\mathcal{F}_x$, $\mathcal{F}_y$, $\mathcal{F}_z$ and the moment tensor elements describing doublets in the horizontal plane (M_{xx}, M_{xy}, M_{yx}, M_{yy}) we will just need a traction discontinuity, but the remaining elements representing couple and dipoles with a z component (M_{xz}, M_{zx}, M_{yz}, M_{zy}, M_{zz}) will contribute to a displacement discontinuity.

An alternative approach is to regard the source as giving rise to a discontinuity in the wavevector $\mathbf{v}$. Such an approach has been used by Haskell (1964) and Harkrider (1964) to specify their sources.

We regard the source as lying in a locally uniform region about the source plane z_S, and can then relate the motion-stress vectors $\mathbf{b}(z_S-)$, $\mathbf{b}(z_S+)$ immediately above and below the source plane to their up and downgoing wave parts as

$$\mathbf{b}(z_S-) = \mathbf{D}_S\mathbf{v}(z_S-), \qquad \mathbf{b}(z_S+) = \mathbf{D}_S\mathbf{v}(z_S+). \tag{14.4.1}$$

The jump in the wavevector across the source level is

$$\mathbf{v}(z_S+) - \mathbf{v}(z_S-) = \Sigma(k, m, z_S, \omega) = \mathbf{D}_S^{-1}\left[\mathbf{b}(z_S-) - \mathbf{b}(z_S+)\right]. \tag{14.4.2}$$

The contribution Σ arising directly from the source will correspond to upgoing radiation $\boldsymbol{U}^S$ into z_S- and downward radiation $\boldsymbol{D}^S$ into z_S+, so that

$$\Sigma = [-\boldsymbol{U}^S, \boldsymbol{D}^S]^T. \tag{14.4.3}$$

In terms of the jump in displacement components $\boldsymbol{S}_W$ and the traction jump $\boldsymbol{S}_T$, we can use the explicit form for $\mathbf{D}_S^{-1}$ (12.1.43) to generate expressions for the radiated components

$$\begin{aligned}\boldsymbol{U}^S &= \mathrm{i}[\boldsymbol{m}_{DS}^T\boldsymbol{S}_T - \boldsymbol{n}_{DS}^T\boldsymbol{S}_W],\\ \boldsymbol{D}^S &= \mathrm{i}[\boldsymbol{m}_{US}^T\boldsymbol{S}_T - \boldsymbol{n}_{US}^T\boldsymbol{S}_W].\end{aligned} \tag{14.4.4}$$

These expressions for the radiation components can be used in a region whose properties vary with depth by splitting the medium at the source level z_S and then considering each of the two halves of the stratification to be extended by uniform half spaces with the properties at z_S. This approach corresponds directly to our treatment of reflection and transmission problems and the radiation components U^S, D^S enter into a compact physical description of the seismic wavefield in combination with reflection and transmission terms.

14.5 Inclusion of source radiation

We now turn our attention to the generation of a full physical description of the seismic wavefield in the frequency/slowness domain in terms of the sequential action of reflection and transmission matrices.

The first stage of generating the representation of the wavefield is to include the description of the radiation from a seismic source. Then we can build in the effects of reflection from below the source, and multiples generated at the free surface. The final stage is to include transmission to the source and the influence of near-receiver structure.

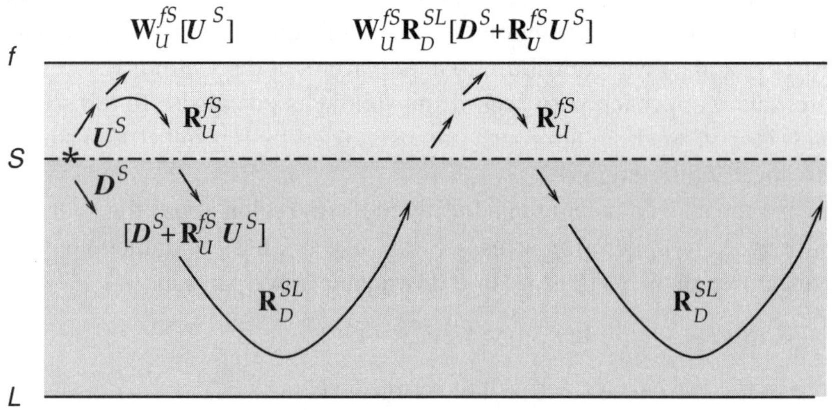

Figure 14.5. Schematic presentation of the action of reflection and transmission operators in the representation of the seismic wavefield.

The description of the modelling scheme takes into account the sequential action of the propagation processes above and below the source, as illustrated in figure 14.5. We consider a structural model occupying the region between the free surface and the plane z_L. Beneath L the velocity structure is assumed to be uniform so that we can apply a radiation condition of purely downgoing waves propagating into this region. We will also find it convenient to introduce an intermediate level J which we will use to separate near-surface and deeper propagation effects.

14.5.1 The full wavefield

We consider a source beneath the surface, at a level z_S, and will be able to generate a description of a surface source (if necessary) by taking the limit as the source depth tends to zero. We consider the source to be placed in a locally uniform medium, so that, as discussed above, we can make an unambiguous separation of the radiation into upward $[\boldsymbol{U}^S]$ and downward $[\boldsymbol{D}^S]$ parts, simply by their relation to the plane z_S.

We now regard the region above the source to abut the uniform medium and to be irradiated by the upgoing field $\boldsymbol{U}^S$. Similarly the region below the source is taken to have an incident downward field $\boldsymbol{D}^S$ from the uniform medium. We now have just the situation for which we have defined our reflection and transmission matrices and so may represent the seismic wavefield in terms of matrices for the regions fS and SL, above and below the source.

We start with the energy radiated upward from the source. At surface receivers there will be a direct wave described by

$$\mathbf{W}_U^{fS}\boldsymbol{U}^S, \tag{14.5.1}$$

where $\mathbf{W}_U^{fS}$ is the matrix which generates surface displacement from an upgoing field at S introduced in Section 14.3. The upgoing wave $\boldsymbol{U}^S$ will be reflected back by the region above the source (especially the free surface) and this will give a downgoing wave contribution at S of

$$\mathbf{R}_U^{fS}\boldsymbol{U}^S. \tag{14.5.2}$$

The total downgoing wave field at S is therefore composed of the source radiation $\boldsymbol{D}^S$ and the surface "ghost" $\mathbf{R}_U^{fS}\boldsymbol{U}^S$ which includes the depth phases such as *pP*, *sP*, *sS* and *pS*, with conversion occurring in the reflection process at the surface. The downward field

$$\boldsymbol{D}^S + \mathbf{R}_U^{fS}\boldsymbol{U}^S, \tag{14.5.3}$$

will now be incident on the lower part of the model SL containing the structures of interest. The once-reflected field is then represented by

$$\mathbf{R}_D^{SL}[\boldsymbol{D}^S + \mathbf{R}_U^{fS}\boldsymbol{U}^S]. \tag{14.5.4}$$

The reflection matrix $\mathbf{R}_D^{SL}$ would normally be constructed by building up the response of various parts of the model using, e.g., the recursion scheme discussed in Section 14.2. The action of $\mathbf{R}_D^{SL}$ will therefore contain all internal multiples in the region of the model below the source level.

At this stage, with only a single reflection back to the source level, we have an upward field at S of $\mathbf{R}_D^{SL}[\boldsymbol{D}^S + \mathbf{R}_U^{fS}\boldsymbol{U}^S]$, and at surface receivers a displacement contribution

$$\mathbf{W}_U^{fS}\mathbf{R}_D^{SL}[\boldsymbol{D}^S + \mathbf{R}_U^{fS}\boldsymbol{U}^S]. \tag{14.5.5}$$

However, we still have to take account of the contribution from multiples generated at the free surface. Each successive reflection above and below the source level requires the application of the combination $\mathbf{R}_D^{SL}\mathbf{R}_U^{fS}$ representing reflection above and below the level of the source.

For the seismic disturbance $\boldsymbol{w}_o$ at the surface, this gives a propagation sequence including free-surface multiples of the form

$$\begin{aligned}\boldsymbol{w}_o &= \mathbf{W}_U^{fS}\boldsymbol{U}^S \\ &\quad + \mathbf{W}_U^{fS}\left(\mathbf{I} + \mathbf{R}_D^{SL}\mathbf{R}_U^{fS} + \mathbf{R}_D^{SL}\mathbf{R}_U^{fS}\mathbf{R}_D^{SL}\mathbf{R}_U^{fS} + \dots\right)\mathbf{R}_D^{SL}[\boldsymbol{D}^S + \mathbf{R}_U^{fS}\boldsymbol{U}^S],\end{aligned} \tag{14.5.6}$$

We can recognise the reverberation sequence in terms of an inverse operator and so we can produce the formal expression

$$\boldsymbol{w}_o = \mathbf{W}_U^{fS}\boldsymbol{U}^S + \mathbf{W}_U^{fS}\left(\mathbf{I} - \mathbf{R}_D^{SL}\mathbf{R}_U^{fS}\right)^{-1}\mathbf{R}_D^{SL}[\boldsymbol{D}^S + \mathbf{R}_U^{fS}\boldsymbol{U}^S]. \tag{14.5.7}$$

The inverse operator $(\mathbf{I} - \mathbf{R}_D^{SL}\mathbf{R}_U^{fS})^{-1}$ includes all multiple interactions between the regions above and below the source. In the representation (14.5.7) for the wavefield we have the reverberation operator for the entire structure combined with reflection from below the source level. Since we have the operator identity

$$\left(\mathbf{I} - \mathbf{R}_D^{SL}\mathbf{R}_U^{fS}\right)^{-1}\mathbf{R}_D^{SL} = \mathbf{R}_D^{SL}\left(\mathbf{I} - \mathbf{R}_U^{fS}\mathbf{R}_D^{SL}\right)^{-1}, \tag{14.5.8}$$

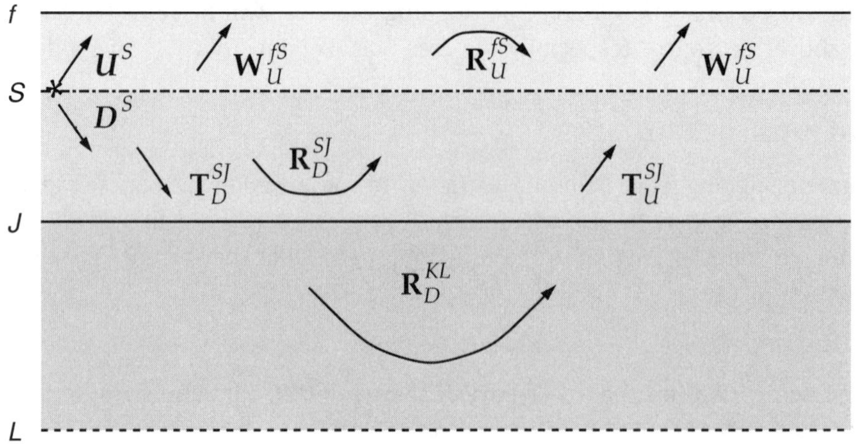

Figure 14.6. Schematic presentation of the action of reflection and transmission operators for the representation of the wavefield with the introduction of a separation surface at the level J below the source level S.

we can view the reverberation sequence as occurring on either the source or receiver sides of the main reflection $\mathbf{R}_D^{SL}$.

For a surface source, we can envisage the level S brought right up to the surface so that $\boldsymbol{D}^S + \mathbf{R}_U^{fS}\boldsymbol{U}^S$ will just be the net downward radiation from the source.

14.5.2 Components of the seismic wavefield

Although (14.5.7) gives a complete description of the surface disturbance due to the source it still does not meet all of our needs. In many practical situations the source lies in a zone of relatively low seismic wave velocities underlain by material with higher wave velocities. This is the case for most seismic prospecting situations and, on a larger scale, for sources in the earth's crust. Since the principal interest in seismology is on energy returned by the structure of the Earth at depth, we would like to separate the effects of propagation in the shallow waveguide from the energy which has penetrated deep into the Earth.

14.5.2.1 Shallow sources

We introduce an interface J which lies below all the low-velocity material, and will show how we can split the representation (14.5.7) for the wavefield into a number of pieces with a clear physical interpretation (figure 14.6). We can begin to separate the portion of the seismic wavefield which is confined to the upper part of the model by splitting the reflection matrix $\mathbf{R}_D^{SL}$ at the separation level J using (14.2.11):

$$\mathbf{R}_D^{SL} = \mathbf{R}_D^{SJ} + \mathbf{T}_U^{SJ}\left(\mathbf{I} - \mathbf{R}_D^{JL}\mathbf{R}_U^{SJ}\right)^{-1}\mathbf{R}_D^{JL}\mathbf{T}_D^{SJ}. \tag{14.5.9}$$

We will write (14.5.9) temporarily in a form in which we isolate the part which includes the reflection from beneath J:

$$\mathbf{R}_D^{SL} = \mathbf{R}_D^{SJ} + \mathbf{T}_U^{SJ}\mathbf{R}_D^{KL}\mathbf{T}_D^{SJ}. \tag{14.5.10}$$

The operator $\mathbf{R}_D^{KL}$ includes all deep reflection effects, both directly through $\mathbf{R}_D^{JL}$ and also any modulations due to multiples in the zone beneath the source level through $(\mathbf{I} - \mathbf{R}_D^{JL}\mathbf{R}_U^{SJ})^{-1}$.

We would like to make use of the expansion (14.5.9) to separate out multiple reflections in the zone above J from the full reverberation operator $(\mathbf{I} - \mathbf{R}_D^{SL}\mathbf{R}_U^{fS})^{-1}$ which can be written as

$$\left(\mathbf{I} - \mathbf{R}_D^{SJ}\mathbf{R}_U^{fS} - \mathbf{T}_U^{SJ}\mathbf{R}_D^{KL}\mathbf{T}_D^{SJ}\mathbf{R}_U^{fS}\right)^{-1}. \tag{14.5.11}$$

We now make use of the operator identity

$$(\mathbf{I} - \mathbf{A} - \mathbf{B})^{-1} = (\mathbf{I} - \mathbf{A})^{-1} + (\mathbf{I} - \mathbf{A})^{-1}\mathbf{B}(\mathbf{I} - \mathbf{A} - \mathbf{B})^{-1}, \tag{14.5.12}$$

which leads to an expansion in multiple powers of the operator combination $(\mathbf{I} - \mathbf{A})^{-1}\mathbf{B}(\mathbf{I} - \mathbf{A})^{-1}$.

With the identification of the terms associated with shallow propagation with $\mathbf{A}$,

$$\mathbf{A} = \mathbf{R}_D^{SJ}\mathbf{R}_U^{fS}, \tag{14.5.13}$$

and the terms involving deep propagation with $\mathbf{B}$,

$$\mathbf{B} = \mathbf{T}_U^{SJ}\mathbf{R}_D^{KL}\mathbf{T}_D^{SJ}\mathbf{R}_U^{fS} = \mathbf{T}_U^{SJ}\left(\mathbf{I} - \mathbf{R}_D^{JL}\mathbf{R}_U^{SJ}\right)^{-1}\mathbf{R}_D^{JL}\mathbf{T}_D^{SJ}\mathbf{R}_U^{fS}, \tag{14.5.14}$$

we can develop an expansion of the full reverberation operator to separate propagation in the shallower zone above J. Thus,

$$\begin{aligned}\left(\mathbf{I} - \mathbf{R}_D^{SL}\mathbf{R}_U^{fS}\right)^{-1} &= \left(\mathbf{I} - \mathbf{R}_D^{SJ}\mathbf{R}_U^{fS}\right)^{-1} \\ &\quad + \left(\mathbf{I} - \mathbf{R}_D^{SJ}\mathbf{R}_U^{fS}\right)^{-1}\mathbf{T}_U^{SJ}\mathbf{R}_D^{KL}\mathbf{T}_D^{SJ}\mathbf{R}_U^{fS}\left(\mathbf{I} - \mathbf{R}_D^{SL}\mathbf{R}_U^{fS}\right)^{-1},\end{aligned} \tag{14.5.15}$$

and we can recognise $(\mathbf{I} - \mathbf{R}_D^{SJ}\mathbf{R}_U^{fS})^{-1}$ as the reverberation operator for the upper zone.

With the introduction of the separation level J and the splitting up of the reflections from beneath the source in (14.5.9), we are able to expand the representation for the wavefield (14.5.7) into a form where we can recognise different classes of wave interaction. The surface field

$$\boldsymbol{w}_0 = \mathbf{W}_U^{fS}\boldsymbol{U}^S \tag{14.5.16}$$

$$+\mathbf{W}_U^{fS}\left(\mathbf{I} - \mathbf{R}_D^{SJ}\mathbf{R}_U^{fS}\right)^{-1}\mathbf{R}_D^{SJ}[\boldsymbol{D}^S + \mathbf{R}_U^{fS}\boldsymbol{U}^S] \tag{14.5.17}$$

$$+\mathbf{W}_U^{fS}\left(\mathbf{I} - \mathbf{R}_D^{SJ}\mathbf{R}_U^{fS}\right)^{-1}\mathbf{T}_U^{SJ}\left(\mathbf{I} - \mathbf{R}_D^{JL}\mathbf{R}_U^{fJ}\right)^{-1}\mathbf{R}_D^{JL}\mathbf{T}_D^{SJ}\left(\mathbf{I} - \mathbf{R}_U^{fS}\mathbf{R}_D^{SJ}\right)^{-1}[\boldsymbol{D}^S + \mathbf{R}_U^{fS}\boldsymbol{U}^S]. \tag{14.5.18}$$

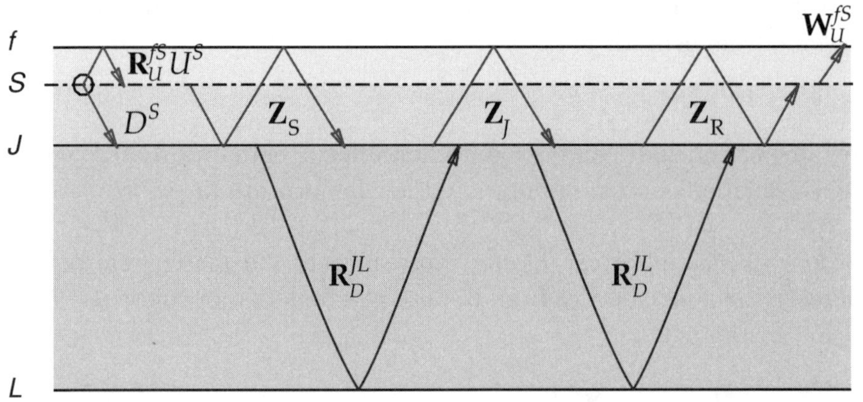

Figure 14.7. The wave propagation phenomena associated with reflection from depth can be isolated by splitting the response of the model at a surface J lying below the source and the near-surface zone of low velocity.

The first two terms in (14.5.16)–(14.5.18) represent propagation confined to the region above the separation level J (note that they have the form of (14.5.7) with L replaced by J). The contribution (14.5.16) corresponds to the direct wave transmitted upward from the source level. The part (14.5.17) represents energy travelling in the upper zone with multiple interaction with the free surface and the increase in velocity with depth. The shallow propagating disturbances represented by (14.5.17) will appear to have horizontal propagation paths when recorded at larger offsets. Such arrivals are shallowly propagating surface waves, as well as wide angle reflections from shallow reflectors.

All reflections from beneath the separation level J are included in the expression (14.5.18) which is presented in a diagrammatic form in figure 14.7. We can recognise the effective source term including the surface ghost, $\boldsymbol{D}^S + \mathbf{R}_U^{fS} \boldsymbol{U}^S$, and the transmission term from the source level to the surface $\mathbf{W}_U^{fS}$ as representing the first and last stages of the propagation process. The representation (14.5.18) for the reflection wavefield includes the possibility of reverberations in the shallow zone between J and the surface in the neighbourhood of the source and the receiver (indicated by $\mathbf{Z}_S$ and $\mathbf{Z}_R$ in figure 14.7). The remaining part of (14.5.18),

$$\mathbf{T}_U^{SJ} \left(\mathbf{I} - \mathbf{R}_D^{JL} \mathbf{R}_U^{fJ}\right)^{-1} \mathbf{R}_D^{JL} \mathbf{T}_D^{SJ}, \tag{14.5.19}$$

provides a description of transmission between the source level S and the separation surface J and the reflections from below J. The expression (14.5.19) includes internal multiples, and long-lag surface multiples involving paths with more than one reflection from beneath J.

We can therefore take (14.5.18) as the formal representation of the portion of the seismic wavefield which has propagated to depth, and we will rewrite this in a compact form that emphasise the nature of the processes:

$$\boldsymbol{w}_o^{\mathrm{de}} = \mathbf{W}_U^{fS}\mathbf{Z}_R\mathbf{T}_U^{SJ}\mathbf{Z}_J\mathbf{R}_D^{JL}\mathbf{T}_D^{SJ}\mathbf{Z}_S[\boldsymbol{D}^S + \mathbf{R}_U^{fS}\boldsymbol{U}^S]. \tag{14.5.20}$$

The shallow reverberation operators

$$\begin{aligned} \mathbf{Z}_{\mathrm{R}} &= (\mathbf{I} - \mathbf{R}_D^{SJ}\mathbf{R}_U^{fS})^{-1}, \\ \mathbf{Z}_{\mathrm{S}} &= (\mathbf{I} - \mathbf{R}_U^{fS}\mathbf{R}_D^{SJ})^{-1}, \end{aligned} \tag{14.5.21}$$

represent the interaction of the seismic waves with shallow structure. Reflection processes involving multiple reflections between the free surface and the structure at depth are contained in

$$\mathbf{Z}_J = (\mathbf{I} - \mathbf{R}_D^{JL}\mathbf{R}_U^{fJ})^{-1}. \tag{14.5.22}$$

Although the expressions (14.5.16)-(14.5.18) are valid for any separation surface J lying beneath the source, it is advantageous to choose this surface to achieve the maximum separation between the different classes of propagation. For sources in the crust, it is therefore desirable to take J a little below the crust-mantle interface so that reflections from the boundary are included in the shallow terms, even though this means that $\mathbf{R}_U^{fJ}$ itself will include contributions at the crust-mantle boundary as well as dominantly from the free surface.

14.5.2.2 Sources at depth

For a deeper source lying below the separation level z_J a convenient form for the respect is obtained by direct use of (14.5.7), so that

$$\begin{aligned} \boldsymbol{w}_o &= \mathbf{W}_U^{fS}\boldsymbol{U}^S && (14.5.23) \\ &\quad + \mathbf{W}_U^{fS}\left(\mathbf{I} - \mathbf{R}_D^{SL}\mathbf{R}_U^{fS}\right)^{-1}\mathbf{R}_D^{SL}[\boldsymbol{D}^S + \mathbf{R}_U^{fS}\boldsymbol{U}^S]. && (14.5.24) \end{aligned}$$

(14.5.23) represents the direct transmission to the surface and (14.5.24) the combination of the direct phases and their accompanying surface reflections, which are reflected back from below the source level as represented by the action of the reflection term $\mathbf{R}_D^{SL}$. The full sequence of surface multiples emerges in the expanded form

$$\left(\mathbf{I} - \mathbf{R}_D^{SL}\mathbf{R}_U^{fS}\right)^{-1}\mathbf{R}_D^{SL} = \mathbf{I} + \mathbf{R}_D^{SL}\mathbf{R}_U^{fS}\mathbf{R}_D^{SL} + \mathbf{R}_D^{SL}\mathbf{R}_U^{fS}\mathbf{R}_D^{SL}\mathbf{R}_U^{fS}\mathbf{R}_D^{SL} + \dots. \tag{14.5.25}$$

Reverberations in the shallow structure are included in the transfer operator to the surface $\mathbf{W}_U^{fS}$. When we make a separation at the level z_J,

$$\mathbf{W}_U^{fS} = \mathbf{W}_U^{fJ}(\mathbf{I} - \mathbf{R}_D^{JS}\mathbf{R}_U^{fJ})^{-1}\mathbf{T}_U^{JS}, \tag{14.5.26}$$

and from (14.3.5), (14.3.6)

$$\mathbf{W}_U^{fJ} = \mathbf{W}_F(\mathbf{I} - \mathbf{R}_D^{0J}\mathbf{R}_F)^{-1}\mathbf{T}_U^{0J},$$
$$\mathbf{R}_U^{fJ} = \mathbf{R}_D^{0J} + \mathbf{T}_D^{0J}(\mathbf{I} - \mathbf{R}_D^{0J}\mathbf{R}_F)^{-1}\mathbf{T}_U^{0J}. \qquad (14.5.27)$$

It is convenient to take the separation level z_J to lie below the zone of lowered wavespeeds near the surface, for example, just below the base of the crust. In this case $\mathbf{R}_D^{JS}$ will generally be small, although reflection may be associated with mantle discontinuities, However, $\mathbf{R}_D^{0J}$ can be much larger because of the reflections from the base of the surface zone, leading to significant reverberations represented through $(\mathbf{I} - \mathbf{R}_D^{0J}\mathbf{R}_F)^{-1}$.

An alternative formulation for the response can be made by working with a set of equivalent sources at the level z_J which generate the same wavefield as the true source at z_S (Kennett, 1983, §9.2):

$$\boldsymbol{w}_o = \mathbf{W}_U^{fJ}\left(\mathbf{I} - \mathbf{R}_D^{JL}\mathbf{R}_U^{fJ}\right)^{-1}\left\{\mathbf{T}_U^{JS}\left(\mathbf{I} - \mathbf{R}_D^{SL}\mathbf{R}_U^{JS}\right)^{-1}[\boldsymbol{U}^S + \mathbf{R}_D^{SL}\boldsymbol{D}^S]\right\}. \qquad (14.5.28)$$

The terms in braces can be reorganised to emphasise the upward transmitted portion and the waves reflected from depth

$$\boldsymbol{w}_o = \mathbf{W}_U^{fJ}\left(\mathbf{I} - \mathbf{R}_D^{JL}\mathbf{R}_U^{fJ}\right)^{-1}\left\{\mathbf{T}_U^{JS}\boldsymbol{U}^S + \mathbf{T}_U^{JS}\left(\mathbf{I} - \mathbf{R}_D^{SL}\mathbf{R}_U^{JS}\right)^{-1}\mathbf{R}_D^{SL}[\boldsymbol{D}^S + \mathbf{R}_U^{JS}\boldsymbol{U}^S]\right\}, \qquad (14.5.29)$$

but unlike (14.5.24) we cannot make a direct identification of the 'depth phases' reflected at the surface.

14.6 Approximations to the response

We will illustrate the way in which we can develop approximations for the seismic response by using the representation (14.5.16)-(14.5.18) for a shallow source with a split of the model at a level z_J:

$$\boldsymbol{w}_o = \mathbf{W}_U^{fS}\boldsymbol{U}^S + \mathbf{W}_U^{fS}(\mathbf{I} - \mathbf{R}_D^{SJ}\mathbf{R}_U^{fS})^{-1}\mathbf{R}_D^{SJ}[\boldsymbol{D}^S + \mathbf{R}_U^{fS}\boldsymbol{U}^S] \qquad (s)$$
$$+\mathbf{W}_U^{fS}(\mathbf{I} - \mathbf{R}_D^{SJ}\mathbf{R}_U^{fS})^{-1}\mathbf{T}_U^{SJ}(\mathbf{I} - \mathbf{R}_D^{JL}\mathbf{R}_U^{fJ})^{-1}\mathbf{R}_D^{JL}\mathbf{T}_D^{SJ}(\mathbf{I} - \mathbf{R}_U^{fS}\mathbf{R}_D^{SJ})^{-1}[\boldsymbol{D}^S + \mathbf{R}_U^{fS}\boldsymbol{U}^S]. \qquad (d)$$

When we construct such approximations we have to be guided by the nature of the wavefield and the particular wavespeed model which is being employed. There are three classes of approximation:

1. definition of the class of physical processes which are included in the response;
2. choice of the dominant wavetype to be followed;
3. the use of numerical integration over restricted integrals in slowness to produce results appropriate to particular distance ranges.

We will consider the first and second class of approximation here and return to the third in Section 16.2.

14.6.1 Choice of propagation processes

The selection of particular classes of propagation processes can be used to try to understand the way in which different features of the wave propagation contribute to the seismogram, or alternatively, to avoid extraneous features and so reduce the computational effort required for calculating a theoretical seismogram.

With the representation (s), (d) for the wavefield response we have a hierarchy of approximations depending on how we treat the upper and lower zones of the model.

14.6.1.1 Inclusion of the shallow response

The terms (s), $\mathbf{W}_U^{fS}\boldsymbol{U}^S + \mathbf{W}_U^{fS}(\mathbf{I} - \mathbf{R}_D^{SJ}\mathbf{R}_U^{fS})^{-1}\mathbf{R}_D^{SJ}[\boldsymbol{D}^S + \mathbf{R}_U^{fS}\boldsymbol{U}^S]$, correspond to propagation purely within the zone above z_J. This region typically would have reduced wavespeed compared with the deeper portions of the model. In consequence the wavetrains associated with (s) are composed of multiple reflections with considerable mutual interference. Such guided waves, trapped between the free surface and the increasing wavespeeds near z_J, can build up to considerable amplitude and play an important role on seismograms. Large "ground-roll" trains are commonly seen in exploration work with surface sources and can have multiple components. Guided waves in the crust (*Pg*, *Sg*, *Lg*) are also important in short-period seismograms at regional ranges out to 1200 km or beyond.

When the offset between the source and receiver is of the same order as the total depth of the layering the shallow response terms (s) should be retained in the representation of the response. However, if the receivers are well separated from the source, the relatively low group velocities of the guided phases mean that the time intervals, which include arrivals from depth as well as shallow propagation, will need to be rather long. It can therefore be convenient to evaluate the shallow contribution separately and add it to the final seismogram.

14.6.1.2 Shallow reverberations

The reverberation operator $(\mathbf{I} - \mathbf{R}_D^{SJ}\mathbf{R}_U^{fS})^{-1}$ for the zone fSJ plays an important role in the response $\boldsymbol{w}_0$ because it appears in both the near-receiver and near-source reverberations as well as being involved with multiple deep reflections.

We can make a number of different approximations for the behaviour in the upper zone:

(i) Modification of the surface boundary condition: if we make the surface transparent, $\mathbf{R}_F = 0$, and so no free surface reflections are included; $\mathbf{R}_U^{fS}$ then only includes the effects of internal boundaries. The surface amplification corrections $\mathbf{W}_U^{fS}$ can be applied if desired to the modified response. Similarly, depth phases from the source would be suppressed.

(ii) Free-surface reflections are included but shallow reverberations are suppressed: the reverberation term for fSJ is replaced by the identity matrix; this leaves surface reflections near the source for the depth phases so that (d) becomes

$$\boldsymbol{w}_o^{d1} = \mathbf{W}_U^{fS}\mathbf{T}_U^{SJ}(\mathbf{I} - \mathbf{R}_D^{JL}\mathbf{R}_U^{fJ})^{-1}\mathbf{R}_D^{JL}\mathbf{T}_D^{SJ}[\boldsymbol{D}^S + \mathbf{R}_U^{fS}\boldsymbol{U}^S]. \tag{14.6.1}$$

(iii) Allowance for free-surface multiples: The reverberation operator for the zone fSJ is reinstated and multiple reflections in the upper zone are included. If the full numerical inverse $(\mathbf{I}-\mathbf{R}_D^{SJ}\mathbf{R}_U^{fS})^{-1}$ is employed then all shallow reverberations are included. In perfectly elastic models or where Q is high these reverberations will often decay rather slowly and so it can be difficult to avoid aliasing in time in the final seismograms. Such problems can be minimised by using a truncated expansion of the inverse. The simplest form is

$$(\mathbf{I} - \mathbf{R}_D^{SJ}\mathbf{R}_U^{fS})^{-1} \approx \mathbf{I} + \mathbf{R}_D^{SJ}\mathbf{R}_U^{fS}, \tag{14.6.2}$$

where one extra surface reflection is included at each appearance of the inverse. The approximation (14.6.2) needs no extra computation since the matrix product is needed to construct the full inverse.

14.6.1.3 *Multiple reflections from depth*

The contribution (d) to the response includes multiple reflections with reflection from below z_J through the reverberation operator $(\mathbf{I} - \mathbf{R}_D^{JL}\mathbf{R}_U^{fJ})^{-1}$. As for shallow multiples we can control the style of multiples both by changing the surface boundary condition and by varying the expansion level of the operator.

When attention is to be concentrated on direct P or S phases, the deep multiple term can be dropped to give the approximation

$$\boldsymbol{w}_o^{de} = \mathbf{W}_U^{fS}(\mathbf{I} - \mathbf{R}_D^{SJ}\mathbf{R}_U^{fS})^{-1}\mathbf{T}_U^{SJ}\mathbf{R}_D^{JL}\mathbf{T}_D^{SJ}(\mathbf{I} - \mathbf{R}_U^{fS}\mathbf{R}_D^{SJ})^{-1}[\boldsymbol{D}^S + \mathbf{R}_U^{fS}\boldsymbol{U}^S]. \tag{14.6.3}$$

This form of the response is very suitable for the generation of teleseismic records and will be considered further in section 16.1. The original *reflectivity* method of Fuchs & Müller (1971) was based on (14.6.3) with only transmission through the upper zone,

$$\boldsymbol{w}_o^{re} = \mathbf{W}_U^{fS}\mathbf{T}_U^{SJ}\mathbf{R}_D^{JL}\mathbf{T}_D^{SJ}\boldsymbol{D}^S. \tag{14.6.4}$$

When the surface multiples are to be included, we have to recognise the long delays associated with the reflection process from beneath z_J. For many purposes it is sufficient to use the truncation of the multiple sequence after one surface interaction, so that the terms associated with reflection from below z_J are

$$\mathbf{R}_D^{JL} + \mathbf{R}_D^{JL}\left(\mathbf{R}_U^{SJ} + \mathbf{T}_D^{SJ}\mathbf{R}_U^{fS}[\mathbf{I} - \mathbf{R}_D^{SJ}\mathbf{R}_U^{fS}]^{-1}\mathbf{T}_U^{SJ}\right)\mathbf{R}_D^{JL}, \tag{14.6.5}$$

where we have expanded $\mathbf{R}_U^{fJ}$ in terms of the shallow reverberation operator. The first term in (14.6.5) includes the P and S phases. The second terms contains *PP, PS, SP* and *SS* as well as possible shallow interactions depending on the treatment of the operator for the zone fSJ and the choice of the separation depth z_J.

14.6.1.4 Control of internal reflections

The recursive schemes used to construct the reflection and transmission response of regions from individual layer contributions can be used to provide controls on the styles of internal multiples which are included in the response. Consider a generic reflection system of the form (14.2.22),

$$\mathbf{R} = \mathbf{R}^C + \mathbf{T}^B\mathbf{R}^B[\mathbf{I} - \mathbf{R}^A\mathbf{R}^B]^{-1}\mathbf{T}^C. \qquad (14.6.6)$$

If we choose to suppress the interaction between A and B we get

$$\mathbf{R} = \mathbf{R}^C + \mathbf{T}^B\mathbf{R}^B\mathbf{T}^C; \qquad (14.6.7)$$

whereas if we allow a single interaction with A we have

$$\mathbf{R} = \mathbf{R}^C + \mathbf{T}^B\mathbf{R}^B[\mathbf{I} + \mathbf{R}^A\mathbf{R}^B]\mathbf{T}^C. \qquad (14.6.8)$$

These approximate forms can be very useful when it is necessary to understand the provenance of a particular feature in the seismic wavetrain. By suppressing different classes of propagation process and examining their influence on the seismogram it is often possible to isolate the feature in the model responsible for the feature.

14.6.2 Choice of wavetype

We have cast the expression for the surface response in a general matrix form. For *SH* waves in isotropic media, the results can be applied directly in terms of the *HH* reflection and transmission coefficients. For the *P-SV* system, or in the more general case of anisotropy, the expressions require the multiplication of 2×2 or 3×3 matrices. It is possible to develop specific approximations which concentrate on one wavetype with limited allowance for conversion (14.2.21), (14.2.23). However, when we wish to maintain maximum flexibility it can be more convenient to use the full matrix development for the propagation terms and impose a choice of wavetype at the source and receiver.

14.6.2.1 Source radiation

The upgoing and downgoing radiation terms $\boldsymbol{U}^S(p, \omega)$ and $\boldsymbol{D}^S(p, \omega)$ can be split into a contribution for each wavetype. The most general form is to include *P*, *SV* and *SH* terms.

Frequently, however, we know what are the propagation paths of particular interest and so can make a specific choice of wavetype. However the choice must be made carefully. For teleseismic *P* waves the obvious contributions are the direct wave and the *pP* surface reflection, but for earthquake sources a very prominent phase is *sP* which leaves the source upwards as *S* and is converted to *P* by reflection at the free surface. Although less common, conversion to *P* can also arise from downgoing waves.

It is therefore often most effective to make a wavetype selection from the composite source term

$$\boldsymbol{D}^S + \mathbf{R}_U^{fS}\boldsymbol{U}^S = \begin{pmatrix} [\boldsymbol{D}^S]_P + [\mathbf{R}_U^{fS}]_{PP}[\boldsymbol{U}^S]_P + [\mathbf{R}_U^{fS}]_{PS}[\boldsymbol{U}^S]_S \\ [\boldsymbol{D}^S]_S + [\mathbf{R}_U^{fS}]_{SS}[\boldsymbol{U}^S]_S + [\mathbf{R}_U^{fS}]_{SP}[\boldsymbol{U}^S]_P \\ [\boldsymbol{D}^S]_H + [\mathbf{R}_U^{fS}]_{HH}[\boldsymbol{U}^S]_H \end{pmatrix}. \quad (14.6.9)$$

The choice of wave type can help in the analysis of a seismic record, by allowing the association of different features and phases with the particular wavetypes radiated from the source.

14.6.2.2 Receiver response

The free-surface amplification matrix $\mathbf{W}_F$ acts on the upgoing wavefield to generate the vertical and horizontal components of displacement. The elements in one column of the matrix correspond to a particular wavetype, (13.1.21),

$$\begin{pmatrix} w_Z \\ w_R \\ w_T \end{pmatrix} = \begin{pmatrix} W_{ZP} & W_{ZS} & 0 \\ W_{RP} & W_{RS} & 0 \\ 0 & 0 & W_{TH} \end{pmatrix} \begin{pmatrix} P_U \\ S_U \\ H_U \end{pmatrix}. \quad (14.6.10)$$

Thus by choosing one column of $\mathbf{W}_F$ attention can be restricted to a particular wavetype at the receiver and so a particular group of propagation paths.

14.6.2.3 Deep reflections

When free surface reflections are ignored as in (14.6.3), the reflection matrix $\mathbf{R}_D^{JL}$ from the lower zone appears only once in the approximate response. The dominant mode of propagation below z_J can then be imposed by selecting only one reflection coefficient, such as $[\mathbf{R}_D^{JL}]_{SS}$ from the matrix, which will still include internal conversions. Alternatively if attention is concentrated on P waves we can exclude the SS coefficient and thereby avoid the long time delays associated with purely S wave propagation.

14.7 Generalized rays

We have seen how we can assemble the response of a medium from the reflection and transmission properties of portions of the stratification, and also generate approximations based on this representation.

For models which are composed of uniform layers and for piecewise smooth models, in a high frequency approximation, we can go further and express the total response as a sum of terms of the form

$$\boldsymbol{w}_o(p, m, \omega) = M(\omega) \sum_I \boldsymbol{g}_I(p, m) \exp[i\omega\tau_I(p)]. \quad (14.7.1)$$

We have assumed a common source spectrum $M(\omega)$ and characterised each of the 'generalized rays' in the sum (14.7.1) by an amplitude factor $\boldsymbol{g}_I(p,m)$ and a phase delay term $\tau_I(p)$. The expression $\boldsymbol{g}_I(p,m)$ factors into two parts. The first part $\boldsymbol{f}_I(p,m)$ represents the way in which the source and receiver terms depend on slowness p. The dependence on angular order m enters from the source; the receiver terms would include free-surface amplification factors. The second part is the product of all reflection and transmission coefficients for slowness p encountered along the ray path. Thus

$$\boldsymbol{g}_I(p) = \boldsymbol{f}_I(p,m) \prod_j \mathrm{T}_j \prod_k \mathrm{R}_k, \tag{14.7.2}$$

where T_j is the transmission coefficient at slowness p for an interface traversed by the ray, and R_k is the reflection coefficient for a situation where the ray changes direction. Reflection may be at an interface, or alternatively by total reflection from a gradient zone for which the reflection coefficient would be $\exp(-\mathrm{i}\pi/2)$ (14.2.49), and the phase delay associated with the turning leg would appear in $\tau_I(p)$.

The accumulated phase delay for the Ith ray,

$$\tau_I(p) = \sum_r \Big\{ n_r \int_{z_r}^{z_{r+1}} \mathrm{d}\zeta\, q_r(p,\zeta) + 2n_r^* \int_{z_r}^{Z_r^*} \mathrm{d}\zeta\, q_r(p,\zeta) \Big\} \tag{14.7.3}$$

where the sum is taken over all the layers $\{r\}$ traversed by the ray. Legs with different wavetypes are counted as distinct. The second term in the sum in (14.7.3) only arises for gradient zones: n_r^* is the number of legs with a turning point at level $Z_r^*(p)$ in layer r.

14.7.1 Generalized Ray Expansion

We can extract the generalized ray expansion from the full response by representing each reflection and transmission element in terms of the complete set of propagation processes for each wavetype. The full response can be written in the form

$$\boldsymbol{w}_0 = \mathbf{W}_U^{fS} \left(\mathbf{I} - \mathbf{R}_D^{SL}\mathbf{R}_U^{fS}\right)^{-1} \mathbf{R}_D^{SL}[\boldsymbol{D}^S + \mathbf{R}_U^{fS}\boldsymbol{U}^S] + \mathbf{W}_U^{fS}\boldsymbol{U}^S, \tag{14.7.4}$$

and each of the free surface terms has a further expansion isolating the free surface reflection,

$$\begin{aligned} \mathbf{W}_U^{fS} &= \mathbf{W}_F(\mathbf{I} - \mathbf{R}_D^{0S}\mathbf{R}_F)^{-1}\mathbf{T}_U^{0S}, \\ \mathbf{R}_U^{fS} &= \mathbf{R}_U^{OS} + \mathbf{T}_D^{0S}\mathbf{R}_U^{0S}(\mathbf{I} - \mathbf{R}_D^{0S}\mathbf{R}_F)^{-1}\mathbf{T}_U^{0S}. \end{aligned} \tag{14.7.5}$$

The reflection for the region below the source, $\mathbf{R}_D^{SL}$ will take the form of a cascaded set of contributions from the different layers, e.g., with intermediate interfaces at levels F,G:

$$\begin{aligned} \mathbf{R}_D^{SL} &= \mathbf{R}_D^{SF} + \mathbf{T}_U^{SF}\mathbf{R}_D^{FL}(\mathbf{I} - \mathbf{R}_U^{SF}\mathbf{R}_D^{FL})^{-1}\mathbf{T}_D^{SF}, \\ \mathbf{R}_D^{FL} &= \mathbf{R}_D^{FG} + \mathbf{T}_U^{FG}\mathbf{R}_D^{GL}(\mathbf{I} - \mathbf{R}_U^{FG}\mathbf{R}_D^{GL})^{-1}\mathbf{T}_D^{FG}. \end{aligned} \tag{14.7.6}$$

Each of the sub-elements can then be further decomposed into the contribution from individual layers. The reverberation operators represented by the inverse terms derive from the infinite sequence of multiple reflections (14.2.12), and concentrating on a single wavetype there is a further set of expansions. Consider, for example, the representation for the *PP* reflection coefficient introduced in Section 14.2

$$\mathrm{R}_{PP} = \mathrm{R}^{C}_{PP} + \mathrm{T}^{B}_{PP}\mathrm{R}^{B}_{PP}[\mathrm{I} - \mathrm{R}^{A}_{PP}\mathrm{R}^{B}_{PP}]^{-1}\mathrm{T}^{C}_{PP}, \tag{14.7.7}$$

we can expand the P reverberation operator in the series

$$\mathrm{R}_{PP} = \mathrm{R}^{C}_{PP} + \mathrm{T}^{B}_{PP}\mathrm{R}^{B}_{PP}[\mathrm{I} + \mathrm{R}^{A}_{PP}\mathrm{R}^{B}_{PP} + \mathrm{R}^{A}_{PP}\mathrm{R}^{B}_{PP}\mathrm{R}^{A}_{PP}\mathrm{R}^{B}_{PP} + ...]\mathrm{T}^{C}_{PP}. \tag{14.7.8}$$

At the individual layer level the reflection and transmission terms have the functional form $h(p)\exp[i\omega y(p)]$ where h is a reflection or transmission coefficient and y an associated phase delay. This form is exact for a medium composed of a stack of uniform layers and represents a high frequency approximation for smooth gradient zones. When the full set of cascading propagation processes are combined with the layer representation we obtain an infinite sequence of generalized ray terms of the form (14.7.1).

However, the merit of the expansion is that attention can be directed to a subset of the generalized rays which represent a particular phenomena. In general, the significance of higher order multiple reflection terms diminishes. A reasonable criterion for the number of terms L to be retained to achieve an accuracy level ϵ in a expansion such as (14.7.8) is that

$$[\bar{\mathrm{R}}^{A}\bar{\mathrm{R}}^{B}]^{L} \le \epsilon, \tag{14.7.9}$$

where $\bar{\mathrm{R}}^{A}$, $\bar{\mathrm{R}}^{B}$ are the absolute values of the largest reflection coefficients in $\mathbf{R}^{A}$, $\mathbf{R}^{B}$ (Kennett, 1974). When incidence on a boundary is near grazing, $\bar{\mathrm{R}}^{B}$ would approach unity and for high accuracy L should be quite large (Müller, 1970). This poses a problem when a stack of uniform layers is used to simulate a gradient zone, since a turning ray is represented by the superposition of sequences of multiple reflections at near grazing incidence in the layers close to the turning level. In order to keep the number of generalized rays within reasonable limits, often only the first order multiples are retained (see e.g. Helmberger, 1968).

When the generalized ray expansion is used for smooth gradient zones we have to recognise the limitations of the high frequency approximation. The generalized ray expansion will not represent the full behaviour when turning points lie close to the boundaries of a gradient zone, since this requires the use of the Airy function forms. Further, the restriction to real phase delays means that tunnelling phenomena into low velocity zones such as the Earth's core (Richards, 1973) cannot be described by the expansion (14.7.1).

14.7.2 Enumerating generalized rays

The nested expansions of (14.7.6) are cumbersome and commonly rays are generated directly by enumeration of possible propagation paths. This process needs to be undertaken with care. Systematic methods exist for generalized rays for a single wavetype (see e.g Hron, 1972; Vered & BenMenahem, 1974). However, when wavetype conversion is allowed the computational mathematics is rather complicated. The solution proposed by Vered & BenMenahem (1974) is to specify the interfaces at which conversion can occur. Rays are constructed to these interfaces, and then the ray generation process is started again with the converted waves.

Commonly many different paths through the model will have the same total phase delay $\tau_k(p)$, because they traverse the same group of layers the same number of times. Those rays with the same phase delay $\tau_k(p)$ are termed members of the same *kinematic group* $\{k\}$. Because the sequence of layer transits will differ, $\boldsymbol{g}_I(p)$ will not be the same for all these rays. But, there are normally subgroups within $\{k\}$ which have a common amplitude term $\boldsymbol{g}_I(p)$ and form a *dynamic group* $\{d\}$. The ray sum can then be written as

$$\boldsymbol{w}_o(p, m, \omega) = M(\omega) \sum_k \Big\{ \sum_d N_{dk} \boldsymbol{g}_d(p, m) \Big\} \exp[i\omega\tau_k(p)], \tag{14.7.10}$$

in terms of the multiplicity of elements N_{dk} in the dynamic group $\{d\}$. The frequency dependent portions are then the same for each kinematic group, the inner sum over dynamic groups accounts for different propagation processes with the same phase delays. For accurate results using generalized rays it is important that N_{dk} is correctly specified; enumerating N_{dk} can be a significant undertaking when multiple internal reflection processes are important.

15

Constructing the Wavefield

In the previous sections we have established the properties of plane elastic waves and examined the way in which such waves interact with contrasts in elastic properties. In this chapter, we show how the full wavefield can be represented in terms of plane wave contributions which have a direct physical significance for stratified media.

The representation of the wavefield is extended to a cylindrical coordinate system for plane stratification, where the response is built from cylindrical waves characterised by horizontal slowness p modulated by an angular dependence imposed by the source. The cylindrical coordinate system is commonly used in the synthesis of the seismic wavefield and so we indicate how the physical wavefield can be extracted from the frequency-slowness response.

For spherical stratification, the wavefield is built up from a sum over discrete components associated with the vector spherical harmonics. But, at high frequencies this representation can be cast into a form which closely resembles the case for plane stratified media.

15.1 Representation of the wavefield

For a general seismic wavefield $v(x_1, x_2, x_3, t)$ recorded across a plane ($x_3 = 0$, say), we can take a Fourier transform with respect to the horizontal variables x_1, x_2 and time t to give

$$\bar{v}(k_1, k_2, x_3, \omega) = \frac{1}{8\pi^3}\int_{-\infty}^{\infty} \mathrm{d}t \int_{-\infty}^{\infty} \mathrm{d}x_1 \int_{-\infty}^{\infty} \mathrm{d}x_2\, v(x_1, x_2, x_3, t)\mathrm{e}^{-\mathrm{i}[k_1x_1+k_2x_2-\omega t]}, \tag{15.1.1}$$

as a function of the horizontal wavenumbers k_1, k_2 and angular frequency ω. The inverse Fourier transform yields the wavefield as a superposition of plane wave components

$$v(x_1, x_2, x_3, t) = \int_{-\infty}^{\infty} \mathrm{d}\omega \int_{-\infty}^{\infty} \mathrm{d}k_1 \int_{-\infty}^{\infty} \mathrm{d}k_2\, \bar{v}(k_1, k_2, x_3, \omega)\mathrm{e}^{\mathrm{i}[k_1x_1+k_2x_2-\omega t]}. \tag{15.1.2}$$

We now introduce the horizontal slownesses p_1, p_2 such that

$$k_1 = \omega p_1, \quad k_2 = \omega p_2, \tag{15.1.3}$$

and then the plane wave representation takes the form

$$v(x_1, x_2, x_3, t) = \int_{-\infty}^{\infty} \mathrm{d}\omega\, \omega^2 \int_{-\infty}^{\infty} \mathrm{d}p_1 \int_{-\infty}^{\infty} \mathrm{d}p_2\, \bar{v}(p_1, p_2, x_3, \omega) \mathrm{e}^{\mathrm{i}\omega[p_1x_1+p_2x_2-t]}. \tag{15.1.4}$$

An alternative decomposition of the wavefield is provided by carrying out the integration with respect to frequency ω on the right hand side of (15.1.4) to give

$$v(x_1, x_2, x_3, t) = -\int_{-\infty}^{\infty} \mathrm{d}p_1 \int_{-\infty}^{\infty} \mathrm{d}p_2\, \partial_{tt}\check{v}(p_1, p_2, x_3, \tau), \tag{15.1.5}$$

where the auxiliary time variable τ is given by

$$\tau = t - p_1x_1 - p_2x_2, \tag{15.1.6}$$

which represents a projection onto the normal to a wavefront with horizontal slowness $(p_1,\ p_2)$. $\check{v}(p_1, p_2, x_3, t)$ represents the inverse transform to time of the frequency slowness response

$$\check{v}(p_1, p_2, x_3, t) = \int_{-\infty}^{\infty} \mathrm{d}\omega\, \bar{v}(p_1, p_2, x_3, \omega)\mathrm{e}^{-\mathrm{i}\omega t}. \tag{15.1.7}$$

The Fourier transform operation (15.1.1) can be applied to any well-behaved wavefield, and so the field representations (15.1.2)-(15.1.5) are quite general. These decompositions of the wavefield are expressed in terms of contributions with the functional form of propagating plane waves but they may well have no direct physical significance.

However, when we deal with stratified media, in which the elastic properties are only a function of depth (x_3), the function $\bar{v}(p_1, p_2, x_3, \omega)$ can be interpreted as the weight to be applied to a set of plane waves characterised by horizontal slowness $(p_1,\ p_2)$ and angular frequency ω. At a horizontal interface between dissimilar media we have seen in Chapter 13 that Snell's law corresponds to the conservation of the horizontal slowness across the interface. This conservation property extends to continuously stratified media so that $(p_1,\ p_2)$ do not vary with x_3. The entire response of a stratified medium to source excitation can therefore be built up by superposing the slowness components with weighting functions which represent the excitation and propagation characteristics of plane waves.

When the medium does not deviate too far from stratification the plane wave superposition gives a good idea of the physical behaviour, even though a full representation of the wavefield requires computations which involve coupling between different slowness components. For a stratified medium we can achieve a further simplification of the representation of the wavefield when the wavefield has a vertical axis of symmetry. Such a condition can be achieved when the source can be approximated by a point. With a cylindrically symmetric source, such as a single explosion or vibrator with a circular plate, we can recast the plane wave expansions (15.1.2), (15.1.3) in terms of

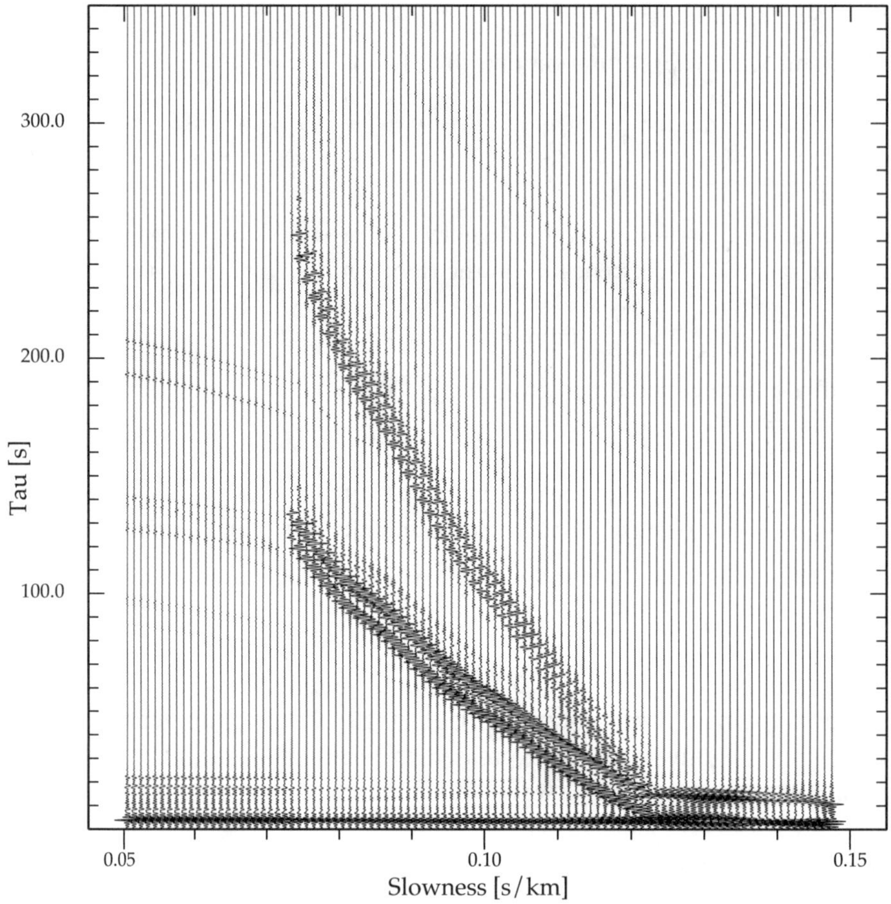

Figure 15.1. The P-response for the upper mantle, generated by a double couple source at 25 km, in the slowness-time domain, traces are normalised to the same maximum amplitude to enhance the reflections at small slowness.

cylindrical waves spreading out from the source. Mathematically, we transform from cartesian components (p_1, p_2) to polar slownesses (p, ϕ). After the change of variables (15.1.4) becomes

$$v(x_1,x_2,x_3,t) = \int_{-\infty}^{\infty} d\omega\, \omega^2 \int_0^{\infty} dp\, pJ_0(\omega pX)\tilde{v}(p_1,p_2,x_3,\omega)e^{-i\omega t}, \quad (15.1.8)$$

where X is the radial distance from the source, $X = (x_1^2 + x_2^2)^{1/2}$, and $J_0(\omega pX)$ is the zeroth order Bessel function.

The response of the medium in the slowness frequency (p_1, p_2, ω) domain depends on the horizontal slowness $p = (p_1^2 + p_2^2)^{1/2}$ and any angular dependence arises from the nature of the source. For example a moment tensor source can excite contributions with no variation with azimuth ϕ as illustrated in (15.1.8), or

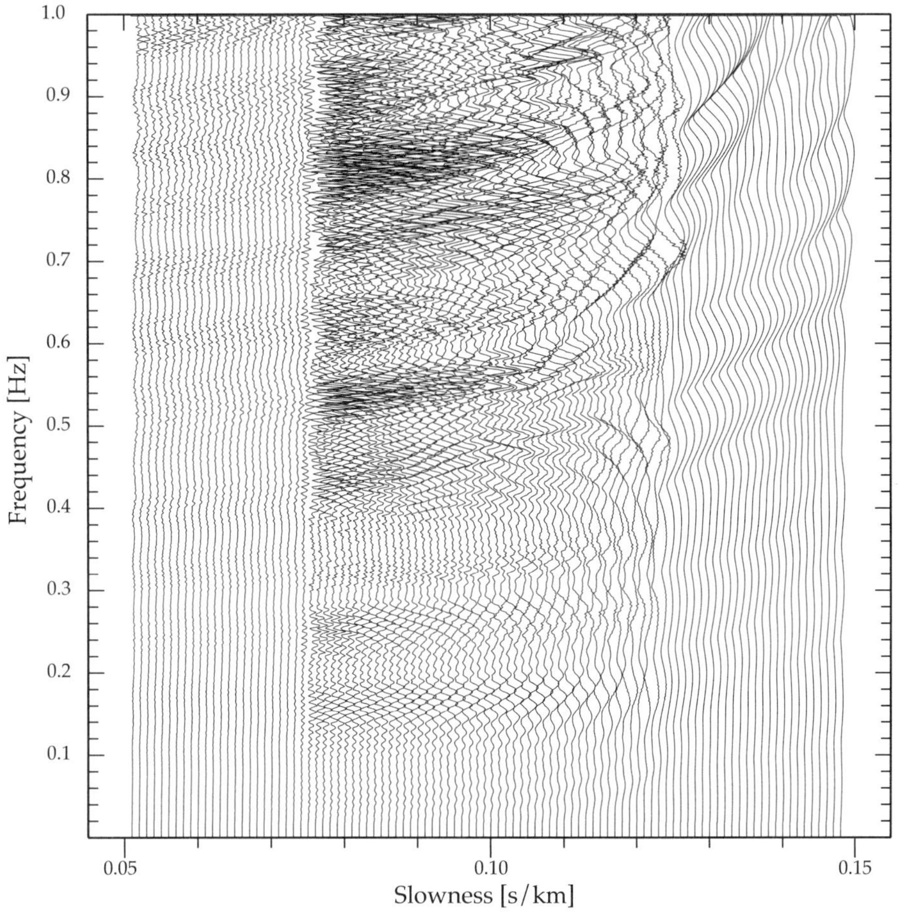

Figure 15.2. The amplitude of the slowness-frequency response for P waves in the upper mantle generated by a double couple source at 25 km.

alternatively with $\cos\phi, \sin\phi$ and $\cos 2\phi, \sin 2\phi$ dependence according to the nature of the source. The higher order angular dependence is accompanied by the appropriate higher order Bessel function.

The character of the seismic wavefield components are very different in the two styles of transform domain: slowness-time $(p - \tau)$ used in (15.1.5) and slowness frequency $(p - \omega)$ used in (15.1.8). The differences in the character have a strong effect on the numerical properties of the various integrals.

We illustrate the differences in the nature of the two domains in figures 15.1, 15.2 by considering P wave propagation in the upper mantle of the AK135 model, using the approximations discussed in Section 14.6 to extract a portion of the full response. We have taken a flattened velocity structure down to 1000 km and have allowed for the possibility of a single surface multiple for energy reflected back from beneath the separation depth of 60 km. A broad frequency band with a flat response to velocity from 0.05 - 0.75 Hz was used with a perfectly elastic model. The *P* wave response at the receiver is synthesised for a $M_{13} + M_{31}$ double couple source including the possibility of shallow multiples. The slowness-time ($p - \tau$) traces are constructed by taking a Fast Fourier transform over time of the complex traces in slowness-frequency ($p - \omega$).

The individual traces in the slowness-time display (figure 15.1) can be regarded as the result of illumination by a single slowness component. The patterns of arrivals map out the expected trajectories for mantle phases (cf. figure 5.7). Traces have been normalised to enhance the reflections at smaller slownesses. The dominant contributions come from *P* and *PP* waves returned from the upper mantle and the *sP*, *sPP* branches about 12 s later. The *P* surface reflections (*pP*, *pPP*) are somewhat smaller. The reflections from the upper mantle discontinuities at 410 and 660 km (accompanied by their depth phases) can be seen as weak arrivals at small slowness and are more distinct for the branches associated with *PP*. At larger slowness ($p > 0.124$ s/km) *P* propagation is largely confined to the crust and hence is concentrated at small τ values.

It is more difficult to display the character of the complex traces in the slowness-frequency domain. In figure 15.2 we show the amplitude of the slowness traces, and need to recall that there is also a phase component which varies even more rapidly than the amplitude. The amplitude behaviour has an interesting character and we can single out different patterns corresponding to distinct propagation regimes. At higher frequencies there can be very rapid variations with slowness in the frequency response, with little coherence between traces. The transition from reflected waves to refracted *P* and *PP* waves at 0.73 s/km has a marked effect on the character of the traces.

15.2 Wavefield representation for cylindrical coordinates

We have seen in the previous section how a plane wave superposition in the form of the Fourier transform over the horizontal cartesian coordinates can be mapped into a Hankel transform over the J_0 Bessel function for an axisymmetric situation, by a change to cylindrical coordinates in both real and slowness space.

It is convenient to take the $\hat{\mathbf{x}}$ axis to the North, $\hat{\mathbf{y}}$ axis to the East, and the $\hat{\mathbf{z}}$ axis vertically downwards. The azimuthal angle ϕ to the x axis in the cylindrical coordinate system then follows the geographical convention.

A general representation of the seismic wavefield in cylindrical coordinates (X, ϕ, z) is obtained by working with vector surface harmonics. We introduce three orthogonal vector harmonics R, S, T such that

$$\mathbf{R}_k^m = \mathbf{e}_z Y_k^m, \quad \mathbf{S}_k^m = k^{-1}\nabla_\perp Y_k^m, \quad \mathbf{T}_k^m = -\mathbf{e}_z \wedge \mathbf{S}_k^m, \tag{15.2.1}$$

where $\mathbf{e}_z$, $\mathbf{e}_X$ and $\mathbf{e}_\phi$ are unit vectors along the cylindrical coordinates and $\nabla_\perp$ is the operator

$$\nabla_\perp = \mathbf{e}_X \partial_X + \mathbf{e}_\phi \frac{1}{X} \partial_\phi. \tag{15.2.2}$$

The dependence on the horizontal coordinates arises through

$$Y_k^m(X, \phi) = J_m(kX) e^{im\phi}, \tag{15.2.3}$$

which depends on the radial distance X through the Bessel function $J_m(kX)$ modulated by the complex exponential azimuthal term. The orthogonality properties between the vector surface harmonics are

$$\int_0^\infty dX\, X \int_{-\pi}^{\pi} d\phi\, \mathbf{R}_k^m \cdot [\mathbf{S}_\kappa^\mu]^* = 0, \tag{15.2.4}$$

where the asterisk denotes a complex conjugate. The individual harmonics have the orthonormality property:

$$\int_0^\infty dX\, X \int_{-\pi}^{\pi} d\phi\, \mathbf{R}_k^m \cdot [\mathbf{R}_\kappa^\mu]^* = (k\kappa)^{-1/2}\, 2\pi \delta_{m\mu} \delta(k - \kappa), \tag{15.2.5}$$

with similar results for $\mathbf{S}_k^m$, $\mathbf{T}_k^m$.

We can represent the displacement field as a sum of contributions from the different harmonics in the form (Takeuchi & Saito, 1972)

$$\mathbf{u}(X, \phi, z, t) = \frac{1}{2\pi} \int_{-\infty}^{\infty} d\omega e^{-i\omega t} \int_0^\infty dk\, k \sum_m [U\mathbf{R}_k^m + V\mathbf{S}_k^m + W\mathbf{T}_k^m], \tag{15.2.6}$$

the summation over the angular order m will be restricted to the range of values needed to represent the source. For a point moment tensor source, $|m| \leq 2$. The wavenumber k may alternatively be expressed in terms of the horizontal slowness p through $k = \omega p$.

The harmonic expansion (15.2.6) can be rewritten to show the explicit dependence of the individual components as

$$u_z(X, \phi, z, t) = \frac{1}{2\pi} \int_{-\infty}^{\infty} d\omega e^{-i\omega t} \int_0^\infty dk\, k \sum_m U(m, z) J_m(kX) e^{im\phi}, \tag{15.2.7}$$

$$\begin{aligned} u_r(X, \phi, z, t) = {} & \frac{1}{2\pi} \int_{-\infty}^{\infty} d\omega e^{-i\omega t} \int_0^\infty dk\, k \\ & \times \sum_m \left[V(m, z) \frac{\partial J_m(kX)}{\partial (kX)} + W(m, z) \frac{im}{kX} J_m(kX) \right] e^{im\phi}, \end{aligned} \tag{15.2.8}$$

$$\begin{aligned} u_\phi(X, \phi, z, t) = {} & \frac{1}{2\pi} \int_{-\infty}^{\infty} d\omega e^{-i\omega t} \int_0^\infty dk\, k \\ & \times \sum_m \left[V(m, z) \frac{im}{kX} J_m(kX) - W(m, z) \frac{\partial J_m(kX)}{\partial (kX)} \right] e^{im\phi}. \end{aligned} \tag{15.2.9}$$

Note that the V and W components are linked on the near field terms which decay more rapidly with distance.

This representation of the displacement $\mathbf{u}$ may be regarded as a superposition of cylindrical waves whose order dictates the nature of their azimuthal modulation. At each frequency and angular order the radial contribution is obtained by superposing all horizontal wavenumbers k from 0 to infinity, thereby including all propagating waves for all wave types at the level z within the stratification. Thus we span from vertically travelling waves to purely horizontal propagation, as well as the whole spectrum of evanescent waves out to infinite wavenumber. At any particular distance X from the source, the relative contributions of the wavenumbers are imposed by the radial phase functions $J_m(kX)$.

We can make a comparable representation of the traction field in terms of the vector harmonics

$$\mathbf{t}(X,\phi,z,t) = \frac{1}{2\pi}\int_{-\infty}^{\infty} d\omega e^{-i\omega t}\int_0^{\infty} dk\, k \sum_m [P\mathbf{R}_k^m + S\mathbf{S}_k^m + T\mathbf{T}_k^m]. \tag{15.2.10}$$

The depth evolution of the coefficients of the harmonics for displacement and traction is described by the systems of coupled equations (12.1.11), (12.1.12), which we can represent in summary form as

$$\partial_z \mathbf{b} = \omega \mathbf{A}\mathbf{b} + \mathbf{F}, \tag{15.2.11}$$

in terms of a motion-stress vector $\mathbf{b}$ and a forcing term $\mathbf{F}$ associated with sources. For *P-SV* waves

$$\mathbf{b} = [U,V,P,S]^T, \tag{15.2.12}$$

and for *SH* waves

$$\mathbf{b} = [W,T]^T. \tag{15.2.13}$$

For small sources we will use the point source representation in terms of a force $\mathcal{E}$ and moment tensor M_{jk}, lying on a vertical axis through the origin of the coordinate system. The cartesian components of the equivalent force system are then

$$f_j = \mathcal{E}_j \delta(\mathbf{x}-\mathbf{x}_S) - \partial_k\{M_{jk}\delta(\mathbf{x}-\mathbf{x}_S)\}. \tag{15.2.14}$$

It is this system of forces which will now appear in the equations of motion and ultimately determine the forcing terms in the differential equations for the stress-displacement vector $\mathbf{b}$, (12.1.11)-(12.1.12). For larger source regions we may simulate the radiation characteristics by the superposition of a number of point source contributions separated in space and time to handle propagation effects. Alternatively we can perform a volume integral over the source region, in which case each volume element $d^3\boldsymbol{\eta}$ has an associated force $\boldsymbol{\epsilon} d^3\boldsymbol{\eta}$ and moment tensor $m_{ij} d^3\boldsymbol{\eta}$.

In each case we have the problem of finding the coefficients F_z, F_V, F_H in a vector harmonic expansion, cf. (15.2.6),

$$\mathbf{f} = \frac{1}{2\pi}\int_{-\infty}^{\infty} d\omega\, e^{-i\omega t}\int_0^{\infty} dk\, k \sum_m [F_z\mathbf{R}_k^m + F_V\mathbf{S}_k^m + F_H\mathbf{T}_k^m]. \tag{15.2.15}$$

If we choose the z axis of our cylindrical coordinate system to pass through the source point $\mathbf{x}_S$, the cartesian components of the point source f_x, f_y, f_z will all be singular at the origin in the horizontal plane $z = z_S$. The coefficients F_z, F_V, F_H will only appear at the source depth z_S and may be evaluated by making use of the orthonormality of the vector harmonics, so that, e.g.,

$$F_V = \frac{1}{2\pi}\int_{-\infty}^{\infty} \mathrm{d}t\, e^{i\omega t} \int_0^{\infty} \mathrm{d}X\, X \int_0^{2\pi} \mathrm{d}\phi\, [\mathbf{S}_k^m]^* \cdot \mathbf{f}. \qquad (15.2.16)$$

To evaluate the integrals over the horizontal plane we have to make use of the expansion of $J_m(kX)e^{im\phi}$ near the origin. It is convenient to work in cartesian coordinates for which

$$F_V = \frac{1}{2\pi}\int_{-\infty}^{\infty} \mathrm{d}t\, e^{i\omega t} \int_{-\infty}^{\infty} \mathrm{d}x_1 \int_{-\infty}^{\infty} \mathrm{d}x_2 [\mathbf{S}_k^m]^* \cdot \mathbf{f}. \qquad (15.2.17)$$

In this cartesian representation the surface harmonics take the form

$$\mathbf{R}_k^m = \left[0, 0, Y_k^m\right]^T, \quad \mathbf{S}_k^m = \frac{1}{k}\left[\partial_1 Y_k^m, \partial_2 Y_k^m, 0\right]^T, \quad \mathbf{T}_k^m = \frac{1}{k}\left[\partial_2 Y_k^m, -\partial_1 Y_k^m, 0\right]^T. \qquad (15.2.18)$$

Near the origin Y_k^m can be approximated by the leading term in a power series expansion

$$Y_k^m(kX) = \frac{k^m (Xe^{i\phi})^m}{2^m m!}, \quad Xe^{i\phi} = x_1 + ix_2, \qquad (15.2.19)$$

so that the integrals in (15.2.17) can be evaluated analytically.

The integrations leading to F_z, F_V, F_H will leave the z dependence of the source terms unaffected. Thus we anticipate that the force components $\mathcal{I}_x$, $\mathcal{I}_y$, $\mathcal{I}_z$ and the moment tensor elements describing doublets in the horizontal plane (M_{xx}, M_{xy}, M_{yx}, M_{yy}) will have a $\delta(z - z_S)$ dependence. The remaining moment tensor elements (M_{xz}, M_{zx}, M_{yz}, M_{zy}, M_{zz}) will appear with a $\delta'(z - z_S)$ term.

For each angular order m, the total forcing term **F**, will therefore have a z dependence,

$$\mathbf{F}(k, m, z, \omega) = \mathbf{F}_1(k, m, \omega)\delta(z - z_S) + \mathbf{F}_2(k, m, \omega)\delta'(z - z_S). \qquad (15.2.20)$$

When we solve for the stress-displacement vector **b** in the presence of the general point source excitation, there will be a discontinuity in **b** across the source plane $z = z_S$

$$\begin{aligned} \mathbf{b}(k, m, z_S+, \omega) - \mathbf{b}(k, m, z_S-, \omega) &= \mathbf{S}(k, m, z_S, \omega), \\ &= \mathbf{F}_1 + \omega\mathbf{A}(p, z_S)\mathbf{F}_2. \end{aligned} \qquad (15.2.21)$$

The action of the $\delta'(z - z_S)$ term is to extract a derivative at the source level and we use (15.2.11) to represent this in terms of the matrix **A**. The influence of the δ' terms arising from couples and dipoles with respect to z is to introduce discontinuities in the displacement components through the action of the matrix **A** on $\mathbf{F}_2$ in addition to the traction discontinuities arising from $\mathbf{F}_1$. The harmonic coefficients $\mathbf{F}_{1,2}$ have to be evaluated from integrals such as (15.2.16) or (15.2.17).

For a representation through a point source, the jump in the components of the stress-displacement vector **b** across z_S depends strongly on angular order. The displacement jumps

$$\begin{aligned} [U]_-^+ &= M_{zz}(\rho\alpha^2)^{-1}, & m &= 0, \\ [V]_-^+ &= \tfrac{1}{2}[\pm M_{xz} - iM_{yz}](\rho\beta^2)^{-1}, & m &= \pm 1, \\ [W]_-^+ &= \tfrac{1}{2}[\pm M_{yz} - iM_{xz}](\rho\beta^2)^{-1}, & m &= \pm 1, \end{aligned} \qquad (15.2.22)$$

and the traction jumps

$$
\begin{aligned}
[P]_-^+ &= -\omega^{-1}\mathcal{E}_z, && m = 0\\
&= \tfrac{1}{2}p[\mathrm{i}(M_{zy} - M_{yz}) \pm (M_{xz} - M_{zx})], && m = \pm 1,\\
[S]_-^+ &= \tfrac{1}{2}p(M_{xx} + M_{yy}) - pM_{zz}(1 - 2\beta^2/\alpha^2) && m = 0,\\
&= \tfrac{1}{2}\omega^{-1}(\mp\mathcal{E}_x + \mathrm{i}\mathcal{E}_y), && m = \pm 1,\\
&= \tfrac{1}{4}p[M_{yy} - M_{xx}) \pm \mathrm{i}(M_{xy} + M_{yx})], && m = \pm 2,\\
[T]_-^+ &= \tfrac{1}{2}p(M_{xy} - M_{yx}), && m = 0,\\
&= \tfrac{1}{2}\omega^{-1}(\mathrm{i}\mathcal{E}_x \pm \mathcal{E}_y), && m = \pm 1,\\
&= \tfrac{1}{4}p[\pm\mathrm{i}(M_{xx} - M_{yy}) + (M_{xy} + M_{yx})], && m = \pm 2.
\end{aligned}
\qquad (15.2.23)
$$

For our point equivalent source (15.2.14), the vector harmonic expansion (15.2.6) for the displacement will be restricted to azimuthal orders $|m| < 2$. These results for a general moment tensor generalise Hudson's (1969a) analysis of an arbitrarily oriented dislocation.

We recall that for an *indigenous* source the moment tensor is symmetric and thus has only six independent components, and further $\boldsymbol{\mathcal{E}}$ will then vanish. This leads to a significant simplification of these results. In particular, for a point source the stress variable P will always be continuous, excitation for $m = \pm 1$ is confined to the horizontal displacement terms and T will only have a jump for $m = \pm 2$.

In an isotropic medium, or a transversely isotropic medium with a vertical axis of symmetry, the only azimuthal dependence in the displacement and traction quantities U, V, W, P, S, T arises from the azimuthal behaviour of the source. For an indigenous source we can therefore associate the azimuthal behaviour of the displacement field in the stratification with certain combinations of the moment tensor elements.

(a) *No variation with azimuth:* For the *P-SV* wavefield this is controlled by the diagonal elements $(M_{xx} + M_{yy})$, M_{zz} and is completely absent for *SH* waves.

(b) $\cos\phi$, $\sin\phi$ *dependence:* This angular behaviour arises from the presence of the vertical couples $M_{x}z$, M_{yz}. The term M_{xz} leads to $\cos\phi$ dependence for *P-SV* and $\sin\phi$ for *SH*, whilst M_{yz} gives $\sin\phi$ behaviour for *P-SV* and $\cos\phi$ for *SH*.

(c) $\cos 2\phi$, $\sin 2\phi$ *dependence:* This behaviour is controlled by the horizontal dipoles and couples M_{xx}, M_{yy}, M_{xy}. The difference $(M_{xx} - M_{yy})$ leads to $\cos 2\phi$ behaviour for *P-SV* and $\sin 2\phi$ for *SH*. The couple M_{xy} gives $\sin 2\phi$ dependence for *P-SV* and $\cos 2\phi$ for *SH*.

These azimuthal dependencies do not rest on any assumptions about the nature of the propagation path through the medium and so hold for both body waves and surface waves.

As we have seen in Section 14.4 we can use the jumps in the motion-stress vector to derive expressions for the radiation components $\boldsymbol{U}^S$, $\boldsymbol{D}^S$ which can then be linked to propagation processes. The radiation components will carry with them the azimuthal dependencies imposed by the contributions of the moment tensor components.

15.3 Evaluation of the wavefield response

We can present the representation of the wavefield by superposition of slowness-frequency components as a double integral and summation of angular components:

$$\mathbf{u}(X,\phi,0,t) = \frac{1}{2\pi}\int_{-\infty}^{\infty} d\omega e^{-i\omega t}\omega^2 \int_0^{\infty} dp\, p \sum_m \mathbf{w}_0^T(p,m,\omega)\mathbf{T}_m(\omega pX), \tag{15.3.1}$$

where $\mathbf{T}$ represents the horizontal phase term of order m. For *P-SV* waves this is a tensor field

$$\mathbf{T}_m(\omega pX) = \left[\mathbf{R}_k^m, \mathbf{S}_k^m\right]^T, \tag{15.3.2}$$

and for *SH* we can equate $\mathbf{T}_m$ to the remaining vector harmonic

$$\mathbf{T}_m(\omega pX) = \mathbf{T}_k^m. \tag{15.3.3}$$

We can exploit the derivative properties of the Bessel functions

$$J'_m(x) = J_{m-1}(x) - mJ_m(x)/x, \tag{15.3.4}$$

to recast the vector surface harmonics $\mathbf{R}_k^m$, $\mathbf{S}_k^m$, $\mathbf{T}_k^m$ in a form that does not contain any derivatives

$$\begin{aligned}
\mathbf{R}_m(\omega pX) &= \left[\mathbf{e}_z J_m(\omega pX)\right] e^{im\phi},\\
\mathbf{S}_m(\omega pX) &= \left[\mathbf{e}_r J_{m-1}(\omega pX) - (\mathbf{e}_r - i\mathbf{e}_\phi)\frac{mJ_m(\omega pX)}{\omega pX}\right] e^{im\phi},\\
\mathbf{T}_m(\omega pr) &= \left[-\mathbf{e}_\phi J_{m-1}(\omega pX) + (\mathbf{e}_\phi + i\mathbf{e}_r)\frac{mJ_m(\omega pX)}{\omega pX}\right] e^{im\phi}.
\end{aligned} \tag{15.3.5}$$

These forms are most convenient for $m \geq 0$, but the values for $m < 0$ are easily obtained from

$$J_{-m}(x) = (-1)^m J_m(x). \tag{15.3.6}$$

For the horizontal components, for azimuthal orders $|m| > 0$ we have 'near-field' components depending on $mJ_m(\omega pX)/\omega pX$ which decay more rapidly than the contributions oriented along the coordinate vectors. These 'near-field' terms couple the radial and tangential components of motion so that there is no clear separation by component of *SV* and *SH* motion at small distances from the source.

To evaluate the double integral in (15.3.1) we have to choose the order in which the frequency and slowness integrals are undertaken. If the slowness integral is calculated first then the intermediate result is the complex frequency spectrum $\bar{\mathbf{u}}(X,\phi,0,\omega)$ at a particular location. This approach may therefore be designated as a *spectral method* and has been used in most attempts to calculate theoretical seismograms by numerical integration of the complete medium response (e.g., Kind 1978, Kennett 1980, Wang & Herrmann 1980). When, alternatively, the frequency integral is evaluated first the intermediate result is a time response for each slowness p, corresponding to the illumination of the medium by a single slowness component. The final result is obtained by an integral over slowness and we follow Chapman (1978) by calling this approach the *slowness method.*

15.3.1 Spectral methods

We can illustrate the spectral approach by considering the spectrum of the surface displacement in the form of a slowness integral with a summation over azimuthal components,

$$\bar{\mathbf{u}}(X, \phi, 0, \omega) = \omega^2 \int_0^\infty \mathrm{d}p\, p \sum_m \boldsymbol{w}_0^T(p, m, \omega)\mathbf{T}_m(\omega p X). \tag{15.3.7}$$

The representations of the wavefield $\boldsymbol{w}_0(p, m, \omega)$ discussed in section 14.5 acquire their azimuthal dependence from the radiation components $\boldsymbol{U}^S$, $\boldsymbol{D}^S$ at the source level. These upgoing and downgoing wave terms derive their dependence on the azimuthal order m through the source jump term $\mathbf{S}(z_S)$ (15.2.21), and a different slowness dependence is introduced for each order. $\boldsymbol{w}_0(p, m, \omega)$ is an odd function of p if m is even, and an even function of p if m is odd.

We recall that the elements of $\mathbf{T}_m(\omega p X)$ depend on $J_m(\omega p X)\mathrm{e}^{\mathrm{i}m\phi}$ and so we may make a decomposition of this 'standing wave' form into a travelling wave representation in terms of the Hankel functions $H_m^{(1)}(\omega p X)$, $H_m^{(2)}(\omega p X)$. Now

$$\begin{aligned} J_m(\omega p X) &= \tfrac{1}{2}[H_m^{(1)}(\omega p X) + H_m^{(2)}(\omega p X)] \\ &= \tfrac{1}{2}[H_m^{(1)}(\omega p X) - \mathrm{e}^{\mathrm{i}m\pi} H_m^{(1)}(-\omega p X)] \end{aligned} \tag{15.3.8}$$

in terms of just the outgoing components. We can use (15.3.8) to rewrite the expressions for the vector surface harmonics and will use the notation $\mathbf{T}_m^{(1)}(\omega p X)$ for the harmonic terms corresponding to outgoing waves from the origin.

With the aid of the symmetry properties of $\boldsymbol{w}_0$ and (15.3.8), we can express (15.3.7) as an integral along the entire slowness axis

$$\bar{\mathbf{u}}(X, \phi, 0, \omega) = \tfrac{1}{2}\omega|\omega| \int_0^\infty \mathrm{d}p\, p \sum_m \boldsymbol{w}_0^T(p, m, \omega)\mathbf{T}_m^{(1)}(\omega p X), \tag{15.3.9}$$

where the contour of integration in the p plane is taken above the branch point for $H_m^{(1)}(\omega p X)$ at the origin. This form shows explicitly that we are only interested in waves which diverge from the source.

For large values of the argument, $H_m^{(1)}(\omega p X)$ may be replaced by its asymptotic form

$$H_m^{(1)}(\omega p X) \sim \left(\frac{2}{\pi \omega p X}\right)^{1/2} \exp\left[\mathrm{i}\omega p X - \mathrm{i}(2m+1)\frac{\pi}{4}\right], \tag{15.3.10}$$

and to the same approximation the vector harmonics take on the character of fields directed along the orthogonal coordinate vectors $\mathbf{e}_z$, $\mathbf{e}_r$, $\mathbf{e}_\phi$. Thus the tensor field $\mathbf{T}_m^{(1)}(\omega pr)$ is approximated by

$$\mathbf{T}_m^{(1)} \sim \begin{pmatrix} \mathbf{e}_z \\ \mathrm{i}\mathbf{e}_r \end{pmatrix} \left(\frac{2}{\pi\omega pX}\right)^{1/2} \exp\left[\mathrm{i}\omega pX - \mathrm{i}(2m+1)\frac{\pi}{4}\right], \tag{15.3.11}$$

and the tangential (*SH*) harmonic

$$\mathrm{T}_m^{(1)}(\omega pr) \sim -\mathrm{i}\mathbf{e}_\phi \left(\frac{2}{\pi\omega pX}\right)^{1/2} \exp\left[\mathrm{i}\omega pX - \mathrm{i}(2m+1)\frac{\pi}{4}\right]. \tag{15.3.12}$$

In this asymptotic limit we are faced with the same slowness and distance dependence in the integrand of (15.3.10) for all three components of displacement.

The symmetry properties we have described will be shared by approximations to the complete response $\boldsymbol{w}_0$, and so we will always have the possibility of employing a standing wave expression (15.3.7) which is very effective for small distances X, or a travelling wave representation (15.3.9) which is most effective for larger distances in the asymptotic regime of the Hankel functions.

For a stratified isotropic medium underlain by a halfspace beneath $z = z_L$, the full response $\boldsymbol{w}_0(p,\omega)$ as a function of slowness p at fixed frequency ω will have branch points at $p = \pm\alpha_L^{-1}$ for the *P-SV* case and $p = \pm\beta_L^{-1}$ in all cases. These arise from the imposition of a radiation condition of only downgoing or decaying waves in $z > z_L$.

For both *P-SV* and *SH* wave contributions we have a sequence of poles in $\boldsymbol{w}_0$ in the region $\beta_L^{-1} < |p| < \beta_{\min}^{-1}$, where $\beta_{\min}$ is the smallest shear wavespeed anywhere in the half space - this is normally attained at the surface. The pole locations do not depend on on azimuthal order m, but will be a function of frequency ω reflecting the dispersion of surface waves.

For *SH* waves, higher mode Love wave poles lie in the region $\beta_L^{-1} < |p| < \beta_{\min}^{-1}$. The poles will lie on the real p axis in a perfectly elastic medium. Whereas for *P-SV* waves, we have higher mode Rayleigh poles whose locations are close to, but not identical to, the Love poles. In addition for $|p| > \beta_{\min}^{-1}$ there is an additional pole associated with the fundamental Rayleigh mode in which evanescent *P* and *SV* waves are coupled in the half space. The limiting slowness for the fundamental Rayleigh mode at high frequency is p_{R0}, corresponding to the non-dispersive Rayleigh wave on a uniform half space with the elastic properties at the surface.

The set of singularities for the *P-SV* wave case at fixed frequency is sketched in figure 15.3; the contour of integration for $\omega > 0$ runs just below the singularities for $p > 0$ and just above for $p < 0$. This contour may be justified by allowing for slight attenuation of seismic waves within the half space (which we may well want on physical grounds) in which case the poles move into the first and third quadrants of the complex p plane. The path of the branch cuts from α_L^{-1}, β_L^{-1} is not critical provided that the conditions on the radicals $\mathrm{Im}(\omega q_{\alpha L}) \geq 0$, $\mathrm{Im}(\omega q_{\beta L}) \geq 0$, are maintained on

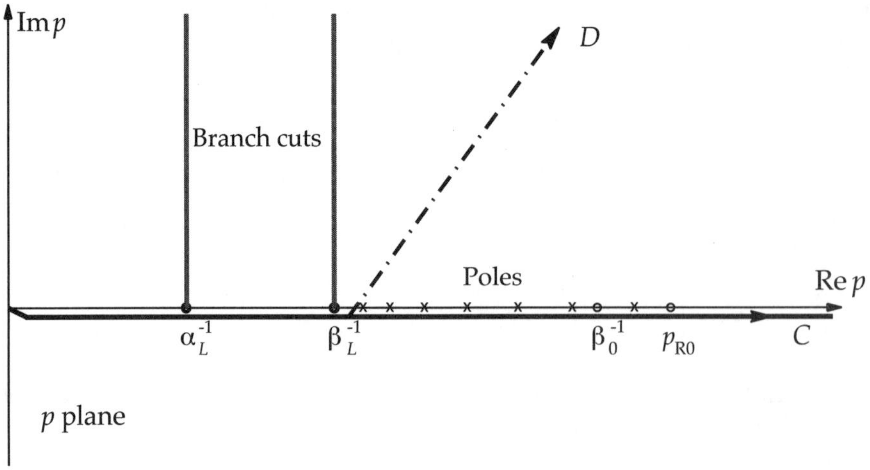

Figure 15.3. The singularities, branch cuts and integration contours in the p plane for the full response of a half space at fixed frequency ω.

the real p axis. It is convenient therefore to follow Lamb (1904) by taking cuts parallel to the imaginary p axis.

The most direct approach to the evaluation of expressions like (15.3.9) is to perform a direct numerical integration along the real p axis, but for a perfectly elastic medium the presence of the poles on the contour is a major obstacle to such an approach.

However, if the contour of integration in (15.3.9) is deformed into the upper half plane to D, we can pick up the polar residue contributions from all the poles to the right of β_L^{-1}. Convergence at infinity is ensured by the properties of the outgoing Hankel function $H_0^{(1)}(\omega pX)$. The result of deforming the contour of integration is to provide a representation of the displacement spectrum as a sum of a contour integral and a polar residue series:

$$\bar{\mathbf{u}}(X,\phi,0,\omega) = \tfrac{1}{2}\omega|\omega| \int_D \mathrm{d}p\, p \sum_m \boldsymbol{w}_0^T(p,m,\omega)\mathbf{T}_m^{(1)}(\omega pX)$$
$$+\pi \mathrm{i}\omega^2 \sum_{j=0}^{N(\omega)} p_j \sum_m \mathrm{Res}_j[\boldsymbol{w}_0^T \mathbf{T}_m^{(1)}(\omega p_j X)], \tag{15.3.13}$$

where Res_j represents the residue at the pole p_j. The contour integration in (15.3.13) consists of a real axis slowness integral from $-\infty$ up to β_L^{-1}, and then a line segment off into the first quadrant where $H_m^{(1)}(\omega pX)$ is a decaying function of complex p. For large ranges X the contribution from negative slownesses is very small and has often been neglected.

As we shall see in Chapter 16, the distribution of poles depends strongly on frequency; at low frequency only a few poles occur in $|p| > \beta_L^{-1}$ whilst at higher frequencies there are a great many. At even moderate frequencies the number of modal residue contributions $N(\omega)$ becomes very large indeed, and locating all the poles is a major computational problem.

The poles with largest slowness give the major contribution to what would generally be regarded as the surface wave train, with relatively low group velocities. The summation of modes with smaller slownesses provides a means of synthesising S body wave phases by modal interference. A residue sum taken over a portion of the real p axis can provide good results for the S wave coda (Kerry 1981), and by forcing α_L, β_L to be very large even P waves can be synthesised (Harvey 1981).

An alternative to the contour deformation procedure we have just described is to arrange to move the poles in the response off the contour of integration. For an attenuative medium, the poles will lie in the first and third quadrants away from the real p axis although their influence is strongly felt on the contour of integration. In most applications we anticipate that at least some loss will occur in seismic wave propagation and so introducing small loss factors is very reasonable.

The pattern of singularities will be modified when approximations to the response is made but direct numerical integration in slowness for an attenuative medium will usually give good results.

15.3.2 Slowness methods

An alternative approach to numerical integration over slowness for each frequency followed by Fourier transformation over frequency, is to perform the frequency integration first to give a representation of the wavefield in the slowness-time domain. Consider the mth azimuthal component

$$\check{\mathbf{u}}(p,X,m,t) = \frac{1}{2\pi}\int_{-\infty}^{\infty} \mathrm{d}\omega\, \mathrm{e}^{-\mathrm{i}\omega t}\omega^2 \mathbf{w}_0^T(p,m,\omega)\mathbf{T}_m(\omega pX), \tag{15.3.14}$$

and this Fourier transform of a product can be expressed as a convolution in time

$$\check{\mathbf{u}}(p,X,m,t) = -\partial_{tt}\left\{\check{\mathbf{w}}_0(p,m,t) \star \frac{1}{pX}\check{\mathbf{T}}_m\left(\frac{t}{pX}\right)\right\}, \tag{15.3.15}$$

where $\check{}$ indicates the inverse Fourier transform with respect to frequency.

The inverse transform of the vector surface harmonics depend on the time transform of $J_m(\omega pX)$ which has a square root singularity at $t = \pm pX$. We can illustrate the approach using the representation at the surface $z = 0$,

$$v(X,0,t) = \int_0^{\infty} \mathrm{d}p\, p \int_{-\infty}^{\infty} \mathrm{d}\omega\, \omega^2 J_0(\omega pX)\tilde{v}(p,0,\omega)\mathrm{e}^{-\mathrm{i}\omega t}. \tag{15.3.16}$$

The radial dependence arises from $J_0(\omega pX)$ and its time transform is given by

$$\pi\left(\frac{1}{pX}\right)\check{J}_0\left(\frac{t}{pX}\right) = \frac{H(t-pX)-H(t+pX)}{(p^2X^2-t^2)^{1/2}}. \tag{15.3.17}$$

On carrying out the inverse Fourier transform of (15.1.8) we obtain,

$$v(X,0,t) = -\frac{1}{\pi}\int_0^\infty \mathrm{d}p\, p\partial_{tt}\left[\check{v}(p,0,t) \star \frac{\{H(t-pX) - H(t+pX)\}}{(p^2X^2 - t^2)^{1/2}}\right],$$
$$= -\frac{1}{\pi}\int_0^\infty \mathrm{d}p\, p\partial_{tt}\int_{-pX}^{pX} \mathrm{d}s \frac{\check{v}(p,0,t-s)}{(p^2X^2 - s^2)^{1/2}}. \qquad (15.3.18)$$

For large pX we can factor the kernel of the integral into two separated singular terms,

$$\frac{H(pX-t) - H(t+pX)}{(p^2X^2 - t^2)^{1/2}} \approx \frac{-1}{(2pX)^{1/2}}\left[\frac{H(pX-t)}{(pX-t)^{1/2}} + \frac{H(t+pX)}{(pX+t)^{1/2}}\right]. \qquad (15.3.19)$$

With this approximation, which is equivalent to using asymptotic forms for the Hankel functions, (15.3.18) becomes

$$v(X,0,t) = -\frac{1}{\pi(2X)^{1/2}}\partial_{tt}\int_0^\infty \mathrm{d}p\, p^{1/2}$$
$$\int_{-\infty}^\infty \mathrm{d}s\, \check{v}(p,0,t-s)\left[\frac{H(pX-t)}{(pX-t)^{1/2}} + \frac{H(pX+t)}{(pX+t)^{1/2}}\right]. \qquad (15.3.20)$$

When there is a common time dependence $M(t)$ for the source components so that $\check{v}(p,0,t) = M(t) \star \check{V}(p,0,t)$, the double integral for $v(X,0,t)$ can be written as

$$v(X,0,t) = -\frac{1}{\pi(2X)^{1/2}}\int_{-\infty}^\infty \mathrm{d}s\, \mathcal{M}(t-s)$$
$$\int_0^\infty \mathrm{d}p\, p^{1/2}\left\{\hat{V}(p,s-pX) + \check{V}(p,s+pX)\right\}. \qquad (15.3.21)$$

where $\hat{V}(p,0,t)$ is the Hilbert transform of $\check{V}(p,0,t)$ and

$$\mathcal{M}(t) = \int_0^t \mathrm{d}\tau\, \partial_{\tau\tau}M(\tau)\frac{H(t-\tau)}{(t-\tau)^{1/2}}, \qquad (15.3.22)$$

may be regarded as an 'effective source' (cf Chapman, 1978). In general the main contribution in (15.3.22) comes from $\hat{V}(p,s-pX)$ and the incoming part $\check{V}(p,s+pX)$ can be neglected.

The dominant contribution will come from the neighbourhood of the line $\tau = t - pX$ and a reasonably good approximation for the wavefield is provided by

$$v(X,0,t) = -\frac{1}{\pi(2X)^{1/2}}\frac{H(t)}{t^{1/2}} \star \int_{-\infty}^\infty \mathrm{d}p\, p^{1/2}\hat{v}(p,0,t-pX), \qquad (15.3.23)$$

which is close in form to the cartesian representation (15.1.5) but includes weightings appropriate to the cylindrical wavefronts.

The inverse transform for the response vector

$$\check{\boldsymbol{w}}_0(p,m,t) = \frac{1}{2\pi}\int_{-\infty}^\infty \mathrm{d}\omega\, \mathrm{e}^{-\mathrm{i}\omega t}\boldsymbol{w}_0(p,m,\omega), \qquad (15.3.24)$$

depends strongly on slowness p.

Since we are interested in sources which start at $t = 0$, $\boldsymbol{w}_0(p, m, \omega)$ is analytic in the upper half plane $\mathrm{Im}\,\omega > 0$. The exponential term $\mathrm{e}^{-\mathrm{i}\omega t}$ enables us to deform the contour, if necessary, into the lower half plane for $t > 0$.

The quantity $\check{\boldsymbol{w}}_0(p, m, t)$ can be thought of as the time response of the half space to irradiation by a single slowness component. The radiation condition into the underlying half space means that for the full half-space response $\boldsymbol{w}_0(p, m, \omega)$, as a function of ω, there are no pole singularities for $0 < p < \beta_L^{-1}$. There will be a branch point at $\omega = 0$, and the branch cut can be conveniently taken along the negative imaginary ω axis. In this slowness range we get individual pulse-like arrivals corresponding to the major phases with a shape determined by the source time function. The pattern of arrivals across the band of slowness gets repeated in time with delays associated with multiple surface reflections. The radiation leakage of S waves into the underlying uniform medium means that each successive surface multiple set will be of smaller amplitude, and the decay of the multiples will be enhanced by the presence of attenuation in the medium. However, a long time series is needed to include all surface multiples and this can create difficulties when $\check{\boldsymbol{w}}_0(p, m, t)$ is evaluated numerically.

For larger slownesses $p > \beta_L^{-1}$, both P and S waves are evanescent in the underlying half space and we have poles in $\boldsymbol{w}_0(p, m, \omega)$. The poles will lie on the real ω axis for perfectly elastic media. The lowest frequency pole will correspond to fundamental mode surface waves. For an attenuative structure the poles move off the real axis into the lower half plane. Just at the branch point at β_L^{-1} we get the maximum density of poles along the ω axis (see figures 3.22, 16.8) and the spacing between the poles expands as p increases to $\beta_{\mathrm{min}}^{-1}$. For $p > \beta_{\mathrm{min}}^{-1}$ we have only one pole for the *P-SV* case corresponding to the fundamental Rayleigh mode. Since the poles in ω are symmetrically disposed about the imaginary axis, the residue contribution to $\check{\boldsymbol{w}}_0(p, m, t)$ takes the form (for $p < \beta_{\mathrm{min}}^{-1}$)

$$\mathrm{Re}\{\sum_{k=0}^{\infty} \mathrm{e}^{-\mathrm{i}\omega_k t}\mathrm{Res}_k[\boldsymbol{w}_0^T(p, m, \omega_k)]\}, \tag{15.3.25}$$

and this representation can prove convenient in surface wave studies. There will also be a continuous spectrum contribution from the sides of the branch cut along the negative imaginary ω axis.

Once we have found the inverse transform of the medium response and performed the convolution (15.3.15) we need to carry out the slowness integral and summation over angular order to recover the full response in space and time

$$\mathbf{u}(X, \phi, 0, t) = \sum_m \int_0^\infty \mathrm{d}p\, p\check{\mathbf{u}}(p, X, m, t), \tag{15.3.26}$$

to generate the seismograms for a particular range X. This forms the basis of the numerical procedure employed by Fryer (1980).

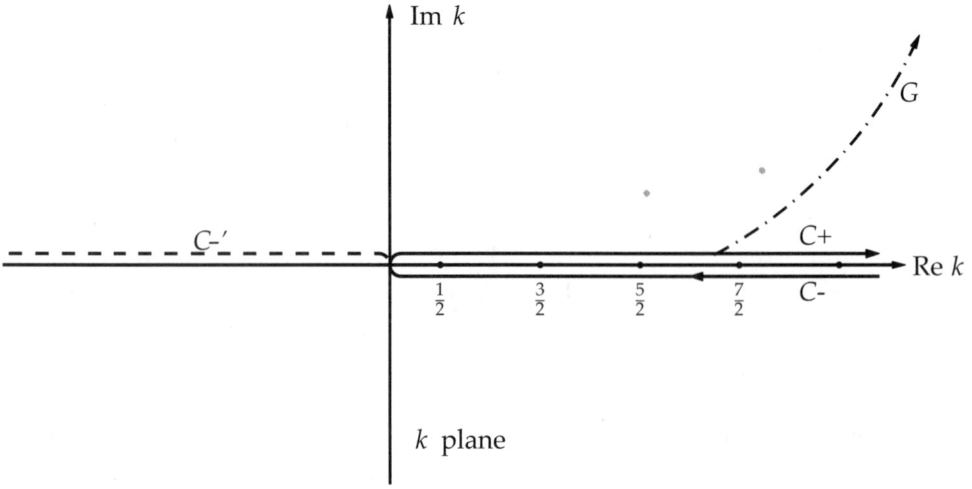

Figure 15.4. The contour employed for the Watson transformation (15.4.3) at fixed frequency ω.

15.4 Spherical stratification

As we have shown in Section 12.2, the displacement response for a spherically stratified medium can be expressed as a sum of discrete angular components in the frequency domain associated with vector tesseral harmonics $\mathbf{P}_l^m$, $\mathbf{B}_l^m$, $\mathbf{C}_l^m$. The spatial and temporal response can be written as

$$\mathbf{u}(r,\theta,\phi,t) = \frac{1}{2\pi}\int_{-\infty}^{\infty} \mathrm{d}\omega\, \mathrm{e}^{-i\omega t} \sum_{l=0}^{\infty} \sum_{m=-l}^{l} \Big[U(r,l,m,\omega)\mathbf{P}_l^m + V(r,l,m,\omega)\mathbf{B}_l^m + W(r,l,m,\omega)\mathbf{C}_l^m\Big]. \tag{15.4.1}$$

P-SV waves are represented through the U, V terms and *SH* waves by W.

As in the cylindrical coordinate system, the effect of a source can be introduced through a discontinuity in the motion stress vector $\mathfrak{b}$ across the source radius r_S. For a point source the simplest representation of the resulting displacement field is achieved by taking the coordinate axis $\theta = 0$ to lie through the source location, with the result that the θ coordinate can be interpreted as the epicentral distance Δ. The sum over azimuthal orders is then restricted to $|m| < 2$.

The convergence of the sum over angular order l is slow for higher frequencies. For body waves with periods of a few seconds, several thousand terms are needed in the l-summation to give an accurate result. An alternative is then to transform the summation over l into an integral by means of the Watson transformation (Watson 1918) or the equivalent Poisson sum formula (Morse & Feshbach, 1953, p466).

We will illustrate this approach for a source without angular variation, such as an explosion, so that $m = 0$. The frequency spectrum of the response can then be written in the forms

$$\bar{f}(\omega) = \sum_{l=0}^{\infty}(l+\tfrac{1}{2})P_l(\cos\Delta)f_l, \tag{15.4.2}$$

$$= \frac{1}{2\mathrm{i}}\int_C \mathrm{d}k\,\frac{k}{\cos k\pi}P_{k-\frac{1}{2}}(\cos\Delta)f(k)\mathrm{e}^{\mathrm{i}k\pi}, \tag{15.4.3}$$

where the Watson transformation is taken over the contour C in figure 15.4. The integrand in (15.4.3) has simple poles at the half-integral values of k along the positive real axis ($k = \frac{1}{2}, \frac{3}{2}, \frac{5}{2}, \ldots$). The equivalence of (15.4.2) and (15.4.3) can be demonstrated by evaluating the contour integral using the Cauchy residue theorem, with the requirement that

$$f(k) = f_l \quad \text{when} \quad k = l + \tfrac{1}{2}. \tag{15.4.4}$$

The extension of the definition of the Legendre functions to continuous argument can be made either in terms of hypergeometric functions (BenMenahem & Singh, 1981, p629) or via an integral representation (Dahlen & Tromp, 1998, p865).

For $\mathrm{Im}(k) > 0$, $(\cos k\pi)^{-1}$ can be expanded as

$$\frac{1}{\cos k\pi} = \frac{2\mathrm{e}^{\mathrm{i}k\pi}}{\mathrm{e}^{2\mathrm{i}k\pi}+1} = 2\mathrm{e}^{\mathrm{i}k\pi}\sum_{q=0}^{\infty}(-1)^q\mathrm{e}^{2\mathrm{i}qk\pi}, \tag{15.4.5}$$

with a similar form for $\mathrm{Im}(k) < 0$. With this expansion the transformation (15.4.3) can be shown to be equivalent to the use of the Poisson sum formula,

$$\bar{f}(\omega) = \sum_{q=\infty}^{\infty}(-1)^q\int_0^{\infty}\mathrm{d}k\,kP_{k-\frac{1}{2}}(\cos\Delta)f(k)\mathrm{e}^{2\mathrm{i}qk\pi}, \tag{15.4.6}$$

where $f(k)$ has been extended as an even function of complex wavenumber k. The summation over q corresponds to the number of circuits of the globe (see, e.g., Gilbert 1976).

The standing wave representation (15.4.6) can be recast in terms of the travelling wave terms $Q^{(1,2)}$ introduced in (12.2.15). $Q^{(1)}(\Delta)$ is a wave that propagates in the direction of increasing Δ and $Q^{(2)}(\Delta)$ travels in the opposite sense. Away from the source ($\Delta = 0$) and its antipole ($\Delta = \pi$), the response spectrum can be written as

$$\bar{f}(\omega) = \sum_{q=1,3,5,\ldots}^{\infty}(-1)^{(q-1)/2}\int_0^{\infty}\mathrm{d}k\,kQ^{(1)}_{k-\frac{1}{2}}(\Delta)f(k)\mathrm{e}^{2\mathrm{i}(q-1)k\pi}$$
$$+ \sum_{q=2,4,6,\ldots}^{\infty}(-1)^{q/2}\int_0^{\infty}\mathrm{d}k\,kQ^{(1)}_{k-\frac{1}{2}}(\Delta)f(k)\mathrm{e}^{2\mathrm{i}qk\pi}, \tag{15.4.7}$$

The terms $q = 1$ corresponds to direct propagation from the source along the minor arc and $q = 2$ to the major arc (in the opposite direction along the great circle). The remaining terms correspond to multi-orbit waves that have circled the Earth more than once before arriving at the receiver.

For a more general source when we consider just the first orbit, the representation for the vertical displacement (i.e., along the radius vector) takes the form

$$\bar{f}(\omega) = \int_0^\infty \mathrm{d}k\, k P^m_{k-\frac{1}{2}}(\cos\Delta) f(k) \mathrm{e}^{\mathrm{i}m\phi}, \tag{15.4.8}$$

and at high frequencies,

$$P^m_{k-1/2}(\cos\Delta)\mathrm{e}^{\mathrm{i}m\phi} \sim \left(\frac{\Delta}{\sin\Delta}\right)^{1/2} J_m(kX)\mathrm{e}^{\mathrm{i}m\phi}, \tag{15.4.9}$$

in terms of the horizontal range $X = r_e\Delta$ at the surface $r = r_e$. We can represent the wavenumber k in terms of a horizontal slowness p through $k = \omega p$, and in terms of the spherical slowness wp introduced in Chapter 12, $p = \wp/r_e$. Thus, asymptotically, we recover the expressions (15.3.7) for the surface displacements, with a scaling factor $(\Delta/\sin\Delta)^{1/2}$ to compensate for sphericity. The travelling wave representations (15.4.7) for the sphere provide the link to the outgoing wave forms used in (15.3.9).

Dahlen & Tromp (1998) provide a detailed treatment of the construction of the response for a spherically stratified medium. It should be noted that the convention for Fourier transforms over time used in that book (which is common in normal modal studies) is the complex conjugate of the one used here; as a result, the roles of $\mathrm{Im}(k)\gtrless 0$ are interchanged at fixed frequency ω.

16

Body Waves and Surface Waves

The combination of the methods of constructing the spatial and temporal response with subsets of the full representation in the frequency-slowness domain provides a wide range of different approaches to the construction of specific features of the wavefield.

By suitable choices we can concentrate on the distinctive phase arrivals associated with body waves, or on the later-arriving surface waves where the character of the record is expressed through frequency dispersion. The sequence of free-surface multiples for S waves grades into the wavetrain of the surface waves, and this portion of the seismogram is amenable to different styles of realisation.

16.1 Body waves

Those seismic waves propagating into the body of the Earth carry high frequency energy to substantial distances and are recorded as distinct phase arrivals (or groups of arrivals) on surface seismometers. These body wave phases can be extracted from the full seismic response by suitable approximations to the numerical integrals, coupled to targeted representations in the (p, ω) domain constructed from reflection and transmission elements for portions of the stratification. The approximation to the response or the interval of slowness integration will be such that we encounter no difficulties from surface wave poles.

We will build on the approximations developed in Sections 14.6 and 14.7. We will either exploit the isolation of a specific group of similar propagation processes or use *generalized ray* methods where a portion of the seismic response is expressed as a sum of contributions for which the inversion integrals take a simple form.

16.1.1 Reflectivity and associated approximations

Fuchs & Müller (1971) introduced a direct numerical integration technique over the transform domain to construct synthetic seismograms for a portion of the wavefield. In their 'reflectivity' technique attention was concentrated on reflection from beneath a level z_J and only transmission was allowed for in the upper region. This approximation (14.6.4) for the partial response can be written in the form

$$\boldsymbol{w}_o^{re} = \mathbf{W}_F \mathbf{T}_U^{0J} \mathbf{R}_D^{JL} \mathbf{T}_D^{SJ} \boldsymbol{D}^S. \tag{16.1.1}$$

The first numerical integral to produce the seismogram spectrum at range X is

$$\bar{\mathbf{u}}(X,0,t) = \tfrac{1}{2}\omega|\omega|M(\omega) \int_{p_a}^{p_b} \mathrm{d}p\, p[\boldsymbol{w}_o^{re}]^T \mathbf{T}^{(1)}(\omega p X), \tag{16.1.2}$$

where the integration is taken over a finite range of slowness along the real axis, cf. (15.3.13). There is no free-surface, so there are no complications from surface wave poles, but there are branch points at the slownesses for P and S waves for the surface, the source level z_S and also the top and bottom of the reflection zone JL. Fuchs & Müller (1971) restricted the integral to real angles of incidence at z_J, and performed the integral over angle rather than slowness. Most subsequent work has employed a slowness integral and allowed for the possibility of evanescent waves, since these reduce the effects of truncating the integral to a limited range of slowness.

The integral (16.1.2) includes an oscillatory component through the Bessel functions contained in $\mathbf{T}^{(1)}(\omega p X)$, and also commonly quite rapid variation in $\boldsymbol{w}_o^{re}$ as a function of slowness p. The evaluation of the oscillatory integral can be accomplished in a number of ways. When $\omega p X$ is not large the behaviour of $\boldsymbol{w}_o^{re}$ tends to be important and a trapezoidal rule integration will generally give good results if the increment in slowness δp is small enough. For large $\omega p X$ the oscillatory Bessel functions dominate and then there are significant advantages in using adaptations of Filon's method to evaluate the integrals, with a local representation of $\boldsymbol{w}_o^{re}(p,\omega)$ by polynomials. Frazer (1988) provides a useful account of the different classes of procedures which can be used to reduce the computational cost of the slowness integration.

All of the methods we have described for generating the response of a model to source excitation depend on the reflection and transmission properties of portions of the stratification and can be regarded as generalisations of the 'reflectivity' method.

Whether we use the full-response of the medium to simulate reflected events at short distances or a targeted approximation such as (14.6.3) for the main phases at distant stations, we have to evaluate the double integrals over slowness and frequency:

$$\mathbf{u}(X,\phi,0,t) = \frac{1}{2\pi}\int_{-\infty}^{\infty} \mathrm{d}\omega\, \mathrm{e}^{-\mathrm{i}\omega t}\omega^2 \int_0^{\infty} \mathrm{d}p\, p \sum_m \boldsymbol{w}_o^T(p,m,\omega)\mathbf{T}_m(\omega p X), \tag{16.1.3}$$

for some class of representation $\boldsymbol{w}_o$. The commonest approach is to use a spectral approach, and usually just the outgoing Hankel function is used as in (16.1.2). However, when the ranges are small it is better to use the 'standing-wave' forms in terms of the Bessel functions $J_m(\omega p X)$ to avoid singularities at the origin. In this case the suppression of apparently incoming arrivals can be used as a measure of the accuracy of the slowness integration.

The choice of the integration interval in slowness is important, since quite large arrivals can arise from the truncation of the integral. These numerical arrivals can be identified because they appear at one of the limiting slownesses. The edge effects can be muted by applying a taper to $\boldsymbol{w}_o$ near the limits p_a, p_b. Alternatively the linearity of the problem can be exploited by adding integrals for additional segments of the slowness axis to shift the interference from the numerical arrivals. If a further interval (p_b, p_c) is added, the edge effect at p_b cancels. Often p_c can be chosen so that the truncation arrival does not cut a significant portion of the record. The choice of multiple panels in slowness also allows a variable integration increment and can reduce the computational effort required to produce accurate seismograms.

The integration to time from the spectra of the seismograms is normally accomplished using the Fast Fourier transform (FFT), which gives a fixed length of time window. If an arrival occurs beyond the bounds of the window, the periodicity of the FFT means that it will be wrapped back into the window as an aliased arrival. The inclusion of physical attenuation and a well-chosen approximation for the slowness-frequency response to include the arrivals of interest (see Section 14.6) can help to alleviate such problems of aliasing. One device that can also be useful is to introduce a constant imaginary part (ε) so that the calculated seismogram includes a damping $e^{-\varepsilon t}$ (see, e.g., Rosenbaum, 1974; Bouchon, 1979; Temme & Müller, 1982). The amplitudes of arrivals falling beyond the FTT window are reduced and so the impact of aliasing is reduced. After inversion to time the true amplitude can be recovered by applying an $e^{\varepsilon t}$ weighting. However this reweighting amplifies any numerical noise in the later part of the window and so the procedure need to be applied judiciously.

In order to get the maximum benefit from a fixed time interval at varying ranges it is convenient to work in reduced time $t - p_{\mathrm{red}}X$ (Fuchs & Müller, 1971), where p_{red} is chosen so that arrivals can be followed over a range of distances. The necessary time shift is introduced by multiplying the spectrum at range X by $\exp[-\mathrm{i}\omega p_{\mathrm{red}}X]$ before taking the Fourier transform over frequency.

When calculations are carried out close to the source the *P-SV* and *SH* wave parts of the seismogram should be constructed at the same time, because the near-field contributions can then be represented correctly. The *P-SV* and *SH* wave parts each have non-causal, non-propagating arrivals which cancel when the two parts are added (Wang & Herrman, 1980). For long period waves, near-field terms are significant to moderate ranges. A useful working rule is that a far-field representation is adequate for ranges larger than three wavelengths for the longest period waves being considered.

Although the use of a *slowness* approach to the evaluation of the integrals (16.1.3) is less common it offers some advantages, particularly with regard to the isolation of specific arrivals in time. The integral over frequency is evaluated first, and this will require long time windows to avoid aliasing, and then the slowness integral follows (see Section 15.3.2). The integrand in the slowness-time domain is generally well behaved and so sampling in slowness can be reduced compared with the *spectral* integral (Fryer, 1980).

The form and depth extent of the model employed for the construction of the synthetic seismograms will have a significant influence on the total computational effort. It is possible to represent smoothly varying wavespeed profiles by a stairstep structure of uniform layers, but the layers must be thin enough that the internal multiples in the layers do not appear as distinct arrivals. For weak gradients it can be more efficient to use methods based on the Langer approximations (see Section 14.2.2). However, when the wavespeed gradients are strong, the correction terms are very important and it is generally most effective to use a cascade of thin uniform layers.

16.1.2 Teleseismic arrivals

At distant recording sites the time separation between different arrivals whose propagation path penetrates deep into the mantle is long enough that we can concentrate just on a related group of phases, such as S and its associated depth phases pS, sS.

The character of the onset of P and S is used in studies of the mechanism and depth of seismic events in a variety of different inversion schemes (see, e.g., Langston & Helmberger, 1975; Pearce, 1980; Sipkin, 1994; Goldstein & Dodge, 1999). At long periods the influence of the shallow structure near the source and receiver is not too important and relatively simple calculation schemes are effective. As frequency increases the crustal reverberations become more important and can tend to mask small amplitude depth phases. A better comparison can be made with broad-band and short-period records when the influence of structure at source and receiver is included (Douglas, Hudson & Blamey, 1973). A suitable representation of the seismic response is provided by an approximation with only a single reflection from below a separation level z_J (14.6.3) including shallow reverberations (see figure 16.1),

$$\begin{aligned} \boldsymbol{w}_o^{tel} &= \boldsymbol{C}_R \mathbf{R}_D^{JL} \boldsymbol{C}_S, \\ &= \mathbf{W}_U^{fS} (\mathbf{I} - \mathbf{R}_D^{SJ} \mathbf{R}_U^{fS})^{-1} \mathbf{T}_U^{SJ} \mathbf{R}_D^{JL} \mathbf{T}_D^{SJ} (\mathbf{I} - \mathbf{R}_U^{fS} \mathbf{R}_D^{SJ})^{-1} [\boldsymbol{D}^S + \mathbf{R}_U^{fS} \boldsymbol{U}^S], \end{aligned} \tag{16.1.4}$$

which we have written in a form that emphasises the different parts of the propagation process. The element $\boldsymbol{C}_S$ represents the source and the multiples and conversions in the structure near the source, $\mathbf{R}_D^{JL}$ extracts a single reflection below z_J and the element $\boldsymbol{C}_R$ is the contribution from the structure in the neighbourhood of the receiver.

The choice of the separation level z_J depends on the nature of the structure but by choosing it to lie some way below the base of the crust (16.1.4) can be used for all shallow events.

At teleseismic ranges between 3000 km and 9500 km, the PP, SS or HH elements of the mantle reflection matrix $\mathbf{R}_D^{JL}$ can normally be approximated in terms of a simple process of total reflection from the lower mantle so that, including anelastic effects,

$$[\mathbf{R}_D^{JL}]_{PP} \sim \exp\{i\omega\tau_M(p) - i\frac{\pi}{2}\}\mathcal{Q}(\omega), \tag{16.1.5}$$

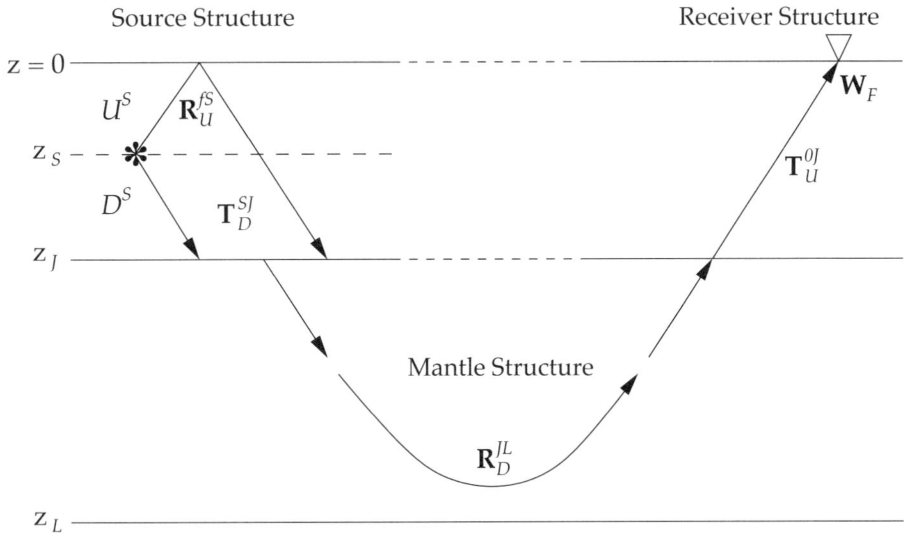

Figure 16.1. Schematic representation of teleseismic propagation processes, shallow reverberations are not indicated.

where $\tau_M(p)$ is the phase delay for the mantle below the separation level z_J,

$$\tau_M(p) = 2\int_{z_J}^{Z_p} \mathrm{d}z\, q_\alpha(z), \tag{16.1.6}$$

and a phase delay of $\exp(-\mathrm{i}\pi/2)$ is associated with the turning point Z_p (14.2.49).

The effect of attenuation in the mantle $\mathcal{Q}(\omega)$ can be included via an empirical operator based on the assumption that for a particular wave type the product of travel-time and overall loss factor is a constant t^*. In terms of frequency

$$\mathcal{Q}(\omega) \approx \exp\left\{-\mathrm{i}\omega t^* \frac{1}{\pi}\ln\left(\frac{\omega}{2\pi}\right)\right\}\exp\left\{-\tfrac{1}{2}|\omega|t^*\right\}, \tag{16.1.7}$$

including the effects of velocity dispersion. For P waves $t_\alpha^* = T_\alpha Q_\alpha^{-1}$ is frequently taken to be 1.0 but for low loss paths can be about 0.3. For S waves t_β^* is rather higher, e.g., Langston & Helmberger (1975) have assumed $t_\beta^* = 3.0$.

In order to obtain an accurate representation of the teleseismic wavefield, core reflections and refractions should also be included, as these phases can interfere with the main P and S arrivals at larger teleseismic distances. For example, *SKS* cuts across the S arrival near 82° and the interference of the main phases, and their associated depth phases, needs to be taken into consideration.

The additional phases can easily be included by modifying the mantle reflection matrix $\mathbf{R}_D^{JL}$. The core reflections (*PcP*, *ScS*, and *ScP*) require the mantle model to be extended to the core-mantle boundary, with the inclusion of the appropriate reflection coefficients. For the refracted core phase *SKS* out to 100°, the model needs to extend to the inner core boundary and the reflection terms involve two-way transmission at the core-mantle boundary, with a turning point approximation for P energy in the fluid outer core.

To recover the seismograms at a given range X we have to evaluate an integral of the form (16.1.2) over a restricted range of positive slowness with $\boldsymbol{w}_0^{tel}$ (16.1.4) as the response term i.e.

$$\bar{\mathbf{u}}(X,0,\omega) = \tfrac{1}{2}\omega|\omega|M(\omega)\int_\Gamma \mathrm{d}p\, p\, [\boldsymbol{w}_0^{tel}]^T \mathbf{T}^{(1)}(\omega p r), \tag{16.1.8}$$

where we have extracted a common source spectrum $M(\omega)$. For simplicity the summation over angular order m to include the influence of source radiation has been suppressed.

The dominant contribution to (16.1.8) will come from the neighbourhood of the saddle point associated with the direct wave. This saddle lies at the geometric slowness p_g given by

$$X + \partial_p\{\tau_M(p_g) + \tau_C(p_g)\} = 0, \tag{16.1.9}$$

where $\tau_C(p_g)$ is the phase delay in the crust corresponding to the transmission terms $\mathbf{T}_U^{0J}\mathbf{T}_D^{SJ}$. The expression in braces is the geometrical ray theory expression for the range $-X(p_g)$ for slowness p_g. The direct waves and their surface reflections and reverberations have slightly different slownesses. With an expansion of the reverberation operators in (16.1.4) there is a sequence of terms corresponding to the surface reflections and crustal multiples each of which is associated with a subsidiary saddle point close to that for the direct arrival.

16.1.2.1 Saddle Point Approximation

An asymptotic approximation to the slowness integral in (16.1.8) for high frequencies can be obtained by the method of steepest descent across the saddle point at p_g.

For P waves the frequency spectrum of the seismogram at range X is then approximated as

$$\bar{\mathbf{u}}(X,0,\omega) = \begin{pmatrix} \mathbf{e}_z C_{RS}^Z(p_{gP},\omega) \\ \mathrm{i}\mathbf{e}_X C_{RS}^R(p_{gP},\omega) \\ 0 \end{pmatrix} \frac{-\mathrm{i}\omega M(\omega)(p_{gP})^{1/2}}{[X\,|\partial_p X(p_{gP})|]^{1/2}} \mathrm{e}^{\mathrm{i}\omega T(p_{gP})} \mathcal{Q}_P(\omega) \tag{16.1.10}$$

where p_{gP} is the geometric slowness and $T(p_{gP})$ the travel-time for the direct P wave. C_{RS} includes amplitude and phase terms associated with the crustal propagation at source and receiver.

For S waves the spectrum takes the form,

$$\bar{\mathbf{u}}(X,0,\omega) = \begin{pmatrix} \mathbf{e}_z C_{RS}^Z(p_{gS},\omega) \\ \mathrm{i}\mathbf{e}_X C_{RS}^R(p_{gS},\omega) \\ -\mathrm{i}\mathbf{e}_\phi C_{RS}^T(p_{gS},\omega) \end{pmatrix} \frac{-\mathrm{i}\omega M(\omega)(p_{gS})^{1/2}}{[X\,|\partial_p X(p_{gS})|]^{1/2}} \mathrm{e}^{\mathrm{i}\omega T(p_{gS})} \mathcal{Q}_S(\omega) \tag{16.1.11}$$

where the various quantities are evaluated at the slowness p_{gS} for the direct S wave.

This approximation has been justified by treating the radiation leaving the source crust as seen at large ranges as a plane wave, using ray theory in the mantle and a plane wave amplification factor $\mathbf{W}_U^{fJ}$ at the receiver (see e.g. Hudson, 1969b). For comparison with long-period records the main interest is in the interference of the direct wave and the surface reflected phases. Langston & Helmberger (1975) have introduced a simple approximation based on a composite source term to model these effects. For *P* waves, a composite downward radiation term is used including the surface reflections,

$$D_P^C = [\boldsymbol{D}^S]_P + [\mathbf{R}_F]_{PP}\mathrm{e}^{\mathrm{i}\omega\Delta\tau_1}[\boldsymbol{U}^S]_P + [\mathbf{R}_F]_{PS}\mathrm{e}^{\mathrm{i}\omega\Delta\tau_2}[\boldsymbol{U}^S]_S, \tag{16.1.12}$$

where $\Delta\tau_1$ is the phase lag of *pP* relative to the direct wave and $\Delta\tau_2$ is the phase lag of *sP*. The change in wavefront divergence on conversion at the surface is included implicitly in (16.1.12) through our use of energy normalised reflection coefficients $[\mathbf{R}_F]_{PP}$, $[\mathbf{R}_F]_{PS}$. With this composite source a single slowness is used to calculate receiver effects and mantle propagation is included only through the attenuation operator. Thus the receiver displacement is approximated by

$$\bar{\mathbf{u}}(r,0,\omega) = -[\mathbf{W}_F]_P D_P^C(p_g,\omega)\mathcal{Q}(\omega)\mathrm{i}\omega M(\omega), \tag{16.1.13}$$

where p_g is the geometric slowness given by (16.1.9). Langston & Helmberger suggest the use of a far-field source time function consisting of a trapezoid of unit height described by three time parameters, which allows relative time scaling of rise time, fault duration and stopping time.

16.1.2.2 Adaptive reflectivity method

The saddle-point approximation is based on the use of a single slowness and neglects the contributions from the subsidiary saddle points associated with the surface reflections and crustal multiples. The range of slownesses occupied by the saddles is not large and we can achieve a more accurate result by using a bundle of slownesses clustered around p_g rather than the single slowness in (16.1.10), (16.1.11). This provides an improved representation of amplitude effects due to conversion, and also of the decay of the crustal reverberations.

For teleseismic *SV* waves conversion to *P* at the Moho is important and using a range of slownesses we can also model shear coupled *PL* waves. Also for closer ranges than 3000 km we need to make a more accurate representation of the reflection terms in $\mathbf{R}_D^{JL}$ to account for the detailed structure in the upper mantle. The presence of triplications in the travel time curves means that a band of slownesses is needed to represent the response.

A restricted slowness integral, tuned to a particular range, can be used to evaluate the spectral contribution

$$\bar{\mathbf{u}}(X,\phi,0,\omega) = \tfrac{1}{2}\omega|\omega|M(\omega)\int_{p_a}^{p_b} \mathrm{d}p\, p\, h(p)\sum_m [\boldsymbol{w}_0^{tel}]^T\mathbf{T}_m^{(1)}(\omega pX), \tag{16.1.14}$$

where the interval (p_a, p_b) covers a range of slownesses clustered around the geometric slowness for the direct wave, and a weighting function $h(p)$ is included to mute truncation artifacts (Marson-Pidgeon & Kennett, 2000a). Once the sum over azimuthal order in (16.1.14) has been carried out to to include source radiation effects, the numerical integration over p can be performed over a limited bundle of slownesses using a trapezium rule or the modified Filon form (Frazer 1988). Since the integration is over a small slowness interval, numerical arrivals at the limiting slownesses are often produced; these can be muted by applying cosine tapers in the weighting function $h(p)$ at the ends of the slowness interval.

The use of the adaptive reflectivity ('slowness bundle') approach provides a very effective representation of the expected arrivals with an appropriate treatment of amplitude effects due to conversion, and the decay of the crustal reverberations.

Figure 16.2 shows a comparison of a number of simulations of S wave teleseismic seismograms at an epicentral distance of 60° using both the single slowness approximation and the slowness bundle approach. In the restricted slowness integral we have employed a bundle of 100 slownesses spanning the interval $[0.95p_g, 1.05p_g]$ about the geometric slowness p_g. For comparison, we also include seismograms calculated using generalized ray theory, by modeling S, pS and sS, using (16.1.12), (16.1.13) adapted to S wave propagation.

The single slowness calculations are performed using a saddle-point approximation at the geometric slowness p_g for direct S, and illustrate the problems associated with using a single slowness approach. Significant energy arriving after the main S wave packet can be seen in figure 16.2(a). The later arriving energy appears to be a combination of regional P waves and their corresponding surface reflections, which have then undergone conversion to S near the source, and waves which have undergone reflection and conversion at the free surface near the receiver (such as SP_n). These additional waves with slowness p_g would not be expected to arrive at the same range. These extra arrivals are associated with the use of the full reverberations in the source and receiver structures and can be muted by calculating a restricted response. At the source we set $[\boldsymbol{D}^S + \mathbf{R}_U^{fS}\boldsymbol{U}^S]_P = 0$, which ensures that there are no downward traveling P waves below the source level, thus eliminating regional P waves. At the receiver end we impose the constraints $[\mathbf{R}_D^{0J}]_{PP} = [\mathbf{R}_D^{0J}]_{PS} = 0$, to ensure that there are no P waves generated due to reflection from a downward incident wave in the receiver end, thus eliminating such phases as SP_n. The traces in figure 16.2(b) were calculated by applying these restrictions, and it can be seen that the late arriving energy has been muted.

This problem of unwanted arrivals does not occur when a slowness bundle approach is used as can be seen in figure 16.2(c). The addition of a set of responses over a range of slownesses with phase terms depending on epicentral distance, tends to cancel the late arrivals associated with different epicentral distances.

We note a number of differences between the single slowness results and the slowness bundle results, e.g., the amplitude of the sS arrival on the vertical component of the single slowness restricted response is underestimated compared to the slowness bundle response. A prominent arrival is evident between S and pS on the vertical

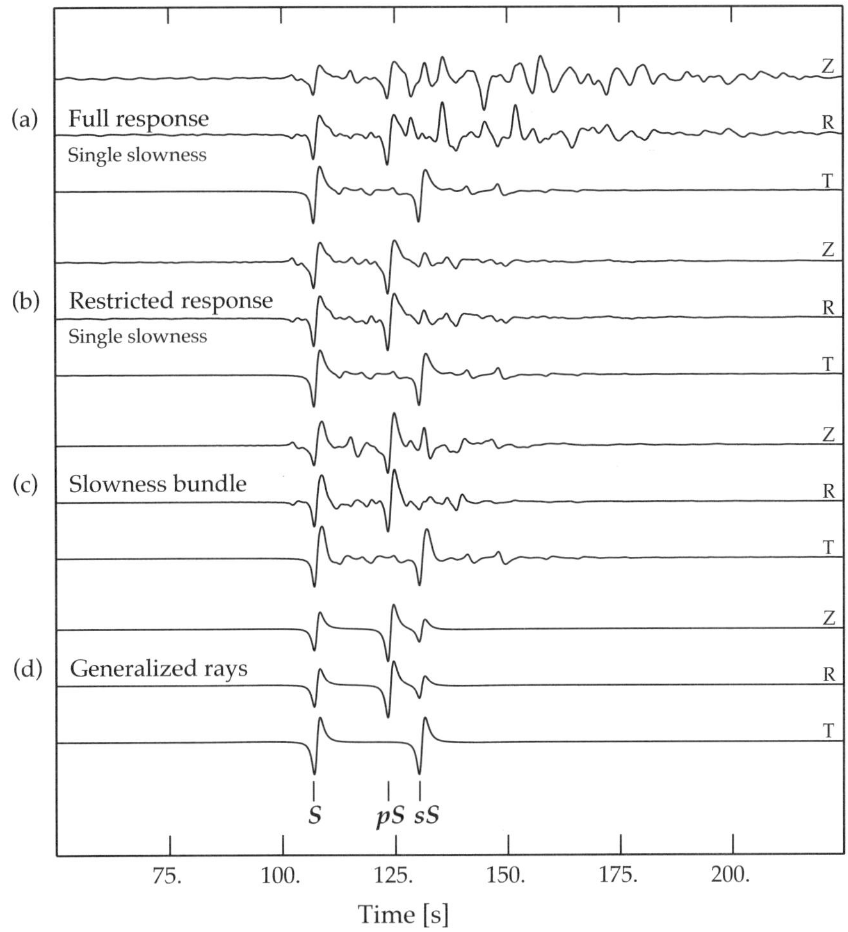

Figure 16.2. *S* wave synthetic seismograms at a distance of 60° for a source depth of 50 km in the *ak135* model. (a) The 3-component traces are calculated using a single slowness and the full response. (b) The 3-component traces are calculated using a single slowness and a restricted response. (c) The 3-component traces are calculated using a bundle of 100 slownesses. (d) The 3-component traces are calculated using generalized ray theory.

component of the slowness bundle traces, which is barely noticeable on the trace with a restricted response for a single slowness. However, this arrival can clearly be seen on the vertical component of the single slowness seismograms with the full response. Thus the constraints used to remove the late arriving energy in the upper traces are a little too restrictive.

The generalized ray results in figure 16.2(d) capture the main arrivals quite well but cannot account for the effects of crustal structure since these are not included in (16.1.13).

16.1.3 Approximations for specific seismic phases

The class of representation (16.1.4) employed for the study of the onset of teleseismic P and S waves is also suitable for studying specific phases with a more sophisticated treatment of the reflection elements in $\mathbf{R}_D^{JL}$.

Suitable forms for the partial response can be built up using, for example, the reflection and transmission properties of gradient zones using the Langer approximations (Section 14.2.2). In this way the role of evanescent waves can be recognised and captured so that tunnelling phenomena can be included. The combination of such partial response terms with a contour integral in the complex p plane to evaluate the spectral response forms the basis of the 'full-wave' method (Richards, 1973; Choy, 1977; Cormier & Richards, 1977; Aki & Richards, 1980 chapter 9). The contour Γ needs to be chosen with a knowledge of the singularities of the partial response in the complex slowness plane. The development of the approximate response terms also needs to take into account situations where a reverberation sequence is needed to cover the range of significant processes, e.g., underside reflections at the top of the core. Cormier and Richards (1988) provide an account of the computational implementation of the 'full-wave' technique.

16.1.4 Generalized rays

As we have seen in Section 14.7, we can expand the frequency-slowness representation for the seismic wavefield as a summation over contributions associated with different styles of propagation path through the medium,

$$\boldsymbol{w}_o(p,m,\omega) = M(\omega)\sum_I \boldsymbol{g}_I(p,m)\mathrm{e}^{\mathrm{i}\omega\tau_I(p)}. \tag{16.1.15}$$

For each of the 'generalized ray' contributions: $\boldsymbol{g}_I(p,m)$ will include excitation by the source, with angular dependence through m, as well as the product of reflection and transmission coefficients for the path, $\omega\tau_I(p)$ will represent the accumulated phase along the path. $M(\omega)$ is the common time dependence for all the moment tensor elements describing the source.

The surface displacement contribution with azimuthal order m from a single generalized ray is given by

$$\mathbf{u}_m^I(X,t) = \frac{1}{2\pi}\int_{-\infty}^{\infty}\mathrm{d}\omega\,\mathrm{e}^{-\mathrm{i}\omega t}\omega^2 M(\omega)\int_0^{\infty}\mathrm{d}p\,\boldsymbol{w}_I(p,m,\omega)\mathbf{T}_m(\omega pX), \tag{16.1.16}$$

with $\boldsymbol{w}_I(p,m,\omega) = \boldsymbol{g}_I(p,m)\mathrm{e}^{\mathrm{i}\omega\tau_I(p)}$. We now follow the *slowness* treatment of Section 15.3.2 and perform the frequency integral first so that we express (16.1.16) as in (15.3.15) as a convolution over time linked to an integral over p:

$$\boldsymbol{u}_m^I(r,t) = \partial_{tt}M(t) * \int_0^{\infty}\mathrm{d}p\,p\left\{\check{\boldsymbol{w}}_I(p,m,t) * \frac{1}{pX}\check{\mathbf{T}}_m\left(\frac{t}{pX}\right)\right\}. \tag{16.1.17}$$

We need therefore to evaluate the Fourier inverse of $\boldsymbol{w}_I(p,m,\omega)$. The final seismograms must be real time functions and so $\mathbf{u}(X,\omega) = \mathbf{u}^*(X,-\omega)$. With the choice of physical Riemann sheet we have made for the radicals appearing in $\boldsymbol{g}_I(p,m)$, $\tau_I(p)$ we require

$$\boldsymbol{g}_I(p,m) = \boldsymbol{g}_I'(p,m) + \mathrm{i}\,\mathrm{sgn}(\omega)\boldsymbol{g}_I''(p,m),$$
$$\tau_I(p) = \tau_I'(p) + \mathrm{i}\,\mathrm{sgn}(\omega)\tau_I''(p). \tag{16.1.18}$$

We may now perform the inverse time transform for $\boldsymbol{w}_I$ to obtain (Chapman, 1978)

$$\boldsymbol{w}_I(p,m,t) = \frac{1}{\pi}\mathrm{Im}\left[\frac{\boldsymbol{g}_I(p,m)}{t-\tau(p)}\right]. \tag{16.1.19}$$

As the imaginary part of τ_I tends to zero and there is just a real phase delay, and $\boldsymbol{w}_I$ tends to a delta function:

$$\mathrm{Im}[t-\tau(p)-0\mathrm{i}]^{-1} \to \delta(t-\tau'(p)) \quad \text{as} \quad \tau'' \to 0. \tag{16.1.20}$$

As in Section 15.3.2 we now restrict attention to the case of azimuthal symmetry and consider just the vertical component of displacement where the radial dependence arises from $J_0(\omega p X)$ The displacement contribution can be found from the convolution of the two slowness dependent terms (16.1.19), (15.3.17),

$$u_{z0}^I(X,t) = \tag{16.1.21}$$
$$\partial_{tt}M(t) * \int_0^\infty \mathrm{d}p\, p\left\{\frac{1}{\pi^2}\int_{-pX}^{pX}\mathrm{d}s\,\mathrm{Im}\left[\frac{G_I(p)}{t-s-\tau_I(p)}\right]\frac{1}{(p^2X^2-s^2)^{1/2}}\right\},$$

where $G_I(p)$ is the vertical component of $\boldsymbol{g}_I(p,0)$. The Im operator can be brought to the front of the slowness integral because only $G_I(p)$ and $\tau_I(p)$ may be complex along the real p axis.

The time and slowness elements in (15.3.17) are common to all the expressions for $\check{\mathbf{T}}_m(t/pr)$ and so the form of the integral in (16.1.22) is modified for other components or angular orders by the addition of well behaved functions. Various methods of calculating theoretical seismograms can now be generated by using different techniques to evaluate the slowness integral in (16.1.22).

Chapman & Orcutt (1985, §4.6) provide an extensive discussion of a group of problems where the contribution to the response can be cast in 'generalized ray' form and for which an analytic approximation can be extracted; these include direct and reflected rays, turning rays, head waves, interface waves, caustics and shadows.

16.1.4.1 The Cagniard method

The application of Cagniard's method to a model composed of a stack of uniform layers was first made by Helmberger (1968), and was also investigated by Müller (1970). Chapman (1976) has shown how this approach can be used with an iterative development to study waves in smoothly varying media.

We examine the inversion of a single generalized ray contribution, without the source time function. We construct

$$u_I(X,t) = \frac{1}{2\pi}\int_0^\infty dp\, p \int_{-\infty}^{\infty} d\omega\, e^{-i\omega t} G_I(p) e^{i\omega\tau_I(p)} J_0(\omega p X) \tag{16.1.22}$$

$$= \frac{1}{4\pi}\int_0^\infty dp\, p \int_{-\infty}^{\infty} d\omega\, e^{-i\omega t} G_I(p) e^{i\omega\tau_I(p)} \{H_0^{(1)}(\omega p X) - H_0^{(1)}(-\omega p X)\}$$

Consider, first, the part which depends on $H_0^{(1)}(\omega p X)$. We distort the contour of integration in the p-plane to the Cagniard path Υ such that

$$\mathrm{Im}\{\theta_I(p,X)\} = \mathrm{Im}\{\tau_I(p) + pX\} = 0. \tag{16.1.23}$$

The contour Υ lies in the fourth quadrant and leaves the real p-axis at the point where θ is equal to the geometric travel time. Along this contour we have a real time variable and so the Fourier transform over ω may be inverted to give

$$u_{I+}(X,t) = \frac{1}{2\pi i}\int_\Upsilon dp\, p G_I(p) \frac{H(t-\tau(p)-pX)}{[(t-\tau(p))^2 - p^2X^2]^{1/2}}. \tag{16.1.24}$$

For the second part, which depends on $H_0^{(1)}(-\omega p X)$, the contour of integration is taken over the mirror image of Υ in the real p-axis. Using the Schwarz reflection principle u_{I-} can be related to the complex conjugate of u_{I+},

$$u_{I-}(X,t) = -[u_{I+}(X,t)]^*. \tag{16.1.25}$$

The generalized ray contribution thus takes the form

$$u_I(X,t) = \frac{1}{\pi}\mathrm{Im}\int_\Upsilon dp\, p G_I(p) \frac{H(t-\tau(p)-pX)}{[(t-\theta)(t-\theta+2pX)]^{1/2}}. \tag{16.1.26}$$

in terms of the time variable θ (16.1.23). We can cast (16.1.26) in the form of a convolution integral in time by changing the variable of integration to θ. Then

$$u_I(X,t) = \frac{1}{\pi}\mathrm{Im}\int_0^t d\theta\, p G_I(p) \left[\frac{\partial p}{\partial \theta}\right] \frac{H(t-\tau(p)-pX)}{[(t-\theta)(t-\theta+2pX)]^{1/2}}, \tag{16.1.27}$$

where p is an implicit function of θ. To reinstate the source time function we must convolve $u_I(X,t)$ with $\partial_{tt}M(t)$. The detailed behaviour of (16.1.27) will depend on the nature of the product of reflection and transmission terms in $G_I(p)$ and the phase delay $\tau(p)$.

For all but the closest ranges, the early part of the contribution from the generalized ray can be obtained by replacing $(t-\theta+2pX)^{1/2}$ in (16.1.26) by $(2pX)^{1/2}$. When we include the source time function, the resulting approximation for the contribution to the seismogram from the Ith generalized ray is

$$u_I(X,t) = \mathcal{M}(t) \star \mathrm{Im}\left\{G_I(p)\left[\frac{\partial p}{\partial \theta}\right]\left(\frac{p}{2X}\right)^{1/2}\right\}, \tag{16.1.28}$$

in terms of the 'effective' source $\mathcal{M}(t)$ defined in (15.3.22).

The properties of the contribution from a generalized ray are controlled by the character of the Cagniard path Υ, in the complex p plane together with the positions of

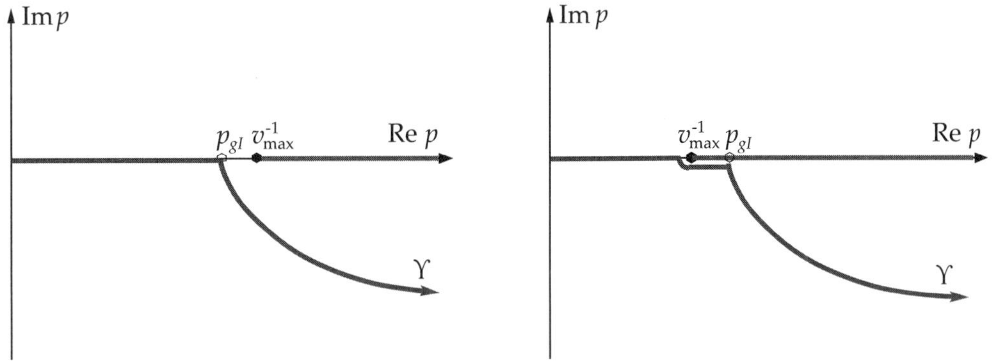

Figure 16.3. The Cagniard path Υ in the complex p-plane, (a) only reflected contributions, (b) head wave segment in additions to reflections.

the branch points for the phase delay $\tau_I(p)$ and the amplitude factor $\boldsymbol{g}_I(p)$. The path Υ is defined by $\text{Im}[pX + \tau_I(p)] = 0$, and is shared by all members of a kinematic group $\{k\}$ (14.7.10); members of a dynamic group $\{d\}$ share the same amplitude distribution along the path.

The Cagniard path Υ starts off along the real axis from $p = 0$ but turns away at right angle at the saddle point corresponding to the geometrical slowness $p_{gI}(X)$ for range X at which $\partial_p\theta_I = 0$. The branch point with smallest slowness lies at the inverse of the maximum wavespeed along the ray path v^{-1}_{max} (including the wavespeed on the deeper side of the lowest interface encountered). If v^{-1}_{max} lies closer to the origin than the saddle point p_{gI} then there will be a head-wave contribution (see figure 16.3). Such head wave segments occur moderately frequently for pure P paths and are very common for generalized rays with a number of S wave legs.

The contribution from a particular generalized ray path can be represented in a convenient form (Vered & BenMenahem, 1974). The head wave contribution is

$$\text{Im}\int_{t_h}^{t_X} \mathrm{d}\theta_I \frac{G_I(p)}{\partial_p\theta_I} U(p,\theta_I), \tag{16.1.29}$$

where $t_h = \theta_I(v^{-1}_{\text{max}}, X)$, $t_X = \min[t, \theta_I(p_{gI}, X)]$ and $U(p,t)$ represents the remainder of the integrand in (16.1.27).

After the geometrical arrival time $t_g = \theta_I(p_{gI}, X)$ we can write the response as

$$\text{Im}\int_{t_g}^{t} \mathrm{d}\theta_I \frac{G_I(p_{gI})}{\partial_p\theta_I} U(p,\theta_I) + \text{Im}\int_{t_g}^{t} \mathrm{d}\theta_I \frac{[G_I(p) - G_I(p_{gI})}{\partial_p\theta_I} U(p,\theta_I). \tag{16.1.30}$$

The first term represents the contribution from reflection along the geometric ray path, and since $\partial_p\theta_I$ vanishes at $p = p_{gI}$ the main contribution to the beginning of the reflected wave will come from the neighbourhood of p_{gI}. The second term represents arrivals with non-least-time paths, and since it vanishes at $\theta(p_{gI}, r)$ makes only a small contribution to the reflection.

The Cagniard method is particularly well suited to models with a limited number of homogeneous layers for which all significant rays can be readily evaluated. The application to gradient zones requires a truncated ray expansion and significant effort in finding the Cagniard contours (see Pao & Gajewski, 1977).

16.1.4.2 The Chapman method

The expression (16.1.22) employs the full form of the transform of the J_0 Bessel function and we can make a further approximation as in (15.3.19) to represent the kernel of the integral in terms of two isolated singularities along the lines $t = \pm pX$. We retain the outgoing term associated with the singularity along $t = pX$,

$$u^I_{z0}(X,t) \approx \partial_{tt}M(t) \star \frac{1}{\pi(2X)^{1/2}} \int_0^\infty \mathrm{d}p\, p^{1/2} \left\{ \frac{1}{\pi} \int_{-\infty}^{\infty} \mathrm{d}s\, \mathrm{Im}\left[\frac{G_I(p)}{s - \theta_I(p,X)} \right] \frac{H(s-t)}{(s-t)^{1/2}} \right\}, \tag{16.1.31}$$

and we recall $\theta_I(p,X) = \tau_I(p) + pX$. The expression (16.1.31) is very suitable for imaginary $G_I(p)$, such as for a turning ray. However, when we consider the general situation of a generalized ray including the possibility of post-critical reflection from interfaces as well as turning ray components we need a form which is suitable for general complex $G_I(p)$. Chapman (1978) has proposed the use of a combination of (16.1.31) and its Hilbert transform (appropriate for real $G_I(p)$) to give the form

$$u^I_{z0}(X,t) \approx \partial_{tt}M(t) \star \frac{1}{\pi^2(2X)^{1/2}} \int_0^\infty \mathrm{d}p\, p^{1/2} \mathrm{Im} \left\{ \int_{-\infty}^{\infty} \mathrm{d}s\, \Lambda(t) G_I(p) \mathrm{Im}\left[\frac{1}{s - \theta_I(p,X)} \right] \right\}. \tag{16.1.32}$$

Here $\Lambda(t)$ is the analytic time function

$$\Lambda(t) = \frac{H(t)}{t^{1/2}} + \frac{H(-t)}{(-t)^{1/2}}, \tag{16.1.33}$$

and combines the inverse square root operator and its Hilbert transform.

For a perfectly elastic medium, $\tau(p)$ is real for those slownesses for which all legs of the generalized ray path involve propagating waves. In this case we recall that from (16.1.20) there will be a δ-function contribution from each slowness value at which $t = \theta_I(p,X)$. We can therefore evaluate (16.1.32) by splitting up the slowness range into intervals spanning each root p_j of $t = \theta_I(p,X)$, corresponding to a particular geometrical ray arrival, and then changing the variable of integration to θ. The contribution from the various arrivals takes the form

$$u^I_{z0}(X,t) \approx \partial_{tt}M(t) \star \frac{1}{\pi^2(2X)^{1/2}} \mathrm{Im} \left\{ \Lambda(t) \star \sum_j \frac{p_j^{1/2} G_I(p_j(t))}{\partial_p \theta_I(p_j(t),X)} \right\}. \tag{16.1.34}$$

which Chapman (1978) has termed the *WKBJ seismogram*. This approximation omits contributions from evanescent waves.

With a locally quadratic approximation to $\theta_I(p,X)$, the results of geometric ray theory can be extracted for an isolated turning ray. The WBKJ seismogram has the

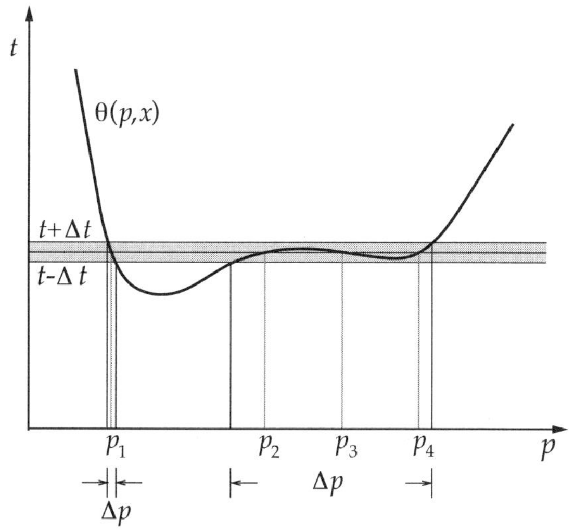

Figure 16.4. The construction of the WKBJ seismogram by smoothing over an interval $(t-\Delta t, t+\Delta t)$ around the desired time t.

distinct advantage that it is still usable at caustics and shadow boundaries where the geometric ray theory fails.

For applications to realistic wavespeed models, Dey-Sarkar & Chapman (1978) demonstrate the value of band-limited frequency filtering which smooths and stabilises the result (16.1.34) by minimising the influence of small fluctuations in θ_I. The stabilising effect of smoothing is illustrated in figure 16.4. For an isolated arrival near p_1, over the time interval $(t - \Delta t, t + \Delta t)$ the associated increment in slowness Δp_1 is small. However, for the small triplication $\{p_2, p_3, p_4\}$ which cannot be resolved at the discretisation interval Δt the effective increment Δp is quite long. Chapman, Chu & Lyness (1988) provide a discussion of the computational implementation of the WKBJ seismogram algorithm.

The merit of the WKBJ seismograms rests with the speed of calculation and the robustness of the results which do not depend on the details of the interpolation of the wavespeed model. With a knowledge of the travel-time and reflection characteristics of the model it is possible to get a rapid idea of the character of contributions to the wavefield; some care has to be taken to avoid truncation artifacts associated with the truncation of the slowness integrals.

16.1.4.3 Inclusion of attenuation

In order to achieve the simplifications of the inverse transforms in both Cagniard's and Chapman's methods for determining the response of a generalized ray we have had to assume that the medium is perfectly elastic.

We can model the effects of attenuation on seismic propagation by letting the elastic wavespeeds become complex (see section 8.3), and in general the wavespeeds will need to be frequency dependent because of the frequency dispersion associated with causal attenuation. In an attenuative medium for a path with just P wave legs, to a first-order approximation,

$$\tau(p) \to \tau_R(p) - \frac{1}{\pi}\ln\left|\frac{\omega}{2\pi}\right| t^*_\alpha(p) + \mathrm{i}\,\mathrm{sgn}\omega\tfrac{1}{2}t^*_\alpha(p) \tag{16.1.35}$$

where $\tau_R(p)$ is constructed from the real part of the velocity distribution. The integrated effect of the attenuation can be found as a sum over ray path elements,

$$t^*_\alpha(p) = \sum_r n_r \int_{z_r}^{z_{r+1}} \mathrm{d}z\, \frac{1}{\alpha^2 Q_\alpha q_{\alpha R}}, \tag{16.1.36}$$

with suitable allowance for turning points, as in (14.7.3).

We see from (16.1.36) that for attenuative media the separation of frequency and slowness effects that has formed the basis of the generalized ray representation cannot be sustained.

For Cagniard's method attenuation can be introduced into the final ray sum by applying an attenuation operator, such as (16.1.7), to each ray contribution allowing for the nature of the path. Burdick & Helmberger (1978) have compared the use of attenuation corrections for each generalized ray to the results obtained by applying a single attenuation operator to the full ray sum. They suggest that often the simpler approximation is adequate.

Chapman et al (1988) discuss the application of the WKBJ seismogram method to slightly attenuative media and suggest the convolution of an attenuation operator $\hat{\mathcal{Q}}(t)$ to the sum over the contributions from θ_I. The operator $\hat{\mathcal{Q}}(t)$ is obtained by taking the inverse Fourier transform of (16.1.7),

$$\hat{\mathcal{Q}}(t) = \mathcal{F}^{-1}\left[\exp\left\{-\mathrm{i}\omega t^* \frac{1}{\pi}\ln\left|\frac{\omega}{2\pi}\right|\right\}\exp\left\{-\tfrac{1}{2}|\omega| t^*\right\}\right] \tag{16.1.37}$$

using the t^* value calculated along the ray path as in (16.1.36).

Usually $t^*(p)$ is a slowly varying function of p and so the approximation of a single attenuative operator for a particular generalized ray contribution at the range X will be justified. However, in complex triplications such as those associated with the upper mantle transition zone it may be be necessary to apply a separate attenuation operator to each contribution to the WKBJ seismogram.

16.2 Ray-mode duality

A significant feature of the seismic wavefield is the presence of guided waves trapped in the zone of low velocity near the surface. These waves are multiply reflected between the free surface and shallow sub-surface structure. The interference of many multiple reflections generated complex wavetrains in which the apparent slowness varies with frequency so that an alternative representation is provided by the superposition of many modes of the near-surface waveguide associated with the increase in seismic wavespeed with depth.

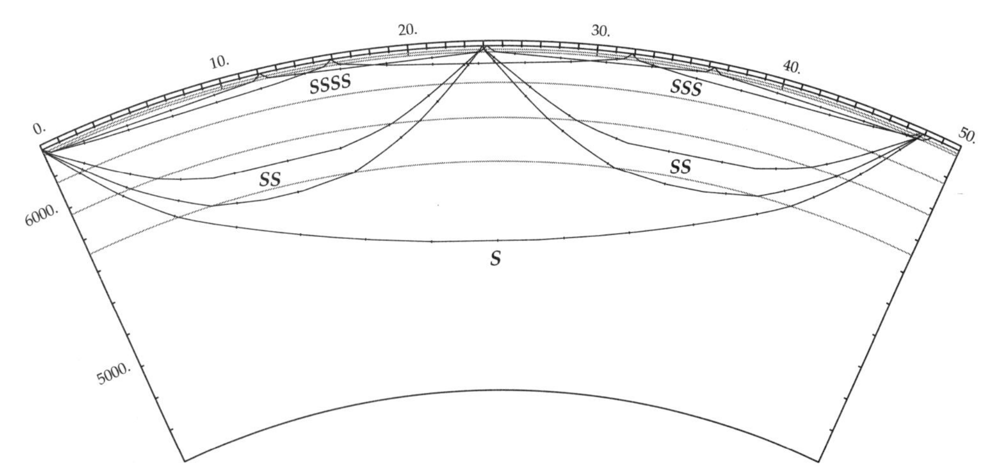

Figure 16.5. *S* propagation paths to a receiver at an epicentral distance of 48°. Note the separation of the *S*, *SS* phases from the sequence of multiples *SS*, *SSSS*, ... which can equally well be described by a few surface wave modes.

We illustrate in figure 16.5 the configuration of *S* waves reaching a receiver at 48° epicentral distance in the AK135 model. The direct *S* waves and the first surface multiple *SS* have turning points well separated in depth from the higher order multiples. The set of multiple reflections *SSS*, *SSSS*, ... crowd into a zone near the surface trapped between the free surface and the zone of rapid increase of *S* wavespeed with depth. The full sequence of *S* multiples is described by the expansion of the operator $(\mathbf{I} - \mathbf{R}_D^{0L}\mathbf{R}_F)^{-1}\mathbf{R}_D^{0L}$ and we can write this expansion in a form where we can separate out the *S* and *SS* contributions by isolating the *SS* or *HH* elements of the reflection terms:

$$(\mathbf{I} - \mathbf{R}_D^{0L}\mathbf{R}_F)^{-1}\mathbf{R}_D^{0L} = \mathbf{R}_D^{0L} + \mathbf{R}_D^{0L}\mathbf{R}_F\mathbf{R}_D^{0L} + (\mathbf{I} - \mathbf{R}_D^{0L}\mathbf{R}_F)^{-1}\mathbf{R}_D^{0L}\mathbf{R}_F\mathbf{R}_D^{0L}\mathbf{R}_F\mathbf{R}_D^{0L}. \tag{16.2.1}$$

The time separation of the successive multiples in the last term in (16.2.1) will be small and, at moderate frequencies, this will lead to a complex interference of overlapping pulses in which it is not possible to distinguish the individual contributions from the multiple *S* phases to the seismograms. The representation (16.3.1) shows how the higher order multiples can be treated as a composite entity. The shift of view to working with surface waves is to concentrate on the singularities of the reverberation operator $(\mathbf{I} - \mathbf{R}_D^{0L}\mathbf{R}_F)^{-1}$ between the free surface and the whole of the wavespeed models. The pole singularities arise because it is possible to find slowness-frequency combinations for which waves can simultaneously satisfy the surface condition of vanishing traction and the requirement for decay at depth. The surface condition enters through $\mathbf{R}_F$ and the decay condition by our restriction to downgoing waves in the region below L through $\mathbf{R}_D^{0L}$. For slowness larger than β_L^{-1} this imposes an evanescent decay in $z > z_L$.

The relation between the propagating wave representation in terms of sequences of multiple reflection and the surface wave form described by standing waves in depth can be well illustrated by comparing the response in the slowness-time ($p - \tau$) and slowness-frequency ($p - \omega$) domains. In figures 16.6 and 16.7 we show pairs of displays of the slowness behaviour, in both time and frequency, for the full response from a double couple source at 25 km depth in the AK135 upper mantle including the effects of attenuation. The calculations were carried out for a frequency band from 0.005 - 0.125 Hz with a time sample of 2 s, in a flattened model with 90 layers to 1000 km depth. Figure 16.6, shows the vertical component of motion and thus the *P-SV* wavefield corresponding to Rayleigh wave modes. The *SH* wavefield is displayed in figure 16.7 through the tangential component, corresponding to Love waves. The upper panel of each of the figures 16.6, 16.7 shows the slowness-time traces derived by a Fast Fourier transform of the p-ω response. The lower panel shows the amplitude in the slowness-frequency domain. All the displays are normalised to a common maximum amplitude on each trace.

For the AK135 model the surface wavespeeds are 5.80 km/s for P and 3.46 km/s for S, so that P waves become evanescent at 0.172 s/km and S waves at 0.289 km/s. Comparison of figures 16.6 and 16.7 shows the substantial differences between the *P-SV* wavefield and the *SH* wavefield. The sequence of branches associated with multiple S propagation in the upper mantle are very clear for *SH* waves (figure 16.7) for slowness less than 0.22 s/km; the progressive loss of higher frequencies for successive branches shows the influence of mantle attenuation. By contrast, for *P-SV* waves (figure 16.6) the multiple branches are less distinct and there is a major loss of amplitude once P waves can propagate in the crust ($p < 0.172$ s/km). The regime of distinct body wave arrivals for slowness less than 0.22 s/km is replaced by quasi-harmonic trains for larger slownesses which extend beyond the surface slowness for S (0.289 s/km) for the *P-SV* wavefield. These decaying oscillatory traces arise from the superposition of very large numbers of multiple S phases with propagation largely confined to the crust, and as we see in the lower panels have an alternative representation in terms of just one or two modes with a well defined trajectory in the slowness-frequency domain.

The simplicity of the frequency response in which we see a very clear set of distinct mode branches is in stark contrast to the complexity of the slowness-time response. Each of the branches in the p-ω domain corresponds to a different pole singularity of the reverberation operator $(\mathbf{I} - \mathbf{R}_D^{0L}\mathbf{R}_F)^{-1}$ and can be classified by the number of modes in the associated eigenfunction as a function of depth (see Section 16.3). The fundamental mode of Love waves (figure 16.7) has a high frequency asymptote at the surface S slowness β_0^{-1} (0.289 s/km). The fundamental Rayleigh mode (figure 16.6) extends to larger slowness, linking evanescent P and S waves, and we note that the modal branch is interrupted at the surface S slowness due to the effect of the surface amplification term $\mathbf{W}_F$.

In Section 15.1, we have displayed a comparison of the slowness - time and slowness - frequency response for P waves in the upper mantle with the inclusion of a single

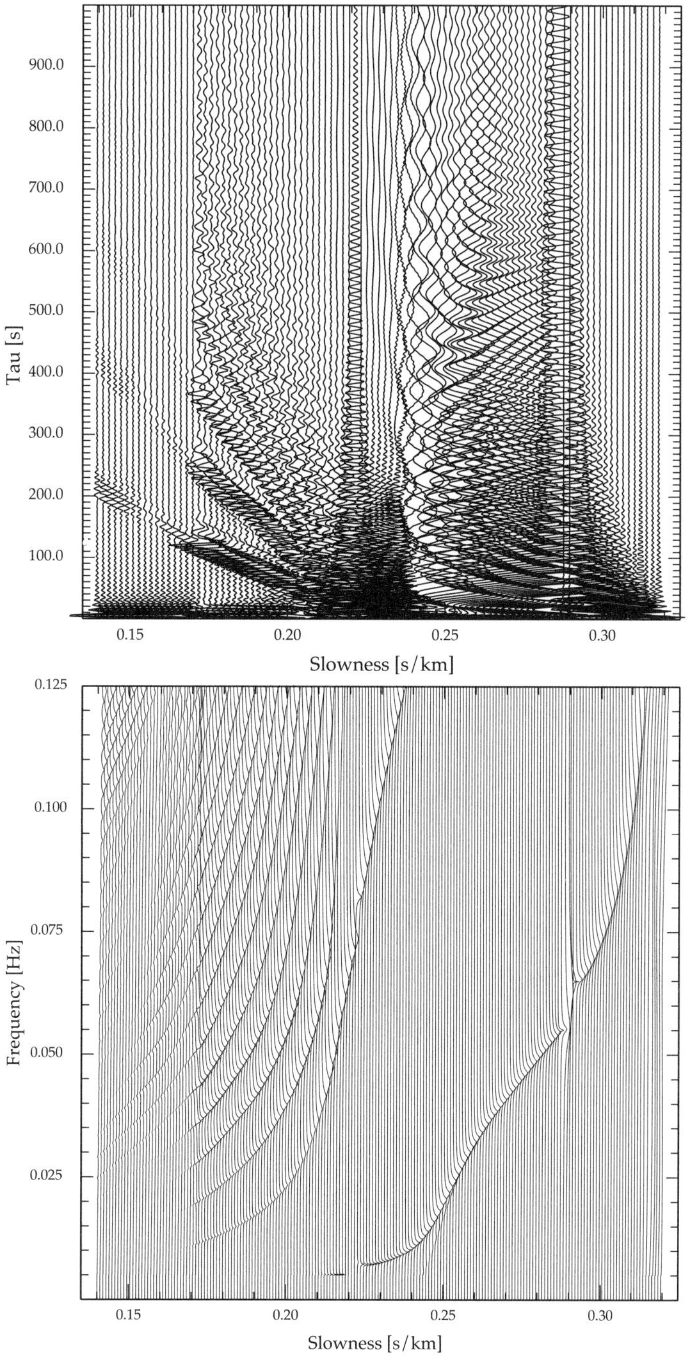

Figure 16.6. Vertical component of the full response of the AK135 model to 1000 km to excitation by a double couple source at 25 km depth. (a) slowness-time domain, (b) slowness-frequency domain.

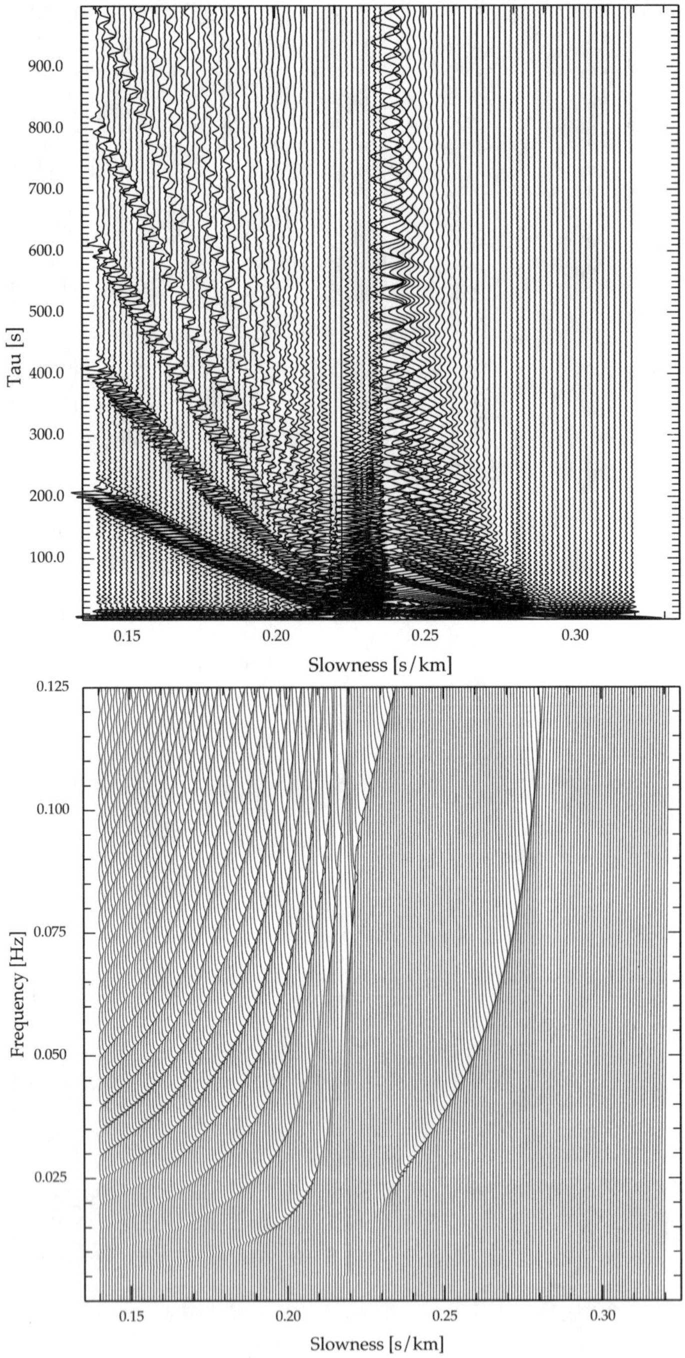

Figure 16.7. Tangential component of the full response of the AK135 model to 1000km to excitation by a double couple source at 25 km depth. (a) slowness-time domain, (b) slowness-frequency domain.

surface reflection (figures 15.1, 15.2). In that case with simple body wave propagation, the p-τ response was simple and the corresponding p-ω representation appeared as a complex interference of harmonics. However, once we allow for the full range of wavefield complexity with the superposition of the complete set of S wave multiples, and associated P waves rising from wavetype conversions, the behaviour is reversed between the slowness-time and slowness-frequency domains. For larger slownesses, the simple organised character in the slowness-frequency domain means that the most economical description is in terms of a limited number of surface wave modes. For intermediate slowness (0.14–0.29 s/km in figures 16.6 and 16.7) a significant number of terms are required to describe the behaviour in terms of either a body-wave or a surface wave. We have a very clear ray-mode duality between the time and frequency representatives. Approximations to the wavefield will take a different character in the two domains. For body waves, we would control the number of mantle reflections, which can be thought of as a form of time windowing. In terms of surface waves we would normally restrict the number of mode branches and so give the effect of a variable filter in frequency.

16.3 Surface waves

The modal description of the guided waves is based on the superposition of the contributions from the poles in the slowness-frequency response, previously encountered in Section 15.3. These poles in the response arise because it is possible to find waves which simultaneously satisfy the boundary condition of vanishing traction at the free surface and decay of the waves at depth. Their location in the frequency-slowness domain can be explained with the aid of representations in terms of reflection and transmission processes.

The phase dependence of a plane wave in a uniform medium takes the form

$$\exp[i\omega q_\beta z] \quad \text{with} \quad q_\beta = (\beta^{-2} - p^2)^{1/2}. \tag{16.3.1}$$

When the slowness p exceeds $1/\beta$ we have a evanescent wave and with the convention we have adopted for the radicals $\mathrm{Im}(\omega q_\beta) \geq 0$ we have

$$\exp[i\omega q_\beta z] \to \exp[-\omega |q_\beta| z], \tag{16.3.2}$$

so that a downgoing wave maps to evanescent decay at depth.

Consider now an upgoing wave $\boldsymbol{v}_U$ incident on the free-surface. In order to satisfy the boundary condition of vanishing traction at the surface, we must have a reflected wave

$$\boldsymbol{v}_D = \mathbf{R}_F \boldsymbol{v}_U. \tag{16.3.3}$$

If the stratified region is ultimately underlain by a uniform halfspace beneath z_L, we can satisfy the condition of decay at depth for $p > \beta_L{}^{-1}$ by generating an upgoing field by the action of $\mathbf{R}_D^{0L}$

$$\boldsymbol{v}'_U = \mathbf{R}_D^{0L}\mathbf{R}_F\boldsymbol{v}_U \tag{16.3.4}$$

The combination of reflection processes will satisfy both boundary conditions in the absence of a source if $\boldsymbol{v}'_U = \boldsymbol{v}_U$, so that

$$\left(\mathbf{I} - \mathbf{R}_D^{0L}\mathbf{R}_F\right)\boldsymbol{v}_U = 0. \tag{16.3.5}$$

The free surface reflection term $\mathbf{R}_F(p)$ is frequency independent and depends on slowness alone, but $\mathbf{R}_D^{0L}(p,\omega)$ depends on both frequency and slowness. The relation (16.3.5) can thus be written as

$$\left(\mathbf{I} - \mathbf{R}_D^{0L}(p,\omega)\mathbf{R}_F(p)\right)\boldsymbol{v}_U(p,\omega) = 0, \tag{16.3.6}$$

which defines a set of trajectories ('dispersion curves') in frequency and slowness for which surface waves can exist. Such points represent singularities of the reverberation operator

$$\left(\mathbf{I} - \mathbf{R}_D^{0L}(p,\omega)\mathbf{R}_F(p)\right)^{-1}, \tag{16.3.7}$$

for the whole stratified region and it is in this guise that the surface waves appear as poles in the slowness domain.

We can make a comparable development by starting from a level z_S to produce an alternative but equivalent representation of the surface wave trajectories

$$\left(\mathbf{I} - \mathbf{R}_D^{SL}(p,\omega)\mathbf{R}_U^{fS}(p,\omega)\right)\boldsymbol{v}_U^{s}(p,\omega) = 0. \tag{16.3.8}$$

which corresponds directly to the singularities in our expression (5.3.7) for the full response of the stratified medium.

16.3.1 Love waves

For slowness $p > \beta_L^{-1}$, the dispersion relation for *SH* waves is

$$1 - \mathrm{R}_D^{HH}(p,\omega)\mathrm{R}_F^{HH}(p) = 0, \tag{16.3.9}$$

where for brevity we have written R_D^{HH} for the *HH*-component of $\mathbf{R}_D^{0L}$. The free surface reflection coefficient is unity for *SH* waves and so the trajectory of the surface wave poles in slowness and frequency is dictated by the requirement that

$$\mathrm{R}_D^{HH}(p,\omega) = 1. \tag{16.3.10}$$

Thus the reflection from the entire region below the surface for *SH* waves is required to have modulus unity, and the dispersion branches occur when there is a match in phase. Introducing the phase of the reflection coefficient, χ_D^{HH}, we can rewrite (16.3.10) as

$$\mathrm{R}_D^{HH}(p,\omega) = \exp[\mathrm{i}\chi_D^{HH}(p,\omega)] = \exp[2n\pi\mathrm{i}] = 1, \tag{16.3.11}$$

and then extracting the phase

$$\chi_D^{HH}(p,\omega) = 2n\pi, \tag{16.3.12}$$

which for fixed slowness p provides a sequence of branches in frequency ω characterised by a mode number n ($n = 0, 1, 2, \ldots$). The example of a single layer over a half-space we discussed in Section 3.2 is in precisely in the form (16.3.12), which represents a constructive interference condition for successive multiple *SH* reflections from the entire structure.

For a perfectly elastic medium, the relation (16.3.12) can be used as the basis of a recursive procedure for the rapid determination of dispersion curves (Kennett & Nolet, 1979; Kennett & Clarke, 1983a). We suppose that the medium has a sequence of interfaces $z_1, z_2, \ldots$ separating uniform media or smoothly varying gradient zones. The reflection coefficient at $z = 0$ can be represented in terms of that at z_1 as

$$\mathrm{R}_D^{HH}(p,\omega) = \mathrm{e}^{\mathrm{i}\chi_D^{HH}(p,\omega)} = \mathrm{e}^{2\mathrm{i}\omega\tau_0(p)}\mathrm{R}_D^{HH}(z_{1-}) = \mathrm{e}^{2\mathrm{i}\omega\tau_0(p)}\mathrm{e}^{\mathrm{i}\chi_1(p,\omega)}, \tag{16.3.13}$$

where $\tau_0(p)$ is the phase delay in the topmost layer. This representation is exact for a uniform medium and a good high frequency approximation for a gradient zone. For a turning ray we would take $\chi_1(p,\omega) = -\pi/2$. For *SH* waves, $\mathrm{R}_D^{HH}(z_1-)$ is related to the reflection coefficient just above the second interface by

$$\mathrm{R}_D(z_1-) = \exp[\mathrm{i}\chi_1] = \frac{r_D^1 + \mathrm{R}_D(z_1-)\mathrm{e}^{2\mathrm{i}\omega\tau_1(p)}}{1 + r_D^1\mathrm{R}_D(z_1-)\mathrm{e}^{2\mathrm{i}\omega\tau_1(p)}} \tag{16.3.14}$$

where we have used the recursion relation (14.2.11) and exploited the properties of the *HH* reflection coefficients. The interface coefficient

$$r_D^1 = \frac{1-y^1}{1+y^1}, \quad \text{with} \quad y^1 = \frac{\mu_1\eta_{\beta 1}}{\mu_1\eta_{\beta 1}} \tag{16.3.15}$$

in terms of the generalized vertical slownesses $\eta_{\beta 1,2}$. Note that for a uniform medium $q_\beta = \eta_\beta$. We can use the exponential representation of a tangent to recast (16.3.14) in the form

$$\tan\left(\tfrac{1}{2}\chi_1\right) = -\mathrm{i}y^1\frac{\mathrm{R}_D(z_1-)\mathrm{e}^{2\mathrm{i}\omega\tau_1(p)} - 1}{\mathrm{R}_D(z_1-)\mathrm{e}^{2\mathrm{i}\omega\tau_1(p)} + 1} \tag{16.3.16}$$

When *SH* waves are propagating on both sides of interface 1, the quantity y^1 is real, and $\mathrm{R}_D(z_2-)$ must also have modulus unity, $\mathrm{R}_D(z_2-) = \exp[\mathrm{i}\chi_2]$, so that

$$\tan\left(\tfrac{1}{2}\chi_1\right) = |y^1|\tan\left(\omega\tau_1 + \tfrac{1}{2}\chi_2\right). \tag{16.3.17}$$

With the aid of trigonometric identities (16.3.17) can be recast directly in terms of the phases

$$\tfrac{1}{2}\chi_1 = \omega\tau_1 + \tfrac{1}{2}\chi_2 - \tan^{-1}\left\{\frac{r_D^1\sin(2\omega\tau_1 + \chi_2)}{1 + r_D^1\cos(2\omega\tau_1 + \chi_2)}\right\}. \tag{16.3.18}$$

A turning point is included by setting $\chi_2 = -\pi/2$. When there is an evanescent wave below z_1, then $\mathrm{R}_D(z_2-)$ and $\mathrm{i}y^1$ are real; and then

$$\tan\left(\tfrac{1}{2}\chi_1\right) = |y^1|\frac{\mathrm{R}_D(z_1-)\mathrm{e}^{-2\omega|\tau_1|} - 1}{\mathrm{R}_D(z_1-)\mathrm{e}^{-2\omega|\tau_1|} + 1}, \tag{16.3.19}$$

at very high frequencies the evanescent decay reduces the right hand side of (16.3.18) to $-|y^1|$.

The relations (16.3.15)-(16.3.19) and their counterparts for deeper interfaces can be used to build an iterative method for calculating surface wave dispersion. For fixed slowness p we set a trial frequency (ω_n^t) for the nth mode. We start from the base of the layering and use the recursive scheme (16.3.14) to carry the *HH* reflection coefficient to successively higher interfaces through the zone of evanescence. As soon as an interface z_p is reached above which SH waves are propagating for slowness p, we calculate the phase χ_p of the reflection coefficient:

$$\exp[i\chi_p] = \mathrm{R}_D(z_p-, p, \omega). \tag{16.3.20}$$

For higher interfaces we increment the phase term via (16.3.18) as:

$$\tfrac{1}{2}\chi_j(p,\omega) = \omega\tau_j(p) + \tfrac{1}{2}\chi_{j+1}(p,\omega) - \tan^{-1}\left\{\frac{r_D^j \sin(2\omega\tau_j + \chi_{j+1})}{1 + r_D^j \cos(2\omega\tau_j + \chi_{j+1})}\right\}, \tag{16.3.21}$$

for each interface up to z_1. Then the total phase at z_1 the surface is

$$\tfrac{1}{2}\chi_D^{HH}(p,\omega) = \omega\tau(p) + X(p,\omega), \tag{16.3.22}$$

where

$$\begin{aligned} \tau(p) &= \sum_{j=0}^{p-1} \omega\tau_j(p), \\ X(p,\omega) &= -\sum_{j=1}^{p-1} \tan^{-1}\left\{\frac{r_D^j \sin(2\omega\tau_j + \chi_{j+1})}{1 + r_D^j \cos(2\omega\tau_j + \chi_{j+1})}\right\} + \chi_p(p,\omega). \end{aligned} \tag{16.3.23}$$

and we have used (16.3.13) to relate χ_D^{HH} to χ_1.

From the dispersion relation (16.3.12) we require

$$\omega_n\tau(p) + X(p,\omega_n) = n\pi, \tag{16.3.24}$$

at the combination (p, ω_n) corresponding to the nth mode. Thus, if the trial frequency ω_n^t were correct (16.3.24) would be satisfied directly, otherwise we can derive an improved estimate from

$$\omega_n' = \frac{n\pi - X(p,\omega_n^t)}{\tau(p)} \tag{16.3.25}$$

and then use this new estimate of the mode frequency to start the cycle of phase estimation once again. A convenient starting frequency is obtained by setting $X(\omega_n^t) = \pi/4$ in (16.3.25), and then the iterative procedure usually converges rapidly. The actual number of iterations depends strongly on the structure and the accuracy that is demanded. For example, for an upper mantle model and an absolute accuracy of 10^{-6} in ω between 4 and 10 iterations are normally required. Higher numbers of iterations are needed if X approaches $\pi/2$. The large reflection at the Moho will provide a significant contribution to the phase response (Kennett & Nolet, 1979).

This iterative scheme represents a generalisation of the approach pioneered by Tolstoy (1955) for scalar wave problems by the inclusion of evanescent waves and thereby improved convergence.

16.3.2 Rayleigh waves

In the *P-SV* case both $\mathbf{R}_F$ and $\mathbf{R}_D^{0L}$ are 2×2 matrices and the dispersion relation takes the form

$$\det\left(\mathbf{I} - \mathbf{R}_D^{0L}(p,\omega)\mathbf{R}_F(p)\right) = 0, \tag{16.3.26}$$

which once again has multiple roots for a stratified medium. However, the analysis is somewhat more complicated because of the link between *SV* and *P* waves through the free surface boundary condition.

A special case exists for a uniform half space; the free surface reflection matrix

$$\mathbf{R}_F = \boldsymbol{n}_D^{-1}(p)\boldsymbol{n}_U(p), \tag{16.3.27}$$

is singular when $\det \boldsymbol{n}_D = 0$, which occurs for a slowness p_{R0} close to $1.1\beta_0^{-1}$. This slowness p_{R0} corresponds to the situation first described by Lord Rayleigh (1885) in which evanescent *P* and *S* fields are coupled by the vanishing traction condition at the free surface. The Rayleigh slowness p_{R0} for a perfectly elastic medium satisfies

$$(2p_{R0}^2 - \beta_0^{-2})^2 - 4p_{R0}^2|q_{\alpha 0}(p_{R0})||q_{\beta 0}(p_{R0})| = 0, \tag{16.3.28}$$

for a half space with wavespeeds α_0, β_0.

In terms of the components of the free surface reflection matrix $\mathbf{R}_F$ and the reflection matrix $\mathbf{R}_D^{0L}$ from the structure beneath the surface, the dispersion relation (16.3.26) can be written as

$$\begin{aligned}[1 - \mathrm{R}_D^{PP}\mathrm{R}_F^{PP} - \mathrm{R}_D^{PS}\mathrm{R}_F^{SP}][1 - \mathrm{R}_D^{SP}\mathrm{R}_F^{PS} - \mathrm{R}_D^{SS}\mathrm{R}_F^{SS}]\\ -[\mathrm{R}_D^{PP}\mathrm{R}_F^{PS} - \mathrm{R}_D^{PS}\mathrm{R}_F^{SS}][\mathrm{R}_D^{SP}\mathrm{R}_F^{PP} - \mathrm{R}_D^{SS}\mathrm{R}_F^{SP}] = 0,\end{aligned} \tag{16.3.29}$$

where for brevity we have used a compressed notation so that

$$\mathrm{R}_F^{PP} = [\mathbf{R}_F]_{PP}, \quad \mathrm{R}_D^{PS} = [\mathbf{R}_D^{0L}]_{PS}, \quad \text{etc.} \tag{16.3.30}$$

As it stands (16.3.29) does not represent a physically feasible set of wave propagation processes, because we have *P* legs juxtaposed with *S* wave legs. However, with some manipulation, we are able to recast (16.3.29) into a form where the physical character of the propagation processes are revealed. We introduce the reverberation term

$$[X^{PP}]^{-1} = [1 - \mathrm{R}_D^{PP}\mathrm{R}_F^{PP} - \mathrm{R}_D^{PS}\mathrm{R}_F^{SP}]^{-1}, \tag{16.3.31}$$

which represents all processes which involve multiple *P* waves, including the possibility of wavetype conversion both at the surface and from the structure. We then divide (16.3.29) by X^{PP} and regroup the terms to emphasis the terms depending on S waves arriving at the surface (R_D^{SS}, R_D^{SP}) to produce

$$1 - \mathrm{R}_D^{SS}\left\{\mathrm{R}_F^{SS} + \mathrm{R}_F^{SP}[X^{PP}]^{-1}\mathrm{R}_D^{PP}\mathrm{R}_F^{PS} + \mathrm{R}_F^{SP}[X^{PP}]^{-1}\mathrm{R}_D^{PS}\mathrm{R}_F^{SS}\right\}$$
$$-\mathrm{R}_D^{SP}\left\{\mathrm{R}_F^{PS} + \mathrm{R}_F^{PP}[X^{PP}]^{-1}\mathrm{R}_D^{PP}\mathrm{R}_F^{PS} + \mathrm{R}_F^{PP}[X^{PP}]^{-1}\mathrm{R}_D^{PS}\mathrm{R}_F^{SS}\right\} = 0. \quad (16.3.32)$$

Now each of the expressions in braces represent a physically realisable sequence of propagation processes linking to the reflections from the structure in R_D^{SS}, R_D^{SP}.

We can interpret (16.3.32) as a constructive interference condition for SV waves incorporating all the different ways in which it is possible to sustain multiples, including conversion to P either at the surface or from the structure itself. The additional complexity compared with the equivalent relation (16.3.9) for SH waves arises from the wide range of propagation options once wave conversion is allowed. Equation (16.3.32) can be simplified somewhat by using the symmetries of the reflection matrices, but at the expense of losing the physical interpretation of the terms (see, e.g., Kennett, 1983, Chapter 11; Kennett & Clarke, 1983b).

16.3.2.1 P and S waves evanescent

When $p > \beta_0^{-1}$, both P and S waves are evanescent throughout the stratification. For perfectly elastic media, all the components of $\mathbf{R}_D^{0L}$ are real and less than unity. Both P and S waves will decay with depth away from the surface, but the rate of decay of the P waves will be much more rapid than for S. In consequence, at moderate to high frequencies only R_D^{SS} has significant amplitude and R_D^{PS}, R_D^{PP} are negligible. The dispersion relation (16.3.32) then reduces to

$$1 - \mathrm{R}_D^{SS}(p,\omega)\mathrm{R}_F^{SS}(p) = 0, \quad (16.3.33)$$

so that the small amplitude of R_D^{SS} has to be compensated by the growth of $\mathrm{R}_F^{SS}(p)$ in this evanescent regime. In terms of the explicit representation of R_F^{SS} we can rewrite (16.3.33) as

$$(2p^2 - \beta_0^{-2})^2(1 + \mathrm{R}_D^{SS}(p,\omega)) - 4p^2|q_{\alpha 0}(p)||q_{\beta 0}(p)|(1 - \mathrm{R}_D^{SS}(p,\omega)) = 0, \quad (16.3.34)$$

The dispersion relation (16.3.34) has a single root and as the frequency increases $\mathrm{R}_D^{SS} \to 0$, and so (16.3.34) will tend to the equation for the Rayleigh slowness p_{R0} in terms of the wavespeeds just at the surface. We can therefore recognise (16.3.34) as a dispersion relation for the *fundamental* Rayleigh mode.

For lower frequencies R_D^{SS} will be larger because the evanescent decay is less, and so the slowness of the fundamental mode will depart significantly from p_{R0} (figures 16.5, 16.8). The modal eigenfunction for this fundamental mode penetrates deeply into the structure (figure 16.11) and the dispersive waves interact with the greater wavespeeds at depth. The phase slowness of the fundamental Rayleigh mode thus decreases with frequency, and hence the phase speed increases with frequency.

For the lowest frequencies, particularly as p approaches β_0^{-1}, R_D^{SP} and R_D^{SS} will be small but no longer negligible and a slight correction is needed to (16.3.34).

16.3.2.2 S propagating, P evanescent

For $\alpha_0^{-1} < p < \beta_0^{-1}$, S waves have travelling wave character at the surface and a turning level above z_L; whilst P waves are still evanescent throughout the structure. For high frequency propagation the decay in evanescent P means that R_D^{PP}, R_D^{SP} are small compared to R_D^{SS}, which for perfectly elastic media has amplitude unity $|\mathrm{R}_D^{SS}| = 1$.

At moderate frequencies we can reduce the dispersion relation (16.3.32) to the approximate form

$$1 - \mathrm{R}_D^{SS}\mathrm{R}_F^{SS} - \left[\mathrm{R}_D^{SS}\mathrm{R}_F^{SP}\mathrm{R}_D^{PP}\mathrm{R}_F^{PS} + \mathrm{R}_D^{SS}\mathrm{R}_F^{SP}\mathrm{R}_D^{PS}\mathrm{R}_F^{SS} + \mathrm{R}_D^{SP}\mathrm{R}_F^{PS}\right] = 0. \tag{16.3.35}$$

where the expression in brackets contains terms with only a single evanescent component. As the frequency increase R_D^{SP}, R_D^{PP} will diminish rapidly and so

$$1 - \mathrm{R}_D^{SS}(p,\omega)\mathrm{R}_F^{SS}(p) \approx 0, \qquad \text{for } \omega \gg 1; \tag{16.3.36}$$

which corresponds to the functional form of the Love wave result (16.3.9). For perfect elasticity R_D^{SS}, R_F^{SS} are both unimodular and we can recast (16.3.36) in terms of the phase of the reflection coefficients

$$\mathrm{R}_D^{SS}(p,\omega) = \mathrm{e}^{\mathrm{i}\psi(p,\omega)}, \qquad \mathrm{R}_F^{SS}(p) = \mathrm{e}^{\mathrm{i}\psi_F(p)}. \tag{16.3.37}$$

Then for a surface wave pole, the constructive interference condition is,

$$\psi(p,\omega) = 2n\pi - \psi_F(p), \tag{16.3.38}$$

with just a slowness dependent phase shift of $\psi_F(p)$ from the Love wave result. The dispersion curves for Love and Rayleigh waves in the slowness-frequency domain are very similar at high frequencies (see figures 16.8,16.9). In the interval $\alpha_0^{-1} < p < \beta_0^{-1}$ where S propagation dominates, the character of higher Rayleigh modes is akin to *SV* Love modes (Kennett & Clarke, 1983b). The high frequency asymptote for the phase slowness of the higher Rayleigh modes is β_0^{-1}. Only the fundamental Rayleigh mode continues into the region where S is evanescent.

The approximate dispersion relation (16.3.36) can be used to develop an iterative scheme to estimate mode frequencies at fixed slowness p based on recursion for the phase $\psi(p,\omega)$ (Kennett & Clarke, 1983a). The structure is similar to the Love wave case but includes allowances for evanescent *PP* and *PS* contributions, as well as the greater complexity of the *P-SV* wave reflection and transmission coefficients.

16.3.2.3 Propagating P and S waves

For $\beta_L^{-1} < p < \alpha_0^{-1}$, both P and S waves will have travelling wave character at the surface, but will turn within the structure above the level z_L. We now need to employ the full form (16.3.32) for the dispersion relation, including the influence of conversions and P reverberations.

We can gain some insight into the nature of the dispersion results by considering a situation in which no conversion is allowed from the structure, i.e., $\mathrm{R}_D^{SP} = \mathrm{R}_D^{PS} \equiv 0$. Under these restrictive conditions (16.3.32) reduces to

$$1 - R_D^{SS}\left\{R_F^{SS} + R_F^{SP}[X^{PP}]^{-1}R_D^{PP}R_F^{PS}\right\} = 0. \tag{16.3.39}$$

In terms of the phase of R_D^{SS}

$$\psi(p,\omega) = 2n\pi - \arg\mathcal{A}^{SS}(p,\omega), \tag{16.3.40}$$

where $\mathcal{A}^{SS}$ is the term in braces in (16.3.39), which represents those free surface reflection effects linking incident and reflected *SV* waves (Levshin, 1981). $\mathcal{A}^{SS}$ includes the simple surface reflection R_F^{SS} and also conversion at the surface to *P*, followed by reflection from the near-surface portion of the stratification with possible *P* reverberation and eventual re-conversion to *SV* at the surface.

When conversion from the structure can be neglected we can use a simple asymptotic representation of the reflection coefficients R_D^{PP}, R_D^{SS}:

$$\begin{aligned} R_D^{PP} &\sim \exp[2i\omega\int_0^{Z_\alpha} dz\, q_\alpha - i\pi/2], \\ R_D^{SS} &\sim \exp[2i\omega\int_0^{Z_\beta} dz\, q_\beta - i\pi/2], \end{aligned} \tag{16.3.41}$$

with integration to the turning points Z_α, Z_β for *P* and *S* waves at slowness p. Note that $Z_\beta \gg Z_\alpha$ because of the shallow penetration of *P* waves. With the approximations (16.3.41), the dispersion relation (16.3.40) can be expressed as (Kennett & Woodhouse, 1978; Brodskiĭ & Levshin, 1979)

$$\tan\left[\omega\int_0^{Z_\alpha} dz\, q_\alpha - \pi/4\right]\tan\left[\omega\int_0^{Z_\beta} dz\, q_\beta - \pi/4\right] = \frac{(2p^2-\beta_0^{-2})}{4p^2 q_{\alpha 0} q_{\beta 0}}. \tag{16.3.42}$$

In general, however, conversion between *P* and *S* does occur within the structure, e.g., at the crust-mantle boundary, and R_D^{PS}, R_D^{SP} are non-zero. With the inclusion of the conversions we can write the dispersion relation (16.3.32) as

$$1 - R_D^{SS}(p,\omega)\left\{R_F^{SS}(p)+\mathcal{B}^{SS}(p,\omega)\right\} - R_D^{SP}(p,\omega)\left\{R_F^{PS}(p)+\mathcal{B}^{PS}(p,\omega)\right\} = 0, \tag{16.3.43}$$

where $\mathcal{B}^{SS}(p,\omega)$ represents the full sequence of interactions between the structure and the surface which link *SV* waves to *SV* waves, and $\mathcal{B}^{PS}(p,\omega)$ is the similar set with conversion from *SV* to *P*.

For slowness p just less than α_0^{-1} the size of R_F^{SS} is well below the unit value attained when *P* is evanescent, and so the significance of the converted terms in (16.3.43) is enhanced relative to the *SS* reflections. There is a smooth continuation of the dispersion curves across the evanescent point α_0^{-1}; however, the presence of propagating *P* waves superimposes a modulation on the spacing of the dispersion curves. The influence appears through the reverberation term $[X^{PP}]^{-1}$ and leads to 'ghost' dispersion branches, controlled by the near-surface *P* wavespeed distribution. The effects are only visible because of a slight change in the spacing of the dispersion branches which is primarily controlled by *S* wave effects.

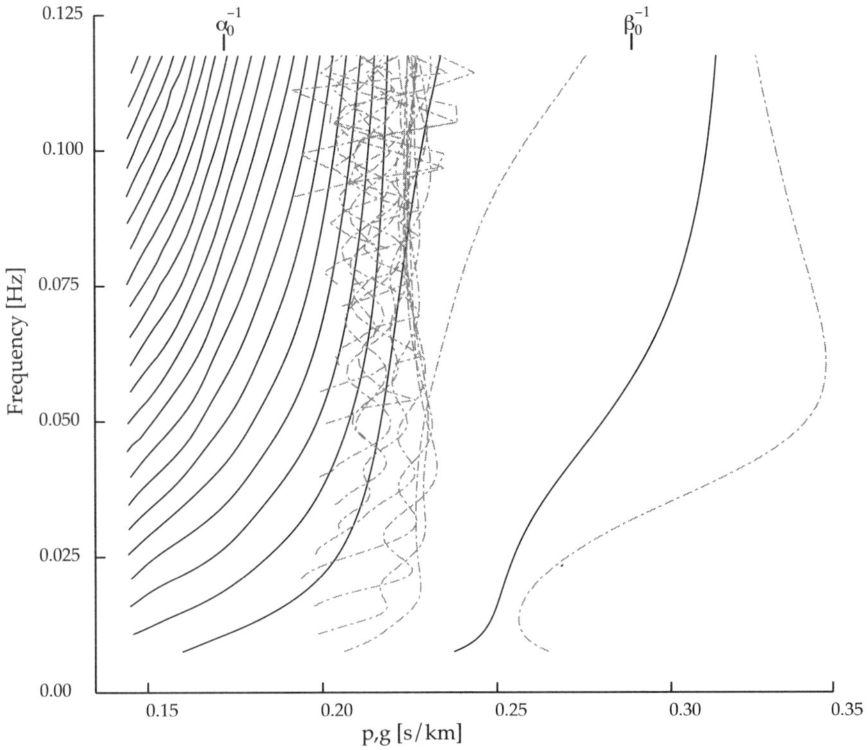

Figure 16.8. Rayleigh wave dispersion curves for the upper mantle of model AK135 in the slowness-frequency domain. Phase slowness is shown as solid lines, group slowness as chain dotted lines in tone.

16.3.3 Dispersion curves and modal shapes

We have seen that the dispersion relation for surface waves can be related to constructive interference of *SH* and *SV* waves between the free surface and the structure at depth. The possibility of conversion to *P* waves means that the Rayleigh wave system is more complicated than the Love wave case for *SH* waves. In particular, the fundamental mode of Rayleigh waves can have a lower phase speed than the shear wave velocity at the surface. We now look at the character of the dispersion for the different mode branches which is one of the major factors influencing seismograms created by modal summation.

We can illustrate the behaviour of the Rayleigh and Love mode dispersion by again using the upper mantle of the AK135 model. In figures 16.8 and 16.9 we compare the behaviour of the dispersion for *P-SV* waves and *SH* waves in terms of slowness for frequencies up to 0.120 Hz. The dispersion calculations for a spherical model include an allowance for anelastic structure and the modal branches can be compared directly with the results in Section 16.2 obtained by working with the wavefield in a flattened model. Except at the lowest frequencies where the influence of density is significant,

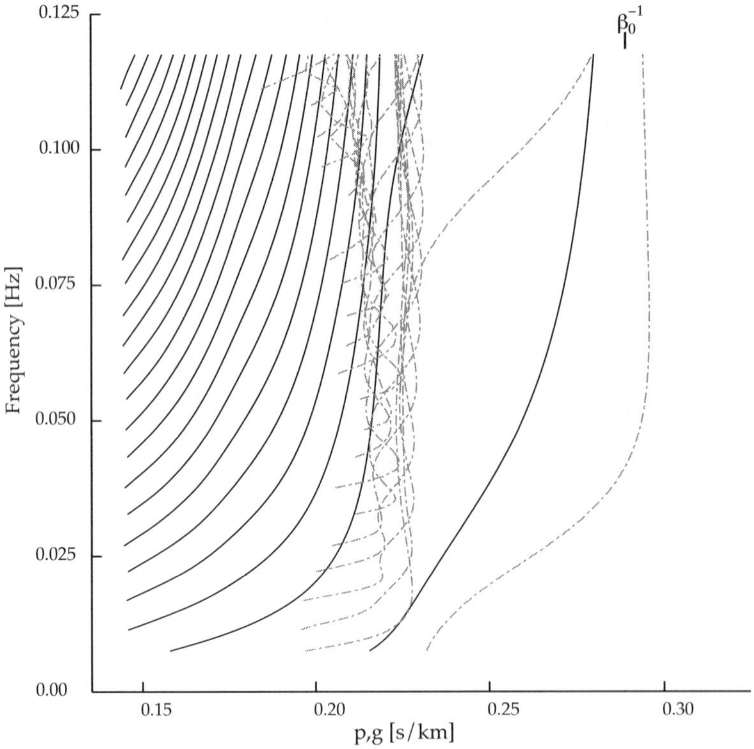

Figure 16.9. Love wave dispersion curves for the upper mantle of model AK135 in the slowness-frequency domain. Phase slowness is shown as solid lines, group slowness as chain dotted lines in tone.

the transform domain results in figures 16.6, 16.7 provide an excellent representation of the dispersion behaviour.

The slowess boundaries between the various dispersion domains have been superimposed on figures 16.8 and 16.9. For the largest slownesses we have, as expected, only the fundamental Rayleigh mode; and at 0.120 Hz we can begin to see the approach to the asymptote p_{R0}. The limiting slowness for all Love waves and all higher Rayleigh modes is β_0^{-1}, and we can see the similar behaviour between the Love and Rayleigh Modes. For slownesses greater than 0.222 s/km, S waves are reflected in the crust or at the Moho and the frequency spacing of these crustal modes is large, so that the higher modes do not appear for these large slownesses in figures 16.8, 16.9. For smaller slownesses the frequency spacing is much tighter. From (16.3.24) we can see the spacing in frequency between mode branches is approximately $1/\{2\tau(p)\}$ and so diminishes quickly as the phase delay $\tau(p)$ grows for smaller slownesses p.

The phase slowness of each of the modes changes with frequency so that if we build a pulse for a single mode it will change its character as its propagates horizontally. A wave packet will travel with the group slowness,

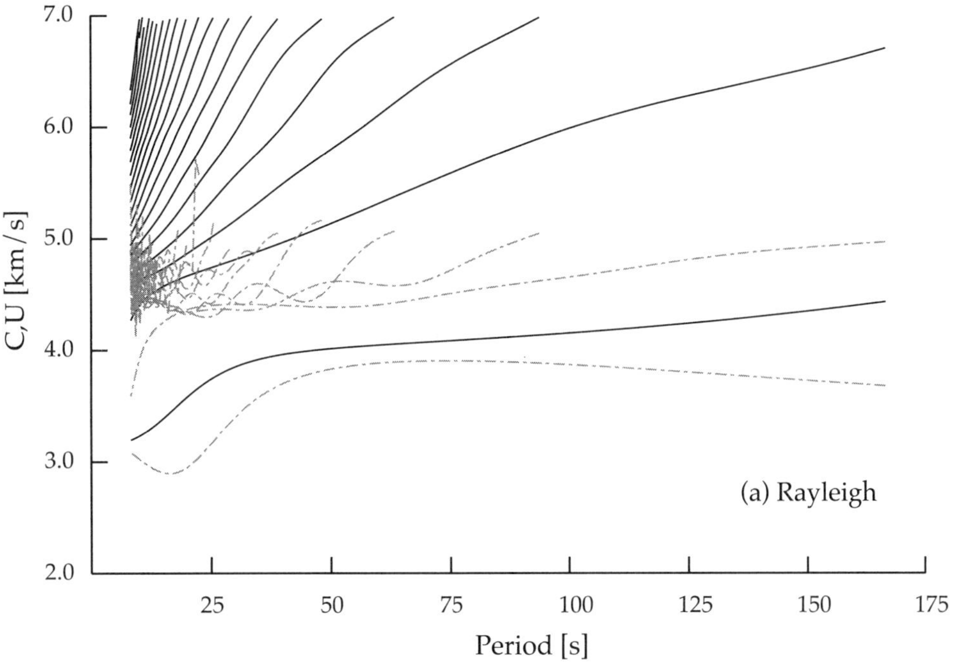

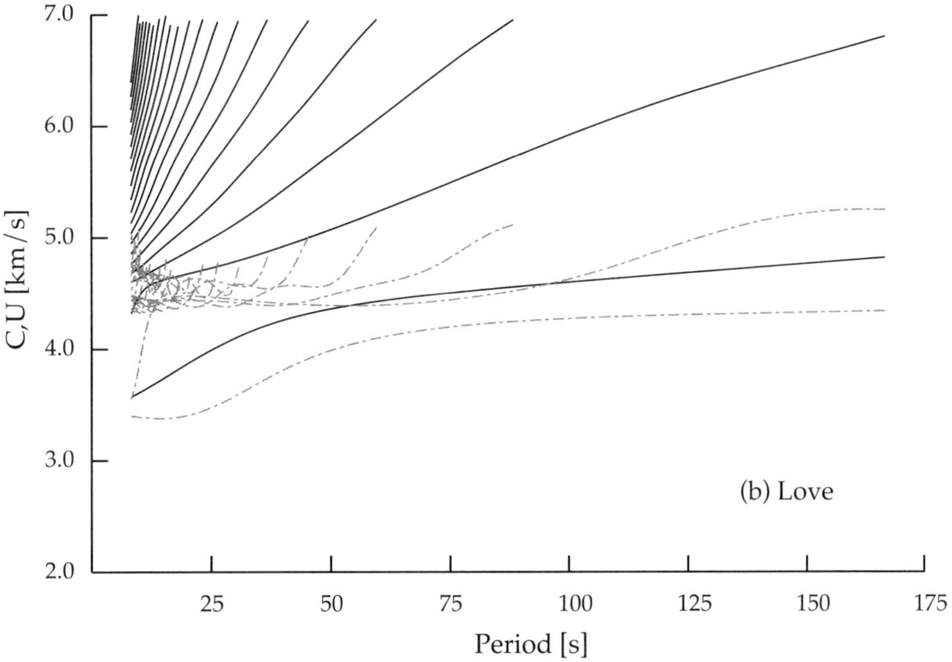

Figure 16.10. Dispersion curves for phase velocity C and group velocity U for the upper mantle of model AK135 as a function of period T: (a) Rayleigh waves, (b) Love waves. Phase velocity is shown as solid lines, group velocity as chain dotted lines in tone.

$$g(\omega) = \frac{\partial}{\partial \omega}(wp) = p + \omega \frac{\partial p}{\partial \omega}, \tag{16.3.44}$$

as discussed in section 3.2. The arrival times of different frequency components will then be $g(\omega)\Delta$ for epicentral distance Δ.

We display the group slowness behaviour as a function of frequency in figures 16.8, 16.9 using chain dotted lines in tone. The fundamental and first higher modes separate to higher group slowness (slower arrivals), but the higher modes form systems with similar group slownesses even though there is considerable variation in phase slowness. Where the group slowness shows an extremum there will tend to be larger amplitudes as a number of frequency components will arrive in close proximity in time. For a maximum in group slowness we will have both higher and lower frequency energy arriving before the components with the maximum slowness giving rise to an Airy phase. The abrupt termination of the dispersive train makes an Airy phase very distinct on a seismogram. Examples are commonly seen for the fundamental mode Rayleigh wave associated with the group slowness maximum near 0.05 Hz (figure 16.8). The group of superimposed Airy phases associated with the first few higher Rayleigh modes for frequencies up to 0.05 Hz is associated with a distinct low-frequency pulse, *Sa* which is often well excited by intermediate depth earthquakes (~ 120 km deep). At higher frequencies we see a large number of modes with a similar group slowness close to 0.22 s/km (4.54 km/s) associated with *Sn* propagation at the top of the mantle. The same modes have group slowness minima near 0.208 s/km (4.8 km/s) arising from trapping above the 410 km discontinuity. The two sets of extrema are very clearly seen in the Love wave dispersion (figure 16.9).

The slowness domain provides a convenient link between body wave and surface wave viewpoints. However, in the study of surface waves it is more common to work with phase velocity $C - 1/p$ and group velocity $U = 1/g$ as a function of period $T = 2\pi/\omega$. The dispersion information from figures 16.8, 16.9 is therefore replotted in figure 16.10 in terms of velocity and period for both Rayleigh and Love waves. We have retained the same convention of plotting the phase velocity in black and the group velocity in tone. In this presentation the higher modes crowd together and the nearly uniform spacing of modes in frequency at fixed phase speed C is not apparent. However, the behaviour of the group velocity at longer periods is very clear.

For a fixed slowness p or phase speed C the sequence of modes is characterised by a mode number relating to the depth dependence of the associated displacement field. The twin boundary conditions of vanishing free surface traction and decay of the displacement field at depth are satisfied by functions with increasing levels of oscillation as frequency increases. For Love waves the mode number is just the number of zero crossings as a function of depth, for Rayleigh waves the situation is a little more complex but the mode number normally is close to the number of zero crossings for the vertical component. The solution of the elastodynamic equation subject to the boundary condition determines only the shape of the eigenfunctions for displacement

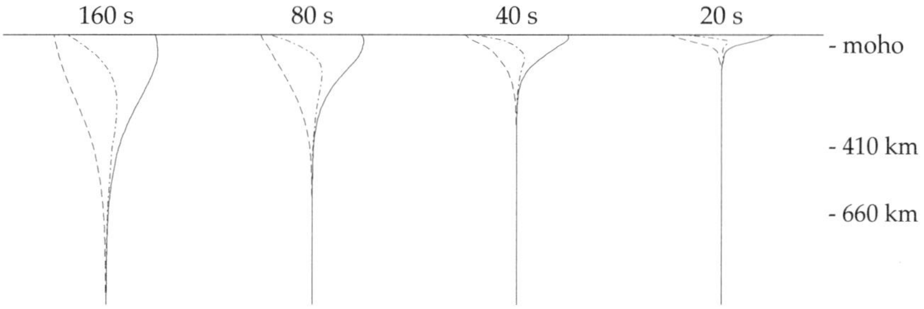

Figure 16.11. Eigenfunctions for the fundamental modes of Rayleigh and Love waves as a function of period. The vertical component U for Rayleigh is shown as a solid line and the horizontal component V by a chain-dotted line. The horizontal component for Love waves W is indicated by the long dashed lines.

and traction and not their amplitude. As we shall see shortly, the eigenfunctions play an important role in the representation of seismograms by modal summation.

The general character of the eigenfunctions is preserved along a mode branch but depth penetration is much greater at longer periods. The behaviour is illustrated in figure 16.11 for the fundamental modes of Rayleigh and Love waves across a broad band of frequencies. At low frequencies, even though the amplitude is decaying away from the surface, the fundamental modes penetrate deep into the structure and so are sensitive to the wavespeed distribution at depth. However by 20 s period (0.05 Hz) the displacement in the fundamental modes is confined quite close to the surface.

We can gain further insight into the nature of the modal eigenfunctions by looking at comparisons of modes at fixed frequency or fixed slowness (figures 16.12, 16.13). We consider both Rayleigh and Love modes, and displace the eigenfunctions for Love waves slightly to the right so that the similarities in behaviour are easier to see.

At fixed period 14.3 s [0.07 Hz], the fundamental modes are confined above 100 km, but even the third higher mode penetrates nearly to the 410 km discontinuity (figure 16.12). The progressive increase in penetration with mode number is related to the decrease in slowness p. S waves become evanescent below the turning point which is increasing function of phase speed C (and decreasing phase slowness p). The close correspondence between the vertical component of the eigenfunction for Rayleigh mode 3 and the tangential component for the third higher Love mode is very striking.

For fixed phase speed 4.8 km/s [0.208 s/km], the penetration depth is comparable for all modes but the frequency increases steadily with mode number (figure 16.13). At this phase speed we are beyond the end of the calculated fundamental mode branches (because we have truncated the AK135 model at 1000 km depth), and so the lowest frequency mode is the first higher mode (mode 1) for both Rayleigh and Love waves. As the mode number increases we see how extra oscillations are introduced into the same depth interval.

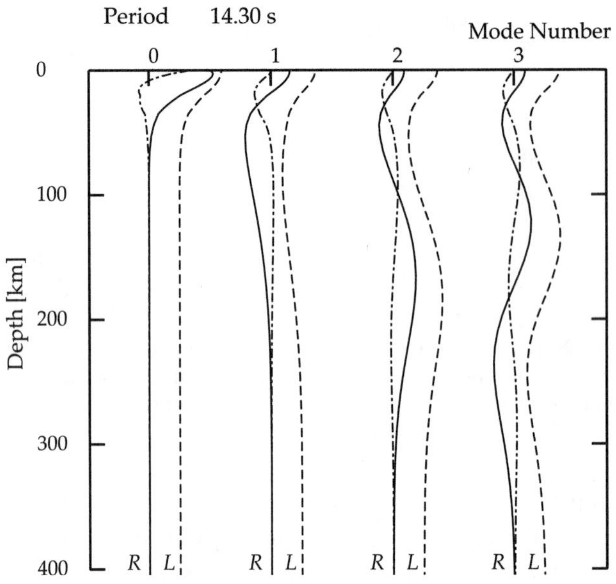

Figure 16.12. Modal eigenfunctions for Rayleigh and Love waves at fixed period. The vertical component U for Rayleigh is shown as a solid line and the horizontal component V by a chain-dotted line. The horizontal component for Love waves W is indicated by the long dashed lines.

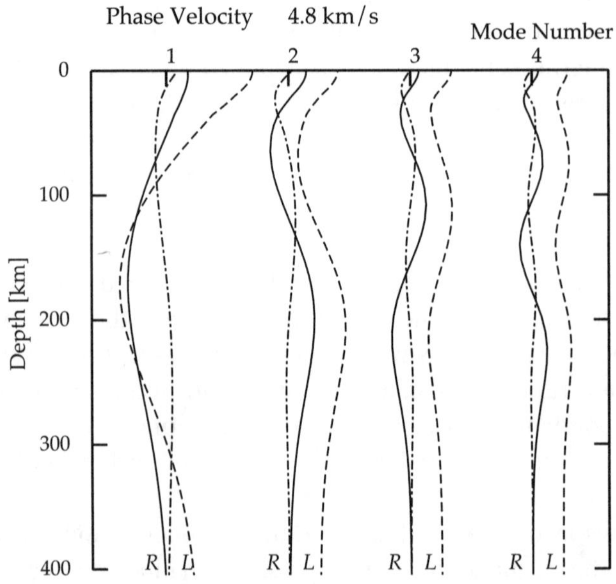

Figure 16.13. Modal eigenfunctions for Rayleigh and Love waves at fixed phase velocity. The vertical component U for Rayleigh is shown as a solid line and the horizontal component V by a chain-dotted line. The horizontal component for Love waves W is indicated by the long dashed lines.

16.4 Seismograms by modal summation

As we have seen in Section 15.3, the representation of the seismic response to a source as a function of angular frequency (ω) and slowness (p) has singularities in the form of poles in the complex $\omega - p$ plane corresponding to normal modes of the stratified structure. The residue contributions from these sets of poles constitute the surface wave field.

The modal contribution to the surface displacement at a range X and an azimuth ϕ from the source for Rayleigh waves can be represented as

$$\mathbf{u}_S(X,\phi,t) = \\ \tfrac{1}{2}\int_{-\infty}^{\infty} d\omega e^{-i\omega t}\, i\omega^2 \sum_m e^{im\phi}\left[\sum_{j=0}^{N(\omega)} p_j \operatorname*{Res}_{\omega,p=p_j}[\boldsymbol{w}_0^{\mathrm{T}}(p,m,\omega)]\mathbf{T}_m^{(1)}(\omega p_j X)\right]. \tag{16.4.1}$$

Here $\boldsymbol{w}_0(p,m,\omega)$ represents the spectral response for the coupled *P-SV* wave system in the ω-p domain and $N(\omega)$ is the number of Rayleigh modes for frequency ω. The angular variation in radiation from the source is represented through the summation over the angular order m of the vector surface harmonics contained in the tensor field $\mathbf{T}_m^{(1)}(\omega p_j X)$ (15.3.2). These surface-harmonic terms include the horizontal phase dependence of the displacement field. In terms of the displacement quantities U, V and the vector harmonics **R**, **S**,

$$\boldsymbol{w}_0^{\mathrm{T}}(p,m,\omega)\mathbf{T}_m^{(1)}(\omega p_j X) = U(p,m,\omega)\mathbf{R}_m^{(1)}(\omega p_j X) + V(p,m,\omega)\mathbf{S}_m^{(1)}(\omega p_j X). \tag{16.4.2}$$

A comparable form can be written for Love waves in terms of the spectral displacement W and the vector surface harmonic **T** which lies wholely in the horizontal plane,

$$\mathbf{u}_H(X,\phi,t) = \\ \tfrac{1}{2}\int_{-\infty}^{\infty} d\omega e^{-i\omega t}\, i\omega^2 \sum_m e^{im\phi}\left[\sum_{l=0}^{M(\omega)} p_l \operatorname*{Res}_{\omega,p=p_l}[W_0(p,m,\omega)]\mathbf{T}_m^{(1)}(\omega p_l X)\right], \tag{16.4.3}$$

where, now, $M(\omega)$ is the number of Love modes at frequency ω.

The azimuthal variation in the spectral representation of the response comes from the expansion of the source contribution which can be conveniently represented in terms of discontinuities in displacement and traction across the source level $z = z_s$,

$$\mathbf{S}(p,m,\omega,z_s) = [\boldsymbol{S}_W^m(z_s), \boldsymbol{S}_T^m(z_s)]^{\mathrm{T}} \tag{16.4.4}$$

At a pole corresponding to a surface-wave mode, a single eigen-displacement field $\boldsymbol{w}_e$ satisfies both the surface and radiation conditions. The residue at the pole can be extracted in terms of the source jumps and this eigen-displacement field

$$\operatorname*{Res}_{\omega,p=p_j}[\boldsymbol{w}_0^{\mathrm{T}}(p,m,\omega)] = \frac{g_j}{2\omega I_j}[\boldsymbol{t}_{ej}^{\mathrm{T}}(z_s)\boldsymbol{S}_W^m - \boldsymbol{w}_{ej}^{\mathrm{T}}(z_s)\boldsymbol{S}_T^m]\boldsymbol{w}_{ej}(0), \tag{16.4.5}$$

where g_j is the group slowness for the particular mode, and I_j is related to the kinetic energy content in the eigen-wavefield

$$I_j = \int_0^\infty \mathrm{d}z\,\rho\,\boldsymbol{w}_{ej}^{\mathrm{T}}\boldsymbol{w}_{ej}. \tag{16.4.6}$$

The residue contribution can be separated into a part which is linked to the source $[\boldsymbol{t}_{ej}^{\mathrm{T}}(z_s)\boldsymbol{S}_W^m - \boldsymbol{w}_{ej}^{\mathrm{T}}(z_s)\boldsymbol{S}_T^m]$, and a displacement contribution at the receiver $\boldsymbol{w}_{ej}(0)$.

With this expression for the residue contribution from each mode, the surface displacement is

$$\mathbf{u}_S(X,\phi,t) = \int_{-\infty}^{\infty} \mathrm{d}\omega \mathrm{e}^{-\mathrm{i}\omega t}\,\mathrm{i}\omega \sum_m \mathrm{e}^{\mathrm{i}m\phi}$$
$$\left[\sum_{j=0}^{N(\omega)} \frac{p_j g_j}{4I_j}[\boldsymbol{t}_{ej}^{\mathrm{T}}(z_s)\boldsymbol{S}_W^m - \boldsymbol{w}_{ej}^{\mathrm{T}}(z_s)\boldsymbol{S}_T^m]\boldsymbol{w}_{ej}(0)\mathbf{T}_m^{(1)}(\omega p_j X)\right]. \tag{16.4.7}$$

The source and receiver contributions to the displacement can be expressed in ways which have more direct physical content (cf chapter 14). For the coupled *P-SV* system, the source dependence can be written as

$$\boldsymbol{S}_{\mathrm{R}}^m(p_j) = [\boldsymbol{t}_{ej}^{\mathrm{T}}(z_s)\boldsymbol{S}_W^m - \boldsymbol{w}_{ej}^{\mathrm{T}}(z_s)\boldsymbol{S}_T^m] = -\mathrm{i}[\boldsymbol{U}_m^S + \mathbf{R}_D^{SL}\boldsymbol{D}_m^S], \tag{16.4.8}$$

where $\boldsymbol{U}_m^S$ is the upgoing radiation from the source in angular order m and $\boldsymbol{D}_m^S$ is the corresponding downgoing radiation. We note that since (16.4.8) involves the processes of reflection from beneath the source level through $\mathbf{R}_D^{SL}$, the contribution involving the source is not localised at the source itself but will involve the full propagation path. The receiver term includes the processes of transmission from the source level to the surface

$$\mathcal{R}(p_j) = \boldsymbol{w}_{ej}(0) = \mathbf{W}_F[\mathbf{I} - \mathbf{R}_D^{0S}\mathbf{R}_F]^{-1}\mathbf{T}_U^{0S}, \tag{16.4.9}$$

where $\mathbf{W}_F$ allows for the amplification of displacement due to interaction with the free surface and the instrument response. $\mathbf{R}_F$ is the free-surface reflection matrix. The reflection and transmission terms $\mathbf{R}_D^{0S}$, $\mathbf{T}_U^{0S}$ involve the region between the source and the surface.

For long propagation paths or higher frequencies we can use asymptotic approximations for the vector surface harmonics. The tensor field $\mathbf{T}_m^{(1)}(\omega p_j X)$, which appears in the Rayleigh wave representation, is approximated by

$$\mathbf{T}_m^{(1)}(\omega p_j X) \sim \begin{pmatrix} \mathbf{e}_z \\ \mathrm{i}\mathbf{e}_X \end{pmatrix} \left(\frac{2}{\pi\omega p_j X}\right)^{1/2} \exp[\mathrm{i}\omega p_j X - \mathrm{i}(2m+1)\pi/4], \tag{16.4.10}$$

where, as in Section 15.2, $\mathbf{e}_z$, $\mathbf{e}_X$ are unit vectors in the vertical and radial direction. This asymptotic form separates the phase contribution from the passage through the distance X away from the source from the decay in amplitude associated with spreading out of the waves across the surface. The equivalent form for the tangential (*SH*) harmonic is

$$\mathbf{T}_m^{(1)}(\omega p_l r) \sim -\mathrm{i}\mathbf{e}_\phi \left(\frac{2}{\pi\omega p_l X}\right)^{1/2} \exp[\mathrm{i}\omega p_l X - \mathrm{i}(2m+1)\pi/4], \tag{16.4.11}$$

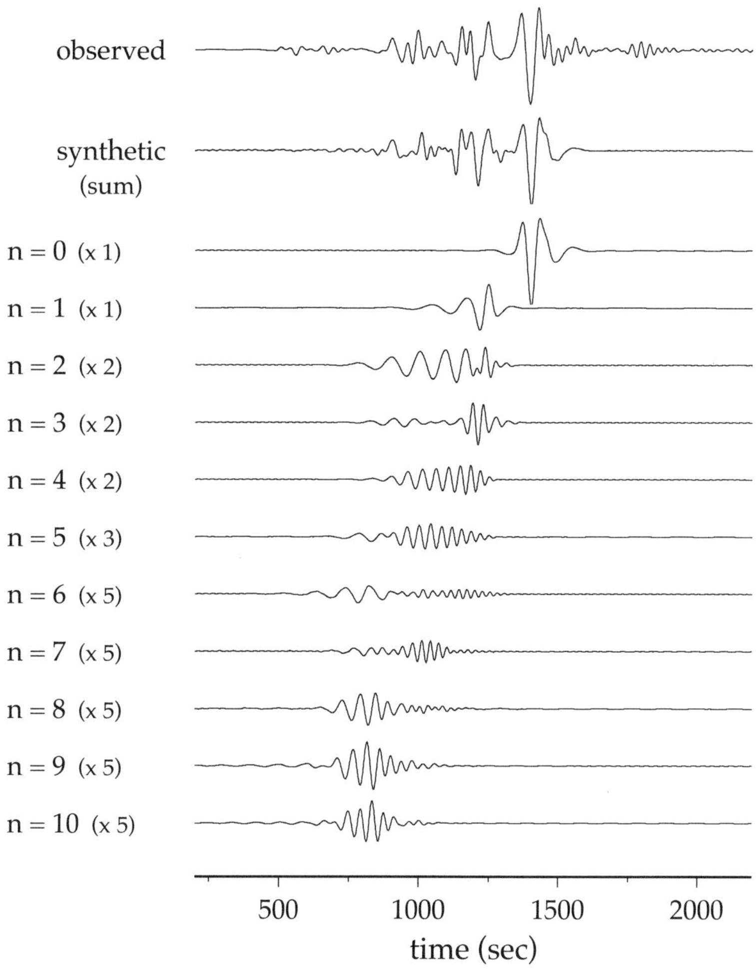

Figure 16.14. The assembly of a seismogram by modal contributions. The vertical component at NWAO from an event in Vanuatu (15.75°S, 167.19°E, 171 km depth) is compared to a synthetic seismogram built by mode summation for the PREM model at an epicentral distance of 48.75°. The seismograms are filtered in the band 0.008-0.030 Hz and the contributions of the first 10 Rayleigh modes are shown separately (together with their magnifications).

where $\mathbf{e}_\phi$ lies in the local transverse direction to the path.

When the argument of the phase function is large for all possible surface wave modes (i.e., $\omega\beta_L^{-1}r \gg 1$), the asymptotic approximations (16.4.10) and (16.4.11) can be used to represent the surface wave part of the wavefield.

For Rayleigh waves,

$$\mathbf{u}_S(X,\phi,t) = \begin{pmatrix} \mathbf{e}_z \\ \mathrm{i}\mathbf{e}_X \end{pmatrix} \sum_m \mathrm{e}^{\mathrm{i}m\phi} \int_{-\infty}^{\infty} \mathrm{d}\omega \qquad (16.4.12)$$

$$\mathrm{i} \sum_{j=0}^{N(\omega)} \sqrt{\frac{2\omega p_j}{\pi X} \frac{g_j}{4I_{\mathrm{R}j}}}\, \boldsymbol{S}_{\mathrm{R}}^m(p_j)\, \boldsymbol{\mathcal{R}}_{\mathrm{R}}(p_j)\, \mathrm{e}^{[\mathrm{i}\omega(p_j X - t) - \mathrm{i}(2m+1)\pi/4]}.$$

We should remember that there is an implied dependence on frequency ω for the phase and group slownesses p_j, g_j and all the terms such as $\boldsymbol{S}$, $\boldsymbol{\mathcal{R}}$ which depend on the properties of individual mode branches. Although we have identified source and receiver terms above, we have already seen that these have non-local properties.

The equivalent representation for Love waves is

$$\mathbf{u}_H(X,\phi,t) = \mathbf{e}_\phi \sum_m \mathrm{e}^{\mathrm{i}m\phi} \int_{-\infty}^{\infty} \mathrm{d}\omega \qquad (16.4.13)$$

$$\sum_{l=0}^{M(\omega)} \sqrt{\frac{2\omega p_l}{\pi X} \frac{g_l}{4I_{\mathrm{L}l}}}\, S_{\mathrm{L}}^m(p_l)\, \mathcal{R}_{\mathrm{L}}(p_l)\, \mathrm{e}^{[\mathrm{i}\omega(p_l X - t) - \mathrm{i}(2m+1)\pi/4]}.$$

In both (16.4.12) and (16.4.13) the major phase contribution comes from the horizontal propagation term but there will also some influence from the source and instrumental response.

In figure 16.14 we show the way in which a seismogram can be built up from modal contributions using the example of an intermediate depth event recorded at 5400 km from the source. The contributions from ten Rayleigh modes in the PREM model are illustrated, with amplification so that the traces are more clearly visible. The modal contributions are added to give the summed synthetic which is a good representation of the observed seismogram at the NWAO station.

16.5 Radiation patterns for surface waves

The source terms $\boldsymbol{S}_{\mathrm{R}}^m$, S_{L}^m contain the radiation patterns imposed by the source. For a point moment tensor, we can sum over the angular orders ($|m| < 2$) to extract the dependence on the moment tensor components that depend on the modal eigendisplacements and tractions through (16.4.8).

In terms of the eigendisplacements and the vertical derivatives at the source depth, we can express the Rayleigh wave radiation term as

$$\begin{aligned} \mathcal{F}^{\mathrm{R}}(p_j,\phi,z_S,\omega) &= p_j V_e(p_j,z_S,\omega)[M_{xx}\cos^2\phi + M_{xy}\sin 2\phi + M_{yy}\sin^2\phi] \\ &\quad + \mathrm{i}p_j \left(U_e(p_j,z_S,\omega) + \partial_z V_e(p_j,z_S,\omega)\right)[M_{xz}\cos\phi + M_{yz}\sin\phi] \\ &\quad - \partial_z U_e(p_j,z_S,\omega) M_{zz}. \end{aligned} \qquad (16.5.1)$$

The dependence on frequency appears through the eigenfunctions and the spectral content of the moment tensor terms from the time history of the source. $\mathcal{F}^{\mathrm{R}}$ depends on the same combinations of moment tensor terms as appear in the radiation patterns

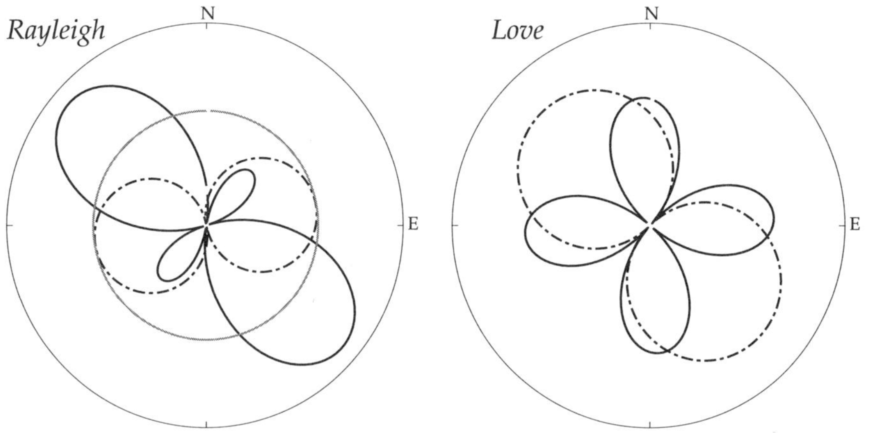

Figure 16.15. The azimuthal dependence of the moment tensor combinations appearing in the Rayleigh and Love wave radiation patterns for an event in southern Xinjiang, China (Table 4.1). For Rayleigh waves: $[M_{xx}\cos^2\phi + M_{xy}\sin 2\phi + M_{yy}\sin^2\phi]$, solid line, $[M_{xz}\cos\phi + M_{yz}\sin\phi]$, chain dotted line, M_{zz} (isotropic), grey tone. For Love waves: $[(M_{yy} - M_{xx})\sin\phi\cos\phi + M_{xy}\cos 2\phi]$ solid line, $[M_{yz}\cos\phi - M_{xz}\sin\phi]$ chain dotted line.

for *P-SV* body waves: $\mathcal{F}^{\mathrm{P}}$ (4.3.4) and $\mathcal{F}^{\mathrm{SV}}$ (4.3.6). The angular dependence of these moment tensor combinations in the radiation pattern for Rayleigh waves is shown in the left hand panel of figure 16.15. We use the moment tensor for the Xinjiang source (Table 4.1) previously used to illustrate body wave radiation.

The Love wave radiation pattern takes a comparable form

$$\mathcal{F}^{\mathrm{L}}(p_l,\phi,z_S,\omega) = \mathrm{i}p_l W_e(p_l,z_S,\omega)[(M_{yy} - M_{xx})\sin\phi\cos\phi + M_{xy}\cos 2\phi] \\ -\partial_z W_e(p_l,z_S,\omega)[M_{yz}\cos\phi - M_{xz}\sin\phi], \qquad (16.5.2)$$

and again incorporates the same moment-tensor combinations as the *SH* radiation for body waves $\mathcal{F}^{\mathrm{SH}}$ (4.3.8). The angular dependence of the moment tensor terms appearing in the radiation pattern for Love waves is shown in the right hand panel of figure 16.15.

For both Rayleigh and Love waves the full radiation pattern for a particular event is controlled by the relative amplitudes of the weighting terms that depend on the particular forms of the modal eigenfunctions and the source depth.

Appendix: Table of Notation

Stress and Strain

$\mathbf{x}$	-	position vector
x_i	-	position coordinates
$\boldsymbol{\xi}$	-	initial position vector
ξ_i	-	initial position coordinates
$\mathbf{u}$	-	displacement vector
$\mathbf{v}$	-	velocity vector
$\mathbf{f}$	-	acceleration vector
$\mathbf{g}$	-	external force vector
$\mathbf{n}$	-	normal vector
$\mathbf{s}$	-	polarisation vector
τ_{ij}	-	incremental stress tensor
σ_{ij}	-	stress tensor (including pre-stress)
e_{ij}	-	strain tensor
$\boldsymbol{\sigma}, \mathbf{t}$	-	traction vectors
$\boldsymbol{\tau}$	-	traction vector at a surface
$\mathbf{n}$	-	normal vector
c_{ijkl}	-	elastic modulus tensor
C_{ijkl}	-	anelastic relaxation tensor
κ	-	bulk modulus
μ	-	shear modulus
ρ	-	density
λ	-	Lamé modulus
R_λ, R_μ	-	relaxation functions
A, C, F, L, N, H	-	moduli for transversely isotropic media
ψ	-	gravitational potential
p, p'	-	pressure, incremental pressure

Waves and Rays

z	-	depth
r	-	radius
t	-	time
ω	-	angular frequency
θ, ϕ	-	coordinate angles
α	-	P wavespeed
β	-	S wavespeed
v	-	general wavespeed
a	-	P slowness $(1/\alpha)$
b	-	S slowness $(1/\beta)$
c	-	wavespeed of sound waves in a fluid
c, C	-	phase velocity
v_g, U	-	group velocity
p	-	horizontal slowness, phase slowness
g	-	group slowness
$\mathbf{n}$	-	direction of travel
$\mathbf{s}$	-	ray direction
$\mathbf{k}$	-	wavenumber vector
k_i	-	wavenumber components
$\mathbf{p}$	-	slowness vector
p_i	-	slowness components
T	-	travel time
X	-	horizontal distance
Δ	-	epicentral angle (for propagation in a sphere)
p	-	ray parameter (horizontal stratification) $(= \mathrm{d}T/\mathrm{d}X)$,
$\wp$	-	ray parameter (radial stratification) $(= \mathrm{d}T/\mathrm{d}\Delta)$
q	-	auxiliary ray parameter
τ	-	intercept time, delay time
q_α	-	vertical slowness for P waves
q_β	-	vertical slowness for S waves
i	-	angle of incidence for P waves
j	-	angle of incidence for S waves
Q^{-1}	-	loss factor
ϑ	-	wavefront/phase function
g_{ik}	-	Christoffel matrix
J	-	ray spreading function
η	-	radial wavespeed variable $(v(r)/r)$
ζ	-	$\mathrm{d}\ln v/\mathrm{d}\ln r$
ξ	-	$\mathrm{d}\ln r/\mathrm{d}\ln \eta$

Sources

R	-	distance from source
$\boldsymbol{\gamma}$	-	unit direction vector from source
γ_i	-	direction cosines
$\mathbf{h}$	-	source position
$\mathbf{R}$	-	receiver location
$\mathbf{S}$	-	source location
$\mathbf{n}$	-	normal vector
$\boldsymbol{\nu}$	-	slip vector
ϕ_s	-	strike angle
δ	-	dip angle
λ	-	rake angle
$\boldsymbol{\epsilon}$	-	force vector
$\boldsymbol{\mathcal{E}}$	-	summary force vector
G_{ij}	-	Green's tensor
H_{ijp}	-	stress tensor derived from Green's tensor
M_{ij}	-	Moment tensor
$m_{ij}(\mathbf{x}, t)$	-	moment tensor density
$\mathbf{c}, \mathbf{d}$	-	auxiliary vectors
$\hat{\mathbf{V}}, \hat{\mathbf{H}}$	-	unit vectors for *SV*, *SH*
$\hat{\mathbf{N}}, \hat{\mathbf{E}}, \hat{\mathbf{D}},$	-	unit vectors for North, East, Down coordinate system
$\mathbf{e}_z, \mathbf{e}_X, \mathbf{e}_\phi$	-	unit vectors in cylindrical courdinate system
$\mathbf{e}_r, \mathbf{e}_\theta, \mathbf{e}_\phi$	-	unit vectors in spherical coordinate system
$\mathbf{Q}, \mathbf{T}$	-	radiation vectors for *P*, *S*

Propagation terms

$\mathbf{A}$	-	coefficient matrix
$\mathfrak{A}$	-	coefficient matrix (spherical)
$\mathbf{b}$	-	motion-stress vector
$\mathfrak{b}$	-	motion-stress vector (spherical)
$\mathbf{C}$	-	transformation matrix
$\mathfrak{C}$	-	transformation matrix (spherical)
$\mathbf{D}$	-	matrix of motion-stress vectors
D	-	motion stress matrix for gradients
$E, \hat{E}$	-	phase matrices for wavetype
E	-	phase matrix
E	-	phase matrix for gradients
$\mathbf{F}$	-	source vector
$\boldsymbol{H}$	-	propagation matrix
$\mathbf{L}, \boldsymbol{L}$	-	correction terms for gradient zones

Propagation terms (cont.)

$\boldsymbol{Q}$	-	interface matrix
$\boldsymbol{t}$	-	traction vector in $p - \omega$ domain
$\boldsymbol{v}$	-	wave vector
$\boldsymbol{w}$	-	displacement vector in $p - \omega$ domain
U, V, W	-	displacement components
P, S, T	-	traction components
k_α, k_β	-	spherical wavenumbers
$\epsilon_\alpha, \epsilon_\beta, \epsilon_H$	-	normalisation factors for wave energy
$\boldsymbol{v}_{U,D}$	-	wave vector components
$\boldsymbol{m}_{U,D}$	-	displacement matrices
$\boldsymbol{n}_{U,D}$	-	traction matrices
$\mathbf{E}_{U,D}$	-	phase delay matrices
$\boldsymbol{U}^S, \boldsymbol{D}^S$	-	upgoing, downgoing waves from source
$\mathbf{R}_F$	-	free-surface reflection matrix
$\mathrm{R}_F^{HH}, \mathrm{R}_F^{HH}$	-	free surface reflection coefficients
$\mathbf{W}_F$	-	free-surface amplification matrix
$\mathbf{R}_U, \mathbf{R}_D$	-	reflection matrices for incident upgoing, downgoing waves
$\mathbf{T}_U, \mathbf{T}_D$	-	transmission matrices for incident upgoing, downgoing waves
$P_{U,D}$	-	weighting coefficients for upgoing, downgoing *P* waves
$S_{U,D}$	-	weighting coefficients for upgoing, downgoing *SV* waves
$H_{U,D}$	-	weighting coefficients for upgoing, downgoing *SH* waves
$\phi_{U,D}$	-	wave amplitudes for upgoing, downgoing *P* waves
$\psi_{U,D}$	-	wave amplitudes for upgoing, downgoing *P* waves
$\chi_{U,D}$	-	wave amplitudes for upgoing, downgoing *P* waves
φ	-	phase of *PP* reflection coefficient
ψ	-	phase of *SS* reflection coefficient
χ	-	phase of *HH* reflection coefficient
$\eta, \hat{\eta}$	-	generalised vertical slownesses
$\gamma_A, \gamma_B, \gamma_C$	-	coupling coefficients
C_R, C_R	-	receiver terms
C_S, C_S	-	source terms
C_{RS}	-	composite source and receiver term
$\mathcal{P}$	-	propagation term
$\mathcal{Q}$	-	attenuation factor
$\mathcal{M}(t)$	-	effective source time function
$\boldsymbol{g}$	-	amplitude weight for generalised ray
Υ	-	Cagniard path in complex p plane
$[X]^{-1}$	-	reverberation operator
Z_R, Z_S, Z_J	-	reverberation operators

Mathematical

a, b, c	-	constants
$\mathcal{H}$	-	Hilbert transform operator
$\mathcal{L}, \mathcal{M}, \mathcal{N}$	-	differential operators
Ai,Bi	-	Airy functions
Aj,Ak,Bj,Bk	-	Airy function wave variables
Ej,Ek,Ej,Ek	-	Airy function wave variables
$H_m^{(1),(2)}(x)$	-	Hankel functions
$J_m(x)$	-	Bessel function
$\mathbf{R}_k^m, \mathbf{S}_k^m, \mathbf{T}_k^m$	-	vector surface harmonics (cylindrical)
$\mathbf{T}_m, \mathbf{T}_m^{(1)}$	-	tensor field of vector harmonics
$j_l(x), h_l^{(1),(2)}(x)$	-	spherical Bessel functions
$P_l(x), P_l^m(x)$	-	Legendre functions
$Q_l^{(1),(2)}$	-	travelling wave form of Legendre function
$\mathbf{P}_l^m, \mathbf{B}_l^m, \mathbf{C}_l^m$	-	vector surface harmonics (cylindrical)
$Y_l^m, \mathcal{Y}_{lm}$	-	surface harmonics on a sphere
Λ	-	analytic time function

Bibliography

Abo-Zena, A.M., 1979. Dispersion function computations for unlimited frequency values, *Geophys. J. R. Astr. Soc*, **58**, 91–105.

Aki, K. & Richards, P.G., 1980. *Quantitative Seismology*, W.H. Freeman, San Francisco.

Aki, K., 1981. Attenuation and scattering of short period seismic waves in the lithosphere, *Identification of Seismic Sources*, ed. E. S. Husebye, Noordhof, Leiden.

Aki, K. & Chouet, B., 1975. Origin of coda waves: source, attenuation and scattering effects, *J. Geophys. Res.*, **80**, 3322–3342.

Alekseev, A.S. & Mikhailenko, B.G., 1980. Solution of dynamic problems of elastic wave propagation in inhomogeneous media by a combination of partial separation of variables and finite difference methods, *J. Geophys.*, **48**, 161–172.

Altman, C. & Cory, H., 1969. The generalised thin film optical method in electromagnetic wave propagation, *Radio Sci.*, **4**, 459–469.

Anderson, D.L., BenMenahem, A. & Archambeau, C.B., 1965. Attenuation of seismic energy in the upper mantle *J. Geophys. Res.*, **70**, 1441–1448

Azbel', I.Ya. & Yanovskaya, T.B., 1972. Approximation of velocity distributions for calculation of *P* times and amplitudes, in *Computational Seismology*, ed .V.I. Keilis-Borok, Consultants Bureau, New York, 62–70.

Azimi, Sh., Kalinin, A.V., Kalinin, V.V. & Pivovarov, B.L., 1968. Impulse and transient characteristics of media with linear and quadratic absorption laws, *Izv. Physics of Solid Earth*, **2**, 88–93.

Babich, V.M., 1961. Ray method for the computation of the intensity of wavefronts in elastic inhomogeneous anisotropic media, *Problems of the Dynamic Theory of Propagation of Seismic Waves*, **5**, Leningrad University Press (in Russian).

Backus, G.E. & Mulcahy, M., 1976a. Moment tensors and other phenomenological descriptions of seismic sources I - Continuous displacements, *Geophys. J. R. Astr. Soc.*, **46**, 341–362.

Backus, G.E. & Mulcahy, M., 1976b. Moment tensors and other phenomenological descriptions of seismic sources II - Discontinuous displacements, *Geophys. J. R. Astr. Soc.*, **47**, 301–330.

Babuška, V. & Cara, M., 1991. *Seismic Anisotropy in the Earth*, Kluwer Academic Publishers, Dordrecht.

Bamford, D., 1977. *Pn* velocity in a continental upper mantle, *Geoph. J. R. Astr. Soc.*, **39**, 29–48.

Ben-Menahem A. & Singh, S.J., 1981. *Seismic Waves and Sources*, Springer Verlag, New York.

Bloxham, J. & Gubbins, G., 1989. Geomagnetic secular variation, *Phil. Trans. R. Soc. Lond.*, **329A**, 415–502.

Boltzman, L., 1876. Zur Theorie der elastichen Nachwirkung, *Pogg. Ann. Erganzungbd.*, **7**, 624–654.

Bouchon, M., 1979. Discrete wave number representation of elastic wave fields in three-space dimensions, *J. Geophys. Res.*, **84**, 3609-3614.

Bowers, D.A. & Hudson, J.A., 1999. Defining the scalar moment of a seismic source with a general moment tensor, *Bull. Seism. Soc. Am.*, **89**, 1390–1394.

Bowman, J.R., 1992. The 1998 Tennant Creek, Northern Territory, earthquakes: a synthesis, *Aust, J. Earth. Sci.*, **39**, 651–669.

Bowman, J.R., Gibson, G. & Jones, T., 1990. Aftershocks of the 22 January 1988 Tennant Creek, Australia intraplate earthquakes: Evidence for a complex fault geometry, *Geophys. J. Int.*, **100**, 87–97.

Braile, L.W. & Smith, R.B., 1975. Guide to the interpretation of crustal refraction profiles, *Geophys. J. R. Astr. Soc.*, **40**, 145–176.

Brekhovskikh, L.M., 1960. *Waves in Layered Media*, Academic Press, New York.

Brennan, B.J. & Smylie, D.E., 1981. Linear viscoelasticity and dispersion in seismic wave propagation, *Rev. Geophys. Space. Science*, **19**, 233–246.

Brodskií, M. & Levshin, A., 1979. An asymptotic approach to the inversion of free oscillation data, *Geophys. J. R. Astr. Soc.*, **58**, 631–654.

Budden, K.G., 1961. *Radio Waves in the Ionosphere*, Cambridge University Press, Cambridge.

Buland, R. & Chapman, C.H., 1983. The computation of seismic travel times, *Bull. Seism. Soc. Am.*, **73**, 1271–1302.

Bullen, K.E., 1965. *An Introduction to the Theory of Seismology*, 3rd edition, Cambridge University Press, Cambridge.

Bullen, K.E. & Bolt, B.A., 1985. *An Introduction to the Theory of Seismology*, Cambridge University Press, Cambridge.

Bullen, K.E., 1975. *The Earth's Density*, Chapman & Hall, London.

Burdick, L.J., & Helmberger, D.V., 1978. The upper mantle P velocity structure of the western United States, *J. Geophys. Res.*, **83**, 1699–1712.

Červený, V., 1972. Seismic rays and ray intensities in inhomogeneous anisotropic media, *Geophys. J. R. Astr. Soc.*, **29**, 1–13.

Červený, V., Molotkov, I.A. & Pšenčík, I., 1977. *Ray Method in Seismology*, Univerzita Karlova, Praha.

Chapman, C.H., 1971. On the computation of seismic ray travel times and amplitudes, *Bull. Seism. Soc. Am.*, **61**, 1267–1274.

Chapman, C.H., 1973. The earth flattening transformation in body wave theory, *Geophys. J. R. Astr. Soc.*, **35**, 55–70.

Chapman, C.H., 1974a. Generalised ray theory for an inhomogeneous medium, *Geophys. J. R. Astr. Soc.*, **36**, 673–704.

Chapman, C.H., 1974b. The turning point of elastodynamic waves, *Geophys. J. R. Astr. Soc.*, **39**, 613–621.

Chapman, C.H., 1976. Exact and approximate generalized ray theory in vertically inhomogeneous media, *Geophys. J. R. Astr. Soc.*, **46**, 201–234.

Chapman, C.H., 1978. A new method for computing synthetic seismograms, *Geophys. J. R. Astr. Soc.*, **54**, 481–518.

Chapman, C.H., 1981. Long period correction to body waves: Theory, *Geophys. J. R. Astr. Soc.*, **64**, 321–372.

Chapman, C.H. & Orcutt J.A., 1985. The computation of body wave synthetic seismograms in laterally homogeneous media, *Rev. Geophys.*, **23**, 105–163.

Chapman, C.H. & Woodhouse J.H., 1981. Symmetry of the wave equation and excitation of body waves, *Geophys. J. R. Astr. Soc.*, **65**, 777–782.

Chapman, C.H., Chu, J-Y & Lyness, D.G., 1988. The WKBJ seismogram algorithm, 47–74 in *Seismological Algorithms*, ed. D.J. Doornbos, Academic Press, London.

Choy, G.L., 1977. Theoretical seismograms of core phases calculated by frequency-dependent full wave theory, and their interpretation, *Geophys. J. R. Astr. Soc.*, **51**, 275–312.

Choy, G.L., Cormier, V.F., Kind, R., Müller, G. & Richards, P.G., 1980. A comparison of synthetic seismograms of core phases generated by the full wave theory and the reflectivity method, *Geophys. J. R. Astr. Soc.*, **61**, 21–40.

Cormier, V.F. & Richards, P.G., 1977. Full wave theory applied to a discontinuous velocity increase: the inner core boundary, *J. Geophys.*, **43**, 3–31.

Cormier, V.F., 1980. The synthesis of complete seismograms in an earth model composed of radially inhomogeneous layers, *Bull. Seism. Soc. Am.*, **70**, 691–716.

Cormier, V.F. & Choy, G.L., 1981. Theoretical body wave interactions with upper mantle structure *J. Geophys. Res.*, **86**, 1673–1678.

Cormier, V.F. & Richards, P.G., 1988. Spectral synthesis of body waves in Earth models specified by vertically varying layers, 3–45, in *Seismological Algorithms*, ed. D.J. Doornbos, Academic Press, London.

Coulomb, J. & Jobert, G., 1972, *Traité de Geophysique Interne*, Masson & Cie, Paris.

Crampin, S., 1981, A review of wave motion in anisotropic and cracked elastic media, *Wave Motion*, **3**, 343–391.

Crampin, S., Evans, R., Uçer, B., Doyle, M., Davis, J.P., Yegorkina, G. V. & Miller, A., 1980. Observations of dilatancy induced polarisation anomalies and earthquake prediction. *Nature*, **286**, 874–877.

Crampin, S. & Zatsepin, S.V., 1997. Modelling the compliance of crustal rock - II. Response to temporal changes before earthquakes, *Geophys. J. Int.*, **129**, 495–506.

Crampin, S., Volti, T. & Stefánsson, R., 1999. A successfully stress-forecast earthquake, *Geophys. J. Int.*, **138**, F1–F5.

Creager, K.C., 1999. Large scale variations in inner core anisotropy *J. Geophys. Res.*, *104*, 23 127–23 139.

Dahlen, F.A. & Tromp, J., 1998. *Theoretical Global Seismology*, Princeton University Press, Princeton.

DeMets, C., Gordon, R.G., Argus, D.F. & Stein, S., 1990. Current plate motions, *Geophys. J. Int.*. **101**, 425–478.

DeMets, C., Gordon, R.G., Argus, D.F. & Stein, S., 1994. Effect of recent revisions to the geomagnetic reversal time scale on estimate of current plate motion, *Geophys. Res. Lett.*, **21**, 2191–2194.

Denman, E.D., 1970. *Coupled Modes in Plasmas, Elastic Media and Parametric Amplifiers*, Elsevier, New York.

Der, Z., 1998. High frequency *P*- and *S*-wave attenuation in the Earth, *Pure Appl. Geophys.*, **153**, 273–310.

Doornbos, D.J., 1981. The effect of a second-order velocity discontinuity on elastic waves near their turning point, *Geophys. J. R. Astr. Soc.*, **64**, 499–511.

Doornbos, D.J., 1988. Asphericity and ellipticity corrections, 75–85, in *Seismological Algorithms*, ed. D.J. Doornbos, Academic Press, London.

Douglas, A., Hudson, J.A., & Blamey, C., 1973. A quantitative evaluation of seismic signals at teleseismic distances III - Computed P and Rayleigh wave seismograms, *Geophys. J. R. Astr. Soc.*, **28**, 345–410.

Durek, J.J. & Ekström, G., 1996. A radial model of anelasticity consistent with long-period surface-wave attenuation, *Bull. Seism. Soc. Am*, **86**, 144–158.

Durek, J.J. & Ekström, G., 1997. Investigating discrepancies among measurements of traveling and standing wave attenuation, *J. Geophys. Res*, **102**, 24 529–24 544.

Durek, J.J., Ritzwoller, M. & Woodhouse, J.H., 1993. Constraining upper mantle anelasticity using surface-wave amplitudes, *Geophys. J. Int.*, **114**, 249–272.

Dziewonski, A.M., Hales, A.L., & Lapwood, E.R., 1975. Parametrically simple Earth models consistent with geophysical data, *Phys. Earth Planet. Inter.*, **10**, 12–48.

Dziewonski, A.M. & Anderson D.L., 1981. Preliminary reference Earth model, *Phys. Earth Planet. Inter.*, **25**, 297–356.

Dziewonski, A.M., Chou, T.-A. & Woodhouse, J.H., 1981. Determination of earthquake source parameters from waveform data for studies of global and regional seismicity, *J. Geophys. Res.*, **86**, 2825–2852.

Dziewonski, A.M. & Woodhouse J.H., 1983. Studies of the seismic source using normal-mode theory, *Earthquakes: Observation, Theory and Interpretation*, Proc. Intl School of Physics "Enrico Fermi", Course LXXXV, pp 45–137, ed H. Kanamori & E. Boschi, North-Holland, Amsterdam.

Dziewonski, A.M., Friedman, A., Giardini, D. & Woodhouse J.H., 1983. Global seismicity of 1982: centroid moment tensor solutions for 308 earthquakes, *Phys. Earth Planet. Inter.*, **33**, 76–90.

Engdahl, E.R., van der Hilst R.D. & Buland, R., 1998. Global teleseismic earthquake relocation with improved travel times and procedures for depth determination, *Bull. Seism. Soc. Am.*, **88**, 722–743.

Faber, S. & Müller, G., 1980. *Sp* phases from the transition zone between the upper and lower mantle, *Bull. Seism. Soc. Am.*, **70**, 487–508.

Felsen, L.B. & Isihara, T., 1979. Hybrid ray-mode formulation of ducted propagation, *J. Acoust. Soc. Am.*, **65**, 595–607.

Frazer, L.N., 1988. Quadrature of wavenumber intergals, 75–85, in *Seismological Algorithms*, ed. D.J. Doornbos, Academic Press, London.

Frohlich, C., 1996. Cliff's nodes concerning plotting nodal lines for *P*, *SH* and *SV*, *Seism. Res. Lett.*, **67**, 16–24.

Fryer, G.J. & Frazer, L.N., 1987. Seismic waves in stratified anisotropic media II - elastodynamic eigensolutions for some anisotropic systems, *Geophys. J. R. Astr. Soc.*, **91**, 73–101.

Fryer, G.J., 1980. A slowness approach to the reflectivity method of seismogram synthesis, *Geophys. J. R. Astr. Soc.*, **63**, 747–758.

Fuchs, K., 1968. Das Reflexions- und Transmissionsvermogen eines geschichteten Mediums mit belieber Tiefen-Verteilung der elastischen Modulor und der Dichte für schragen Einfall Ebener Wellen, *Z. Geophys.*, **34**, 389–411.

Fuchs, K. & Müller, G., 1971. Computation of synthetic seismograms with the reflectivity method and comparison with observations, *Geophys. J. R. Astr. Soc.*, **23**, 417–433.

Fuchs, K., 1975. Synthetic seismograms of *PS*-reflections from transition zones computed with the reflectivity method, *J. Geophys.*, **41**, 445–462.

Furumura, T. & Takenaka H., 1996. 2.5-D modelling of elastic waves using the pseudospectral method, *Geophys. J. Int.*, **124**, 820–832.

Furumura, T., Kennett, B.L.N. & Furumura, M., 1998. Synthetic seismograms for laterally heterogeneous whole earth models using the pseudospectral method, *Geophys. J. Int.*, **135**, 845–860.

Furumura, M., Kennett, B.L.N. & Furumura, T., 1999. Seismic wavefield calculation for laterally heterogeneous earth models - II. The influence of upper mantle heterogeneity, *Geophys.J.Int*, **139**, 623–644.

Gans, R., 1915. Fortpflanzung des Lichtes durch ein inhomogenes Medium. *Ann. Physik*, **47**, 709–732.

Garmany, J., 1989, A student's garden of anisotropy, *Ann. Rev. Earth Planet. Sci.*, **17**, 285–308.

Gilbert, F., 1976. The representation of seismic displacements in terms of travelling waves, *Geophys. J. R. Astr. Soc.*, **44**, 275–280.

Gilbert, F. & Helmberger, D.V., 1972. Generalized ray theory for a layered sphere, *Geophys. J. R. Astr. Soc.*, **27**, 57–80.

Glatzmaier, G.A. & Roberts, P.H., 1996. Rotation and magnetism of earth's inner core, *Science*, **274**, 1887–1891.

Goldstein, P. & Dodge, D., 1999. Fast and accurate depth and source mechanism estimation using P-waveform modelling: a tool for special event analysis, event screening and regional calibration, *Geophys. Res. Lett.*, **26**, 2569–2572.

Gutenberg, B., 1932. *Handbuch der Geophysik, Band IV: Erdbeben*, Gebrúder Bornträger, Berlin.

Gutenberg, B. & Richter C.F., 1949. *Seismicity of the Earth*, Princeton University Press.

Harkrider, D.G., 1964. Surface waves in multi-layered elastic media I: Rayleigh and Love waves from buried sources in a multilayered half space, *Bull. Seism. Soc. Am.*, **54**, 627–679.

Haskell, N.A., 1964. Radiation pattern of surface waves from point sources in a multi-layered medium, *Bull. Seism. Soc. Am.*, **54**, 377–393.

Harvey, D. J., 1981. Seismogram synthesis using normal mode superposition: The locked mode approximation, *Geophys. J. R. Astr. Soc.*, **66**, 37–70.

Helbig, K., 1958. Elastische Wellen in anisotropen Medien, *Gerlands Beitr. Geophysik*, **67**, 177–211.

Helmberger, D.V., 1968. The crust-mantle transition in the Bering Sea, *Bull. Seism. Soc. Am.*, **58**, 179–214.

Helmberger, D.V., 1973. On the structure of the low velocity zone, *Geophys. J. R. Astr. Soc.*, **34**, 251–263.

Helmberger, D.V. & Engen, G., 1980. Modelling the long-period body waves from shallow earthquakes at regional ranges, *Bull. Seism. Soc. Am.*, **70**, 1699–1714.

Helmberger, D.V. & Wiggins, R. A., 1971. Upper mantle structure of the midwestern United States, *J. Geophys. Res.*, **76**, 3229–3245.

Hirn, A., Steimetz, L., Kind, R. & Fuchs, K., 1973. Long range profiles in Western Europe II: Fine structure of the lithosphere in France (Southern Bretagne), *Z. Geophys.*, **39**, 363–384.

Hron, F., 1972. Numerical methods of ray generation in multilayered media, *Methods in Computational Physics*, **12**, ed. B. A. Bolt, Academic Press, New York.

Hudson, J.A., 1962. The total internal reflection of SH waves, *Geophys. J. R. Astr. Soc.*, **6**, 509–531.

Hudson, J.A., 1969a. A quantitative evaluation of seismic signals at teleseismic distances I - Radiation from a point source, *Geophys. J. R. Astr. Soc.*, **18**, 233–249.

Hudson, J.A., 1969b. A quantitative evaluation of seismic signals at teleseismic distances II - Body waves and surface waves from an extended source, *Geophys. J. R. Astr. Soc.*, **18**, 353–370.

Hudson, J.A., 1980. *The Excitation and Propagation of Elastic Waves*, Cambridge University Press.

Hudson, J.A., 1991. Overall properties of heterogeneous media, *Geophys. J. Int.*, **107**, 505–511.

Hudson, J.A., Pearce, R.G. & Rogers, R.M., 1989. Source-type plot for inversion of the moment tensor, *J. Geophys. Res.*, **94**, 765–774.

Illingworth, M.R., 1982. *Seismic Waves in Stratified Media*, Ph.D. thesis, University of Cambridge.

Inoue, H., Fukao, Y., Tanabe, K. & Ogata, Y., 1990. Whole mantle *P*-wave mantle tomography, *Phys. Earth. Planet. Inter.*, **59**, 294–328.

Jackson, D.D. & Anderson, D.L., 1970. Physical mechanisms of seismic wave attenuation, *Rev. Geophys. Sp. Phys.*, **8**, 1–63.

Jackson, I., Paterson M.S. & Fitz Gerald, J.D., 1992. Seismic wave dispersion and attenuation in Åheim Dunite: an experimental study, *Geophys. J. Int.*, **108**, 517–534.

Jackson, I. & Rigden, S.M., 1998. Composition and temperature of the Earth's mantle: seismological models interpreted through experimental studies of earth materials, in *The Earth's Mantle: Structure, Composition and Evolution*, 405–460, ed. I. Jackson, Cambridge University Press.

Jackson, I., 2000. Laboratory measurement of seismic wave dispersion and attenuation: Recent progress, in S.I. Karato, A.M. Forte, R.C. Liebermann, G. Masters & L. Stixrude, eds., *Earth's Deep Interior: Mineral Physics and Tomography from the Atomic to the Global Scale*, AGU Geophysical Monograph Series, **117**, 265–289, American Geophysical Union, Washington, D C.

Jarosch, H. & Aboodi, E., 1970. Towards a unified notation for source parameters, *Geophys. J. R. Astr. Soc*, **21**, 513–529.

Jeffreys, H., 1939. The times of *P*, *S* and *SKS* and the velocities of *P* and *S*, *Mon. Not. R. Astr. Soc, Geophys. Suppl.*, **4**, 498–536.

Jeffreys, H., 1958. A modification of Lomnitz's law of creep in rocks, *Geophys. J. R. Astr. Soc.*, **1**, 92–95.

Jeffreys, H. & Bullen, K.E., 1940. *Seismological Tables*, British Association Seismological Committee, London.

Johnson, L.R., 1967. Array measurements of *P* velocities in the upper mantle, *J. Geophys. Res.*, **72**, 6309–6325.

Jordan, T.H., 1975. The continental tectosphere, *Rev. Geophys.*, **13**, 1–12.

Jordan, T.H., 1978. Composition and development of the continental tectosphere, *Nature*, **274**, 544–548.

Julian, B.R. & Anderson, D.L., 1968. Travel times, apparent velocities and amplitudes of body waves, *Bull. Seism. Soc. Am.*, **58**, 339–366.

Kaiho, Y. & Kennett, B.L.N., 2000. Three-dimensional structure beneath the Australasian region from refracted wave observations, *Geophys. J. Int.*, **142**, 651–688.

Kanamori, H. & Anderson, D.L., 1977. Importance of physical dispersion in surface wave and free oscillation problems: review, *Rev. Geophys. Space. Phys.*, **15**, 105–112.

Karato, S., 1998. A dislocation model of seismic wave attenuation and micro-creep in the Earth: Harold Jeffreys and the rheology of the Earth, *Pure Appl. Geophys.*, **153**, 239–256.

Keith, C. & Crampin, S., 1977. Seismic body waves in anisotropic media: reflection and refraction at a plane interface; propagation through a layer; synthetic seismograms, *Geophys J. R. Astr. Soc.*, **49**, 181–208; 209–223; 225–243.

Kennett, B.L.N., 1983. *Seismic Wave Propagation in Stratified Media*, Cambridge University Press.

Kennett, B.L.N., 1986. Wavenumber and wavetype coupling in laterally heterogeneous media, *Geophys. J. R. Astr. Soc.*, **87**, 313–331.

Kennett B.L.N. 1988a. Systematic approximations to the seismic wave field, 237–259. in *Seismological Algorithms*, ed. D.J. Doornbos, Academic Press, London.

Kennett B.L.N. 1988b. Radiation from a moment-tensor source, 427–441. in *Seismological Algorithms*, ed. D.J. Doornbos, Academic Press, London.

Kennett B.L.N., 1991. *IASPEI 1991 Seismological Tables*, Bibliotech, Canberra.

Kennett, B.L.N., & Woodhouse, J. H., 1978. On high frequency spheroidal modes and the structure of the upper mantle, *Geophys. J. R. Astr. Soc.*, **55**, 333–350.

Kennett, B.L.N., & Nolet, G., 1979. The influence of upper mantle discontinuities on the toroidal free oscillations of the Earth, *Geophys. J. R. Astr. Soc.*, **56**, 283–308.

Kennett, B.L.N. & Illingworth, M.R., 1981. Seismic waves in a stratified half space III - Piecewise smooth models, *Geophys. J. R. Astr. Soc.*, **66**, 633–675.

Kennett, B.L.N. & Clarke, T.J. 1983a. Rapid calculation of surface wave dispersion, *Geophys. J. R. Astr. Soc.*, **72**, 619–631.

Kennett, B.L.N. & Clarke, T.J. 1983b. Seismic waves in a stratified half-space - IV: *P-SV* wave decoupling and surface wave dispersion, *Geophys. J. R. Astr. Soc.*, **72**, 633–645.

Kennett, B.L.N. & Engdahl, E.R., 1991. Traveltimes for global earthquake location and phase identification, *Geophys. J. Int.*, **105**, 429–465.

Kennett, B.L.N., Engdahl, E.R. & Buland, R., 1995. Constraints on seismic velocities in the Earth from travel times, *Geophys. J. Int.*, **122**, 108–124.

Kerry, N.J., 1981. The synthesis of seismic surface waves, *Geophys. J. R. Astr. Soc.*, **64**, 425–446.

Koketsu, K. & Furumura, T., 1998 Specific distribution of ground motion during the 1995 Kobe earthquake and its generation mechanism, *Geophys. Res. Lett.*, **25**, 785–788.

Kostrov, B.V. & Das, S., 1988. *Principles of Earthquake Source Mechanics*, Cambridge University Press, Cambridge.

Kuge, K. & Kawakatsu, H., 1990. Analysis of a deep "non double couple" earthquake using very broadband data, *Geophys. Res. Lett.*, *17*, 227–230.

Kulhánek O., 1990. *Anatomy of Seismograms*, Elsevier Science Publishers, Amsterdam.

Lamb, H., 1904. On the propagation of tremors over the surface of an elastic solid, *Phil. Trans. R. Soc. Lond.*, **203A**, 1-42.

Levshin, A., 1981. On the relation of P and S travel times, phase velocities of shear modes and frequencies of spheroidal oscillation in the radially inhomogeneous Earth, *Computational Seismology*, **13**, 103-109.

Love, A.E.H., 1903. The propagation of wave motion in an isotropic solid medium, *Proc. Lond. Maths. Soc.*, **1**, 291-316.

Love, A.E.H., 1911. *Some Problems of Geodynamics*, Cambridge University Press, Cambridge.

Love, A.E.H., 1927. *A Treatise on the Mathematical Theory of Elasticity*, 2nd edition, Cambridge University Press, Cambridge.

Marson-Pidgeon, K. & Kennett, B.L.N., 2000a. Flexible computation of teleseismic synthetics for source and structural studies, *Geophys. J. Int*, **125**, 229-248.

Marson-Pidgeon, K. & Kennett, B.L.N., 2000b. Source depth and mechanism inversion at teleseismic distances, using a neighbourhood algorithm, *Bull. Seism. Soc. Am.*, **100**, 1369-1383.

McKenzie, D.P. & Brune, J.N., 1972. Melting on fault planes during large earthquakes, *Geophys. J. R. Astr. Soc.*, **29**, 65-78.

Mitchell, B.J., 1995. Anelastic structure and evolution of the continental crust and upper mantle from seismic surface wave attenuation, *Rev. Geophys.*, **33**, 441-462.

Mitchell, B.J. & Cong, L., 1998. *Lg* coda *Q* and its relation to the structure and evolution of the continents: a global perspective, *Pure Appl. Geophys.*, **153**, 655-663.

Montagner, J-P & Kennett, B.L.N., 1996. How to reconcile body-wave and normal-mode reference Earth models?, *Geophys. J. Int*, **125**, 229-248.

Mooney, W., Laske, G. & Masters, G., 1998. CRUST5.1, a global crustal model at $5° \times 5°$, *J. Geophys. Res.*, **103**, 727-747.

Morelli, A. & Dziewonski, A.M., 1993. Body wave traveltimes and a spherically symmetric *P*- and *S*-wave velocity model, *Geophys. J. Int.*, **112**, 178-194.

Morse, P.M. & Feshbach, H., 1953. *Methods of Theoretical Physics*, McGraw-Hill, New York.

Müller, G., 1970. Exact ray theory and its application to the reflection of elastic waves from vertically inhomogeneous media, *Geophys. J. R. Astr. Soc.*, **21**, 261-283.

Müller, G., 1977. Earth flattening approximations for body waves derived from geometric ray theory - improvements, corrections and range of applicability, *J. Geophys.*, **44**, 429-436.

Müller, G., 1985. The reflectivity method: a tutorial, *J. Geophys.*, **58**, 153-174.

Musgrave, M.J.P., 1970. *Crystal Acoustics*, Holden-Day, San Francisco.

Nataf, H.C. & Ricard, Y., 1995. 3SMAC: an *a priori* tomographic model of the upper mantle based on geophysical modeling, *Phys. Earth. Planet. Inter.*, **95**, 101-122.

Nolet G., Grand S. & Kennett B.L.N., 1994. Seismic heterogeneity in the Upper Mantle, *J. Geophys. Res.*, **99**, 23 753-23 766.

Nussenveig, H.M., 1965. High frequency scattering by an impenetrable sphere, *Ann. Phys.*, **34**, 23-95.

O'Brien, P.N.S. & Lucas, A.L., 1971. Velocity dispersion of seismic waves, *Geophys. Prospect.*, **19**, 1-26.

O'Neill, H.St.C. & Palme, H., 1998. Composition of the silicate Earth: implications for accretion and core formation, in *The Earth's Mantle: Structure, Composition and Evolution*, 3-126, ed. I. Jackson, Cambridge University Press.

Olver, F.W.J., 1974. *Asymptotics and Special Functions*, Academic Press, New York.

Owens, T.J., Randall, G.E., Wu, F.T. & Zeng, R.S., 1993. PASSCAL instrument performance during the Tibetan plateau passive seismic experiment, *Bull. Seism. Soc. Am.*, **83**. 1959-1970.

Pao, Y-H., & Gajewski, R. R., 1977. The generalised ray theory and transient responses of layered elastic solids, *Physical Acoustics*, **13**, ed. W. Mason, Academic Press, New York.

Peterson, J., 1993. Observations and modeling of seismic background noise, *USGS Open-File report 93-322*, pp95.

Raitt, R.W., 1969. Anisotropy of the upper mantle, *The Earth's Crust and Upper Mantle*, ed. P. J. Hart, American Geophysical Union, Washington, D.C.

Rayleigh, Lord, 1885. On waves propagated along the plate surface of an elastic solid. *Proc. Lond. Maths. Soc.*, **17**, 4–11.

Redheffer, R., 1961. Difference equations and functional equations in transmission line theory, in *Modern Mathematics for the Engineer (2nd series)*, ed. E.F. Beckenback, McGraw Hill, New York.

Ricard, Y., Nataf, H.C. & Montagner, J.P., 1996. The three-dimensional seismological model *a priori* constrained : confrontation with seismic data, *J. Geophys. Res.*, **101**, 8457–8472.

Richards, P.G., 1973. Calculations of body waves, for caustics and tunnelling in core phases, *Geophys. J. R. Astr. Soc.*, **35**, 243–264.

Richards, P.G., 1974. Weakly coupled potentials for high frequency elastic waves in continuously stratified media, *Bull. Seism. Soc. Am.*, **64**, 1575–1588.

Richards, P.G., 1976. On the adequacy of plane wave reflection/transmission coefficients in the analysis of seismic body waves, *Bull. Seism. Soc. Am.*, **66**, 701–717.

Richards, P.G. & Frasier, C.W., 1976. Scattering of elastic waves from depth-dependent inhomogeneities, *Geophysics*, **41**, 441–458.

Richter, C.F., 1958. *Elementary Seismology*, W. H. Freeman and Company Inc., San Francisco.

Rosenbaum, J.M., 1974. Synthetic microseisms: Logging in porous formations, *Geophysics*, **39**, 14–32.

Saastamoinen, P., 1980. On propagators and scatterers in wave problems of layered elastic media - a spectral approach, *Bull. Seism. Soc. Am.*, **70**, 1125–1135.

Sato, H. & Fehler, M.C., 1998. *Seismic Wave Propagation and Scattering in the Heterogeneous Earth*, American Institute of Physics, New York.

Shearer, P.M., 1991. Constraints on upper-mantle discontinuities from observations of long-period reflected and converted phases, *J. Geophys. Res.*, **96**, 18 147–18 182.

Sheriff, R.E. & Geldart, L.P., 1982. *Exploration Seismology*, Cambridge University Press, Cambridge.

Silver, P.G., 1996. Seismic anisotropy beneath the continents; probing the depths of geology, *Ann. Rev. Earth Planet. Sci.*, **24**, 385-432.

Sipkin, S., 1994. Rapid determination of global moment-tensor solutions, *Geophys. Res. Lett.*, **21**, 1667-1670.

Smith, M.L. & Dahlen, F.A., 1981. The period and Q of the Chandler Wobble, *Geophys. J. R. Astr. Soc.*, **64**, 223–281.

Song, X. & Helmberger, D.V., 1992. Velocity structure near the inner core boundary from waveform modelling, *J. Geophys. Res.*, **97**, 6573–6586.

Stokes, G.G., 1849. Dynamical theory of diffraction, *Trans. Camb. Phil. Soc.*, **9**, 1.

Stoneley, R., 1924. Elastic waves at the surface of separation of two solids, *Proc. R. Soc. Lond.*, **106A**, 416–420.

Stutzmann, E., Roult, G. & Astiz, L., 2000. GEOSCOPE station noise levels, *Bull. Seism. Soc. Am.*, **90**, 690–701.

Sykes, L.R., 1967. Mechanisms of earthquakes and the nature of faulting on mid-ocean ridges, *J. Geophys. Res.*, **72**, 2131–2153.

Takeuchi, H. & Saito, M, 1972. Seismic Surface Waves, in *Methods in Computational Physics*, **11**, ed. B.A. Bolt, Academic Press, New York.

Tan B., Jackson, I. & Fitz Gerald, J., 1997. Shear wave dispersion and attenuation in fine-grained synthetic olivine aggregates: preliminary results, *Geophys. Res. Lett.*, **24**, 1055–1058.

Tazime, K., 1994. *Elements of elastic wave theory*, Makishoten, Tokyo (in Japanese).

Temme, P. & Müller, G., 1982. Numerical simulation of vertical seismic profiling, *J. Geophys.*, **50**, 177–182.

Thomson, C.J. & Chapman, C.H., 1984. On approximate solutions for reflection of waves in a stratified medium, *Geophys. J. R. Astr. Soc.*, *79*, 385–410.

Titchmarsh, E.C., 1937. *An Introduction to the Theory of Fourier Integrals*, Oxford University Press.

Ursin, B., 1983. Review of elastic and electromagnetic wave propagation in horizontally layered media, *Geophysics*, **48**, 1063-1081.

van der Hilst, R.D., Kennett B.L.N., Christie, D. & Grant J., 1994. Project SKIPPY explores the mantle and lithosphere under Australia, *EOS*, **75**, 177, 180-181.

Vered, M. & BenMenahem, A., 1974. Application of synthetic seismograms to the study of low magnitude earthquakes and crustal structure in the Northern Red Sea region, *Bull. Seism. Soc. Am.*, **64** , 1221-1237.

Vinnik, L.P., Makeyeva, L.I., Milev, A. & Usenko, A. Yu., 1992. Global patterns of azimuthal anisotropy and deformations in the continental mantle, *Geophys. J. Int*, **111**, 433-447.

Walton, K., 1974. The seismological effects of prestraining within the Earth, *Geophys. J. R. Astr. Soc.*, **36**, 651-677.

Wang, C.Y., & Herrmann, R.B., 1980. A numerical study of *P*-, *SV*- and *SH*-wave generation in a plane layered medium *Bull. Seism. Soc. Am.*, **70**, 1015-1036.

Watson, G.N., 1918. The diffraction of electric waves by the Earth, *Proc. R. Soc. Lond.*, **95A**, 83-99.

Woodhouse, J.H., 1974. Surface waves in a laterally varying layered structure. *Geophys. J. R. Astr. Soc.*, **37**, 461-490.

Woodhouse, J.H., 1978. Asymptotic results for elastodynamic propagator matrices in plane stratified and spherically stratified earth models, *Geophys. J. R. Astr. Soc.*, **54**, 263-280.

Young, C.J. & Lay, T.J., 1990. Multiple phase analysis of the shear velocity structure in the D'' region beneath Alaska, *J. Geophys. Res.*, **95**, 17 385-17 402.

Yoshida, S., Koketsu, K., Shibazaki, B., Sagiya, T., Kato, T. & Yoshida, Y., 1996, Joint inversion of near- and far-field waveforms and geodetic data for the rupture process of the 1995 Kobe earthquake, *J. Phys. Earth.*, **44**, 437-454.

Yoshizawa, K., Yomogida, K. & Tsuboi, S., 1999. Resolving power of surface wave polarisation data for higher-order heterogeneities, *Geophys. J. Int.*, **138**, 205-220.

Zatsepin, S.V. & Crampin, S., 1997. Modelling the compliance of crustal rock - I. Response of shear-wave splitting to differential stress, *Geophys. J. Int.*, **129**, 477-494.

Index